AF782321

Handbook of Polysaccharides

Handbook of Polysaccharides

Edited by **Erica Young**

Callisto Reference

New York

Published by Callisto Reference,
106 Park Avenue, Suite 200,
New York, NY 10016, USA
www.callistoreference.com

Handbook of Polysaccharides
Edited by Erica Young

© 2015 Callisto Reference

International Standard Book Number: 978-1-63239-409-5 (Hardback)

This book contains information obtained from authentic and highly regarded sources. Copyright for all individual chapters remain with the respective authors as indicated. A wide variety of references are listed. Permission and sources are indicated; for detailed attributions, please refer to the permissions page. Reasonable efforts have been made to publish reliable data and information, but the authors, editors and publisher cannot assume any responsibility for the validity of all materials or the consequences of their use.

The publisher's policy is to use permanent paper from mills that operate a sustainable forestry policy. Furthermore, the publisher ensures that the text paper and cover boards used have met acceptable environmental accreditation standards.

Trademark Notice: Registered trademark of products or corporate names are used only for explanation and identification without intent to infringe.

Printed in China.

Contents

Preface

This book compiles the various aspects of polysaccharides in microorganisms, plants and animals. The diversity of these polysaccharides emerges from the structural alterations and the monosaccharide content which is under genetic control. Their physical as well as chemical properties have made them useful in a variety of food, pharmaceutical and industrial applications. These properties determine their biological activity and their functions in different applications. The use of polysaccharides as therapeutics, the role played by them in protection and preservation of food, as carriers of drugs, nutrients and their ability to communicate with molecules for improving textures of food colloids, as well as for efficient delivery are a few important functions. The book covers two broad sections namely, Sources & Biological Properties of Polysaccharides and Physical & Chemical Characteristics of Polysaccharides.

This book is a comprehensive compilation of works of different researchers from varied parts of the world. It includes valuable experiences of the researchers with the sole objective of providing the readers (learners) with a proper knowledge of the concerned field. This book will be beneficial in evoking inspiration and enhancing the knowledge of the interested readers.

In the end, I would like to extend my heartiest thanks to the authors who worked with great determination on their chapters. I also appreciate the publisher's support in the course of the book. I would also like to deeply acknowledge my family who stood by me as a source of inspiration during the project.

Editor

Section 1

Sources and Biological Properties of Polysaccharides

Yeast (*Saccharomyces cerevisiae*) Glucan Polysaccharides – Occurrence, Separation and Application in Food, Feed and Health Industries

Stefan Kwiatkowski and Stefan Edgar Kwiatkowski

Additional information is available at the end of the chapter

1. Introduction

Energy is stored in plants in form of carbohydrates, including sucrose (i.e., saccharose, a disaccharide) and starch (a polysaccharide). Sucrose and starch are derived from glucose, the product of carbon dioxide and water in photosynthesis. These carbohydrates are not only essential as energy supplies in food and animal feed, but in recent years have also been tapped as a renewable fuel energy source by the fuel ethanol industry. Some plants, like sugar cane or sugar beet, produce juice rich in sucrose, a mixed glucose-fructose disaccharide. Brewer's yeast (*Saccharomyces cerevisiae*) can directly ferment sucrose into ethanol, whereas starch, like that found in corn or potatoes, must first be hydrolyzed, to glucose or fermentable glucose oligosaccharides by an enzyme (e.g., amylase) before fermenting into ethanol. Cellulose, structural polysaccharides in plants, is even more difficult to hydrolyze to glucose. To obtain fermentable sugars, cellulose-containing plant material is usually subjected to a physicochemical pretreatment followed by enzymatic hydrolysis with cellulases. Such pretreatment processes are expensive and not yet used commercially on a large scale.

Because sugar and starch are major components of human food, and cellulose is the principal component of ruminant feed, their large-scale use in fuel alcohol production is already affecting food prices in the world. Brazil began producing fuel ethanol from sugar cane 36 years ago and in 2010 supplied enough ethanol to sell gasoline containing 25% ethanol throughout the country. Similarly, a large-scale, government sponsored program of fuel ethanol production from starch, using maize as its source, was started in the United States in 2007. This industry has created vast quantities of yeast byproducts; including a commercial-scale source of yeast extract, yeast glucose and mannose polysaccharides. Sugar cane juice and sugar beet juice fermentations are especially convenient for yeast collection, because the post-fermentation solution does not contain solids such as non-fermentable residues, like those produced during corn mash fermentation.

The yeast cell wall contains three major constituents: glucan (glucose polysaccharide), mannan (mannose polysaccharide) and a protein fraction. The separation of these natural polymers is simple and inexpensive. However, it is difficult and expensive to obtain more than 65% pure fractions and, therefore, these components are produced and sold at these low levels of purity. Their largest commercial application has been as animal feed nutritional supplements (see, http://www.Alltech.com).

There are three well known types of glucose polysaccharides in yeast: poly-(1→3)(1→6)-β-D-glucopyranose - commercial name: yeast β-D-glucan, yeast glucan; poly-(1→4)-α-D-glucopyranose (commercial name yeast glucogen) and poly-(1→6)(1→3)(1→4)-α-D-glucopyranose a recently "rediscovered" polysaccharide (Arvindecar & Patil, 2002) yet without a commercial name. The prefix "poly" informs about the polymeric nature of the material built of D-glucose cyclic monomers (D-glucopyranose rings) and the letter D informs that glucose belongs to the group of naturally occurring plant sugars with D stereochemistry. The letter α or β refers to the configuration of the glycosidic bonds between the C-1 of a nonreducing ring and the C-6 of a reducing ring or the C-1 of a nonreducing ring and the C-3 of a reducing ring (see Fig. 1, Lindhorst 2003). Figure 1 shows structures of β-D-glucopyranose isomer with a C-1 OH group in the β position (solid line) or a C-1 OH group in the (α) position (dotted line). Penta-(1→6)(1→3)-β-D-glucopyranose drawing shows the location of all OH groups, glycosidic bonds and glucopyranose rings within the structure. (The hydrogen atoms attached to carbon atoms are not shown.)

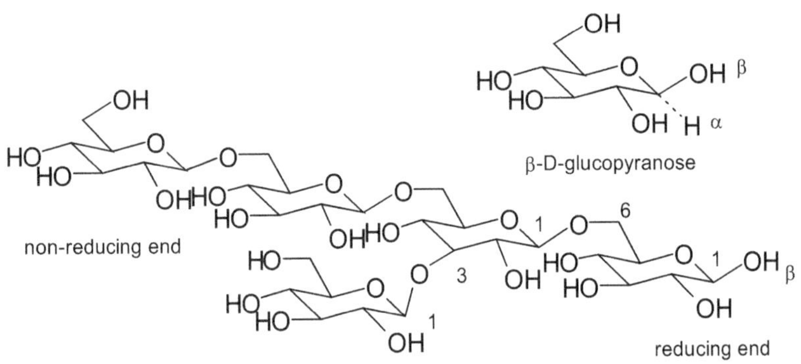

Fig. 1

Figure 1.

Solving polysaccharide structure is not simple. At first, a sample needs to be homogenous and 98-99% pure. Different separation techniques (i.e., size exclusion chromatography, affinity (lectin) chromatography and in some cases high performance liquid chromatography (HPLC) are suitable for the purification of water or solvent soluble samples. Spectroscopic methods, such as ^{1}H and ^{13}C nuclear magnetic resonance (NMR) 1D/2D and mass spectrometry (MS) combined with chemical treatments of the sample, can provide further clues about the kind, sequence and ratios of the monosaccharides present, the type of bonds (α- or β-) between individual sugars, the branching points and reducing and non-reducing ends, and its molecular weight.

The experimental procedures that are most frequently used in the process of solving the polysaccharide structure are:

1. Sample purification

A solid, crude, non-homogenous, polysaccharide must be solubilized in an appropriate solvent (water) or, in the case of insoluble polysaccharides, it must be subjected to chemical or enzymatic degradation (with minimal damage to the original structure) to reduce the polymer size which then allows the sample to become soluble. Sample solubilization makes the use of various chromatographic techniques (i.e., ion exchange chromatography, size exclusion chromatography, affinity-chromatography, sorbent-liquid chromatography) for sample fractionation and purification, possible.

2. Composition analysis

Complete acid hydrolysis is used to cleave all glycosidic bonds in the polysaccharide and to release D-glucose as a single product (in the case of glucans), which is then identified by comparison to a D-glucose standard.

3. Methylation linkage analysis

All of the polysaccharide hydroxyl (OH) groups are methylated using an excess of methylation reagent (dimsyl sodium followed by methyl iodide). Obtained this way, the fully O-methylated derivative is then subjected to acidic hydrolysis of all glycosidic linkages. The released O-methylated derivatives of D-glucose: 2, 3, 4, 6-O-tetramethyl-D-glucose; 2, 3, 4-O-trimethyl-D-glucose; 2, 4-O-dimethyl-D-glucose are identified and quantified using HPLC/MS and appropriate standards. In the case of pentasaccharide (Fig.1) the ratios of the methylation products would be 2:2:1.

4. Spectroscopy

Finally, ^{1}H and ^{13}C NMR and M S are used to establish the sequence of glucopyranose rings, the position of branching points, the number of non-reducing ends and the nature of glycosidic bonds (α or β). If the polysaccharide contains repeating sequences, their ratios can be also established.

5. Enzymatic action

Glycosidases and lectins of known specificity are applied to confirm glucose oligosaccharide structures, from which the complete polysaccharide architecture can be deduced.

The Complex Carbohydrates Research Center located at The University of Georgia (http://www.ccrc.uga.edu) offers services covering all of the aspects of the polysaccharide structural assignments.

2. Fungal and yeast cell wall, D-glucose homo- and hetero-polysaccharides

Fungi constitute an independent evolutionary group of eukaryotic organisms equal in rank to that of plants or animals. They can excrete hydrolytic enzymes that break down biopolymers (polysaccharides and proteins) of plant or animal origin and use sugars and/or amino acids for their nutrition. Yeast is a single-cell fungus and the *Saccharomyces cerevisiae* species (baker's or brewer's yeast) has been utilized by people from early human civilization in food and beverage preparation. Mushrooms are fungal fruiting bodies; edible mushrooms are taste improving, "healthy" food ingredients that are very popular on European and Asian menus. Extracts and dried powders made by processing "medicinal mushrooms" are important ingredients of traditional Eastern medicinal remedies.

Fungal cell walls are made of polysaccharides. Chitin, a homopolymer of (1→4)-β-D-N-acetylglucosamine and various glucans (homopolymers of D-glucose, with α- or β-glycosidyl linkages between C-1 and C-2, C-3, C-4 and C-6 of glucopyranose rings) are the most frequently found building blocks used by nature in construction of the cell wall that separates the fungus from the environment. In yeast, α- and β-glucans and α-mannan are the major polysaccharides (chitin is present at 1-2%) that are utilized in cell wall construction. Their ratio changes with yeast strain and growth stage and is dependent upon growth conditions including oxygen and nutrient availability and the temperature and pH of the medium (Stewart & Russell, 1998). Table 1 shows the average abundance of yeast cell wall components.

Component	Cell wall mass (%, dry weight)
(1→3)-β-D-glucan	50-55
(1→6)-β-D-glucan	5-10
(1→4)-α-(1→3)-β-D-glucan	3-7*
Mannoprotein complex	35-40
Chitin	2

Mannan structure can be represented by the following formula: poly-(1→2) (1→3) (1→6)-α-D-mannopyranose. This polysaccharide is built out of a long backbone chain of (1→6)-α-D-mannopyranose rings with short mannopyranose chains (one to four rings long, connected through (1→2) and (1→3)-α-glucoside bonds (Ballou, 1980).

Table 1. The cell wall components of *Saccharomyces cerevisiae* (Klis et al. 2002; Kath & Kulicke, 1999; Lessage & Bussey, 2006; Kwiatkowski et al., 2009)

The yeast cell wall forms a border that defines the yeast cell's dimensions and separates its organelles from the negative influences of the environment. The individual constituents of the cell wall connect to each other by covalent bonds forming a single supra-molecular biopolymer. To find the nature of these connections, as well as the individual polysaccharides structures, all of the architectural elements must be dissected, solubilized, purified and analyzed. In this process it is almost impossible not to damage particular structural elements and to afterwards distinguish between the original components of the cell wall and soluble polysaccharides that could be present in the cytoplasm or might be trapped (physically absorbed) within cell wall. More than 50 years of research were required to establish the structure of the major, high molecular weight yeast homo-polysaccharides (Bacon & Farmer, 1969; Aimaniada et al., 2009).

The (1→3)-β-D-glucan and (1→6)-β-D-glucan form a single structure in which the (1→3)-β-D-glucan forms the backbone chain with (1→6)-β-D-glucan branches that are attached to branching glucopyranose rings at C-6 (Lessage & Bussey, 2006). The (1→3)-β-D-glucan chains form triple helix tridimensional structures with spring-like mechanical properties, responsible for the yeast cell wall's strength (Klis et al., 2002) and its ability to absorb toxins (Yiannikouris et al., 2004). The (1→6)-β-D-glucan is a linker between (1→3)-β-D-glucan, chitin and mannoproteins (Kaptein et al., 1999; Kollar et al., 1997) that stabilizes the whole structure and is the main cause of yeast cell wall insolubility. The properties and role of a mixed (1→4)-α-(1→3)-β-D-glucan from yeast cells will be discussed in section 3 of this chapter.

Mannoproteins are mostly located on the outside surface of the cell wall (Osumi, 1998). They play a sensory function for nutrients and chemical and bacterial/viral toxins, actively participate in the transport of nutrients and metabolites through the cell wall, and also participate in mating (Klis et al., 2002; Stewart & Russell, 1998). Some cell wall enzymes such as glucanase, mannanase, invertase, alkaline phosphatase and lipase are mannoproteins (Stewart & Russell, 1998) that hydrolyze nutrients and participate in the reconstruction of cell polysaccharides during cell growth and budding. Cell wall mannans are connected with cell wall glucans via covalent bonds, but they can be released under action of an alkaline medium in which they are perfectly soluble. The process of separating yeast cell wall glucans (which stay as an insoluble fraction) from yeast mannans is used on an industrial scale (Sedmak, 2006). The soluble, mannan-rich fraction, can be added back to the cell wall which contains only 10-17% mannan by weight, to increase the content of mannan to 30% by weight. Such a product is sold by Alltech Inc. under the name of Actigen™ to the animal nutrition industry.

The small fraction of chitin present in yeast cell wall is primarily located in the scar rings around buds in budding yeast and secures closure of the gaps in the mother cell and the departing daughter cell (Lessage & Bussey, 2006; Stewart & Russell, 1998).

3. α-D-glucans from baker's yeast (*Saccharomyces cerevisiae*)

The role of (1→3)-β-D-glucan in the maintenance of yeast cell wall shape and rigidity (Lessage & Bussey, 2006; Klis, et al. 2006) and (1→6)-β-D–glucan as a polysaccharide that links together all of the cell wall polysaccharides (Aimaniada et al., 2009; Kollar et al., 1997) is well documented and has been reviewed by Lessage & Bussey (2006) and Klis et al. (2002). The presence of starch-like, "alkali-soluble glycogen, "an "energy storage polysaccharide" in cell cytosol, and "difficult to dissolve acid-soluble," glycogen like (1→4)-α-D-glucan in the cell walls of *Saccharomyces cerevisiae* grown aerobically, was frequently mentioned in early yeast literature (Grba et al., 1976). The two forms of glycogen synthetase have also been identified (Rothman-Denes & Cabib, 1970; Lille & Pringle, 1980) but the yeast literature is still treating the "difficult to dissolve" α-glucan as physically adsorbed cytosol glycogen, trapped within yeast cell wall, and not as an independent component attached by a covalent bond to the other cell wall polysaccharides. In 1973, Manners et al. reported: "β-1, 6-glucan purified after acid extraction had to be exhaustively treated with α-amylase and still showed the presence of glucose and maltose in paper chromatographic analysis, along with gentiobiose and higher gentiooligosaccharides," which are (1→4)-α-D-glucan oligosaccharides formed during enzymatic hydrolysis of the mixed α, β-glucan from the yeast cell wall.

The α-glucan content in the yeast cell wall is reported to vary from as little as 1% (Lille & Pringle, 1980) to as much as 29% (Sedmak, 2006) of the dry weight, depending upon the nutritional status of the cells, the method of isolation, the method of analysis and the phase of growth during which the cells were harvested (Lille & Pringle, 1980). Spectrophotometric analysis of soluble glycogen can be run directly on a water extract, and if properly done, yields reliable results (Quain, 1981). Quantitative analysis of "insoluble" glycogen requires additional, enzymatic release of glucose from its mixed polysaccharide with β-glucan (see, http://www.megazyme.com). The industrially produced brewer's yeast described in Sedmak's US patent application (Sedmak, 2006) contains glucans (α + β) at 28.9% dry weight, which includes 12.4% α-glucan. Hydrolysis of these water-insoluble cells with alkaline protease solubilizes the mannoprotein complex and yields water-insoluble cell wall polysaccharides (including α-glucan) that contain 54.5% dry weight of glucans with more than half of the total weight (29.2%) as α-glucan.

Arvindekar and Patil (2002) proposed an explanation for the presence of α-glucan in the insoluble fraction from yeast cell walls, which has been described by others as "difficult to wash away" yeast glycogen (Gunia-Smith et al., 1977; Manners & Fleet, 1976). They found that the ratio of "soluble" to "insoluble" glycogen in different strains of *Saccharomyces cerevisiae* is in the range of 1:2.5–1:3. When the insoluble fraction was digested with purified lyticase, which possessed only (1→3)-β-D–glucan hydrolyzing activity and no (1→6)-β-D–glucan or (1→4)-α-D-glucan hydrolyzing activities, all of the material dissolved. The affinity chromatography of the solution on a Concanavalin A (ConA) column retained a fraction that contained mixed (1→4)-α-(1→6)-β-D–glucan polysaccharide. The eluent contained exclusively (1→3)-β-D–glucan oligosaccharides. When the (1→4)-α-D-glucan-rich fraction was released from its binding with ConA and treated with amyloglucosidase followed by dialysis to remove glucose, subjecting the resulting solution to affinity chromatography on a ConA column showed none of the material was retained. The material recovered from the solution was then proved to be pure (1→6)-β-D–glucan. These simple experiments showed that the "insoluble glycogen" from yeast cell wall is a mixed glucan in which (1→4)-α-D-glucan is connected to (1→3)-β-D–glucan through a (1→6)-β-D–glucan link. The structure of this mixed yeast cell wall (1→4)-α-(1→6)-β-(1→3)-β-D-glucan was confirmed by our ^{1}H NMR study (Kwiatkowski et al., 2009) and methylation analysis done at CCRC.

We used yeast cell wall α-glucan (prepared in our laboratory) in the development of the enzyme linked immunosorbent assay (ELISA) for yeast cell wall quantitative analysis in a complex polysaccharide matrix (Moran et al., 2011). The antigen was prepared in two stages.

At first a sample of cell wall (1→4)-α-D-glucan (fraction 50-100 kDa separated with the help of an ultra centrifugal filter (Aldrich Inc.) was oxidized mostly at the C-6 carbons using the Albright-Goldman oxidation reagent (Albright & Goldman, 1967; Zekovic et al., 2006) which converted the $-CH_2OH$ groups into aldehyde groups $-CH=O$ in glucopyranose rings. The reaction produced a highly cross-linked polymer, with internal acetal bonds, but still well soluble in DMSO (dimethyl sulfoxide) and water, which was coupled to a bovine serum albumin (BSA), yielding the antigen (glucan-BSA conjugate). To restore the original α-glucan structure, but with $-CH_2$–NH-BSA groups instead of $-CH_2$-OH groups at the glucan's C-6 carbon, the conjugate C-6 imino (-CH=N-BSA) groups were reduced with sodium cyanoborohydride (Baxter & Reitz, 2002). Sodium dodecyl sulfate (SDS) polyacrylamide gel electrophoresis (PAGE) of the conjugate indicated the presence of significant quantities of free BSA (line at 66 kDa) together with the glucan-BSA conjugate of molecular size between 100 and 175 kDa. To remove the excess BSA, the concentrated solution of the conjugate was washed with water of pH 7.0, 8.5 and 4.5 using an ultra centrifugal filter with the 50-kDa membrane. This "purified" conjugate was used then in rabbit immunization. We found that the rabbit polyclonal antibodies were very sensitive to yeast cell wall α-glucan, but (as expected) they were also cross-reactive with BSA. To delete this activity we synthesized the BSA-coated affinity phase (Moran et al., 2011) and removed antibodies specific to BSA from the crude fraction of polyclonal rabbit anti-α-glucan antibodies. We also found that antibodies grown against this α-glucan polysaccharide did not recognize any commercially available starch or glycogen samples, including yeast glycogen. Such specificity confirms structural differences between the cell wall and the cytosol α-glucans from *Saccharomyces cerevisiae*. Because of the high specificity and very low cross reactivity of the "purified" rabbit antibodies with soy proteins and corn carbohydrates, the ELISA cell wall assay detects and quantifies samples of yeast cell wall down to 50 g/ton of feed (Moran et al., 2011). A similar assay was used for quantification of α-D-glucan-protein complex in mushrooms with immunomodulatory activity (Hirotaka et al., 2007).

What is the role of this polysaccharide in yeast cell life? Some remarks in the literature point to its role in yeast flocculation (Patel & Ingledew, 1975), acting as part of a possible killer toxin receptor (Hutchkins & Bussey, 1983) and as being sensitive to environmental conditions and changes in medium (Slaughter & Nomura, 1992; Dake et al., 2010; Jadhav et al., 2008). The cell wall bound α-glucan from fission yeast *Schizosaccharomyces pombe* has been studied in detail (Villar-Tajadura et al., 2008; Vos et al., 2007 ; Garcia et al., 2006; Grün et al., 2005), and systematic identification of the genes affecting glycogen storage in the yeast *Saccharomyces cerevisiae* was attempted (Wilson et al., 2002). Unfortunately, the presence of the two different pools of glycogen was not addressed. One of the possible functions of bound α-glucan might be the temporary formation of mixed α, β-glucan oligosaccharides that become soluble in cytoplasm and can be transported inside the cell wall where the β-glucan fragment is used in the construction of insoluble yeast cell wall β-glucan polysaccharide.

Is it possible that α-glucan polysaccharides from yeast or mushrooms can have medicinal properties? There is no literature concerning the medicinal properties of α-glucan as a separate fraction of the yeast cell wall, which is a consequence of acknowledging it as part of an energy storage cellular polysaccharide-yeast glycogen. Bioactive components containing α-glucan, mixed α, β-glucan or α-glucan-protein complexes are known to be present in extracts from the fruiting bodies of edible medicinal-mushrooms. Like in the case of many other traditional herbal remedies, it is difficult to separate the single bioactive component's (α-glucan) bioactivity from the activities of the other components of the extract, and in some cases, early assignments were incorrect.

Mushrooms with medicinal activities that contain extractable α-glucan in their fruiting bodies, belong to one of two phyla: Ascomycota or Basidiomycota. Hot-water extracts from *Peziza vesiculosa* (Ohno et al., 1985; Suzuki et al., 1982) and *Cordyceps chinensis* (Yan et al., 2011; Holliday & Cleaver, 2008; Khan et al., 2010; Liu et al., 2006; Kiho et al., 1993; Kiho et al., 1996; Li et al., 2006), which are both Ascomycota, have been shown to have antitumor and anti-diabetic activities. The linear α-(1→4)-D-glucan is less active than α-(1→4) (1→6)-D-glucan with 1 to 6 branches on the backbone 1→4 chain with a ratio of 1:8. The biological activity of glycogen-like polysaccharides was found to be molecular-size dependent with smaller ~9.5 kDa molecules being active, whereas 14-24 kDa molecules were not (Kakutani et al., 2007).

More is known about edible mushrooms belonging to Basidiomycota which contain both: α- and β-glucans in their fruiting bodies (Rop et al., 2009). Hot-water extracts from *Peziza vesiculosa* show antitumor activity (Ohno et al., 1985; Suzuki et al., 1982) and extracts from *Grifola frondosa* (Maitake) show glucose suppression (Tanaka et al., 2011) and antidiabetic (Lei et al., 2007) activities. Extracts (Shida & Matsuda, 1974) from *Lentinus edodes* (Shiitake) stimulate the immune system (Shah et al., 2011; Terakawa et al., 2008) and possess antibacterial, antiviral (Mach et al., 2008) and anticancer properties (Shida & Matsuda, 1974; Hyodo et al., 2005; Shah et al., 2011). Extracts from *Tricholoma matsutake* (Matsutake) show antitumor (Ebina et al., 2002) and immunomodulatory (Hirotaka et al., 2007; Hirotaka et al., 2005) activities. Extracts (Smiderle et al., 2010) from *Agaricus bisporus* (Portobello) inhibit breast cancer cell proliferation (Grube et al., 2001) and stimulate the immune system (Ren et al., 2008; Koppada et al., 2009). A glycogen-like polysaccharide from Portobello mushrooms potently activated macrophages, stimulating TNF-α production and phagocytosis of RAW264.7 cells (Kojima et al., 2010). Extracts from *Pleurotus ostreatus* (Oyster mushroom) show antiproliferative and proapoptotic activities on colon cancer cells (Lavi et al., 2006). *Pholiota nameko* (Butterscotch mushroom), which has been in use as a major component of miso soup in Japan, shows antinflammatory activity (Li et al., 2008). In Asian and European culinary traditions, meals containing these mushrooms are believed to be healthy and to heal a variety of ailments.

Whether α-glucan from yeast cell walls can contribute to the known medicinal activities of various extracts from whole yeast is still an open question. Newly developed, simple methods for producing large quantities of this material (Kwiatkowski et al., 2009; Moran et al., 2011) should stimulate new research in this area.

The abundant literature regarding bioactivity of the whole yeast cell wall and its extracts rich in α-mannoprotein or β-glucan, will be discussed in sections 4, 5 and 6 of this chapter.

4. Baker's yeast cell wall β-D-glucan/α-mannoprotein complex

Yeast cell wall biogenesis was studied by Smits (Smits et al, 2001) who found that "The yeast cell wall forms a border that defines the yeast cell's dimensions and separates its organelles from the negative impact of the environment. All of the individual constituents of a yeast cell wall connect to each other by covalent bonds forming a single supra-molecular biopolymer". The (1→3) (1→6)-β-D-glucan and the (1→2) (1→3) (1→6)-α-D-mannan/protein complexes (Vinogradov et al., 1998) are the major components of this supra-molecular biopolymer (Lessage & Bussey, 2006; Kath & Kulicke, 1999). They connect to each other by a covalent bond the nature of which is still under investigation (Kaptein et al., 1999; Kollar et al., 1997). In the process of yeast cell wall isolation the cell wall is first cracked open, with the help of physical, chemical and/or enzymatic treatments (Kath & Kulicke, 1999) and then separated from the soluble yeast cell components by centrifugation (Sedmak, 2006; Jamas et al., 1998). The concentrated yeast cell wall solids containing ~30% dry weight material are then spray-dried. The resulting product is a fine powder with a light tan color. The cell wall polysaccharides can be separated from each other by alkaline extraction, which solubilizes the α-mannoprotein fraction and leaves β-glucan particles in suspension. The β-glucan particulate can be separated from the soluble α-mannoprotein fraction by centrifugation and spray-dried to yield light, yellow colored, fine powder, free of any smell or taste, containing ~65% β-glucan. The α-mannoprotein solution can then be concentrated by using membrane ultrafiltration and spray-dried to produce a light brown, "mannan rich fraction" with ~40% by weight of α-mannan. Yeast cell walls as well as both fractions of polysaccharides are produced on a large scale and have practical applications as animal feed nutritional supplements. Furthermore, this yeast cell wall preparation, sold by Alltech Inc. under the name of Bio-Mos®, was proved (in more than 600 feeding trials) to have a positive impact on the immune system of livestock (Baurhoo et al., 2009; Yang et al., 2008; Rosen et al., 2007; Jacques & Newman, 1994; Morrison et al., 2010) fish (Dimitroglu et al., 2009), and companion animals (Swanson et al., 2002) when added to feed at the rate of 0.5-2.0 kg/ton. Bio-Mos® contains ~17% of α-mannan, which is possibly its bioactive component. An improved form of Bio-Mos® is marketed by Alltech Inc. under the name Actigen™. This improved product contains 30% α-mannan and is manufactured, by mixing yeast cell wall with a mannan-rich fraction. Actigen™ product is four times more active than Bio-Mos® (~13% of it is soluble and therefore, better available to interaction with microbes and animal gut) and its application rate of 250-500 g/ton approaches that of antibiotics. Alpha-D-mannans have a "brush like" structure built out of α-(1→2)- and α-(1→3)-D-mannopyranose branches, 1 to 5 rings long (brush hair), which are attached to a ~120-ring-long α-(1→6)-D-mannopyranose chain, the brush handle (Vinogradov et al., 1998). This structure creates a specific combination of various functionalities that also involve protein conjugates. It can fit with various receptors present on the walls of animal digestive tracts (Mansour & Levitz, 2003) and with the receptors on the membranes of pathogenic bacteria (Wellens et al., 2008). Alpha-D-mannan/protein-conjugates are involved in interactions with animal immune systems and as a result enhance immune system activity (Wismar et al., 2010) and contribute to animal antioxidant and antimutagenic defenses (Krizkova et al., 2006). Immune system strength, blockage of bacterial adhesion to the gut and modification of the gut structure contributes to improved survival and better growth in young animals. The second major yeast cell wall polysaccharide, (1→3)(1→6)-β-D-glucan, might also contribute to yeast cell wall biological properties (section 5 of this chapter), but its major role in animal nutrition is its ability to bind mycotoxins and detoxify animal feed. The (1→3)(1→6)-β-D-glucan is a part of the cell wall's "triple helix tridimensional structure, with spring-like mechanical properties, responsible for yeast cell wall strength and ability to absorb toxins" (Yannikouris et al., 2004) (see section 2, this chapter). Toxins occurring in plant-derived animal feed belong to one of two groups. Mycotoxins (the first group) are the by-products of the secondary metabolism of pathogenic fungi (Bennet & Klich, 2003), whereas chemical toxins (the second group) are incorporated into plant tissue as a result of plants metabolizing agrochemicals from the soil or water used for irrigation (McLean & Bledsoe, 1992). Some of the most problematic mycotoxins in causing human or animal diseases (i.e., aflatoxin, citrinin, ergot alkaloids, fumonisins, ochratoxin A, patulin, trichothecenes and zeralenone; Smith et al., 1995) can be absorbed by yeast cell wall β-glucans (Yiannikouris et al., 2004; Yannikouris et al., 2006). A mixture of cell wall β-glucan with clay (bentonite) sold as Mycosorb® by Alltech Inc. offers a spectrum of mycotoxin absorption superior to that of yeast cell wall glucans alone and also absorbs heavy metals (Brady et al., 1994). As in the case of Bio-Mos® a multitude of feeding trials have demonstrated the efficacy of Mycosorb® as an animal feed detoxifier using companion animals (Leung et al., 2006), horses (Raymond et al., 2003), pigs (Kogan & Kohler, 2007) and poultry (Dvorska et al., 2003; Karaman et al., 2005; Valarezo et al., 1998).

The whole yeast *Saccharomyces cerevisiae* grown in a medium containing inorganic selenium (Demirici & Pometto, 1999; Demirici et al., 1999; Mapelli et al., 2011) is used to produce yet another yeast-based human/animal nutritional supplement SelPlex® (Alltech Inc.). To maintain healthy metabolism, the human body requires 17 μg of selenium a day. Dietary selenium is used to synthesize selenocysteine (in liver) and is then incorporated into more than 25 Se-containing enzymes that play important roles in body's defense against free radical species (Tapiero et al., (2003) and in many other cellular processes including the generation of energy in mitochondria (Rayman, 2000). Because yeast does not possess genes that control selenium metabolism (Rodrigo, 2002), selenium, which has chemical properties extremely similar to sulfur, is metabolized in the same manner as sulfur and randomly incorporated into yeast cytosol, small-molecules like Se-glutathione, Se-adenosylhomocysteine (Uden et al. 2003) and proteins as Se-methionine (Tastet et al., 2008). Unlike yeast, mushrooms can metabolize selenium and can accumulate large quantities of it (Turlo et al., 2007) in the form of selenomethionine and selenocysteine. Animal feeding studies clearly showed that SelPlex® is not just a selenium source, but it also carries a variety of beneficial effects such as increased animal fertility (Rayman, 2000) or improved animal immune system activity (Rayman, 2000). Additionally, feeding SelPlex® to mice has been shown to delay the development of brain tumors from malignant human cancer cells implanted in mice brains (Toborek, 2011) and can significantly limit the deposition of Aβ amyloid plaques in APP/PS1 mouse brains that carry human Alzheimer's disease genes (Lovell et al., 2009).

Only some of these effects can be promoted by regular yeast cell wall preparations. However separating the bioactivities caused by the selenomethionine-containing yeast proteins from the activity caused by the polysaccharide parts of selenized yeast cell walls is difficult. Comparison of genomic activity in tissues taken from animals fed with SelPlex® to those fed with Bio-Mos® spiked with selenomethionine indicates very large differences in the regulation of multiple groups of genes for both treatments (Kwiatkowski et al., 2011), which may indicate that the bioactivity of Sel-Plex® involves cooperation between the selenoprotein and the polysaccharide components of selenized yeast cell walls.

Indeed, selected extracts from selenized yeast/yeast cell walls (Kwiatkowski et al., 2011) show potential as future, possible treatments of diseases such as type 2 diabetes, cancer and Alzheimer's.

5. β-(1→3)(1→6)-D-glucan as valuable by product from yeast fermentation

The commercial source from which the bulk yeast cell wall polysaccharides (including β-glucan) are produced uses the same strains of yeast as are used in fuel alcohol production. Le Saffre/ADM, Lallemand, Enzyme Development (New York) and DSM Life Sciences (Delft) are the largest suppliers of yeast for fuel ethanol producers in the United States and the European Union. Major factors that affect yeast cell wall composition include yeast strain (Hahn-Hagerdal et al., 2005), growth conditions (growth medium, temperature, osmotic pressure, toxic metabolites) and the time of harvesting (Aguilar-Uscanaga & Francois, 2003; Klis et al., 2002; Klis et al., 2006). Fuel alcohol fermentation is a high stress process (Devantier et al., 2005) and the cell walls of the yeast collected as its byproduct contain a high amount of β-glucans (Basso et al., 2008; Knauf & Kraus, 2006; Jones & Ingledew, 1994). In general, yeast strains of *Saccharomyces cerevisiae* that are used in baking (baker's yeast) have a higher β-glucan-to-α-mannan ratio than those that are used for alcohol fermentation (brewer's yeast), therefore it is advantageous to use pure baker's yeast for producing high quality (1→3)(1→6)-β-D-glucan for medicinal applications (Kim et al., 2007). The process of separating various yeast components has been heavily patented. However, the differences in the technologies are minor and in principle do not differ from the methodology described by Manners and Fleet (1976). The process starts from autolysing yeast cells at a temperature between 45°C and 65°C at slightly acidic pH, to release yeast cell walls that are insoluble and denser than the cytoplasmic contents and can be separated by centrifugation (Wheatcroft et al., 2002). These steps can be followed by incubating the yeast cell walls with alkaline protease at a pH of 9 to 10 to solubilize mannans and leave behind insoluble β-glucan (Zapata et al, 2008), which can then be physically separated from the liquid fraction by centrifugation and subsequent washing (Sedmak, 2006). Additional enzymes, like glucoamylase and lipase, can be used to hydrolyze α-glucan from α β-glucan, which is still present in the cell wall material and to solubilize the residue lipids from cell membranes. The final step of β-glucan production is a spray-drying that produces a white-to-maroon colored powder that does not carry any taste or aroma and is useful for feed and food applications. Further alkaline and acidic treatments of the food-grade β-glucan (Kelly,

2001) yields high purity (98.5% β-glucan, <0.1% mannan, 0.4% α-glucan, 0.3% protein, 0.2% chitin) microparticulate β-glucan with reduced molecular weight (from ~1-3 MDa to ~150 kDa) that is much more easily absorbed by the digestive tract and shows improved activity compared with food-grade products containing only ~65% β-glucan. Even further hydrolysis produces soluble yeast β-glucan (Jamas et al., 1998; Lee et al., 2001) that still retains most of the particular β-glucan bioactivities (Janusz et al., 1986; Wakshull et al., 1999).

Yeast *Saccharomyces cerevisiae*, its cell wall and products of its fractionation are generally recognized as safe (GRAS) by the US Food and Drug Administration (FDA, 1997), and they can be legally used as food ingredients but not as food additives. The European Food Safety Authority (EFSA) issued an opinion that yeast β-glucans are a "safe food ingredient" (EFSA, 2011) that can be used as a "food supplement" up to 375 mg/day and in foods for "particular nutritional uses" at dose levels up to 600 mg/day. (The uses of yeast cell wall as an animal feed ingredient were discussed in section 4 of this chapter).

Food-grade yeast β-glucans such as BetaRight® and WGP® (Biothera, Inc.) are used as ingredients in baked foods, beverages, ceral, yogurt, fruit juices, chocolate and as food thickeners in salad dressings, ice cream, mayonnaise, sauces and cheese. The majority of these applications have been patented (Zechner-Krpan et al., 2009; Thammakiti et al., 2004) and a critical review of 300 patented applications is available (Laroche & Michaud, 2007). Yeast β-glucans improve food rheological properties, gelling, water and oil-holding properties, without impacting its taste or odor (Petravic-Tominac et al., 2011). Beta-glucans also add health benefits (Laroche & Michaud, 2007) like antioxidative, bacteriostatic and immunostimmulating activities. Cosmetic products used in skin treatment contain yeast β-glucans as moisturizing and moisture-retaining components that also provide a proper moistening feeling. Because of its emulsion-stabilizing effects, pleasant texture and antioxidant activity yeast β-glucans can prevent skin injuries caused by solar radiation and therefore are used in sun-screens, oils and gels (Michiko & Yutaka, 2007). Deodorants containing yeast β-glucans have proved to be useful in oral preparations, mouthwashes and diapers (Michiko et al., 2005). Acid-treated cell walls (AYC) can be used as new binders in pharmaceutical formulations and, when mixed with traditional fillers like hydroxypropylcellulose or polyvinylpyrrolidone, yield harder pills with very short (~2 min) dissolution times (Yusa et al., 2002). Its adhesive and biological properties can be also utilized in producing coating for surgical instruments (Klein, 2003) and in the manufacture of packaging for the food industry (Cope, 1987). Its antibacterial and antiviral properties have found application in the control of plant pests (Kitagawa, 2007) and viral invasions (Slovakova et al., 1997).

6. Medicinal application of native and chemically modified forms of β-(1→ 6)(1→3)-D-glucan

Approximately 2000 research and review papers covering β-(1→6)(1→3)-D-glucan bioactivity and its medicinal applications have been published since the 1960's, and the majority of the literature has been reviewed (Bohn & BeMiller, 1995; Kogan, 2000; Zekovic et al., 2005; Vetvicka, 2011; http://www.betaglucan.org; and for patented applications see Laroche & Michaud, 2007). Many early assumptions have been confirmed by more rigorous studies, and many others have been disproven. Many of the studies that initially showed promise utilized poorly purified and not sufficiently standardized samples of yeast cell wall glucan (Jaehrig et al., 2007; Vetvicka & Vetvickova, 2007). Highly purified, insoluble, whole-glucan particles (**WGP**) with the size 2-4 μm (Yan et al., 2005), or soluble, yeast poly-(1→6)-β-D-glucopyranosyl-(1→3)-β-D-glucopyranose (**PGG**) with molecular weight ~150 kDa (Gawronski et al., 1999) is presently used in animal and human studies. Glucan samples used in earlier studies could contain significant quantities of (also) bioactive impurities like yeast mannoproteins or α-glucans. The existence of a correlation between the β-(1→6)(1→3)-D-glucan structure and its bioactivity is well established having been confirmed by multiple animal and human studies (Novak & Vetvicka, 2009). The WGP-glucan fate after oral administration and PGG-glucan after intravenous administration has been studied in animals (Vetvicka & Vetvickova, 2008). Mammalian cells lack β-glucan, which is present in the cell wall of many infectious microbes. Thus, mammalian macrophages, which are present "on guard" in the cells lining the digestive track, recognize and bind purified β-glucan, but not the whole yeast cells (which have mannoproteins on the cell surface), via the dectin-1 receptor, which is a glucan-sensing receptor expressed on dectin a transmembrane protein. The macrophages transport β-glucan molecules to the spleen, bone marrow and lymph nodes where they are disintegrated into smaller, soluble, β-glucan oligosaccharides that are released back into the bloodstream. The circulating monocytes, macrophages, neutrophils, natural killer and dendritic cells (Chan et al., 2009) which possesses β-glucan recognizing receptors, including: toll like receptor 2 (TLR-2, Underhill et al., 1999), dectin-1 (Brown et al., 2007), complement receptor 3 (CR-3) (Ross et al., 1987) and lactosyl ceramide receptor, bind to the β-glucan molecules triggering the non-specific-innate immune response, including phagocytosis and production of proinflammatory factors (Qi et al., 2011). This mechanism leads to the elimination of infectious agents (Chan et al., 2009). Cancer cells are recognized by the host immune system, but antibody response is too weak to destroy them. When humanized, anti-cancer monoclonal antibodies are used in cancer therapy the treatments are not uniformly effective. Combined antitumor monoclonal antibodies (mAb's) and β-glucan therapy yields much better results with fewer adverse side effects (Salvador et al., 2008; Liu et al., 2009). The mechanism of action is only partially understood with neutrophils and macrophages playing a role in the process of killing cancer cells (Liu et al., 2009). Strengthening immunity to various diseases (Hofer & Pospisil, 2011) and combating those that are already active in humans and animals are the main applications of yeast β-glucan. In Eastern culinary tradition, eating mushrooms (which contain large quantities of soluble β-glucan in their fruiting bodies) has long been recognized as healthy, and four pharmaceutical preparations, based on mushroom extracts, have been registered as drugs in Japan (Hyodo et al., 2005). In the Western world several clinical trials (Babineu et al., 1994; Weitberg, 2008; Spruijt et al., 2010) proved the beneficial properties of WGP and PGG glucans, which are still waiting to be registered as drugs (Vetvicka, 2011; Lehtovaara & Gu, 2011). The following therapeutic activities of yeast β-glucan are well-established in treating the following conditions: post-surgical infections, hospital pneumonia, acute renal failure (Koc et al., 2011), pressure ulcers (bed sores), wound healing (surgical and as result of injury, Spruijt et al., 2010) and burns caused by heat, UV or X-ray radiation. As an adjuvant yeast β-glucans alleviate stress and improves the mAb's plus PGG anticancer treatment for breast, colorectal, colon, leukemia, lung, ovarian and skin cancers, chemotherapy and radiotherapy; antireumatic drug therapy (Sener et al., 2006), and antifungal therapies. They stimulate bone marrow healing and bone repair (due to injuries), which increases red blood cell count and neutrophil production. Many of these activities were studied in the 1990s and 2000s and have been reviewed. Studies since 2010 have been referred to herein.

Some of the negative side effects that have been observed in response to prolonged or large doses of β-glucan preparations are the consequences of their mode of action as well as their physical and chemical properties. Their therapeutic activity is based upon immune system mobilization and the production of monocytes, macrophages, neutrophils, natural killer and dendritic cells which fight and destroy pathogens. However, prolonged pro-inflammatory changes in cells can cause the development of autoimmune diseases. The human body is missing enzymes that can hydrolyze β-glucan polysaccharides and for their degradation uses less efficient oxidative pathways. This insufficiency extends the cellular half-life of WGP and PPG glucans and can lead to the formation of granulomas in the liver, causing inflammation that can result in liver cirrhosis. Therefore the use of soluble and relatively low molecular weight PGG glucan is recommended. The well documented β-glucan abilities to lower blood sugar and blood pressure justify further study aimed at producing a glucan material with molecular weight lower than that of PGG glucan to preclude accumulation in granulomas, when used as a long-term food additive. The next stage of development of less toxic, soluble β-glucan preparations should include its size fractionation to produce more standardized material for medicinal applications. The majority of *in vivo* studies on β-glucan medicinal applications were done using rodent models, therefore it is easy to foresee veterinary applications of this food/feed ingredient to improve immune systems, facilitate burn and wound healing, stimulate post-trauma bone repair and help fight cancers.

Author details

Stefan Kwiatkowski
Alltech Inc. Nicholasville KY, USA

Stefan Edgar Kwiatkowski
University of Kentucky, USA

Acknowledgement

The author expresses deep gratitude to Dr. Pearse T. Lyons, President, and Dr. Karl Dawson, CEO, Alltech Inc. for their continuing support and inspiration.

7. References

Aimaniada, V., Clavaud, C., Simenel, C., Fontaine, T., Delepierre, M. & Latage, J-P. (2009). Cell wall (1→6)-β-D-glucan of *Saccharomyces cerevisiae* – structural characterization and in situ synthesis, *The Journal of Biological Chemistry* Vol. 284: 13401-13412.

Aguilar-Uscanaga, B., & Francois, J.M. (2003). A study of the yeast cell wall composition and structure in response to growth conditions and mode of cultivation, *Letters in Applied Microbiology* Vol. 37: 268-274.

Albright, J.D. & Goldman, L. (1967). Dimethyl sulfoxide-acid anhydride mixtures for oxidation of alcohols, *Journal of the American Chemical Society* Vol. 89: 2416.

Arvindekar, A.U. & Patil, N.B., (2002). Glycogen – a covalently linked component of the cell wall in Saccharomyces cerevisiae, *Yeast* Vol. 19(No. 2): 131-139.

Babineau, T.J., Marcello, P., Swails, W., Keuler, A., Bistrian, B. & Forse, A. (1994). Randomized Phase I/II Trial of a Macrophage-Specific Immunomodulator (PGG-Glucan) in High Risk Surgical Patients, *Annals of Surgery* Vol. 220(No. 5): 601-609.

Bacon, J.S.D., Farmer, V.C., Jones, D. & Taylor I.F. (1969). The glucan components of the cel wall of Baker's Yeast *(Saccharomyces cerevisiae)* considered in relation to its ultrastructure, *Biochemistry Journal* Vol. 114: 557-567.

Ballou, C.E. (1980). Genetics of mannoprotein biosynthesis" in *Fungal Polysaccharides*, Sandford, P., et al; ACS Symposium Series; American Chemical Society: Washington D.C. (1980) p 1-14.

Basso, L.C., de Amorim, H.V., de Oliveera, A.J. & Lopes, M.L. (2008). Yeast selction for fuel alcohol production in Brazil, *FEMS Yeast Research* Vol. 8(No. 7): 1155-1163.

Baxter, E.W. & Reitz, A.B. (2002). Reductive aminations of carbonyl compounds with borohydride and borane reducing agents, *Organic Reactions* Vol. 59: 1. Wiley: New York 2002.

Baurhoo, B., Ferket, P.R. & Zhao, X. (2009). Effects of diets containing different concentrations of mannooligosaccharides or antibiotics on growth performance, intestinal development, cecal and litter microbial populations, and carcass parameters in broilers, *Poultry Science* Vol. 88(No.11): 2262-2272.

Bennet, J.W. & Klich, M. (2003). Mycotoxins, *Clinical Microbiology Reviews*. Vol. 16(No. 3), 497-516.

Borchers, A.T., Keen, C.L. & Gershwin, M.E.et al. (2004). Mushrooms, tumors and immunity: an update, *Experimental Biology and Medicine* Vol. 229(No. 5): 393-406.

Bohn, J.A. & Be Miller, J.N. (1995). (1→3)(1→6)-β-D-Glucans as biological response modifiers: a review of structure-functional activity relationships, *Carbohydrate Polymers* Vol. 28(No. 1): 3-14.

Brady, D., Stoll, A.D. & Duncan, J.R. (1994). Chemical and enzymatic extraction of heavy metal binding polymers from isolated cell walls of *Saccharomyces cerevisiae, Biotechnol. Bioeng. (USA)* Vol. 44(No.3), 297-302.

Brown, J., O'Callaghan, Ch., Marshall, A.S.J., Gilbert, R.J.C., Siebold, Ch., Gordon, S., Brown, G.D. & Jones, Y.J. (2007). Structure of the fungal β-glucan binding receptor dectin-1: implications for function, *Protein Science* Vol. 16: 1042-1052.

Chan, G.Ch-F., Chan, W.K. & Sze, D.M-Y. (2009). The effects of β-glucan on human immune and cancer cells, *Journal of hematology & oncology* Vol. 2: 25.

Cope, Roy (1987) Adhesive substances from yeast cell wall, *Eur. Pat. Appl.* EP 230378 A2.

Dake, M.S., Jadhv, J.P. & Patil, N.P. (2010). Variations of two pools of glycogen and carbohydrate in *Saccharomyces cerevisiae* grown with various ethanol concentrations, *Journal of Industrial Microbiology & Biotechnology* Vol. 37(No. 7): 701-706.

Demirici, A. & Pometto III, A.L. (1999). Production of organically bound selenium yeast by continuous fermentation, *J. Agric. Food Chem.* Vol. 47(No. 6): 2491-2495.

Demirici, A., Pometto III, A.L. & Cox, D.J. (1999). Enhanced organically bound selenium yeast production by fed-batch fermentation, *J. Agric. Food Chem.* Vol. 47(No. 6): 2496-2500.

Devantier, R., Pedersen, S. & Olsson, L. (2005). Characterization of very high gravity ethanol fermentation of corn mash. Effect of glucoamylase dosage, pre-saccharification and yeast strain, *Appl Microbiol Biotechnol* Vol. 68: 622-629.

Dimitroglu, A., Merrifield, D.A., Moate, R., Davies, S.J., Spring, P., Sweetman, J. & Bradley, G. (2009). Dietary mannan oligosaccharide supplementation modulates intestinal microbial ecology and improves gut morphology of rainbow trout, *Oncorhynchus mykiss (Walbaum), Journal of Animal Science* Vol. 87(No. 10): 3226-3234.

Dvorska, J.E. & Surai, P.F. (2003). Protective effect of modified glucomannans against aurofusarin induced changes in quail egg and embryo" *Asian-Australian Journal of Animal Science* Vol.17(No.3): 434-440.

Ebina, T., Kubota., Ogama, N. & Matsunaga, K. (2002). Antitumoreffect of a peptide-glucan preparation extracted from a mycelium of *Tricholoma matsutake* (S. Ito and Imai) Sing, *Biotherapy* Vol. 16(No. 3): 255-259.

EFSA. (2011) Scientific opinion on the safety of 'yeast β-glucans' as novel food ingredient, *EFSA Journal* Vol. 9(No. 5): 2137-2159.

FDA. (1997) Substances generally recognized as safe, *Federal Register* 62 FR 18938, 17 April 1997.

Garcia, I., Tajadura, V., Martin, V., Toda, T. & Sanchez, Y. (2006). Synthesis of α-glucans in fission yeast spores is carried by three α-glucan synthase paralogs Mok12p, Mok13p and Mok14p, *Molecular Microbiology* Vol. 59(No. 3): 836-853.

Gawronski, M., Park, J.T., Magee, A.S. & Conrad, H. (1999). Microfibrillar structure of PGG glucan in aqueous solution as triple helix aggregates by small angle x-ray scattering, *Biopolymers* Vol. 50: 569-578.

Grba, S., Oura, E. & Soumalainen, H. (1976). Formation of glycogen and trehalose in baker's yeast, *European Journal of Applied Microbiology* Vol. 2(No. 1): 29-37.

Grube B.J., Eng, E.T., Kao, Y.C., Kwon, A. & Chen, S. (2001). White button mushroom phytochemicals inhibit aromatase activity and breast cancer cell proliferation, *J. Nutrition* Vol. 131(No. 12): 3288–93.

Grün, Ch.H., Hochstenbach, F., Humbel, B.M., Verkleij, A.J., Sietsma, J.H., Klis, F.M., Kamerling, J.P. & Vliegenthart, J.F.G. (2005). The structure of cell wall α-glucan from fission yeast, *Glycobiology* Vol. 15(No. 3): 245-257.

Gunia-Smith, Z., Patil, N.B. & Smith, E.E., (1977). Two pools of glycogen in *Saccharomyces*, *Journal of Bacteriology* Vol. 1977: 818-825

Hahn-Hagerdal, B., Karhumaa, K., Larsson, Ch.U., Gorwa-Grauslund, M., Gorgens, J. & van Zyl, W.H. (2005). Role of cultivation media in the development of yeast strains for large scale industrial use, *Microbial Cell Factories* Vol. 4: 31-47.

Hirotaka, H., Yoko, Y., Kenichi, M., Yoko, I. & Tdeshi, Y. (2007). Development of enzyme-linked immunoassay to detect immunomodulatory α-D-glucan-protein complex, MPG-1, in basidiomycete Tricholoma matsutaka and related processed food, *J. Agric. Food Chem.* Vol. 55(No. 21): 8508-8515.

Hirotaka, H., Yoko, Y., Kenichi, M., Yoko, I. & Tdeshi, Y. (2005). Isolation and characterization of a novel immunomodulatory α-glucan-protein complex from the mycelium of *Tricholoma matsutake* in basidiomycetes, *J. Agric. Food Chem.* Vol. 53(No. 23): 8948-8956.

Hofer, M. & Pospisil, M. (2011). Modulation of animal and human hematopoiesis by β glucans: a review, *Molecules* Vol. 16: 7969-7979.

Holliday J., Cleaver, M. (2008). Medicinal Value of the Caterpillar Fungi Species of the Genus Cordyceps (Fr.) Link (Ascomycetes), *International Journal of Medicinal Mushrooms* Vol. 10(No. 3): 219

Huang, D., Farkas, I. & Roach, P.J. (1996). Pho85p, a cyclin-dependent protein kinase, and the Snf1p protein kinase act antagonistically to control glycogen accumulation in *Saccharomyces cerevisiae, Molecular and Cellular Biology* Vol. 16(No. 8): 4357-4365.

Hutchins, K. & Bussey, H. (1983). Cell wall receptor for yeast killer toxin: involvement of (1, 6)-β-D-glucan, *Journal of Bacteriology* Vol. 154: 161-169.

Hyodo I., Amano, N., Eguchi, K., Narabayashi, M., Imanishi, J., Hirai, M., Nakano, T. & Takashima S. (2005). Nationwide survey on complementary and alternative medicine in cancer patients in Japan, *Journal of Clinical Oncology* Vol. 23(No. 12): 2645–54.

Jacques, K.A. & Newman, K.E. (1994). Effect of oligosaccharide supplements on performance and health of Holstein calves pre- and post-weaning, *Journal of Animal Science* Vol. 72(Suppl. 1): 295.

Jadhav, J.P. Dake, M.S. & Patil, N.P. (2008). α, β-Glucan complex in *Saccharomyces carlsbergensis, Asian Journal of Chemistry* Vol. 20(No. 1): 55-65.

Jaehrig, S.C., Rohn, S., Kroh, L.W. Fleisher, L-G. & Kurtz, T. (2007). In vitro potential antioxidant activity of (1→3) (1→6)-β-D-glucan and protein fractions from *Saccharomyces cerevisiae* cell walls, *J. Agric. Food Chem.* Vol. 55(No. 12): 4710-4716.

Jamas, S., Easson Jr, D.D. & Ostroff, G.R. (1997). Glucan preparation, *US Patent* US 5622939, Apr 22, 1997.

Jamas, S., Easson Jr, D.D. & Ostroff, G.R. (1998). Underivatized, aqueous soluble β-(1, 3)-glucan, composition and making some, *US Patent* US 5817643, Oct 6, 1998.

Janusz, M.J., Austen, K.F. & Czop, J.K. (1986). Isolation of soluble yeast beta-glucan that inhibit human monocyte phagocytosis mediated by beta-glucan receptors, The Journal of Immunology Vol.137 (No. 10):3270-3276.

Jones, A.M. & Ingledew, W.M. (1994). Fuel alcohol production: optimization of temperature for very efficient very-high gravity fermentation, *Applied and Environmental Microbiology* Vol. 1994: 1048-1051.

Kakutani, R., Adachi, Y., Kajiura, H., Takata, H., Kuriki, T. & Ohno, N. (2007). Relationship between structure and immunostimulating activity of enzymatically synthesized glycogen, *Carbohydrate Research* Vol. 342: 2371-2379.

Kaptein, J.C., Van Den Ende, H. & Klis, F.M. (1999). The contribution of cell wall proteins to the organization of the yeast cell wall, *Biochimical Biophysical Acta-General Subjects* Vol. 1426(No. 2): 373-383.

Karaman, M., Basmocioglu, H., Ortatatli, M. & Oguz, H. (2005). Evaluation of detoxifying effect of yeast glucomannan on aflatoxicosis in broilers as assessed by gross examination and histopathology, *British Poultry Science* Vol. 46(No. 3): 394-400.

Kath, F. & Kulicke, W-M. (1999). Mild enzymatic isolation of mannan and glucan from yeast *Saccharomyces cerevisiae, Die Angewandte Makromoleculare Chemie* Vol. *268*: 59-68.

Kelly, G.E., (2001) Process for glucan preparation and therapeutic uses of glucan, *US Patent* US 6,242,594 B1.

Khan, M.A., Mousumi, T., Dian-Zheng Zhang & Han-Chun Chen (2010). *Cordyceps* Mushroom: A Potent Anticancer Nutraceutical, *The Open Nutraceutical Journal* Vol. 3: 179–183.

Kiho, T., Hui, J., Yamane A. & Ukai S. (1993). Polysaccharides in fungi. XXXII. Hypoglycemic activity and chemical properties of a polysaccharide from the cultural mycelium of Cordyceps sinensis, *Biol. Pharm. Bull.* Vol. 16(No. 12): 1291–1293.

Kiho, T., Hui, J., Yamane A. & Ukai S. (1996). Polysaccharides in fungi. XXXVI. Hypoglycemic activity of a polysaccharide (CS-F30) from the cultural mycelium of Cordyceps sinensis and its effect on glucose metabolism in mouse liver, *Biol. Pharm. Bull.* Vol. 19(No. 2): 294–296.

Kim, Y-H., Kang, S.W., Lee, J.H., Chang, H-I., Yun, Ch-W., Paik, H-D., Kang, Ch-W. & Seung, W. (2007). High density fermentation of *Saccharomyces cerevisiae* JUL3 in fed-batch culture for the production of β-glucan, *Journal of Industrial and Engineering Chemistry (Seoul)* Vol. 13(No. 1): 153-158.

Kitagawa, T. (2007). Yeast cell wall degradation product as plant pest control agent, *Jpn. Kokai Tokkyo Koho* JP 2007045708 A.

Klein, B (2003). Immunostimulating coating for surgical devices, *US Patent* US 6541678 B2.

Klis, FM., Mol, P., Hellingwerf, K. & Brul S. (2002) Dynamics of cell wall structure in *Saccharomyces cerevisiae, FEMS Microbiology Reviews* Vol. 26(No. 3): 239-256.

Klis, F.M., Boorsma, A. & DeGroot, P.W.J. (2006). Cell wall construction in *Saccharomyces cerevisiae, Yeast* Vol. 23: 185-202.

Knauf, M. & Kraus, K. (2006). Specific yeast developed for modern ethanol production, *Sugar Industry* Vol. 131: 753-758.

Koc, E., Reis, K.A., Ebinc, F.A., Pasaoglu, H., Demirtas, C., Omeroglu, S., Derici, U.B., Erten, Y., Bali M., Arinsov, T. & Sindel, S. (2011). Protective effect of beta-glucan on contrast induced-nephropathy [acute renal failure] and a comparison of beta-glucan with nebivolol and N-acetylcysteine in rats, *Clin. Exp. Nephrol.*Vol. 15(No. 5): 658-665.

Kogan, G. (2000). (1→3, 1→6)-β-D-Glucans of yeast and fungi and their biological activity, *Studies in Natural Products Chemistry* Vol. 23: 107-151/Atta-ur-Rahman (Ed.) Elsevier 2000.

Kogan, G. & Koher, A. (2007). Role of yeast cell wall polysaccharides in pig nutrition and health protection, *Livestock Sci.* Vol. 109: 161-165.

Kogan, G., Pajtinka, M., Babincova M., Miadokova, E., Ranko, P., Slamenova, D. &

Korolenko, T.A., (2008).Yeast Cell Wall Polysaccharides as antioxidants and antimutagens: can they fight cancer? *Neoplasma* Vol. 55: 5.

Kojima, H., Akaki, J., Nakajima, S., Kamei, K. & Tamesada, M. (2010). Structural analysis of glycogen like polysaccharides having macrophage-activating activity in extracts, *J. Nat. Med.* Vol. 64:16-23.

Kollar, R., Reinhold, B.B., Petrakova, E., Yeh, H.J.Ashwell, G., Drgonova, J., Kaptein, J.C.,

Klis, F.M. & Cabib E., (1997). Architecture of the yeast cell wall β-(1-6)-glucan interconnects mannoprotein, β-(1-3) glucan and chitin, *The Journal of Biological Chemistry* Vol. 272(No. 28): 17762-17775.

Koppada, R., Norozian, F. N., Torbati, D., Kalomiris Sophia, Ramachandran, Ch. & Totapally B.R. (2009). Physiological effects of a novel immune stimulator drug (1→4)-α-D-glucan in rats, *Basic and Clinical Pharmacology and Toxicology* Vol. 105(No. 4): 217-221.

Krizkova L., Zitnanova, I., Mislovicova, D., Masarova, J., Sasinkova, V., Durackova, Z. & Krajovic, J. (2006). Antioxidant and antimutagenic activity of mannan neoglycoconjugates: mannan-human serum albumin and mannan-penicilin G acylase, *Mutation Research/Genetic Toxycology and Environmental Mutagenesis* Vol. 606(No. 1-2): 72-79.

Kwiatkowski S., Thielen, U., Glenny, P. & Moran, C. (2009). A study of *Saccharomyces cerevisiae* cell wall glucans, *The Institute of Brewing & Distilling* Vol. 115(No. 2):151-158.

Kwiatkowski, S., Powers, R., Matney, C., Ghorghchian, P.P. & Ostertag, E.M. (2011). Compositions and methods for separating, characterizing and administering selenoglycoproteins, USPTO Application: 20110280948.

Laroche, C. & Michaud, P. (2007). New developments and perspective applications for β- (1, 3)-glucans, *Recent Patents on Biotechnology* Vol. 2007(No. 1): 59-73.

Lavi, I., Friesem, D., Geresh, S., Hadar, Y. & Schwartz, B. (2006). An aqueous polysaccharide extract from the edible mushroom *Pleurotus ostreaus* induces antiproliferative and pro-apoptotic effects on HT-29 colon cancer cells, *Cancer Letters (Amsterdam, Netherlands)* Vol. 244(No. 1): 61-70.

Lee, J.N., Lee, D.Y., Kim, G.E., Kim, H.N., Sohn, J., Kim, S. & Kim, C.W. (2001). Purification of soluble beta-glucan with immune-enhancing activity from cell wall of yeast, *Biosci. Biotechnol. Biochem.* Vol. 65(No. 4): 831-841.

Lei, H., Ma, X. & Wu, W. (2007). Anti-diabetic effect of α-glucan from fruiting body of maitake (*Grifola frondosa*) on KK-Ay mice, *Journal of Pharmacy and Pharmacology* Vol. 59(No. 4): 575-582.

Lehtovaara, B.C. & Gu, F.X. (2011). Pharmacological, Structural, and Drug Delivery Properties and Applications of 1, 3-β-Glucans, *J. Agric. Food Chem.* Vol. 59(No. 5): 6813-6828.

Lessage, G. & Bussey, H. (2006). Cell wall assembly in *Saccharomyces cerevisiae, Microbiology and Molecular Biology Reviews* Vol. 70(No. 2): 317-343.

Leung, M.C.K., Diaz-Llano, G. & Smith, T.K. (2006). Mycotoxins in pet food: a review on worldwide prevalence and preventive strategies, *J. Agric. Food Chem.* Vol. 54 (No. 26): 9623-9635.

Li, S.P., Zhang, G.H., Zeng, Q., Huang, Z.G., Wang, Y.T., Dong, T.T.X & Tsim, K.W.K. (2006). Hypoglycemic activity of polysaccharide, with antioxidation, isolated from cultured *Cordyceps mycelia, Phytomedicine* Vol. 13(No. 6): 428–433.

Li H., Lu, X., Zhang, S., Lu, M. & Liu, H. (2008). Anti-inflammatory activity of polysaccharide from *Pholiota nameko, Biochemistry (Moscow)* Vol. 73(No. 6): 669-675.

Lichtenthaler, F.W. & Peters, S. (1980). Carbohydrates as green raw materials for the chemical industry, *Comptes Rendus Chimie* Vol. 7(No. 2): 65-90.

Lille, S.H. & Pringle, J.R. (1980). Reserve carbohydrate metabolism in *Saccharomyces cerevisiae*: response to nutrition limitation, *Journal of Bacteriology* Vol. 143: 1384-1394.

Lindhorst, T.K. (2003). Essentials of carbohydrate chemistry and biochemistry, Wiley-VCH, Weinheim, ISBN 3-527-30664-1

Liu, J., Gunn, L., Hansen, R. & Yan, J. (2009). Combined yeast derived β-Glucan with antitumor monoclonal antibody for cancer immunotherapy, *Experimental and Molecular Pathology* Vol. 86(No. 3): 208-214.

Liu, W-Ch., Wang, S-Ch., Tsai, M-L., Chen, M-C., Wang, Y-C., Hong, J-H., McBride W. & Chiang, Ch-S. (2006). Protection against radiation-induced bone marrow and intestinal injuries by *Cordyceps sinensis*, a Chinese herbal medicine, *Radiation Research* Vol. 166(No. 6): 900–907.

Lovell, M.A., Xiong, S., Lyubartseva, G. & Markesbery, W.R. (2009). Organo-selenium (Sel Plex) decreases amyloid burden and RNA and DMA oxidative damage in APP/PS1 mice, *Free. Radic. Biol. Med.* Vol. 46(No. 11): 1527-1533.

Mach C.M., Fugii, H., Wakame, K. & Smith, J. (2008). Evaluation of active hexose correlated compound hepatic metabolism and potential for drug interactions with chemotherapy agents, *Journal of the Society for Integrative Oncology* Vol. 6(No. 3): 105–9.

Manners, D.J. & Fleet, G.H. (1976). Isolation and composition of alkali soluble glucan from cell walls of *Saccharomyces cerevisiae, J. Gen. Microbiol.* Vol. 94: 180-192.

Manners, D.J., Masson, A.J., Patterson, J.C., Bjorndal, H. & Lindberg L., (1973) The structure of β-(1→6)-D-glucan from Yeast cell wall, *Biochemical Journal* Vol. 135: 19-30.

Mansour, M.K. & Levitz S.M. (2003). Fungal mannoproteins: the sweet path to immunodominance, *ASM News* Vol. 69 (No. 12): 595-560.

Mapelli, V., Hillestrom, P.R., Kapolna, E., Larsen, E.H. & Olsson, L. (2011). Metabolic and bioprocess engineering for production of selenized yeast with increased content of seleno-methylselenocysteine" *Metabolic Engineering* Vol. 13(No. 3): 282-293.

McLean J.E. & Bledsoe B.E. (1992). Behavior of metals in soils, *EPA Ground Water Issue 1992,* EPA/540S-92/018.

Michiko, Y., Tomohiko, N. & Yutaka, K. (2005). Deodorants containing yeast cell wall fraction, *Jpn. Kokai Tokkyo Koho* JP 2005325065 A.

Michiko, Y. & Yutaka, K. (2007) Skin conditioning composition containing cell wall fractions, JP 2007045714 Feb 22, 2007.

Moran, C., Kwiatkowski, S., Yannikouris, A. & Thielen, U. (2011). Yeast cell wall components and detection thereof, *International Patent Application* WO 2011143613 (A1) 20-11-17.

Morrison S.J., Dawson, S. & Carson, A.F. (2010). The effects of mannan oligosaccharide and *Streptococcus faecium* addition to milk replacer on calf health and performance, *Livestock Science* Vol. 131(No. 2): 292-296.

Novak, M. & Vetvicka, V. (2009). Glucans as biological response modifiers, *Endocrine, Metabolic and Immune Disorders-Drug Targets* Vol. 9: 67-75.

Ohno, N., Mimura, H., Suzuki, I. & Yadomae, T. (1985). Antitumor activity and structural characterization of polysaccharide fractions extracted with cold alkali from a fungus, *Peziza vesiculosa, Chem. Pharm. Bull. (Tokyo)* Vol. 33(No. 6): 2564-8.

Osumi, M. (1998) The ultrastructure of yeast: cell wall structure and formation, *Micron* Vol. 29: 207-233.

Patel, G.B. & Ingledew, W.M. (1975). The relationship of acid soluble glycogen to yeast flocculation, *Can. J. Microbiol.* Vol. 21: 1608-1613.

Petravic-Tominac V., Zechner-Krpan, V., Berkovic, K., Golovic, P. Hercog, Z., Srecec, S. & Spoliaric, I. (2011). Rheological properties, water-holding and oil-binding capacities of particulate β-glucans isolated from spent Brewer's yeast by three different procedures, *Food Technol. Biotechnol.* Vol. 49(No. 1): 56-64.

Qi, C., Cai, Y., Ding, Li. B., Kloecker, G., Qian, K., Vasilakos, J., Saijo, S., Iwakura, Y., Yannelli, J.R. & Yan, J. (2011). Differential pathways regulating innate and adaptive antitumor immune responses by particulate β-glucan, *Blood* Vol. 117(No. 25): 6825-6836.

Quain, D.E. (1981). The determination of glycogen in yeast, *J.Inst.Brew.* Vol. 87: 289-291.

Rayman, M.P. (2000). The importance of selenium to human health, *Lancet* Vol. 356: 233-241.

Raymond, S.L., Smith, T.K. & Swamy H.V.L.N. (2003). Effects of feeding a blend of grains naturally contaminated with *Fusarium* mycotoxins on feed intake, serum chemistry and hematology of horses and the efficacy of a polymeric glucomannan mycotoxin adsorbent, *J. Anim. Sci.* Vol. 81: 2123-2130.

Ren Z., Guo, Z., Meydani, S.N. & Wu, D. (2008). White button mushroom enhances maturation of bone marrow-derived dendritic cells and their antigen presenting function in mice, *J.Nutrition* Vol. 138 (No. 3): 544–50.

Rodrigo, M-J., Moskovitz, J., Salamini, F. & Bartels, D. (2002). Reverse genetic approaches in plants and yeast suggest a role for novel, evolutionary conserved, selenoprotein-related genes in oxidative stress defense, *Mol. Genet. Genomicks* Vol. 267: 613-621.

Rop, O., Mlcek, J. & Jurikova, T. (2009) Beta-glucans in higher fungi and their health effects, *Nutrition Reviews* Vol. 67 (No. 11): 624–31.

Rosen, G.D. (2007). Holo-analysis of the efficacy of Bio-Mos in pig nutrition, *Animal Science* Vol. 82: 683-689.

Ross, G.D., Cain, J.A., Moynes, B.L., Newman, S.L. & Lachman, P.J. (1987). Role of complement receptor type three and serum opsonins in the neutrophil response to yeast, *Complemment Inflamm.* Vol. 4: 61-74.

Rothman-Denes, L.B. & Cabib, E. (1970). Two forms of glycogen synthetase and their role in glycogen accumulation, *P.N.A.S.* Vol. 66: 967-974.

Salvador, C., Li, B., Haqnsen, R., Cramer, D.E., Kong, M & Jun, Y. (2008). Yeast derived β glucan augments the therapeutic efficacy mediated by anti-vascular endothelial growth factor monoclonal antibody in human carcinoma xenograft models, *Clin. Cancer Res.*Vol. 14(No. 4): 1239-1247.

Sedmak, JJ. (2006). Production of β-glucans and Mannans, *US Patent Application* Nov 23, 2006 US 2006/0263415 A1.

Sener, G., Eksioglu-Demiraop, E., Cetiner, M., Ercan, F. & Yegen, B.C. (2006). Beta-glucan ameliorates methotrexate-induced oxidative organ injury via its antioxidant and immunomodulatory effects, *European J Pharmacology* Vol. 542(No. 1-3): 170-178.

Shah, S.K., Walker, P.A., Moore-Olufemi, S.D., Sundaresan, A., Kulkarini, A.D. & Andrassy, R.J. (2011). An evidence-based review of a *Lentinula edodes* mushroom extract as complementary therapy in the surgical oncology patient, *Journal of Parenteral and Enteral Nutrition* Vol. 35(No. 4): 449–58.

Shida, M. & Matsuda, K. (1974). Studies on polysaccharides from *Lentinus edodes* (Shii-ta-ke), *Mushroom Science* Vol. 9(pt1): 531-539.

Slaughter, J.C. & Nomura, T. (1992). Intracellular glycogen and trehalose contents as predictors of yeast viability, *Enzyme and Microbial Technology* Vol. 14(No. 1): 64-67.

Slovakova, L., Subikova, V., Sandula, J. & Huncikova, S. (1997) An agent for preventive protection of plants from viral diseases, SK 278413 Apr 9, 1997.

Smiderle, F.R., Sassaki, G.L., Van Arkel, J., Jacomini, M., Wickers, H.J. & Van Griensven, L.J.L.D. (2010). High molecular weight glucan of the culinary medicinal mushroom *Agaricus bisporus* in an α-glucan that forms complexes with low molecular weight galactan, *Molecules* Vol. 15(No. 8): 5818-5830.

Smith, J.E., Solomons, G., Lewis, C. & Anderson, J.G. (1995. Role of mycotoxins in human and animal nutrition and health, *Natural Toxins* Vol. 3(No. 4): 187-192.

Smits, G.J., van den Ende, H. & Klis, F.M. (2001). Differential regulation of cell wall biogenesis during growth and development in yeast, *Microbiology* Vol. 147: 781-794.

Spruijt, N.E., Visser, T. & Leenen, L.P.H. (2010). A systematic review of randomized controlled trials exploring the effect of immunomodulative interventions on infection, organ failure, and mortality in trauma patients, *Crit. Care* Vol. 14: R150.

Stewart G.G. & Russell I. (1998). Brewer's Yeast, *Brewing Science & Technology Series III* (1998): 28-46, The Institute of Brewing ISBN No. 0900489 13 8.

Suzuki, I., Yadomae, T., Yonekubo, H., Nishijima, M. & Miyazaki, T. (1982). Antitumor activity of an immunomodulating material extracted from a fungus, *Peziza vesiculosa, Chem. Pharm. Bull. (Tokyo)* Vol. 30(No. 3): 1066-8.

Swanson, K.S., Grieshop, C.M., Fuckinger, E.A., Bauer, U., Healy, H-P., Dawson, K.A., Merchen, N.R. & Fahey Jr., G.C. (2002). Supplemental fructooligosaccharides and mannooligosaccharides influence immune function, ileal and total tract nutrient digestibilities, microbial populations and concentrations of protein catabolites in the large bowel of dogs, *The Journal of Nutrition* Vol. 132(No. 5): 980-989.

Tanaka, A. (2011). Blood glucose suppression effect of Maitake mushroom extract, *BRAIN Techno News* Vol. 146: 30-34.

Tapiero, H., Townsend, D.M. & Tew, K.D. (2003). The antioxidant role of selenium and seleno-compounds, *Biomedicine & Pharmacotherapy* Vol. 57: 134-144.

Tastet, L., Schaumloffel, D., Bouisierre, B. & Lobinski, R. (2008). Identification of selenium containing proteins in selenium rich yeast aqueous extract by 2D gel electrophoresis, nanoHPLC-ICP MS and nanoHPLC-ESI MS/MS, *Talanta* Vol. 75: 1140-1145.

Terakawa, N., Matsui, Y. Satoi, S., Yanagimoto, H., Takahashi, K., Yamamoto, T., Yamao, Y.,

Takai, S., Kwon, A-Hon. & Kamiyama, Y. (2008). Immunological effect of active hexose correlated compound (AHCC) in healthy volunteers: a double-blind, placebo-controlled trial, *Nutrition and Cancer* Vol. 60(No. 5): 643–51.

Thammakiti, S., Suphantharika, M., Phaesuwan, T. & Verduyn, C. (2004). Preparation of brewer's yeast β-glucans for potential application in the food industry, *International Journal of Food Science and Technology* Vol. 39(No. 1): 21-29.

Toborek, M. (2011) University of Miami, USA, Private communication.

Turlo, J., Gutkowska, B. & Malinowska, E. (2007). Relationship between the selenium, selenomethionine and selenocysteine in submerged cultivated mycelium of *Lentinula edodes* (Berk.), *Acta Chromatographica* Vol. 18: 36-48.

Tzianabos, A.O. (2000). Polysaccharide immunomodulators as therapeutic agents: structural aspects and biologic function, *Clinical Microbiology Reviews* Vol. 2000: 523-533

Uden, P.C., Boakye, H.T., Kahakachchi, C., Hafezi, R., Nolibos, P., Block, E., Johnson, S Tyson, J.F. (2003). Element selective characterization of stability and selectivity of selenium species in selenized yeast, *J. Anal. At. Spectrom.* Vol. 19: 65-73.

Underhill, D.M., Ozinsky, A., Hajjar, A.M., Stevens, A., Wilson, C.B., Bassetti, M. & Andersen, A. (1999). The Toll like receptor 2 is recruited to macrophage phagosomes and discriminates between pathogens, *Nature* Vol. 401: 811-815.

Valarezo, S. (1998). Comparative effects of antibiotic, mannooligosaccharide and mycotoxin adsorbent on performance of commercial broilers fed pelleted diets, *Poultry Science* Vol. 77(Suppl. 1): 137.

Vetvicka, V. (2011). Glucan-immunostimulant, adjuvant, potential drug, *World Journal of Clinical Oncology* Vol. 2(No. 2): 115-119.

Vetvicka, V. & Vetvickova, J. (2007). Physiological effects of different types of β-glucan, *Biomed. Pap. Med. Fac. Univ. Palacky Olomouc. Czech Repub.* Vol. 151(No. 2): 225-231.

Vetvicka, V. & Vetvickova, J. (2008). A comparison of injected and orally administered β glucans, *JANA* Vol. 11(No. 1): 42-49.

Villar-Tajadura M., Madrid, M., Casando, J., Santos, B. & Perez, P. (2008). Rga2 is a Rho2 GAP that regulates morphogenesis and cell integrity in *Saccharomyces pombe, Molecular Microbiology* Vol. 70(No. 4): 867-881.

Vinogradov E., Petersen, B. & Bock, K. (1998). Structural analysis of intact polysaccharide mannan from *Saccharomyces cerevisiae* yeast using ^{1}H and ^{13}C NMR spectroscopy at 750MHz, *Carbohydrate Research* Vol. 307(No. 1-2): 177-183.

Vos, A., Dekker, N., Distel, B., Leunissen, J.A.M. & Hochstenbach, F. (2007). Role of the synthase domain of Ags1p in cell wall α-glucan boiosynthesis in Fission Yeast, *J. Biol. Chem.* Vol. 282(No. 26): 18969-18979.

Wakshull, E., Brunke-Reese, D., Lindermuth, J., Fisette, L., Nathans, R.S., Crowley, J.J., Tufs, J.C., Zimmerman, J., Mackin, W. & Adams, D.S. (1999). PGG-Glucan, a soluble β-(1, 3)-glucan enhances the oxidative burst response, microbial activity, and activities NF-κB-like factor in human PMN: evidence for glycosphingolipid β-(1, 3)-glucan receptor, *Immunopharmacology* Vol. 41: 89-107.

Weitberg, A.B. (2008). A phase I/II trial of β-(1→3) (1→6)-D-glucan in the treatment of patients with advanced malignances receiving chemotherapy, *J. Exp. Clin. Cancer Res.* Vol. 27: 40-44.

Wheatcroft, R., Kulandai, J., Gilbert, R.W., Sime, K.J., Smith, C.G. & Langeris, W.H., (2002). Production of β-glucan-mannan preparations by autolysis of cells under certain pH, temperature and time conditions, *US Patent* US 6444448 B1 Sep. 3, 2002.

Wellens, A., Garafalo, C., Nguyen, H., Van Gerven, N., Slattegard, R., Hernalsteens, J-P.,

Wyns, L., Oscarson, S., De Greve, H., Hultgren, S. & Bouckaert, J. (2008). Intervening with urinary tract infections using anti-adhesives based on the crystal structure of the FimH-oligomannose-3 complex, *PLoS ONE* Vol. 3(No. 4): e2040, 1-13.

Wilson, W.A., Wang, Z. & Roach, P.J. (2002). Systematic identification of the genes affecting glycogen storage in the yeast *Saccharomyces cerevisiae, Molecular & Cellular Proteomics* Vol. 1(No. 3): 232-242.

Wismar, R., Pedersen, S.B., Frokiaer, H. & Laerke, H.N. (2010). Dietary fibers as immunoregulatory compounds in health and disease, Annals *of the New York Academy of Sciences* Vol. 1190: 70-85.

Yan, J., Allendorf, D.J. & Bradley, B. (2005). Yeast whole glucan particle (WGP) β-glucan in conjunction with antitumor monoclonal antibodies to trat cancer, *Expert Opin. Biol. Ther.* Vol. 5: 691-702.

Yan, J-K., Wan, W-Q., Li, L. & Wu, J-Y. (2011). Physicochemical properties and antitumor activities of two α-glucans isolated from hot water extract of *Cordyceps* (CS-HKI) fungal mycelia, *Carbohydrate Polymers* Vol. 85(No. 4): 753-758.

Yang, Y., Iji, P., Kocher, A., Mikkelsen, L.L. & Choct, M. (2008). Effects of mannooligosaccharide and fructooligosaccharide on the response of broilers to pathogenic *Escherichia coli* challenge, *British Poultry Science* Vol. 49(No. 5): 550-559.

Yannikouris, A., Francois, J., Poughon, L., Dussap, C-G., Bertin, G., Jeminet, G. & Jouany, J. P. (2004). Alkali extraction of β-D-glucans from *Saccharomyces cerevisiae* cell wall and study of their absorptive properties toward Zeralenone, *J. Agric. Food Chem.* Vol. 52(No. 11): 3666-3673.

Yannikouris, A., Andre, G., Poughon, L., Dussap, C-G., Bertin, G., Jeminet, G. & Jouany, J-P. (2006). Chemical and conformational study of the interactions involved in mycotoxins complexation with β-Glucans, *Biomacromolecules* Vol. 7(No. 4): 1147-1155.

Yusa, H., Katsuyama, H., Ozeki, T., Takashima, Y., Okada, H., Kasai, T., Eguchi, T. & Kakiuchi, H. (2002). Application of acid treated yeast cell wall (AYC) as a novel binder, *Yakuzaigaku* Vol. 62(No. 4): 153-160.

Zapata, G.M., Meza, M. & Meza Mendoza, H.A. (2008). Method for obtaining yeast glucan by autolysis of *Saccharomyces cerevisiae* Baker's yeast cells, *PCT Int. Appl.*, WO2008032134 A1.

Zechner-Krpan, V., Petravic-Tominac, V., Panjkota-Krbavicic, I., Grba, S. & Berkovic, K. (2009). Potential application of yeast β-glucans in food industry, *Agriculturae Conspectus Scientificus* Vol. 74(No. 4): 277-282.

Zekovic, D.B., Kwiatkowski, S., Vrvic, M. M., Jakovlievic, D. & Moran, C. (2005). Natural and modified (1→3)-β-D-glucans in health promotion and disease alleviation, *Critical Reviews in Biotechnology* Vol. 25(No. 4): 205-230.

Zekovic, D.B., Radulovic, M., Nastasovic, A., Vrvic, M.M., Jakovlievic, D. & Kogan, G. (2006). Mild Pfitzner-Moffat oxidation of the (1→3)-β-D-glucan from *Saccharomyces cerevisiae, Chemical Papers* Vol. 60(No. 3): 24-53.

Chapter 2

Is Chitosan a New Panacea? Areas of Application

Susana P. Miranda Castro and Eva G. Lizárraga Paulín

Additional information is available at the end of the chapter

1. Introduction

Polysaccharides are extremely common in nature and cellulose is the most common organic compound on the planet. It is said that the second most common polysaccharide in the world after cellulose is chitin. "Chitin is to shellfish what cellulose is to trees".

It's been more than two centuries since chitin was discovered formally and considered very important from the scientific and industrial point of view, as it has many applications in many different areas.

The development of commercial applications for chitin and chitosan has progressed. The first known use of chitosan was a durable, flexible film used as a component in the varnish applied to Stradivarius violins, however new efforts are changing its vision in the market. The emphasis on environmentally friendly technology has stimulated interest in biopolymers, which are more versatile and far more biodegradable than their synthetic counterpart.

The purpose of this chapter is to highlight the basic concepts of chemistry and the application of this polysaccharide that is gaining much interest due to the properties it presents and the many applications in various fields. Thousands of scientific articles have been reported in the last 20 years where companies appeared engaging and exploiting this material worldwide. Through investigation many questions have arisen but have not yet answered, however, this polysaccharide has been very successful in many applications.

Furthermore, this chapter aims to convince young readers to further research on possible technology that tend to care for the environment and health.

2. The origin and discovery of chitin

The universe began about 15 million years ago. Materials with high temperature and density were expanded, released energy, then cooled and gave birth to stars, planets and all living beings. The sun was born 5 billion years and 0.4 million years later gave birth to Earth.

Why talk about the birth of the earth? This is because the chitin could be a constituent of the first living cell. It actually came into existence long before the dinosaurs. In the late Precambrian period, two billion years ago, living cells appeared with nuclei containing chitin around it. In the Silurian period 440 million years, land plants appeared containing cellulose. Fish appeared in the Carboniferous period and later, the arthropods in the Devonian period. The first dinosaurs lived two hundred million years ago and during the second half of the Jurassic period the crab rich in chitin appeared.

After dinosaurs occupied the Earth for 100 million years, from the Jurassic to Cretaceous, they were extinguished by a comet that crashed into the Yucatan Peninsula 65 million years ago, but crabs and small animals escaped this catastrophy.

Since living beings appeared, cellulose and chitin have been beneficial in general and both maintained an ecological balance. Chitin is the animal version of the cellulose and it is the second most abundant in nature, but Professor M Peter has challenged that assumption by saying that Chitin is certainly a very abundant material even if much of it is not readily accessible for industrial use and suggested that hemicelluloses, which occur in conjunction with cellulose in trees and other plants, are actually more abundant than chitin. The hemicellulose component averages about half of the cellulose component, whereas the normal estimate of chitin production is that it is one whole order of magnitude less than that of cellulose. Another possible contender is lignin, which again occurs in conjunction with cellulose in most plants and, like hemicelluloses, averages about half of the cellulose component. A fourth possible contender is starch which like cellulose it is a major component of vegetable matter where it acts as a reserve material rather than a structural component [1].

The English word "chitin" comes from the French word chitine, which first appeared in 1836. These words were derived from the Greek word chitōn, meaning mollusk that is influenced by the Greek word khitōn, meaning "tunic" or "frock". That word may come from the Central Semitic word *kittan, the Akkadian words kitû or kita'um, meaning flax or linen, and the Sumerian word gada or gida. A similar word, "chiton", refers to a marine animal with a protective shell (also known as a "sea cradle" [2].

It is normally accepted as a fact that chitin was first isolated from mushrooms and called "fungine" by the French chemist Henri Bracconot in 1811. Charles Jeuniaux suggested in a paper presented at the 1st International Conference of the European Chitin Society held in Brest in 1995, that chitin had previously been isolated from arthropod cuticle by the English scientist A Hachett in 1795. However, as pointed out by Professor Jeuniaux, Hachett only reported the presence in the cuticle of an organic material particularly resistant to the usual chemical reagents but did not investigate it further. Braconnot on the other hand carried out chemical analysis on his fungical culture, and reported the formation of acetic acid from it on treatment with hot acid, and concluded it was a new material. Braconnot may be considered the discoverer of chitin even though his name for the new material, 'fungine', was soon replaced by its current name "chitin"which was first proposed by Odier [3,4].

Chitin is a big molecule composed of –beta-1,4-N-acetylglucosamine (GlcNAc) monomers. There are three forms of chitin: α, β, and γ chitin. The α-form, is mainly obtained from crab and shrimp. Both α and β chitin/chitosan are commercially available [5].

3. Sources of chitin

In the book "Chitin" published by Muzarelli in 1977, we can find a complete list of organisms that contain chitin: Fungi, Algae, Cnidaria (jellyfish), Aschelminthes (round worm), Entoprocta, Bryozoa (Moss or lace animals, Phoronida (Horseshoe worms), Brachiopoda(Lamp shells), Echiruda, Annelida (Segmented worms), Mollusca, Arthropoda and Ponogophora [6]). Herring, P.J in 1979 wrote that chitin is the main component of arthropod exoskeletons, tendons, and the linings of their respiratory, excretory, and digestive systems, as well as insects external structure and some fungi. It is also found in the reflective material (iridophores) of both epidermis and eyes of arthropods and cephalopods (phylum: Mollusca) and the epidermal cuticle of the vertebrate Paralipophrys trigloides (fish) is also chitinous [7,8]. The main commercial sources of chitin are the shell wastes of shrimp, lobsters, crabs and krill. There are three forms of chitin: α, β, and γ chitin. The α-form, is composed of alternating antiparallel polysaccharide strands and is mainly obtained from crab and shrimp. α -Chitin is by far the most abundant; it occurs in fungal and yeast cell walls, krill, lobster and crab tendons and shells, shrimp shells, and insect cuticle. The rarer β -chitin is composed of parallel strands of polysaccharides, is found in association with proteins in squid pens [9,10] and in the tubes synthesized by pogonophoran and vestimetiferan worms [11,12]. It also occurs in aphrodite chaetae [13] as well as in the lorica, built by some seaweeds or protozoa [14,15,16]. And 2 parallel chains alternating with an antiparallel strand constitute gamma chitin and are found in fungi [15].

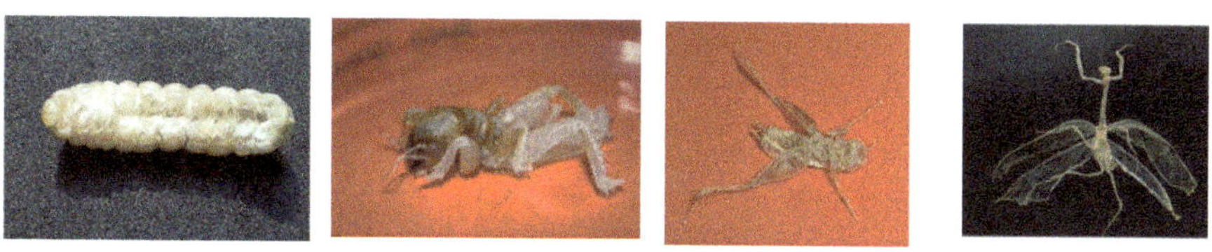

Figure 1. Chitinous structure of worm and insects

4. Chitin from crustacean

Currently most commercial production of chitin is based on extracting it from the exoskeleton of shrimp, prawn, crab and other crustaceans. This source contains a high percentage of inorganic material, primarily $CaCO_3$ and a rough calculation indicates that for every tonne of chitin produced, 0.8 tonne of CO_2 is released into the environment. In view of current concerns about global warming this cannot be considered to be a truly environmentally friendly process [3].

Another source of chitin that is more environmentally friendly, although much more limited in volume, is squid pen. This waste contains very little in the way of inorganic material and very little, if any CO_2 would be released in the extraction and purification process. Another and perhaps more sustainable source in the long run is vegetable chitin from fungal sources such as waste mycelia. There is extensive literature on the topic, but it is only recently that it has become commercially available [3].

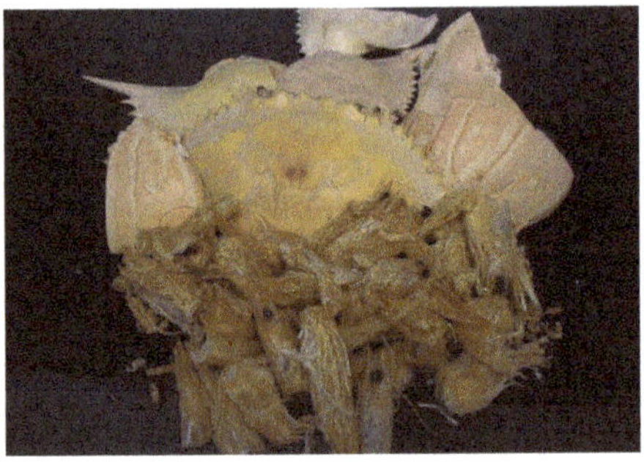

Figure 2. Exosqueleton of crustacea, this is the source of commercial chitin

5. Chitin from fungi

Chitin is widely distributed in fungi, occurring in *Basidiomycetes, Ascomycetes,* and *Phycomycetes,* where it is a component of the cell walls and structural membranes of mycelia, stalks, and spores. The amounts vary between traces and up to 45% of the organic fraction, the rest being mostly proteins, glucans and mannans. However, not all fungi contain chitin, and the polymer may be absent in one species that is closely related to another. Variations in the amounts of chitin may depend on physiological parameters in natural environments as well as on the fermentation conditions in biotechnological processing or in cultures of fungi [4].

The chitin in fungi possesses principally the same structure as the chitin occurring in other organisms. However, a major difference results from the fact that fungal chitin is associated with other polysaccharides which do not occur in the exoskeleton of arthropods. The molecular mass of chitin in fungi is not known. However, it was estimated that bakers' yeast synthesizes rather uniform chains containing 120 ± 170 GlcNAc monomer units which corresponds to 24,000 ± 34,500 Daltons [4].

6. Chemical methods to prepare chitin

Several procedures have been developed through the years to prepare chitin; they are at the basis of the chemical processes for industrial production of chitin and chitosan. Various methods are reported in Muzzarelli's book such as: Method of Rigby (1936 and 1937); Hackman (1954); Foster and Hackman (1957); Horowitz, Roseman and Blumenthal (1957);

Whistler and Be Miller (1962); Takeda and Abe (1962); Takeda and Katsuura (1964); Broussignac (1968); Lovell, Lafleur and Hoskins (1968); Madhavan and Ramachandran (1974) [6]. There is also a review that summarizes methods of preparation of various chitin and its conversion to chitosan [17].

7. Enzymatic methods to prepare chitin

A new process for deproteinization of chitin from shrimp head was studied [18]. Recovery of the protein fraction of the shrimp waste has been widely studied by enzymatic hydrolysis method [19,20].The enzymatic deproteinization process has limited value due to residual small peptides directly attached to chitin molecules ranging from 4.4% to 7.9% of total weight [21]. As these processes are costly because of the use of commercial enzymes, there is now a need to develop an efficient and economical method for extracting proteins from shellfish waste. One interesting new technology for extraction of chitin that offers an alternative to the more harsh chemical methods is fermentation by using microorganisms. Fermentation has been envisaged as one of the most ecofriendly, safe, technologically flexible, and economically viable alternative methods [22-28]. Fermentation of shrimp waste with lactic acid bacteria results in production of a solid portion of chitin and a liquor containing shrimp proteins, minerals, pigments, and nutrients [26,29]. Deproteinization of the biowaste occurs mainly by proteolytic enzyme produced by Lactobacillus [30]. Lactic acid produced by the process of breakdown of glucose, creating the low pH condition of ensilation; suppress the growth of microorganisms involved in spoilage of shrimp waste [31]. The lactic acid reacts with calcium carbonate component in the chitin fraction leading to the fermentation of calcium lactate, which gets precipitated and can be removed by washing. There is now a need to develop an efficient, simpler, eco- friendly, economical, and commercially viable method.

8. Chitosan

Despite the wide spread occurrence of chitin, up to now the main commercial sources of chitin have been crab and shrimp shells. In industrial processing, chitin is extracted from crustaceans by acid treatment to dissolve calcium carbonate followed by alkaline extraction to solubilized proteins. In addition a decolorization step is often added to remove leftover pigments and obtain a colorless product. These treatments must be adapted to each chitin source but by partial deacetylation under alkaline conditions, one obtains "chitosan" [16]. Chitosan is the most important derivate of this naturally occurring polymer being one of the most abundant polysaccharides after cellulose. Chitosan is a copolymer composed of N-acetyl-D-glucosamine and D-glucosamine units. It is obtained in three different ways, thermochemical deacetylation of chitin in the presence of alkali, by enzymatic hydrolysis in the presence of a chitin deacetylase, or naturally found in certain fungi as part of their structure. In chitosan part of the amino groups remain acetylated. It is generally accepted that N-acetylglucosamine residues are randomly distributed along the whole polymer chain. In an acid medium, amino groups are protonated and thus determine the positive charge of the chitosan molecule. Thus, chitosan behaves like a polycation in solution [32]. Properties of chitosan, such as the mean polymerization degree, the degree of N-deacetylation, the positive charge, and the nature of chemical modifications of its molecule, strongly influence its biological activity.

Chitin contains 6–7% nitrogen and in its deacetylated form, chitosan contains 7–9.5% nitrogen. In chitosan, between 60 to 80% of the acetyl groups available in chitin are removed [33]. The chain distribution is dependant on the processing method used to obtain biopolymer [34-36]. It is the N-deacetylated derivative of chitin, but the N-deacetylation is almost never complete [35]. Chitin and chitosan are names that do not strictly refer to a fixed stoichiometry. Chemically, chitin is known as poly-N-acetylglucosamine, and in accordance to this proposed name, the difference between chitin and chitosan is that the degree of deacetylation in chitin is very little, while deacetylation in chitosan occurs to an extent but still not enough to be called polyglucosamine [37].

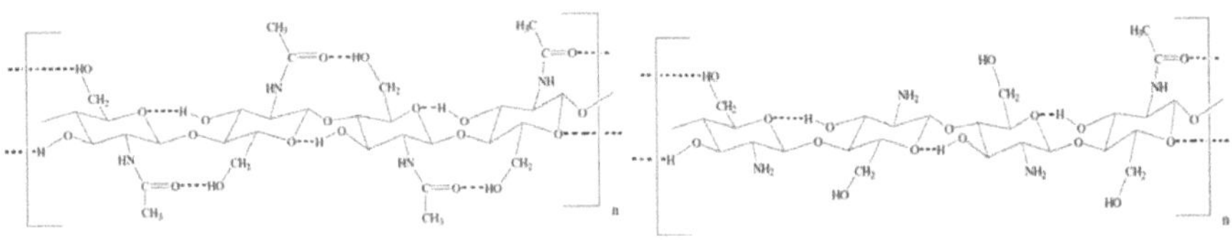

Figure 3. Chitin and chitosan chemical structure

9. Sources of chitosan

Chitosan is commercially produced from deacetylated chitin found in shrimp and crab shell. However, supplies of raw materials are variable and seasonal and the process is laborious and costly [38]. Furthermore, the chitosan derived from such process is heterogeneous with respect to its physiochemical properties [38]. Recent advances in fermentation technology provide an alternative source of chitosan. Fungal cell walls and septa of *Ascomycetes, Zygomycetes, Basidiomycetes* and *Deuteromycetes* contain mainly chitin, which is responsible for maintaining their shape, strength and integrity of cell structure [38]. The production of chitosan from fungal mycelia has a lot of advantages over crustacean chitosans such as the degree of acetylation, molecular weight, viscosity and charge distribution of the fungal chitosan. They are more stable than crustacean chitosans. The production of chitosan by fungus in a bioreactor at a technical scale offers also additional opportunities to obtain identical material throughout the year. The fungal chitosan is free of heavy metal contents such as nickel, copper [39-41]. Moreover the production of chitosan from fungal mycelia gives medium-low molecular weight chitosans ($1–12 \times 10^4$ Da), whereas the molecular weight of chitosans obtained from crustacean sources is high (about 1.5×10^6 Da) [41]. Chitosan with a medium-low molecular weight has been used as a powder in cholesterol absorption [42] and as thread or membrane in many medical-technical applications. For these reasons, there is an increasing interest in the production of fungal chitosan.

There are some examples of chitosan extracted from fungi. Chitosans isolated from *Mucorales* typically show Mw in the range 4 x 10^5 to 1.2 x 10^6 Daltons and FA values between 0.2 and 0.09. Amino acid analysis of chitosan prepared from *Aspergillus niger* reveals covalently bound arginine, serine, and proline. Nadarajah et. al., 2001, studied chitosan production from mycelia of *Rhizopus* sp KNO1 and KNO2, *Mucor sp* KNO3 and *Asperigullus niger* with the highest amount of extractable chitosan obtained at the late exponential phase. *Mucor* sp KNO3, produced the highest amount of 557mg per 2.26 g of dry cell weight /250 ml of culture. Kishore et. al.(1993), examined the production of chitosan from mycelia of *Absidia coerulea, Mucor rouxii, Gongronella butieri, Phvcomyces blakesleeanus* and *Absidia blakesleeana*. Chitosan yields of *A. coerulea, M. rouxii, G. butieri, P. blakesleeanus* and *A. blakesleeana* were 47–50, 29–32, 21–25, 6 and 7 mg/100 mL of medium, respectively. The degree of acetylation of chitosan ranged from 6 to 15%; the lowest was from strains of *A. coerulea*. Viscosity average molecular weights of fungal chitosans were equivalent, approximately 4.5 x 10^5 Daltons. Wei-Ping Wang et.al., (2008) evaluated the physical properties of fungal chitosan from *Absidia coerulea* (AF 93105*), Mucor rouxii* (Ag 92033), and *Rhizopus oryzae (*Ag 92033). Their glucosamine contents and degrees of deacetylation (DD) were over 80%, differences had been observed in their molecular weight (Mw), ranging from 6.6 to 560 kDa. Chitosan was isolated and purified from the mycelia of *Rhizomucor miehei* and *Mucor racemosus* with a degree of deacetylation of 97 y 98 respectively [43-45].

Considerable research has been carried out on using mycelium waste from fermentation processes as a source of fungal chitin and chitosan. It is argued that this would offer a stable non-seasonal source of raw material that would be more consistent in character than shellfish waste, but so far this route does not appear to have been taken up by chitosan producing companies. Currently there is only one commercial source of fungal chitosan and is produced by the company Kitozyme. However their raw material is not mycelium waste from a fermentation process, which is what is normally envisaged when fungal chitosan is referred to, but actually conventional edible mushrooms grown under contract in France and shipped to Belgium for processing. So mycelium waste still remains a vast and as yet untapped potential source of chitosan.

10. Genetic engineering approach to produce chitin

It is difficult to obtain pure carbohydrates, especially chitin, through conventional techniques. Bacterial cells have been engineered in an effort to overcome this problem [46]. *E. coli* has been engineered to produce chitobiose. This method took advantage of NodC, which is a chito-oligosaccharide synthase, and genetically engineered chitinase to make a cell factory with the ability to produce chito-oligosaccharides [47]. Recombinant chito-oligosaccharides have also been obtained using *E coli* cells which expressed nodC or nodBC genes [48]. By expressing different combinations of nod genes in *E. coli*, O-acetylated and sulfated chito-oligosaccharide have been produced [49].

11. Parameters influencing the behavior of the biopolymer

The main parameters influencing the characteristics of chitosan are its degree of deacetylation (DD) and molecular weight (Mw), which affect the solubility, rheological and physical properties. Various grades of chitosan are available commercially, which differ primarily in the degree of deacetylation and molecular weight. Different conditions such as type and concentration of reagents, time and temperature employed throughout the processing can affect the physical characteristics and performance of the final chitosan product [50]. However, both DD and molecular weight can be further modified. For example, DD can be lowered by reacetylation [51-55] and molecular weight can be lowered by acidic or enzymatic depolymerisation [56-58].

12. Degree of Deacetylation (DD)

Deacetylation describes a reaction that removes an acetyl functional group. When the degree of deacetylation of chitin reaches about 50% (depending on the origin of the polymer), it becomes soluble in aqueous acidic media and is called chitosan. The solubilization occurs by protonation of the $-NH_2$ function on the C-2 position of the D-glucosamine repeat unit, whereby the polysaccharide is converted to a polyelectrolyte in acidic media. Chitosan is the only pseudonatural cationic polymer and thus, it finds many applications that follow from its unique character (flocculants for protein recovery, depollution, etc.). Being soluble in aqueous solutions, it is largely used in different applications as solutions, gels, or films and fibers.

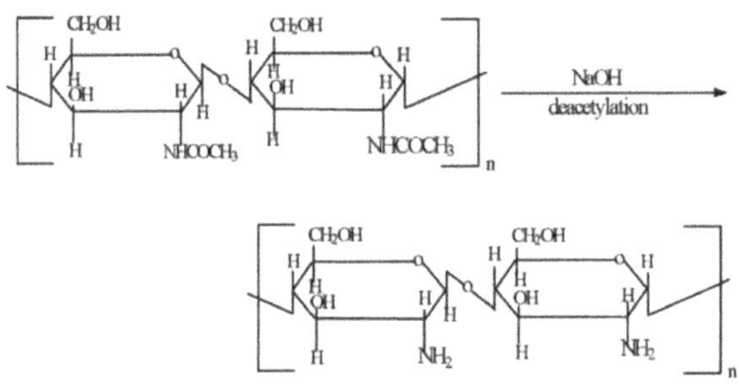

Figure 4. Chitin deacetylation

A highly deacetylated polymer has been used to explore methods of characterization [59]. The solution properties of a chitosan depend not only on its average DA but also on the distribution of the acetyl groups along the main chain in addition of the molecular weight [60-62]. Several methods have been proposed for alkaline deacetylation to obtain chitosan [6,17]. The conditions used in the deacetylation determines the polymer molecular weight and degree of deacetylation (DD).

Chitosan has been largely employed in many areas, such as photography, biotechnology, cosmetics, food processing, biomedical products (artificial skin, wound dressing, contact lens, etc.), system of controlled liberation of medicines (capsules and microcapsules), treatment of industrial effluents for removal of metallic and coloring ions. The amino groups are responsible for the distinct characteristics attributed to this basic polymer (compared to an acidic biopolymer). Therefore, the characterization of the polymer in either chitin or chitosan is extremely important according to the structure-properties relationship, defining a possible industrial application. Thus many techniques are available to determine the degree of deacetylation. Elson Santiago de Alvarenga (2011) published on line describing the most important parameters to be evaluated in chitosan as "deacetylation degree" (DD) [63].

The methods for carrying out the analysis of the degree of deacetylation are: Elemental analysis; Titration; HPLC; Infrared; ^{1}H nuclear magnetic resonance; CP-MAS ^{13}C NMR; CP-MAS ^{15}N NMR; steric exclusion; nitrous acid deamination; thermal analysis.

13. Molecular weight

Another important characteristic to consider for these polymers is the molecular weight and its distribution. The first difficulty encountered in this respect concerns the solubility of the samples and dissociation of aggregates often present in polysaccharide solutions [16, 57, 64, 65, 66]. As to choice of a solvent for chitosan characterization, various systems have been proposed, including an acid at a given concentration for protonation together with a salt to screen the electrostatic interaction. The solvent is important also when molecular weight has to be calculated from intrinsic viscosity using the Mark–Houwink relation.

14. Biological properties of chitosan

14.1. Biocompatibility

Biocompatibility of a biomaterial refers to the extent to which the material does not have toxic or injurious effects on biological systems [67, 68]. One of the present trends in biomedical research requires materials that are derived from nature as natural materials have been shown to exhibit better biocompatibility with humans and because chitosan's monomeric unit, N-acetylglucosamine, occurs in hyaluronic acid, an extracellular macromolecule that is important in wound repair. Additionally, the N-acetylglucosamine moiety in chitosan is structurally similar to glycosaminoglycans (GAGs), heparin, chondroitin sulphate and hyaluronic acid in which they are biocompatible, and hold the specific interactions with various growth factors, receptors and adhesion proteins besides being the biologically important mucopolysaccharides and in all mammals. Therefore, the analogous structure in chitosan may also exert similar bioactivity and biocompatibility [69, 70].

The potential of chitosan stems from its cationic nature and high charge density in solution. An effective approach for developing a clinically applicable chitosan is to modify the surface of the material that already has excellent biofunctionality and bulk properties [71]. Altering the physical and chemical properties of the chitosan in order to improve its medicinal quality will also influence its biocompatibility [69.70].

The excellent biological properties of chitosan can be potentially improved with a variety of additional chemicals such as polyethylene glycol and carboxymethyl N-acyl groups in order to produce biocompatible chitosan derivatives for use as wound dressings [72]. Chitosan's positive surface charge enables it to effectively support cell growth [73]. Chitosan-gelatin sponge wound dressing had demonstrated a superior antibacterial effect. Additionally, chitosan gelatin sponge allowed the wound site to contract markedly and shortened the wound healing time, as compared with sterile Vaseline gauze [74]. Widely used surface modification techniques include coating, oxidation by low temperature plasma for better printing and adhesion and surfactant addition for antistatic. Blends are often used to improve tensile properties and to provide a stronger structural component for separation media that supports the active polymer. The physical properties of a polymer can also be altered by introducing a second polymer that improves the properties of the original polymer in certain aspects, such as hydrophobility, lowered melt temperature, raised glass transition temperature, etc [75]. A thorough understanding of cell and proteins interactions with artificial surfaces is of importance to design suitable implant surfaces and substrates. The surface properties of newly synthesized biomedical grade chitosan derivatives, including surface composition, wettability, domain composition, surface oxidation, surface charge and morphology, may influence protein adsorption and subsequently, the cellular responses to biomaterial implants [76-81].

15. Biodegradability

The claim "biodegradable" is often associated with environmentally friendly products. It is defined as being able to be broken down by natural processes, into more basic components. Products are usually broken down by bacteria, fungi or other simple organisms [82].

An important aspect in the use of polymers as drug delivery systems is their metabolic fate in the body or biodegradation. In the case of the systemic absorption of hydrophilic polymers such as chitosan, they should have a suitable Mw for renal clearance. If the administered polymer's size is larger than this, then the polymer should undergo degradation. Biodegradation (chemical or enzymatic) provides fragments suitable for renal clearance. Chemical degradation in this case refers to acid catalysed degradation i.e. in the stomach. Enzymatically, chitosan can be degraded by enzymes able to hydrolyse glucosamine–glucosamine, glucosamine–N- acetyl-glucosamine and N-acetyl-glucosamine–N-acetyl- glucosamine linkages [83]. Even though depolymerisation through oxidation–reduction reaction [84] and free radical degradation [85] of chitosan have been reported these are unlikely to play a significant role in the in vivo degradation.

Chitosan is thought to be degraded in vertebrates predominantly by lysozyme and by bacterial enzymes in the colon [83, 86]. However, eight human chitinases (in the glycoside hydrolase 18 family) have been identified, three of which have shown enzymatic activity

[87]. A variety of microorganisms synthesises and/or degrades chitin, the biological precursor of chitosan. In general, chitinases in microorganisms hydrolyze N-acetyl-β-1,4-glucosaminide linkages randomly i.e. they are endo-chitinases (EC 3.2.1.14). Chitinases are also present in higher plants, even though they do not have chitin structural components.

Chemical characterisation assays determining the degradation of chitosan commonly use viscometry and/or gel permeation chromatography to evaluate a decrease in Mw [88]. Lysozyme has been found to efficiently degrade chitosan; 50% acetylated chitosan had 66% loss in viscosity after a 4 h incubation in vitro at pH 5.5 (0.1 M phosphate buffer, 0.2 M NaCl, 37 °C) [88]. This degradation appears to be dependent on the degree of acetylation with degradation of acetylated chitosan (more chitin like) showing the faster [89,90].

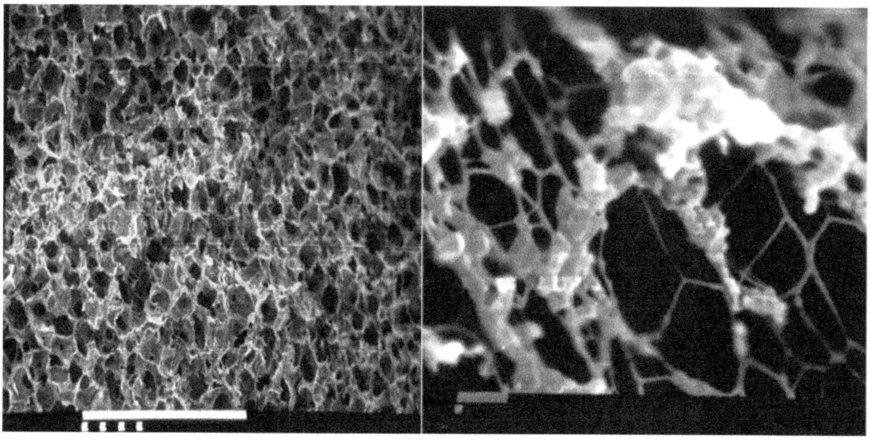

Figure 5. Biodegradation of chitosan thermosensible gel inside the rat's body after 5 days.

16. Safe biomaterial

Chitosan is a potentially biologically compatible material that is chemically versatile (–NH_2 groups and various Mw). These two basic properties have been used by drug delivery and tissue engineering to create a great amount of formulations and scaffolds that show promise in healthcare. It is approved for dietary applications in Japan, Italy and Finland [91] and it has been approved by the FDA for use in wound dressings [92] but is not approved for any product in drug delivery. The term "Chitosan" represents a large group of structurally different chemical entities that may show different biodistribution, biodegradation and toxicological profiles.

The formulation of chitosan with a drug may alter the pharmacokinetic and biodistribution profile.The balancing, or reduction, of the positive charges on the chitosan molecule has effects on its interaction with cells and the microenvironment, often leading to decreased uptake and a decrease in toxicity. The modifications made to chitosan could make it more or less toxic and any residual reactants should be carefully removed. In addition, the route of administration determines the uptake, concentration, contact time and cell types affected.

Chitosan details (DD, MW)	Modification	Assessment	IC50
95% DD, 18.7 kDa	Steric acid conjugation micelle	In vitro ,A549 cells	369±27 μg/ml
95% DD, 18.7 kDa	teric acid conjugation and entrapment in micelle	In vitro ,A549 cells	234±9 μg/ml
97% DD, 65 kDa	N-octyl-O-sulphate	Invitro, primary rat hepatocytes	>200 mg/ml
87% DD, 20, 45, 200, 460 kDa	None, aspartic acid salt	In vitro, Caco-2 cells, pH 6.2	0.67±0.24, 0.61±0.10, 0.65±0.20, 0.72±0.16 mg/ml
87% DD, 20, 45, 200, 460 kDa	None, glutamic acid salt		0.56±0.10, 0.48±0.07, 0.35±0.06, 0.46±0.06 mg/ml
87% DD, 20, 45, 200, 460 kDa	None, Lactic acid salt		0.38±0.13, 0.31±0.06, 0.34±0.04, 0.37±0.08 mg/ml
87% DD, 20, 45, 200, 460 kDa	None, hydrochloride salt		0.23±0.13, 0.22±0.06, 0.27±0.08, 0.23±0.08 mg/ml
78% DD, <50 kDa	None, lactic acid salt	In vitro B16F10 cells	2.50 mg/ml
82% DD, 150–170 kDa	None, lactic acid salt	In vitro B16F10 cells	2.00±0.18 mg/ml
>80% DD, 60–90 kDa	None, glutamic acid salt	In vitro B16F10 cells	2.47±0.14 mg/ml
77% DD, 180–230 kDa	None, lactic acid salt	In vitro B16F10 cells	1.73±1.39 mg/ml
85% DD, 60–90 kDa	None, hydrochloric acid salt	In vitro B16F10 cells	2.24±0.16 mg/ml
81% DD, 100–130 kDa	None, hydrochloric acid salt	In vitro B16F10 cells	0.21±0.04 mg/ml
100% DD, 152 kDa	Glycol chitosan	In vitro B16F10 cells	2.47±0.15 mg/ml
100% DD, 3–6 kDa	20, 44, 55% Trimethyl chitosan, chloride salt	In vitro, MCF7 and COS7 cells, 6 h & 24 h	>10 mg/ml
100% DD, 3–6 kDa	94% Trimethyl chitosan, chloride salt	In vitro, MCF7, 6 h	1.402±0.210 mg/ml
100% DD, 3–6 kDa	94% Trimethyl chitosan, chloride salt	In vitro, COS7, 6 h	2.207±0.381 mg/ml
100% DD, 100 kDa	36% Trimethyl chitosan, chloride sal	In vitro, MCF7, 6 h	0.823±0.324 mg/ml
100% DD, 100 kDa	36% Trimethyl chitosan, chloride sal	In vitro, COS7, 6 h	>10 mg/ml

Chitosan details (DD, MW)	Modification	Assessment	IC50
84.7% DD, 400, 100, 50, 25, 5 kDa	40% Trimethyl chitosan	In vitro, L929 cells, 3 h	30, 70, 90, 270, >1000 μg/ml
84.7% DD, 1.89 MDa	12% PEG modified 40% trimethyl chitosan	In vitro, L929 cells, 3 h	220 μg/ml
84.7% DD, 3.6 MDa	25.7% PEG modified 40% trimethyl chitosan	In vitro, L929 cells, 3 h	370 μg/ml
84.7% DD, 300 kDa	6.44% PEG modified 40% trimethyl chitosan (and all PEG modified TMC with lower Mw)	In vitro, L929 cells, 3 h	>500 μg/ml
97% DD, 65 kDa	N-octyl-O-sulphate	In vivo, IV, mice	102.59 mg/kg
97% DD, 65 kDa	N-octyl-O-sulphate	n vivo, IP, mice	130.53 mg/kg

Table 1. Toxicity of chitosan and chitosan derivatives
Table taken from [94]

In a series of articles are described the effects of chitosans with differing molecular weights and degree of deacetylation in vitro and in vivo. Toxicity was found to be degree of deacetylation and molecular weight dependent. At high DD the toxicity is related to the molecular weight and the concentration, at lower DD toxicity is less pronounced and less related to the molecular weight [93].

A summary of toxicities of chitosan and derivatives assessed through IC_{50} values is presented in the next table [94].

From this table it can be gathered that most chitosans (and derivatives) are not significantly toxic compared to a toxic polymer such as polyethylenimine [94].

It appears that the toxicity of chitosan is related to the charge density of the molecule, toxicity increases with increasing density. It appears that there is a threshold level below which there are too few contact points between a molecule and the cell components to produce a significantly toxic effect. This balance is between 40 and 60% DD, or degree of trimethylation, although any sufficiently small chitosan (<10 kDa) is not appreciably toxic. Modifications that do not increase the charge on the molecule seem to have little effect on the toxicity beyond that of the native molecule [94].

17. Antimicrobial activity

The exact mechanism by which chitosan exerts its antimicrobial activity is currently unknown, it has been suggested that the polycationic nature of this biopolymer that forms from acidic solutions below pH 6.5 is a crucial factor. Thus, it has been proposed that the positively charged amino groups of the glucosamine units interact with negatively charged components in microbial cell membranes, altering their barrier properties, and thereby preventing the entry of nutrients or causing the leakage of intracellular contents [95-100]. Another reported mechanism involves the penetration of low- molecular weight chitosan into the cell, binding to DNA and the subsequent inhibition of RNA and protein synthesis [101]. Chitosan has also been shown to activate several defence processes in plant tissues and it inhibits the production of toxins and microbial growth because of its ability to chelate metal ions [102,103].

El-Ghaouth et. al.(1992) have proposed possible antibacterial actions of chitosan and its derivatives [104]. They asserted that chitosan reacted with the cell surface, altered cell permeability, and further prevented the entry of material or caused the leakage of material. However, no evidence has been provided to demonstrate the relationship between the antibacterial activity of chitosan and the surface characteristics of the bacterial cell wall. Antimicrobial activity of chitosan has been demonstrated against many bacteria, filamentous fungi and yeasts [105-108]. Chitosan has wide spectrum of activity and high killing rate against Gram-positive and Gram-negative bacteria, but lower toxicity toward mammalian cells [109,110]. Variations in chitosan's antimicrobial efficacy arise from various factors. These factors can be classified into four categories as follows: (1) microbial factors, related to microorganism species and cell age; (2) intrinsic factors of chitosan, including positive charge density, Mw, concentration, hydrophilic/hydrophobic characteristic and chelating capacity; (3) physical state, namely water-soluble and solid state of chitosan; (4) environmental factors, involving ionic strength in medium, pH, temperature and reactive time [111].

Although owning a broad spectrum of antimicrobial activity, chitosan exhibits different inhibitory efficiency against different fungi, Gram-positive and Gram-negative bacteria. Chitosan exerts an antifungal effect by suppressing sporulation and spore germination [112]. In contrast, the mode of antibacterial activity is a complicating process that differs between Gram positive and Gram-negative bacteria due to different cell surface characteristics.

Based on the available evidences, bacteria appear to be generally less sensitive to the antimicrobial action of chitosan than fungi. The antifungal activity of chitosan is greater at lower pH values [113]. For a given microbial species, age of the cell can influence antimicrobial efficiency.

Chitosan has a broad spectrum of unique biological activities, including its ability to induce resistance to viral infections in plants, inhibit viral infection in animal cells, and prevent the development of phage infection in infected microbial cultures. High polymeric chitosan, when added to a nutrient medium, prevents the accumulation of the infectious phage progeny in infected cultures of Gram-positive and Gram-negative organisms. The yield of infectious DNA containing phage can decrease by several orders of magnitude in the presence of chitosan [114, 115]. The observed effect also depends on the concentration, degree of polymerization, and molecular structure of chitosan. Thus, chitosans with a polymerization degree of 250 and higher were much more effective in inhibiting coliphage infection than their fragments with a polymerization degree of 15–19. On the other hand, chitosan oligomers were more effective than their polymeric precursor in inhibiting the replication of 1-97A phage in *Bacillus thuringiensis* cultures. Factors determining such strong differences are currently unclear. Anionic derivatives of chitosan, such as 6-O-sulfate and N-succinate-6-O-sulfate, caused no effect on phage infection [114]. This finding showed that the positive charge of a chitosan molecule is important for inhibition of phage infection.

It has been suggested that chitosan can inhibit the replication of bacteriophages by several mechanisms: it can (a) decrease the viability of cultured bacterial cells, (b) neutralize the infectivity of mature phage particles in the inoculum and/or daughter phage particles, and (c) block the replication of the virulent phage [114].

The condition of the phage culture is known to be of paramount importance for the development of phage infection because phages can reproduce only in viable cells. However, there is evidence that chitosan displays a bactericidal activity toward many microbial species including *Escherichia coli* [116, 117] and species of the genus Bacillus [118]. Chitosan, because of its polycationic nature, binds to the external membrane of Gram negative microorganisms by electrostatic forces, which is demonstrated in experiments with core phosphate groups of lipid A, thereby decreasing the potency of endotoxin. It was suggested that the antibacterial effects of chitosan and many other cationic agents are based on their ability to increase the permeability of the outer membrane of Gram negative microorganisms to an extent incompatible with their viability [116-119].

Great amount of literature support the essential importance of polycationic structure in antimicrobial activity. A higher positive charge density leads to strong electrostatic interaction. Therein, the positive charge is associated with DD or degree of substitution (DS) of chitosan or its derivatives, which affect positive charge density. Concerning chitosan derivatives, antimicrobial activity mostly depends on DS of the grafting groups.

There are several works regarding the influence of the molecular weight of chitosan on its antimicrobial properties [120-126]. Some of them have demonstrated that COS (chitooligosaccharides), which are soluble in water, were the least effective in terms of biocide properties [124-126]. Recent work carried out by Qin et al. 2006, on the evaluation of chitosan solutions against the growth of *Candida albicans, Escherichia coli* and *Staphylococcus aureus,* it has shown that only water insoluble chitosan in organic acidic solutions, i.e. chitosonium salts, exhibit efficient biocide properties. On the other hand, a research performed by Fernandes et al, 2008 showed that the growth of *E. coli* was markedly inhibited by COS, and this inhibition decreased slightly as molecular weight increased. In another work performed by Fernandez-Saiz et al. 2009, changes in molecular weight of the chitosan materials tested, i.e. 310–375 and 50–190 kDa, did not lead to significant variations in biocide properties [127, 128,129].

Concerning DD, there are several works that consider this feature, and there is no doubt that the antimicrobial properties of chitosan increased with this variable. The antimicrobial performance tends to increase upon an increase in the DD of chitosan, which is related to an increase in the positive charge of the polymer [130-133].

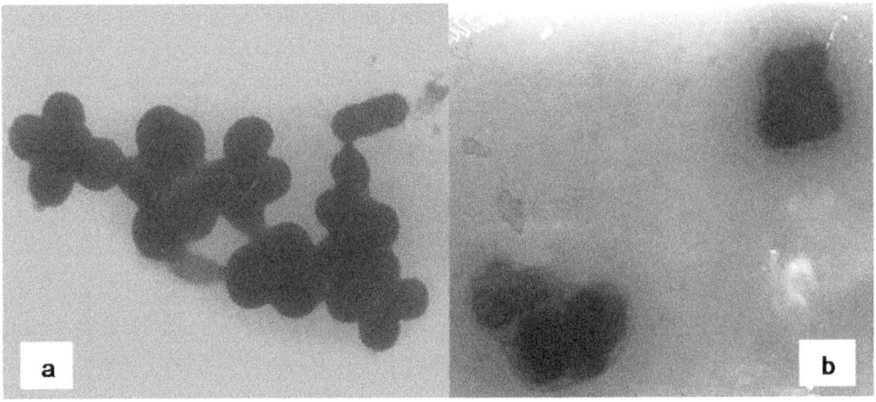

Figure 6. Staphylococcus bacteria (a), covered with chitosan (b)

18. Analgesic activity

Some investigators reported that chitin and chitosan induce analgesia. Allan in 1984 found that chitosan provide a cool and pleasant soothing effect when applied to open wounds. Ohshima et al in 1987, reported that chitin proved excellent pain relief in 83 out of 91 cases who received the agent topically over open wounds such as burns, skin abrasion, skin ulcers, skin graft areas and so on [134-135]. Minami in 1993 and Okamoto in 1993, reported that animals did not feel pain when their wounds were covered with chitin and chitosan [136-137].

Chitin and chitosan have been found to reduce the inflammatory pain due to intraperitoneal administration of acetic acid in rats [138]. When the chitosan suspension was mixed with acetic acid, the amino groups in C2 the position were protonated to NH_3 subsequently the particles resolved in the solution. Bradykinin is one of the main substances related to pain and the levels of this substance decrease in the presence of chitin and chitosan. Chitin absorbs Bradykinin more extensively than chitosan and this could be one of the main analgesic effect [138].

19. Antitumor activity

In some medical applications of chitin/chitosan, as antitumor compounds, for example, their degradation products are preferred, as they have a lower viscosity and a better solubility in water. The antitumor activity of chitin/chitosan is manifested by the stimulation of the immune system (production of lymphokines, including interleukins 1 and 2, stimulation of NK, etc.) [139-141]. Jeon and Kim in 2002, tested the antitumor activity of three kinds of COSs (high molecular weight ranging from 6.5 to 12 kDa – HMWCOS, medium molecular weight ranging from 1.5 to 5.5 kDa -MMWCOS, and low molecular weight ranging from 0.5 to 1.4 kDa – LMWCOS) against Sarcoma 180 solid (S180) and Uterine cervix carcinoma No

14 (U14) [140]. The efficiency of tumor growth inhibition for both types of tumor cells in mice was best in the case of MMWCOS. There are many reports of S180 tumor cells being used for testing the antitumor activity of chitosan [141-142].

Maeda and Kimura 2004, investigated the antitumour effect of three water-soluble low molecular weight chitosans (21 kDa, 46 kDa, 130 kDa) and various doses of 650 kDa chitosan in sarcoma 180-bearing mice. They found that LMWC (21 and 46 kDa) and also smaller oligosaccharides could activate the intestinal immune system of animals, thus preventing tumor growth. But no antitumor effect was observed after the oral administration of chitosan samples, even of low molecular weight (46 kDa). The same authors confirmed that high molecular weight chitosan (650 kDa) prevents the adverse reactions of some cancer chemotherapeutic drugs [141]. Qin et al. in 2002 also tested the antitumor activity of LMWC against sarcoma 180, but they came to the opposite conclusions [143]. They noted that oral administration of LMW chitosan decreases the weight of the tumour [139, 142], although administration by intraperitoneal injection led to a higher inhibitory rate [142]. It was reported the higher the MW of LMWC, the better the tumor inhibitory effect [139]. The introduction of acidic groups as a result of chitosan oxidation has the opposite effect, and an increase in MW decreases antitumor activity. Qin Cai-qin et al in 2002, prepared low molecular weight chitosans by oxidative degradation with H_2O_2. They found that carboxyllc contents increased with decrease in molecular weight (M~). They also found that the introduction of carboxylic group is advantage to water-solubility of chitosan, but more acidic groups decrease the function of amino groups of chitosan against sarcoma 180 tumor. There is a correlation between the activity and the molecular weights of the oxidized chitosans, and a maximum of inhibition was found around 4 100 [143]. The influence of LMWC and COS (including the pentamer, hexamer, and higher oligomers) on the growth inhibition of Ehrlich ascites tumor (EAT) cells and tumor induced neovascularization was investigated [144]. Based on experimental results concerning the inhibition of angiogenesis and the induction of apoptosis, it was confirmed that COSs seem to be more potent angio inhibitory and antitumor compounds. Wang et al. reported that chitin oligosaccharides (DP 1-6) also reduced the number of K562 cells (human erythromyeloblastoid leukemia cell line) [145].

20. Chitosan applications

20.1. Biomedical

Potential applications of chitosan can only be exploited if its usable forms are properly developed and prepared. In solution and gel, it can be used as a bacteriostatic, fungistatic and coating agent. Gels and suspensions may play the role of carriers for slow release or controlled action of drugs, as an immobilising medium and an encapsulation material. Film and membranes are used in dialysis, contact lenses, dressings and the encapsulation of mammal cells, including cell cultures. Chitosan sponges are used in dressings, and to stop bleeding in mucous membranes. Chitosan fibres are used as resorbable sutures, non-wovens for dressings, and as drug carriers in the form of hollow fibres.

Figure 7. These are some ways you can shape the chitosan: films, gels, microspheres, tubes, sponges, powder

21. Artificial skin

The preparation of artificial skin by natural materials such as gelatin, pectin, starch, cellulose, alginate, chitin, collagen, polyamino acids, and dextran has been established to enhance the healing process. The structures of these natural materials are analogs of protein and growth factor structures in the human body that may be more relevant for stimulating the appropriate hysiological responses required for cellular regeneration and tissue reconstructing in wounds [146].

Dressing materials based on chitin, chitosan and derivatives are well-known on the market, and are produced mainly in Japan and the US. JEX KK Co produces composite dressings made of synthetic resins, chitosan and materials of collagen and acetylchitiosan. Eisai Co is manufacture of chitin dressings in sponge form (Chitopack C®) or a PET non-woven modified with chitin (Chitopack C®). The Japanese Unitika Co offers a dressing non-woven of chitosan fibres. The American 3M proposes a chitosan gel preparation (Tegasorb®) and a hydrocolloid (Tegaderm®) designed for the healing of extensive internal wounds.ChorioChit sponge is a biological dressing obtained by lyophilisation of human placenta blended with MCCh.

22. Scaffold for the regeneration of tissue

Chitin and its derivatives have been used as scaffolds for bone and other natural tissue regeneration [147] as well as structures by which three-dimensional formation of tissues are supported [148]. While looking for a good material for a good scaffold, there are at least four important factors that should be taken into account: (1) ability to form temporary matrix, (2) ability to form porous structure for tissue to grow, (3) biodegradability, and finally (4) non-toxic byproducts from the digestion [149,150]. Thus, neither the physical nor biological properties of such biomaterials should be ignored [147]. Chitin and its derivatives have been shown to possess these criteria.

23. Haemostasis and wound healing

Hemostasis through blood coagulation is an important step for wound healing. The main cellular components in blood coagulation are platelets. It has been shown that chitosan has a hemostatic effect [151]. Okamoto et al 2003, reported that chitin is an effective agent for hemostasis maintenance through aggregating platelets, and suggested that the effect of chitin and chitosan is due to both physical and chemical properties of these biopolymers, especially their amino groups [152].

Haemostatic dressings containing chitin and chitosan as bioactive agents are also well known, notably the Syvek patch, RDH (Marine Polymer Technologies), Clo-Sur PAD (Medtronic-Scion), Chito-Seal (Abbot), the M-Patch and Trauma DEX (Medafor).

24. Peripheral nerve prosthesis

Prosthesis is made from various forms of utility polysaccharides. The main objective of research is to develop replacement implants that will not be rejected by the body of the recipient and offering the ability to regenerate damaged nerve. Because chitin has high mechanical strength under physiological conditions (low for chitosan) chitin has the potential to be a good nerve guidance channel. Ferier et al, 2005, used this fact and made chitin tubes that could support nerve cell adhesion and neurite outgrowth [153]. In a research related to nerve regeneration, it was shown that rabbits with the crushed common peroneal nerve exhibit better improvement in peripheral nerve regeneration in the presence of chitooligosaccharide. As a result, chito-oligosaccharide can be used as neuroprotective material with an ability to improve injured peripheral nerve regeneration [154].

25. Immunology

The key property of chitin-derived products for application in various biomedical applications is the immuno-modulating effect [155,156]. Some mechanisms of immuno-enhancement activity of chitin and its derivatives have been reported, for example, chitosan exhibited the ability to boost NO production from macrophages in the presence of interferon-γ (IFN-γ) through the NF-κB signaling pathway [157]. Minami et al. in 1998 found that chitin and chitosan affected C3 and C5 components of complement system and concluded that complement system is activated by chitin and chitosan through the alternative pathway. After activating the complement, C5 is produced followed by an increase in migration of polymorphonuclear cells (PMN) to the injured tissue. This is a normal inflammatory reaction but in the presence of chitin and chitosan, there are no inflammatory symptoms, such as erythema, temperature elevation and abscess formation [158]. The intensity of complement [158] and macrophage [159] activation of chitin is less than chitosan; therefore, chitin is more immunomodulatory.

26. Blood cholesterol control

Chitin and chitosan are among the candidates to battle obesity and hypercholesterolemia. It has been reported that they can reduce the amount of cholesterol in rats [160]. Several mechanisms have been proposed to explain this phenomenon. One is through electrostatic interaction between lipids and aminopolysaccharides [161]. Chitin binds to lipid (cholesterol) micelles and inhibits their absorption. Another proposed mechanism is increasing the excretion of bile acid by which the amount of fecal fat increases [162]. The hypocholesterolemic effect of chitosan has also been found in humans. The proposed cholesterol lowering mechanism of chitosan was that it combines bile acids in the digestive tract, and excretes them into the feces, thus decreasing the resorption of bile acids, so that the cholesterol pool in the body was decreased and the level of serum cholesterol consequently decreased [163].

27. Drug delivery carriers

It is important for a drug delivery carrier to be efficiently removed after delivering drugs. In other words, it must not accumulate in the body nor must it be toxic [164].

Chitosan offers several advantages, and these include its ability to control the release of active agents and avoid the use of hazardous organic solvents while fabricating particles since it is soluble in aqueous acidic solution. Chitosan in the form of colloidal structures can entrap macromolecules by various mechanisms. These associated macromolecules have been shown to transport through mucosa and epithelia more efficiently [165]. Cationic chitosan in combination with other natural polymers has been shown to enhance the drug encapsulation efficiency of liposomes via the layer-by-layer (L-b-L) self-assembly technique [166]. Nanoparticles made of chitosan in association with polyethylene oxide have been used as protein carrier [167]. Moreover, an oral delivery system has been developed by using chitosan and tripolyphosphate. In this system, micro- and nano-particles were entrapped in beads made from chitosan in solution of tripolyphosphate [168].

28. Food

28.1. Chitosan films

Edible films and coatings have received considerable attention in recent years because of their advantages including use as edible packaging materials over synthetic films. This could contribute to the reduction of environmental pollution.

By functioning as barriers, such edible films and coatings can feasibly reduce the complexity and thus improve the recyclability of packaging materials, compared to the more traditional non-environmental friendly packaging materials, and may be able to substitute such synthetic polymer films [169].

Edible films are defined as a thin layer of material which can be consumed and provides a barrier to moisture, oxygen and solute movement for the food. The material can be a complete food coating or can be disposed as a continuous layer between food components [170]. Edible films can be formed as food coatings and free-standing films, and have the potential to be used with food as gas aroma barrier [171].

Chitosans are described in terms of the degree of deacetylation and average molecular weight and their importance resides in their antimicrobial properties in conjunction with their cationicity and their film forming properties [172]. Chitosan can form semi-permeable coatings, which can modify the internal atmosphere, thereby delaying ripening and decreasing transpiration rates in fruits and vegetables [173-176]. Films from aqueous chitosan are clear, tough, flexible and good oxygen barriers [177,178].

29. Bread

Applications of chitosan for extension of shelf life of bread by retarding starch retrogradation and/or by inhibiting microbial growth have been documented. Park and others in 2002 investigated the effect of chitosan (493 kDa) coating on shelf life of baguette [179].

Chitosan coating may offer a protective barrier for moisture transfer through the bread surface, thus reducing weight loss, retarding hardness, retrogradation, inhibiting microbial growth, retarding oxidation [179- 181].

30. Eggs

Several problems are encountered during storage of eggs, such as weight loss, interior quality deterioration, and microbial contamination [182-183]. The movement of carbon dioxide and moisture from the albumen through the shell governs quality changes in albumen and yolk, and weight loss of [184,185].

Chitosan coating may offer a protective barrier for moisture and gas transfer from the albumen through the egg shell, thus extending the shelf life of eggs [182, 186].

31. Fruits and vegetables

The major postharvest losses of fruits are due to fungal infection, physiological disorders, and physical injuries [102, 104, 187]. One of the potential approaches to extend the storability of these perishable commodities is to apply edible coatings on the surface, followed by a cold storage [188]. Edible coatings can be used as a protective barrier to reduce respiration and transpiration rates through fruit surfaces, retard microbial growth and color changes, and improve texture quality of fruits [171]. Coating fruits with semipermeable film has generally been shown to retard ripening by modifying the endogenous CO_2, O_2, and ethylene levels of fruits [102]. Chitosan coating is likely to modify the internal atmosphere without causing anaerobic respiration, since chitosan films are more selectively permeable to O_2 than to CO_2 [189]. Therefore, chitosan coating with its ability to modify internal atmosphere in the tissue and fungistatic property has a potential to prolong storage life and control decay of fruits.

32. Juice and beverages

Processing of clarified fruit juices commonly involves the use of clarifying aids, including gelatin, bentonite, silica sol, tannins, polyvinylpyrrolidone, or combinations of these compounds [190]. Chitosan with a partial positive charge has been shown to possess acid-binding properties [191] and to be effective in aiding the separation of colloidal and dispersed particles from food processing wastes [192,193]. These properties make chitosan an attractive processing aid in fruit juice production.

33. Mayonaise

Few studies have been conducted on the use of chitosan to enhance emulsification in mayonnaise preparation. Lee (1996) reported that addition of chitosan (1500 kDa, 0.1% based on egg yolk weight) increased emulsifying capacity of egg yolk by about 10% and enhanced emulsion stability of mayonnaise by 9.4% compared with those of the control [186]. Kim and Hur (2002) also suggested the use of chitosan as an emulsion stabilizer in commercial mayonnaise preparation [194].

Chitosan possesses a positive ionic charge and has both reactive amino and hydroxyl groups, which give it the ability to chemically bond with negatively charged protein. When pH is less than 6.5, chitosan solution carries a positive charge along its backbone. Because of its polar groups, chitosan also provides additional stabilization due to hydration forces [195]. According to Filar and Wirick in 1978, chitosan functions only in acid systems to show possible utility as a thickener and stabilizer [196].

34. Meat

Meat or meat products are highly susceptible to lipid oxidation, which leads to rapid development of rancid or warmed-over flavor. Chitosan possesses antioxidant and antibacterial capacity [126, 197], and may retard the lipid oxidation and inhibit the growth of spoilage bacteria in meat during storage.

Darmadji and Izumimoto in 1994 observed that addition of 1.0% chitosan to beef decreased the TBA value by about 70% compared to that of the control sample after 3 days of storage at 4 °C. Chitosan has a desirable effect on the development of the red color of beef during storage [198]. Sagoo et al in 2002, demonstrated that chitosan was an effective inhibitor of microbial growth in chilled comminuted pork products and that the effect of chitosan was concentration dependent [199].

35. Milk

A few attempts have been made to evaluate the possibility of using chitosan to improve the quality and shelf life of milk. Ha and Lee in 200, investigated the effectiveness of water-soluble chitosan (0.03%) to minimize the microbial (bacterial and yeast) spoilage of processed milk [200]. Complete inhibition of microbial growth was observed in the banana-flavored milk containing chitosan, in contrast to that observed in control milk (without chitosan), during storage for 15 days at 4 and 10 ◦C. The banana-flavored milk containing chitosan also maintained relatively higher pH than that of control milk during storage for 15 d at both temperatures [200].

36. Sausages

Sodium nitrite is generally used as a curing agent for color and flavor development as well as preservative effect in sausages {201]. However, nitrite reacts with amine in meat and may produce nitrosoamine, a strong toxicant detrimental to human health. Several workers [202] have investigated the possible role of chitosan, in lieu of sodium nitrite, as curing agent in sausage, and found that addition of chitosan could reduce or replace the use of nitrite without affecting preservative effect and color development.

37. Seafoods and seafood products

Seafood products are highly susceptible to quality deterioration due to lipid oxidation of unsaturated fatty acids, catalyzed by the presence of high concentrations of hematin compounds and metal ions in the fish muscle [203]. Furthermore, seafood quality is highly influenced by autolysis, contamination by and growth of microorganisms, and loss of protein functionality [204].

The oxidative stability of fish flesh with added chitosans was compared with those added with conventional antioxidants, butylated hydroxyanisole + butylated hydroxytoluene (BHA + BHT, 200 ppm) and tert butylhydroquinone (TBHQ, 200ppm), during storage at 4 ◦C. Chitosan was most effective in preventing lipid oxidation than the others. The antioxidant capacity of chitosan added to the fish muscle depended on the molecular weight and concentration of chitosan [204]. Similarly, Kim and Thomas in 2007 also observed that the antioxidative effects of chitosan in salmon depended on its molecular weight [205].

38. Chitosan in agriculture

Due to the antifungal, antibacterial and antiviral properties of chitosan, it has been used successfully in agriculture in recent years: in plant protection, like growth promoter, in soil correction, enhancer of secondary metabolites production, and activator of defense mechanisms to mention a few.

39. Seed coating

Chitosan application can be done by different ways: in the seed, in the soil or by foliar way. In seeds, it has been used as a coating material for cereals, nuts, fruits and vegetables [206-208]. It has been shown that this way of application alters permeability of the seed plasma membrane, increasing the concentrations of sugars and proline, and enhancing peroxidase (POD), catalase (CAT), phenylalanine ammonia-lyase (PAL) and tyrosine ammonia-lyase (TAL) activities [207,209]. By this way, germination rates increases significantly [210] and seedlings germinate quicker, better, and vigorously [211-214]. Chitosan is used not only in seed coatings, but also in fruits and vegetables, because it gives more firmness and it promotes diminution of the normal microbiological charge [215] increasing the product life.

40. Leaf coating

Chitosan foliar application increases stomatal conductance and reduces transpiration, without affecting plant height, root length, leaf area or plant biomass [216]. When chitosan is sprayed in leaves, abscisic acid (ABA) content increases [217]. It promotes the activation of defense mechanisms which allow plants to deal with stress and to defend against diseases due to the antiviral, antifungal and antibacterial nature of chitosan [218,219].

41. Fertilizer

By applying chitosan in soil, it has been demonstrated that it stops plant wilting because it acts as a potent fertilizer due to the high concentration of nitrogen content in its molecular structure [220,221]. Also, it has been used as a soil amendment, controlling diseases caused by fungal species like *Fusarium acuminatum, Fusarium sp, Cylindrocladium floridanum* and *Aspergillus flavus* [208, 218, 222].

42. Plants growth promoter

Chitosan acts as plant growth promoter in some crops like Faba bean plant, radish, passion fruit, potato, gerbera, cabbage, soybean and other crops when it is incorporated in solution, increasing plant production and protecting plants against pathogens too. Chitosan has a significant effect on growth rates of roots, shoots, flowering, and number of flowers [219, 223].

43. Plant self defense

Plants react naturally against most of biological and environmental adverse conditions, but sometimes defense has to be induced in order to fight against harder threats. It has been reported that chitosan is a great biopolymer used for this purpose, because it induces defense reactions in some plants, sensitizing them in order to increase their responses against pathogens attack. Some substances that get favored due to the presence of chitin and chitosan are phytoalexines, pathogenesis related proteins (PR), protein inhibitors, chitinases and glucanases, as well as Reactive Oxygen Species (ROS) and hydrogen peroxide generation [224]. This is because chitosan interacts with cellular DNA generating multiple biochemical reactions in the plant, generating a rapid response in the plant against pathogens attack. For this reason, chitosan has been considered as an elicitor, namely a defense mechanism activator in plants, generating a process at cellular level in which plant cells get and transduce biological signals in order to activate defense responses [225]. There are some specific elicitor-binding proteins which act like physiological receptors in signal transduction cascades, varying their specificity depending on the studied system, which allows researchers to find the molecular bases that origin the signal interchanges between host plants and microbial pathogens [225-227].

Not only at biochemical level but also at microbiological level, chitosan is effective on plant protection. It has been found that application of chitosan in plants by the ways mentioned in sections above reduces visibly the damages caused in the plants by pathogenic fungi because of the antibiotic nature of chitosan [215, 218]. Because of being a polysaccharide, chitosan acts as a bioremediator molecule that stimulates the activity of beneficial microorganisms in the soil such as Bacillus, fluorescent, Pseudomonas, Actinomycetes, Mycorrhiza and Rhizobacteria [228-233], which alter the microbial equilibrium in the rhizosphere disadvantaging plant pathogens, making them able to compete through mechanisms such as parasitism, antibiosis, and induced resistance [234,235].

44. Bioinsecticide

Chitosan research has been focused principally in controlling bacterial and fungal burden; nevertheless there are some investigations about the use of chitosan as bioinseciticide. One of the first findings was that chitosan is active against some insects like lepidopterous and homopterous, with a mortality of 80%, and this percentage increases when increasing oligo-chitosan concentration too [236].

Not only chitosan, but also its derivates (as N-acetyl (NAC) and N-benzyl (NBC) chitosan derivatives) had shown significant insecticidal activities superior to those of chitosan itself, particularly against species like Spodoptera littoralis, an important destructive pest of subtropical and tropical agriculture in northern Europe, affecting cotton, vegetable and ornamental crops [237]. Some other insects have been successfully attacked by chitosan derivates, like Helicoverpa armigera (H¨ubn), Plutella xylostella (L), Aphis gossypii (Glover), Metopolophium dirhodum (Walker), Hyalopterus pruni (Geoffroy), Rhopalosiphum padi L, Sitobium avenae (Fabricius) and Myzus persicae (Sulzer) [238].

Active chitinases from chitosan are relevant enzymes for biopesticide control mechanisms, being the hydrolysis of chitin-containing media a common practice to evaluate the efficiency of bioinsecticide organisms. It has been considered to add chitin derivatives to formulations containing these microorganisms to increase biopesticide effectiveness, to provide a favorable developmental environment and resistance against adverse conditions [239]. New chitosan derivatives with insecticidal or fungicidal properties may thus serve as good alternatives for broad-spectrum and highly persistent pesticides because they are non-toxic to vertebrates and humans, and have a biodegradable matrix.

45. Biopesticide

Tricoderma sp. and Bacilus sp. are microorganisms which often increase chitin and chitosan production, enhancing its efficiency to control pathogenic microorganisms and pests [238]. Native populations of biocontrol microorganisms became increased by adding chitin in soils infected with pathogenic agents. Thereafter, these endogenous control strains can be isolated, cultured and potentially used as biological controls. It has also been demonstrated a significant increase in chitinolytic microorganisms even in very infertile soils like in dunes, improving soil microbiota and its properties [239, 241].

46. Bionematicide

Nematodes proliferation can be controlled when chitosan is applied in soil, because chitinolytic microorganisms proliferate destroying nematode eggs and degrading the chitin-containing cuticle of young nematodes [240]. Because of the high content of nitrogen in chitosan and chitin molecules, concentrations of ammonia emissions increase turning toxic to nematodes which principally affect plant roots and shoots [239, 243].

Further research is still required to find more applications of chitosan in agriculture, but nowadays this polymer means to be a cheap and easy material to deal with crop problems pre-harvest, harvest and post-harvest level.

47. Conclusions

Scientific databases reveal thousands of articles and patents related to chitin, chitosan and its derivatives and increasingly opens up new possibilities to produce new derivatives as well as new applications.

The answer to the question if the chitosan is a "new panacea", is given by the multiple applications for this new biopolymer and its predecessor, the chitin. Two hundred years have passed since its discovery and this biopolymer has shown unique qualities that many other polymers do not have, as it can be applied in different areas like in the agricultural and medical field or in related areas such as pharmacy and biomedical.

As seen in this chapter, chitosan's behavior in different applications within diverse areas, is governed by its molecular weight, degree of deacetylation, degree of polymerization and source of obtention. Twenty years ago the articles published did not provide data on the characterization of material but today most papers focus on the properties of the polymer before the application.

Also, through the study of this biopolymer and due to the great demand of chitin and chitosan, it is very important to direct all efforts to seek methods of production through environment-friendly processes and on the other hand, through genetic engineering methods, finding the way to produce a more uniform material with characteristics previously designed, especially for medical or pharmaceutical items.

Acknowledgement

Gabriela Montaraz Miranda, Rosa Guadalupe Valadez Olguin and Ana María Sixto Berrocal for her technical assistance.

To PAPITT Program No. IT220411 Estudio fitopatológico, bioquímico y molecular de la respuesta contra estreses bióticos y abióticos en plántulas de maíz. For the financial support.

To PAPIME Program No. PE203211 Elaboración de material didáctico para las diferentes estrategias utilizadas en la enseñanza de la biotecnología de última generación. For the financial support.

Author details

Susana P. Miranda Castro* and Eva G. Lizárraga Paulín
Laboratorio de Biotecnología, Facultad de Estudios Superiores Cuautitlan, UNAM, Cuautitlan Izcalli, Estado de México, México

48. References

[1] Matsunaga A. (1998) Chitosan: The Ultimate Health Builder. Vantage Press, 134 p. ISBN 0533126290, 9780533126293.
[2] American Heritage dictionary of the English Language: Fourth Edition. 2000. http://www.specialist-online-dictionary.com/ Entry for chiton.
[3] Roberts George, European Chitin Society Newsletter, Editor: George A F Roberts, Truth or Myth? July 2006, N° 21.
[4] Peter M G (2005) Chitin and Chitosan in Fungi. Biopolymers. DOI: 10.1002/3527600035.bpol6005.
[5] Jang MK, Kong BG, Jeong YI, Lee CH, Nah JW (2004) Physicochemical characterization of α-chitin, β-chitin, and γ-chitin separated from natural resources. J. polym. sci. A. polym. chem. 42: 3423–3432. DOI: 10.1002/pola.20176.
[6] Muzzarelli, R.A.A. (1977) Enzymatic synthesis of chitin and chitosan. Occurrence of chitin. In: Chitin (Muzzarelli, R.A.A., ed.), pp. 5–44. Pregamon Press, New York, NY. ISBN 0080203671, 9780080203676
[7] Herring PJ(1979) Marine Ecology and natural products. Pure appl. chem. 51: 1901–1911. Available: http://pac.iupac.org/publications/pac/pdf/1979/pdf/5109x1901.pdf.

* Corresponding Author

[8] Wagner GP, Lo J, Laine R, Almeder M (1993) Chitin in the epidermal cuticle of a vertebrate (Paralipophrystrigloides, Blenniidae, Teleostei). Cell mol. life sci. 4: 317–319. DOI: 10.1007/BF01923410.

[9] Rudall KM, Kenchington W (1973) The chitin system. Biol. rev. 48:597–636. DOI: 10.1111/j.1469-185X.1973.tb01570.x.

[10] Rudall KM (1969) Chitin and its association with other molecules. J. polym. Sci Part C. 28: 83–102. DOI: 10.1002/polc.5070280110.

[11] Blackwell J, Parker KD, Rudall KM (1965) Chitin in pogonophore tubes. J. mar boil. Assoc UK. 45: 659–61. Available: http://sabella.mba.ac.uk/2313/01/Chitin_in_Pogonophore_tubes.pdf.

[12] Gaill F, Persson J, Sugiyama P, Vuong R, Chanzy H (1992) The chitin system in the tubes of deep sea hydrothermal vent worms. J. struct. Boil. 109: 116–28. Available: http://dx.doi.org/10.1016/1047-8477(92)90043-A.

[13] Lotmar W, Picken Ler (1950) A new crystallographic modification of chitin and its distribution. Experientia. 6: 58–59. DOI: 10.1007/BF02174818.

[14] Herth W, Kuppel A, Schnepf E (1977) Chitinous fibrils in the lorica of the flagellate chrysophyte Poterioochromonas stipitata (syn. Ochromonas malhamensis). J. cell biol. 73: 311–21. Available: http://www.ncbi.nlm.nih.gov/pmc/articles/PMC2109911/pdf/jc732311.pdf.

[15] Herth W, Mulisch M, Zugenmaier P (1986) Comparison of chitin fibril structure and assembly in three unicellular organisms. In: Muzzarelli R, Jeuniaux C, Gooday GW, editors. Chitin in nature and technology. New York: Plenum Publishing Corporation. p. 107–20. ISBN 0306422115, 9780306422119.

[16] Rinaudo M (2006) Chitin and chitosan: Properties and applications. Progress in Polymer Science (Oxford), 31 (7), pp. 603-632. http://dx.doi.org/10.1016/j.bbr.2011.03.031.

[17] No HK, Meyers SP (1995) Preparation and Characterization of Chitin and Chitosan—A Review. J. aquat. food prod. technol. 4: 27-52. DOI:10.1300/J030v04n02_03.

[18] Maryam MA and Mahmood A B (2007) A new process for deproteinization of chitin from shrimp head waste. Proceedings of European Congress of Chemical Engineering (ECCE-6) 16–20. Available: http://www.nt.ntnu.no/users/skoge/prost/proceedings/ecce6_sep07/upload/647.pdf

[19] Simpson BK and Haard NF (1985) The use of enzymes to extract carotene protein from shrimp waste. J appl. bioche. 7:212–222.

[20] Cano-Lopez A, Simpson BK, Haard NF (1987) Extraction of carotenoprotein from shrimp process wastes with the aid of trypsin from Atlantic cod. J. food sci. 52:503–504. DOI: 10.1111/j.1365-2621.1987.tb06656.x.

[21] Synowiecki J and Al-Khateeb NAAQ (2000) The recovery of protein hydrolysate during enzymatic isolation of chitin from shrimp Crangon crangon processing discards. Food chemistry, 68:147-152.ISSN: 03088146 http://dx.doi.org/10.1016/S0308-8146(99)00165-X.

[22] Rao MS, Munoz J, Stevens WF (2000) Critical factors in chitin production by fermentation of shrimp biowastes. Appl. microbiol. biotechnol. 54: 808–813. DOI: 10.1007/s002530000449.

[23] Shirai K, Guerrero I, Huerta S, Saucedo G, Castillo A, Gonzalez R (2001) Effect of initial glucose concentration and inoculation level of lactic acid bacteria in shrimp waste ensilation. Enz. microbiol. biotechnol. 28: 446–452. ISSN: 01410229. DOI: 10.1016/S0141-0229(00)00338-0.
[24] Healy M, Green M, Healy A (2003) Bioprocessing of marine crustacean shell waste. Acta biotechnol. 23: 151–160. ISSN: 01384988.
[25] Bhaskar N, Suresh PV, Sakhare PZ, Sachindra NM (2007) Shrimp biowaste fermentation with Pediococcus acidolactici CFR2182: optimization of fermentation conditions by response surface methodology and effect of optimized conditions on deproteination/ demineralization and carotenoids recovery. Enzym. microb. Tech. 40: 1427–1434. ISSN: 01410229. DOI: 10.1016/j.enzmictec.2006.10.019.
[26] Prameela K, Murali M CH, Hemalatha KPJ (2010a) Extraction of pharmaceutically important chitin and carotenoids from shrimp biowaste by microbial fermentation method. J. pharm. Res. 3: 2393–2395. ISSN: 0974-6943. Available: www.jpronline.info
[27] Prameela K, Murali MCH, Smitha PV, Hemalatha KPJ (2010b) Bioremediation of shrimp biowaste by using natural probiotic for chitin and carotenoid production an alternative method to hazardous chemical method. Int. j. appl. biol. pharm. technol. 1: 903–910. ISSN 0976-4550. Available: http://ijabpt.com/pdf/87022-Prameela-Gitam%5B1%5D.pdf.
[28] Prameela K, Murali M CH, Hemalatha KPJ (2010c) Bio- efficiency of Pediococcus acidilactici (ATCC 8042) for recovery of chitin and carotenoids in the fermentation of shrimp biowaste. Int. j. chem.tech. res. 2: 1924–1928. DOI: 10.1007/s00253-011-3651-2.
[29] Rao MS and Stevens WF (2005) Quality parameters of chitosan derived from fermentation of shrimp biomaterial using a drum reaction. J. chem. technol. Biotechnol. 80: 1080–1087. DOI: 10.1002/jctb.1286.
[30] Woods B (1998) Microbiology of fermented foods, vol 1. Blackie, NY. BN 978-0-7514-0216-2. ISBN 0-7514-0216-8
[31] Legarrenta GI, Zakaria Z, Hall GM (1996) Lactic fermentation of prawn waste: comparison of commercial and isolated starter culture. In: Domard A, Jeuniaux C, Muzzarelli RAA, Roberts GAF (eds) Advances in chitin science, vol 1. Jacques Andre, Lyon, pp 399–406 ISBN 2907922408, 9782907922401.
[32] Muzzarelli RAA and Peter MG, Eds.1997. Chitin Handbook,, Grottammare: Atec, ISBN 8886889011, 9788886889018
[33] Mathur NK and Narang CK (1990)Chitin and chitosan, versatile polysaccharides from marine animals. J. chem. educ. 67, 938-942. DOI: 10.1021/ed067p938.
[34] Dodane V and Vilivalam VD (1998) Pharmaceutical applications of chitosan. Pharm. sci. tech. today 1: 246–253. ISSN: 14615347. DOI: 10.1016/S1461-5347(98)00059-5.
[35] Kumar MNVR (2000) A review of chitin and chitosan applications. Reactive & Functional Polymers, 46: 1-27. http://dx.doi.org/10.1016/S1381-5148(00)00038-9.
[36] Khor E and Lim L (2003) Implantable applications of chitin and chitosan. Biomaterials. 24: 2339-2349. ISSN: 01429612. DOI: 10.1016/S0142-9612(03)00026-7.
[37] Muzzarelli RAA Natural (1973) Polymers, Pergamon Press, London. ISBN 0080172350, 9780080172354.

[38] Crestini C, Kovac B and Giovannozzi-Sermanni G (1996) Production and isolation of chitosan by submerged and solid-state fermentation from Lentinusedodes. Biotechnol. bioeng. 50: 207–210. DOI: 10.1002/bit.260500202.

[39] Tan SC, Tan TK, Wong SM, Khor E (1996) The chitosan yield of zygomycetes at their optimum harvesting time Carbohydrate Polymers. 30: 239-242. http://dx.doi.org/10.1016/S0144-8617(96)00052-5.

[40] Arcidiacono S and Kaplan D L (1992) Molecular weight distribution of chitosan isolated from Mucor rouxii under different culture and processing conditions. Biotechnol. bioeng. 39: 281–286. doi: 10.1002/bit.260390305.

[41] Nwe N, Stevens WF(2002) Chitosan isolation from the chitosan-glucan complex of fungal cell wall using amylolytic enzymes. Biotechnology Letters 24:1461-1464. DOI: 10.1023/A:1019898715518.

[42] Ikeda I, Sugano M, Yoshida K, Sasaki E, Iwamoto Y, and Hatano K (1993) Effects of chitosan hydrolyzates on lipid absorption and on serum and liver lipid concentration in rats. J. agric. food chem. 41: 431-435. DOI: 10.1021/jf00027a016.

[43] Nadarajah K. Kader J, Mazmira M and Paul DC (2001) Pakistan Journal of Biological Sciences 4 (3):263-265. Available: http://docsdrive.com/pdfs/ansinet/pjbs/2001/263-265.pdf.

[44] Kishore D. Raneab and Dallas G (1993) Hoove Production of Chitosan by fun. Food Biotechnology .7: 11-33 DOI:10.1080/08905439309549843.

[45] Wei-Ping W, Yu-Min D and Xiao-Ying W (2008) Physical properties of fungal chitosan. World J. Microb. Biot. 24: 2717-2720. DOI: 10.1007/s11274-008-9755-x.

[46] Endo T and Koizumi S. (2000) Large-scale production of oligosaccharides using engineered bacteria. Curr. Opin. Struct. Biol. 10: 536–541. ISSN: 0959440X. DOI: 10.1016/S0959-440X(00)00127-5.

[47] Cottaz S and Samain E (2005) Genetic engineering of Escherichia coli for the production of NI,NII- diacetylchitobiose (chitinbiose) and its utilization as a primer for the synthesis of complex carbohydrates. Metab. Eng. 7: 311–317. http://dx.doi.org/10.1016/j.ymben.2005.05.004.

[48] Samain E, Drouillard S, Heyraud A, Driguez H, Geremia RA (1997) Gram-scale synthesis of recombinant chitooligosaccharides in Escherichia coli. Carbohydr. res. 302: 35–42. ISSN: 00086215. DOI: 10.1016/S0008-6215(97)00107-9.

[49] Samain E, Chazalet V, Geremia RA (1999) Production of O-acetylated and sulfated chitooligosaccharides by recombinant Escherichia coli strains harboring different combinations of nod genes. J. biotechnol. 72: 33–47. ISSN: 01681656. DOI: 10.1016/S0168-1656(99)00048-6.

[50] Khan TA, Peh KK, Ching H (2002) Reporting degree of deacetylation values of chitosan: the influence of analytical methods. J. pharm sci. 5:205–12. Available: (www.ualberta.ca/~csps) 5(3):205-212, 2002.

[51] Shigemasa Y, Matsuura H, Sashiwa H, Saimoto H (1996) Evaluation of different absorbance ratios from infrared spectroscopy for analyzing the degree of deacetylation in chitin. Int. j. biol. macromol. 18: pp. 237-242. ISSN: 01418130. DOI: 10.1016/0141-8130(95)01079-3.

[52] Muzzarelli RAA and Muzzarelli BB (1998) Structural and functional versatility of chitins. In: Structural Diversity and Functional Versatility of Polysaccharides. S. Dumitriu (ed) Marcel Dekker, New York. ISBN 0824754808, 9780824754808
[53] Brugnerotto J, Lizardi J, Goycoolea F, Arguelles-Monal W, Desbrieres J, Rinaudo M (2001) An infrared investigation in relation with chitin and chitosan characterization. Polymer 42: 3569-3580. ISSN: 00323861. DOI: 10.1016/S0032-3861(00)00713-8.
[54] Duarte ML, Ferreira MC, Marvao MR, Rocha J. (2002) An optimised method to determine the degree of acetylation of chitin and chitosan by FTIR spectroscopy. Int. J. biol. macromol. 31: 1-8. ISSN: 01418130. DOI: 10.1016/S0141-8130(02)00039-9.
[55] Cao W, Jing D, Li J, Gong Y, Zhao N and Zhang X (2005) Effects of the Degree of Deacetylation on the Physicochemical Properties and Schwann Cell Affinity of Chitosan Films. J. biomater appl. 20: 157-177. doi:10.1177/0885328205049897
[56] Zhang H, Du Y, Yu X, Mitsutomi M and Aiba SI (1999) Preparation of chitooligosaccharides from chitosan by a complex enzyme. Carbohyd. res. 320: 257-260. DOI: http://dx.doi.org/10.1016/S0008-6215(99)00154-8.
[57] Jia Z and Shen D (2002) Effect of reaction temperature and reaction time on the preparation of low-molecular-weight chitosan using phosphoric acid. Carbohyd polym. 49 :393-396. ISSN: 01448617. DOI: 10.1016/S0144-8617(02)00026-7.
[58] Qin C, Du Y, Zong L, Zeng F, Liu Y, Zhou B (2003) Effect of hemicellulase on the molecular weight and structure of chitosan. Polym. degrad. stabil. 80 :435-441. ISSN: 01413910. DOI: 10.1016/S0141-3910(03)00027-2.
[59] Domard A and Rinaudo M (1983) Preparation and characterization of fully deacetylated chitosan. Int.j. biol macromol. 5: 49–52. http://dx.doi.org/10.1016/0141-8130(83)90078-8.
[60] Kubota N and Eguchi Y (1997) Facile preparation of water-soluble N-acetylated chitosan and molecular weight dependence of its water-solubility. Polym j. 29:123–127. ISSN: 00323896.
[61] Aiba SI (1991) Studies on chitosan: 3. Evidence for the presence of random and block copolymer structures in partially N-acetylated chitosans. Int. J. Biol. Macromol.13 : 40-44. Available: http://dx.doi.org/10.1016/0141-8130(91)90008-I.
[62] Rinaudo M, Domard A (1989) Solution properties of chitosan. In: Skjak-Braek G, Anthonsen T, Sandford P, editors. Chitin and chitosan. Sources, chemistry, biochemistry, physical properties and applications. London and New York: Elsevier; p. 71–86. ISBN 1851663959, 9781851663958.
[63] Elson Santiago de Alvarenga (2011) Characterization and Properties of Chitosan, Biotechnology of Biopolymers, Prof. MagdyElnashar (Ed.). ISBN: 978-953-307-179-4. InTech, Available from: http://www.intechopen.com/books/biotechnology-of-biopolymers/characterization-and-properties-of-chitosan.
[64] Allan G and Peyron M (1995) Molecular weight manipulation of chitosan I: Kinetics of depolymerization by nitrous acid. Carbohyd. res. 277: 257-272. Available: http://dx.doi.org/10.1016/0008-6215(95)00207-A.
[65] Rinaudo M, Pavlov G, Desbrieres J (1999) Influence of acetic acid concentration on the solubilization of chitosan. Polymer. 40:7029-7032 Available: http://dx.doi.org/10.1016/S0032-3861(99)00056-7.

[66] Kasaai M R, Arul J and Charlet G (2000) Intrinsic viscosity–molecular weight relationship for chitosan. J. polym. sci. B Polym. phys. 38: 2591–2598. doi: 10.1002/1099-0488(20001001)38:19<2591::AID-POLB110>3.0.CO;2-6.

[67] Dutta PK (2005) Chitin and chitosan opportunities and challenges. P.K. Dutta (Ed.) SSM International Publication. Contai, Midnapore , India p 6.

[68] Keong LC and Halim A S(2009) In Vitro Models in Biocompatibility Assessment for Biomedical-Grade Chitosan Derivatives in Wound Management. Int. j. mol. sci. 10:1300-1313. doi:10.3390/ijms10031300.

[69] Li Q, Dunn ET, Grandmaison EW, Goosen MFA (1992) Applications and properties of chitosan. J. compat. pol. 7:370–97. doi: 10.1177/088391159200700406.

[70] Suh FJK and Matthew HWT (2000) Applications of chitosan-based polysaccharide biomaterials in cartilage tissue engineering: A review. Biomater. 21: 2589-2598. http://dx.doi.org/10.1016/S0142-9612(00)00126-5.

[71] Ikada Y (1994) Surface modification of polymers for medical applications. Biomaterials.15:725–36. http://dx.doi.org/10.1016/0142-9612(94)90025-6.

[72] Zhang M, Li XH, Gong YD, Zhao NM and Zhang XF (2002) Properties and biocompatibility of chitosan films modified by blending with PEG. Biomaterials 23: 2641-2648. http://dx.doi.org/10.1016/S0142-9612(01)00403-3.

[73] Zileinski BA and Acbischer P (1994) Chitosan as a matrix for mammalian cell encapsulation. Biomaterials. 15: 1049-1056. http://dx.doi.org/10.1016/0142-9612(94)90090-6.

[74] Deng CM, He LZ, Zhao M, Yang D. Liu Y (2007) Biological properties of the chitosan-gelatin sponge wound dressing. Carbohyd. Polym. 69: 583-589. http://dx.doi.org/10.1016/j.carbpol.2007.01.014.

[75] Rathke TD and Hudson SM (1994) Review of chitin and chitosan as fiber and film formers. J Mater Sci: Rev Macromolecular ChemPhys. C34:375–437. DOI:10.1080/15321799408014163.

[76] Karel SJR, Jaromir L, Vera P, Jiřina B, Fu-Tong L, Jiří V, Hans-Joachim G. (1997) Effect of chemical structure of hydrogels on the adhesion and phenotypic characteristics of human monocytes such as expression ofgalectins and other carbohydrate-binding sites. Biomaterials. 18:1009–1014.Available: http://dx.doi.org/10.1016/S0142-9612(97)00037-9.

[77] Den Braber ET, De Ruijter JE, Ginsel LA, Von Recum AF, Jansen JA (1996) Quantitative analysis of fibroblast morphology on microgrooved surfaces with various groove and ridge dimensions.Biomaterials.17:2037-2044. http://dx.doi.org/10.1016/0142-9612(96)00032-4.

[78] Singhvi R, Stephanopoulos G, Daniel IC (1994) Effect of substrate morphology on cell physiology. Biotechnol. bioeng.43: 764–71. DOI: 10.1002/bit.260430811.

[79] Suzuki T and Mizushima Y (1997) Characteristics of silica-chitosan complex membrane and their relationships to the characteristics of growth and adhesiveness of L-929 cells cultured on the biomembrane. J. ferment. Bioeng. 84: 128-132. ISSN: 0922338X. DOI: 10.1016/S0922-338X(97)82541-X.

[80] Hallab NJ, Bundy KJ, O'Connor K, Clark R, Moses RL (1995) Cell adhesion to biomaterials: correlations between surface charge, surface roughness, adsorbed protein and cell morphology. J. long-term defects med impl. 5:209–31. PMID:10172729.
[81] Ruardy TG, Moorlag HE, Schakenraad JM, Van Der Mei HC, Busscher HJ (1997) Growth of fibroblasts and endothelial cells on wettability gradient surfaces. J. Colloid interf- sci. 188:209–217. http://dx.doi.org/10.1006/jcis.1997.4769.
[82] Kean T and Thanou M, (2009) Chitin and chitosan—sources, production and medical applications, in: P.A. Williams, R. Arshady (Eds.), Desk reference of Natural Polymers, their Sources, Chemistry and Applications, Kentus Books, London, pp. 327–361. ISBN: 978-1-84973-351-9. DOI:10.1039/9781849733519-00292
[83] Kean T and Thanou M. (2010) Biodegradation, biodistribution and toxicity of chitosan. Adv. Drug. Deliver. Rev. 62: 3-11. ISSN: 0169409X. DOI: 10.1016/j.addr.2009.09.004.
[84] Hsu SC, Don TM, Chiu WY (2002) Free radical degradation of chitosan with potassium persulfate. Polym. Degrade. Stabil. 75: 73-83. ISSN: 01413910. DOI: 10.1016/S0141-3910(01)00205-1.
[85] Zoldners J, Kiseleva T, Kaiminsh I (2005) Influence of ascorbic acid on the stability of chitosan Carbohyd. Polym. 60: 215-218. ISSN: 01448617. DOI: 10.1016/j.carbpol.2005.01.013.
[86] Saimoto H, Takamori Y, Morimoto M, Sashiwa H, Okamoto Y, Minami S, Matsuhashi A, Shigemasa Y (1977) Biodegradation of Chitin with Enzymes and Vital Components. Macromolecular Symposia. 120:11-18. ISSN: 10221360.
[87] Funkhouser JD, Aronson J NN (2007) Chitinase family GH18: Evolutionary insights from the genomic history of a diverse protein family BMC. Evol. biol. 7, art. no. 96. ISSN: 14712148. DOI: 10.1186/1471-2148-7-96. Available: http://www.biomedcentral.com/1471-2148/7/96.
[88] Onishi H and Machida Y (1999) Biodegradation and distribution of water-soluble chitosan in mice. Biomaterials. 20: 175-182. ISSN: 01429612. DOI: 10.1016/S0142-9612(98)00159-8.
[89] Senel S and McClure SJ (2004) Potential applications of chitosan in veterinary medicine. Adv. drug. deliver. rev.56: 1467-1480. ISSN: 0169409X. DOI: 10.1016/j.addr.2004.02.007.
[90] Zhang H and Neau SH (2002) In vitro degradation of chitosan by bacterial enzymes from rat cecal and colonic contents. Biomaterials, 23: 2761-2766. ISSN: 01429612. DOI: 10.1016/S0142-9612(02)00011-X.
[91] Illum L (1998) Chitosan and its use as a pharmaceutical excipient. Pharm. res. 15: 1326-1331. DOI: 10.1023/A:1011929016601.
[92] Wedmore I, McManus JG, Pusateri AE, Holcomb JB (2006) A special report on the chitosan-based hemostatic dressing: Experience in current combat operations. J. trauma. 60: 655-658. DOI: 10.1097/01.ta.0000199392.91772.44.
[93] Schipper NGM, Varum KM, Artursson P (1996) Chitosans as absorption enhancers for poorly absorbable drugs. 1: Influence of molecular weight and degree of acetylation on drug transport across human intestinal epithelial (Caco-2) cells. Pharm. res. 13:1686-1692. DOI: 10.1023/A:1016444808000.

[94] Bruno Sarmento, José das Neves (2012) Chitosan-Based Systems for Biopharmaceuticals: Delivery, Targeting and Polymer Therapeutics. Bruno Sarmento, José das Neves (eds). John Wiley & Sons. ISBN 111996296X, 9781119962960.

[95] Fernandez-Saiz P, Ocio MJ, Lagaron JM (2010) The use of chitosan in antimicrobial films for food protection, CAB Reviews: Perspectives in Agriculture, Veterinary Science, Nutrition and Natural Resources. Vol 5. ISSN: 17498848. DOI: 10.1079/PAVSNNR20105024.

[96] Ralston GB, Tracey MV, Wrench PM (1964) The inhibition of fermentation in baker's yeast by chitosan. BBA - General Subjects. 93:652-655. ISSN: 03044165. PubMed ID: 14263164.

[97] Helander IM, Nurmiaho-Lassila EL, Ahvenainen R, Rhoades J, Roller S (2001) Chitosan disrupts the barrier properties of the outer membrane of Gram-negative bacteria. Int. j. food. microbiol. 71: 235-244. ISSN: 01681605. DOI: 10.1016/S0168-1605(01)00609-2.

[98] Liu H, Du Y, Xiaohui Wang X, Sun L (2004) Chitosan kills bacteria through cell membrane damage. Int. j. food. Microbial. 95:147–55. PMID:15282127.

[99] Mayachiew P, Devahastin S, Mackey BM, Niranjan K.(2010). Effects of drying methods and conditions on antimicrobial activity of edible chitosan films enriched with galangal extract. Food res. Int. 43(1):125–32. Available: http://dx.doi.org/10.1016/j.foodres.2009.09.006,

[100] Ganan M, Carrascosa V, Martinez-Rodriguez AJ (2009) Antimicrobial activity of chitosan against Campylobacter spp. and other microorganism and its mechanism of action. Journal of Food Protection 8:1735–1738. PMID:19722411.

[101] Hadwiger LA, Kendra DF, Fristensky BW, Wagoner W (1985) Chitosan both activates genes in plants and inhibits RNA synthesis in fungi. In: Muzzarelli RAA, Jeuniaux C, Gooday GW, editors. Chitin in Nature and Technology. Plenum Press, New York; p. 209–22.

[102] El Ghaouth A, Arul J, Ponnampalam R, Boulet M (1991) Chitosan coating effect on storability and quality of fresh strawberries. J Food Sci 56:1618–20. DOI: 10.1111/j.1365-2621.1991.tb08655.x.

[103] Cuero RG, Osuji G, Washington A (1991) N-carboxymethyl chitosan inhibition of aflatoxin production: role of zinc. Biotechnol. lett.13:441–444. Available: http://www.springerlink.com/content/x63116370379vn54/fulltext.pdf.

[104] El Ghaouth A, Arul J, Grenier J, Asselin A (1992a) Antifungal activity of chitosan on two postharvest pathogens of strawberry fruits. Phytopathology 82:398-402. Available: http://www.apsnet.org/publications/phytopathology/backissues/Documents/1992Articles/Phyto82n04_398.pdf

[105] Hirano S and Nagao N, (1989) Effects of chitosan, pectic acid, lysozyme, and chitinase on the growth of several phytopathogens. Agricultural and Biological Chemistry 53, 3065–3066. DOI: http://dx.doi.org/10.1271/bbb1961.53.3065.

[106] Kong M, Chen G, Xing K, Park HJ (2010) Antimicrobial properties of chitosan and mode of action: A state of the art review. Int. j .food microbial. 144:51-63. ISSN: 01681605. DOI: 10.1016/j.ijfoodmicro.2010.09.012.

[107] Uchida Y, Izume M, Ohtakara A (1989) In: Skjak-Braek, G., Anthonsen, T., Sandford, P. (Eds.), Chitin and chitosan. Elsevier, London, UK, p. 373.

[108] Ueno K, Yamaguchi T, Sakairi N, Nishi N, Tokura S (1997) In:Domard, A., Roberts,G.A.F., Varum, K.M. (Eds.), Advances in chitin science. Jacques Andre, Lyon, p. 156.

[109] Franklin TJ and Snow GA (1981) Biochemistry of Antimicrobial Action, 3rd ed. Chapman and Hall, London, p. 175. ISBN 0412302608, 9780412302602.

[110] Kong M, Chen XG, Liu CS, Liu CG, Meng XH, Yu LJ (2008) Antibacterial mechanism of chitosan microspheres in a solid dispersing system against E. coli. Colloid. Surface. B. 65: pp. 197-202.DOI: http://dx.doi.org/10.1016/j.colsurfb.2008.04.003.

[111] Kong M, Chen XG, Xing K, Park HJ (2010)Antimicrobial properties of chitosan and mode of action: A state of the art review. Int. j .food microbial. 144: 51-63. PMID:20951455. Available: http://www.sciencedirect.com/science/article/pii/S0168160510005167.

[112] Hernandez-Lauzardo AN, Bautista-Banos S, Velazquez-del Valle MG, Mendez-Montealvo MG, Sanchez-Rivera MM, Bello-Perez LA (2008) Antifungal effects of chitosan with different molecular weights on in vitro development of Rhizopus stolonifer (Ehrenb.:Fr.) Vuill.Carbohyd. polym. 73: 541-547. ISSN: 01448617. DOI: 10.1016/j.carbpol.2007.12.020.

[113] Roller S, Covill N (1999) The antifungal properties of chitosan in laboratory media and apple juice. Int. j .food microbial. 47: 67-77. ISSN: 01681605. DOI: 10.1016/S0168-1605(99)00006-9.

[114] Kochkina Z M and Chirkov S N (2000a) Influence of chitosan derivatives on the development of phage infection in the Bacillus thuringiensis culture. Microbiology 69: 217-219. DOI: 10.1007/BF02756202.

[115] Kochkina ZM, Surgucheva NA, Chirkov SN (2000b) Influence of chitosan derivatives on the development of phage infection in the Bacillus thuringiensis culture. Microbiology. 69: 217-219. DOI: 10.1007/BF02756202.

[116] Sudarshan NR, Hoover DG, Knorr D(1992) Antibacterial action of chitosan. Food biotechnol. 6: 257-272. ISSN: 08905436

[117] Tsai GJ and Su W (1999) Antibacterial Activity of Shrimp Chitosan against Escherichia coli .J. Food Prot. 62:239–243.

[118] Barrette J, Champagne C, Goulet J (1999) Development of bacterial contamination during production of yeast extracts. J. appl. environ. microbiol. 65: 3261–3263. ISSN: 00992240.

[119] Vaara M (1992) Agents that increase the permeability of the outer membrane. Microbiol. Rev. 56: 395-411. ISSN: 01460749 PubMed ID: 1406489.

[120] Jeon Y-J, Park P-J, Kim S-K(2001) Antimicrobial effect of chitooligosaccharides produced by bioreactor Carbohyd. Polym. 44: 71-76. ISSN: 01448617. DOI: 10.1016/S0144-8617(00)00200-9

[121] Gerasimenko DV, Avdienko ID, Bannikova GE, Zueva OY, Varlamov VP (2004) Antibacterial effects of water-soluble lowmolecular- weight chitosan on different

microorganisms. Appl. Biochem. Micro. 40:253–7. DOI: 10.1023/B:ABIM.0000025947.84650.b4.

[122] Liu N, Chen X, Park H, Liu C, Liu C, Meng X, Yu L (2006) Effect of MW and concentration of chitosan on antibacterial activity of Escherichia coli Bioreactor. Carbohyd.polym.64:60–65. ISSN: 01448617 DOI: 10.1016/j.carbpol.2005.10.028.

[123] Liu X, Guan Y, Yang D, Li Z, Yao K (2001) Antibacterial action of chitosan and carboxymethylated chitosan. Journal of Applied Polymer Science. 79(7):1324–35. ISSN: 00218995. DOI: 10.1002/1097-4628(20010214)79:7<1324::AID-APP210>3.0.CO;2-LDocument

[124] Zivanovic S, Basurto C, Chi S, Davidson P, Weiss J (2004) Molecular weight of chitosan influences antimicrobial activity in oil-in-water emulsions. J. food protect. 67:952–959. ISSN: 0362028X.

[125] Uchida Y, Izume M, Ohtakara A (1989) Preparation of chitosan oligomers with purified chitosanase and its application. In: Skja˚ k-Brk G, Anthonsen T, Sandford P, editors. Chitin and Chitosan: Sources, Chemistry, Biochemistry, Physical Properties and Applications. Elsevier Applied Science, London. p. 373–82.

[126] No HK, Park NY, Lee SH, Meyers SP (2002) Antibacterial activity of chitosans and chitosan oligomers with different molecular weights. Int J FoodMicrobiol 74:65–72. DOI: http://dx.doi.org/10.1016/S0168-1605(01)00717-6.

[127] Qin C, Li H, Xiao Q, Liu Y, Zhu J, Du Y (2006) Water-solubility of chitosan and its antimicrobial activity. Carbohyd.polym. 63: 367-374. DOI: http://dx.doi.org/10.1016/j.carbpol.2005.09.023.

[128] Fernandes C, Tavaria F, Soares J, Ramos O, João Monteiro J, Pintado M, Xavier Malcata F (2008) Antimicrobial effects of chitosan and chitooligosaccharides, upon Staphylococcus aureus and Escherichia coli, in food model systems. Food microbial. 25:922–928. ISSN: 07400020 DOI: 10.1016/j.fm.2008.05.003.

[129] Fernandez-Saiz P, Lagaron J, Ocio M (2009) Optimization of the biocide properties of chitosan for its application in the design of active films of interest in the food area. Food hydrocolloid. 23(3):913–921. ISSN: 0268005X DOI: 10.1016/j.foodhyd.2008.06.001.

[130] Takahashi T, Imai M, Suzuki I, Sawai J (2008) Growth inhibitory effect on bacteria of chitosan membranes regulated with deacetylation degree. Biochem. Eng. j. 40:485–491. ISSN: 1369703X. DOI: 10.1016/j.bej.2008.02.009.

[131] Chen Yen-Men, Chung Ying-Chien, Woan Wang Li, Chen Kung-Tung, Li Shyh-Yuan(2002) Antibacterial properties of chitosan in waterborne pathogen. J. environ. sci. heal. A. 37: 1379-1390. ISSN: 1093-4529. DOI: DOI: 10.1081/ESE-120005993.

[132] Hongpattarakere T, Riyaphan O(2008) Effect of deacetylation conditions on antimicrobial activity of chitosans prepared from carapace of black tiger shrimp (Penaeusmonodon). Songklanakarin Journal of Science and Technology. 30:1–9. Available: http://rdo.psu.ac.th/sjstweb/journal/30-Suppl-1/0125-3395-30-S1-1-9.pdf.

[133] Tsai G-J, Su W-H, Chen H-C, Pan C-L(2002) Antimicrobial activity of shrimp chitin and chitosan from different treatments and applications of fish preservation. Fisheries sci. 68:170-177. ISSN: 09199268. DOI: 10.1046/j.1444-2906.2002.00404.x.

[134] Allan G (1984) Biochemical applications of chitin and chitosan. In Zikakis J, editor. Chitin, Chitosan and related enzymes. Orland, FL: Academic Press. pp 119-133. ISBN 0127809503, 9780127809502.

[135] Ohshima Y, Nishino K, Yonekura Y, Kishimoto S and Wakabayashy S (1987). Clinical application of chitin non-woven fabric as a wound dreesing. Eur. J. plast. surg. 10:66-69. DOI: 10.1007/BF00578375.

[136] Minami S, Okamoto Y, Tanioka S, Sashiwa H,Saimoto H, Matsuhashi A, Shigemasa Y (1993) Effects of chitosan on wound healing. In M. Yalpani, Carbohydrates and carbohydrate polymers: analysis, biotechnology, modification, antiviral, biomedical and other applications. pp. 141-152. Vol 1Mount Prospect: ATL Press. ISBN 1882360400, 9781882360406.

[137] Okamoto Y, Minami S, Matsuhashi A ,Sashiwa H, Saimoto H, Shigemasa Y, Taniagua T, Tanaka Y, and Tokura S (1993) Application of polymeric N-acetyl-D-glucosamine (chitin) to veterinary practice. J. vet. med. sci. 55, 743-747. (PMID:8286525).

[138] Okamoto Y, Kawakami K, Miyatake K, Morimoto M, Shigemasa Y, Minami S (2002) Analgesic effects of chitin and chitosan. Carbohyd. Polym. 49:249-252. ISSN: 01448617. DOI: http://dx.doi.org/10.1016/S0144-8617(01)00316-2,.

[139] Qin CQ, Du YM, Xiao L, Li Z and Gao XH (2002a) Enzymic preparation of water soluble chitosan and their antitumor activity. Int. j. biol. macromol. 31:111-117. ISSN 0141-8130. DOI http://dx.doi.org/10.1016/S0141-8130(02)00064-8.

[140] Jeon Y-J and Kim S-K (2002) Antitumor Activity of Chitosan Oligosaccharides Produced In Ultrafiltration Membrane Reactor System. J. microbiol. biotechnol. 12:503-507. ISSN 1017-7825.

[141] Maeda Y and Kimura Y (2004) Antitumor Effects of Various Low-Molecular-Weight Chitosans Are Due to Increased Natural Killer Activity of Intestinal Intraepithelia lymphocytes in sarcoma 180-bearing mice. J. nutr. 134:945-950.

[142] Qin CQ, Zhou B, Zeng L, Zhang Z, Liu Y.; Du, Y.M. & Xiao, L. (2004). The physicochemical properties and antitumor activity of cellulose-treated chitosan. Food Chem., 84, 107-115, ISSN 0308-8146. doi:10.1016/S0308-8146(03)00181-X.

[143] Qin CQ, Du YM, Xiao L, Gao XH, Zhou JL and Liu HL (2002b) Effect of Molecular Weight and Structure on Antitumor Activity of Oxidized Chitosan. Wuhan univ. j. nat. sci., 7, 231-236, ISSN 1007-1202. DOI: 10.1007/BF02830325.

[144] Harish Prashanth KV and Tharanathan RN (2005) Depolymerized products of chitosan as potent inhibitors of tumor-induced angiogenesis. Biochim. Biophys. Acta, 1722, 22-29, ISSN 0006-3002. DOI: 10.1016/j.bbagen.2004.11.009.

[145] Wang S-L, Lin T-Y, Yen Y-H, Liao H-F and Chen Y-J (2006) Bioconversion of shellfish chitin wastes for the production of Bacillus subtilis W-118 chitinase. Carbohydr. res., 341, 2507–2515, ISSN 0008-6215. DOI: 10.1016/j.carres.2006.06.027.

[146] Parvez S, Rahman MM, Khan MA, Khan MAH, Islam JMM, Ahmed M, Rahman MF, Ahmed B (2012) Preparation and characterization of artificial skin using chitosan and gelatin composites for potential biomedical application. Polymer Bulletin: 1-17. Article in Press. DOI: 10.1007/s00289-012-0761-7.

[147] Brandl F, Sommer F, Goepferich A (2007) Rational design of hydrogels for tissue engineering: Impact of physical factors on cell behavior. Biomaterials. 28:134–146. ISSN: 01429612. DOI: 10.1016/j.biomaterials.2006.09.017.

[148] Tsioptsias C, Tsivintzelis I, Papadopoulou L, Panayiotou C A.(2009) novel method for producing tissue engineering scaffolds from chitin, chitin-hydroxyapatite, and cellulose. Mater. Sci. Eng. C. 29: 159–164. ISSN: 09284931. DOI: 10.1016/j.msec.2008.06.003.

[149] Drury JL and Mooney DJ(2003) Hydrogels for tissue engineering: scaffold design variables and applications. Biomaterials 24, 4337–4351. ISSN: 01429612. DOI: 10.1016/S0142-9612(03)00340-5.

[150] Khor E and Lim LY (2003) Implantable applications of chitin and chitosan. Biomaterials 24: 2339–2349. ISSN: 01429612 DOI: 10.1016/S0142-9612(03)00026-7.

[151] Klokkevold PR, Fukayama H, Sung EC, Bertolami CN (1999) The effect of chitosan (poly-N-acetyl glucosamine) on lingual hemostasis in heparinized rabbits. J. oral maxillofac. surg. 57: 49–52. PII: S0278-2391(99)90632-8

[152] Okamoto Y, Yano R, Miyatake K, Tomohiro I, Shigemasa Y, Minami S (2003) Effects of chitin and chitosan on blood coagulation. Carbohyd. polym. 53: 337-342. ISSN: 01448617. DOI: 10.1016/S0144-8617(03)00076-6.

[153] Freier T, Montenegro R, Shan Koh H, Shoichet S (2005) Chitin-based tubes for tissue engineering in the nervous system. Biomaterials. 26: 4624–4632. ISSN: 01429612. DOI: 10.1016/j.biomaterials.2004.11.040.

[154] Gong Y, Gong L, Gu X, Ding F (2009) Chitooligosaccharides promote peripheral nerve regeneration in a rabbit common peroneal nerve crush injury model. Microsurgery. 29(8): 650–656. ISSN: 07381085 DOI: 10.1002/micr.20686

[155] Aam B, Heggset E, Norberg A, SØrlie M, Vårum K, Eijsink V (2010)Production of Chitooligosaccharides and Their Potential Applications in Medicine. Mar. Drugs. 8(5)1482–1517. ISSN: 16603397 DOI: 10.3390/md8051482

[156] Muzzarelli RAA(2010) Chitins and chitosans as immunoadjuvants and non-allergenic drug carriers. Mar. Drugs, 8, 292–312. ISSN: 16603397. DOI: 10.3390/md8020292.

[157] Jeong HJ, Koo HN, Oh EY, Chae HJ, Kim HR, Suh SB, Kim CH, Cho KH, Park BR, Park ST, Lee YM, Kim HM (2000)Nitric oxide production by high molecular weight water-soluble chitosan via nuclear factor-kappaB activation. Int. j. immunopharmacol. 22: 923–933. http://dx.doi.org/10.1016/S0192-0561(00)00055-2.

[158] Minami S, Suzuki H, Okamoto Y, Fujinaga T, Shigemasa Y (1998) Chitin and chitosan activate complement via the alternative pathway. Carbohyd. polym. 36: 151-155. ISSN: 01448617. http://dx.doi.org/10.1016/S0144-8617(98)00015-0.

[159] Freier T, Montenegro R, Shan Koh H, Shoichet MS (2005) Chitin-based tubes for tissue engineering in the nervous system. Biomaterials. 26: 4624–4632. ISSN: 01429612. DOI: 10.1016/j.biomaterials.2004.11.040.

[160] Razdan A and Pettersson D (1994) Effect of chitin and chitosan on nutrient digestibility and plasma lipid concentrations in broiler chickens. Br. j. nutr. 72: 277–288. ISSN: 00071145 . DOI: 10.1079/BJN19940029.

[161] Muzzarelli RAA and Muzzarelli C (2006) Chitosan, a dietary supplement and a food technology commodity. In Functional Food Carbohydrates; Biliaderis, C.G., Izydorczyk,

M.S., Eds.; Francis and Taylor: Orlando, FL, USA, pp. 215–248. eBook ISBN: 978-1-4200-0351-2. DOI: 10.1201/9781420003512.ch6.

[162] Gallaher CM, Munion J, Hesslink R Jr, Wise J, Gallaher DD (2000) Cholesterol Reduction by Glucomannan and Chitosan Is Mediated by Changes in Cholesterol Absorption and Bile Acid and Fat Excretion in Rats. J. nutr. 130, 2753–2759. ISSN: 00223166 . PubMed ID: 11053517.

[163] Maezaki Y, Tsuji K, Nakagawa Y, Kawai Y, Akimoto M, Tsugita T, Takekawa W, Terada A, Hara H, Mitsuoka T (1993) Hypocholesterolemic Effect of Chitosan in Adult Males. Biosci. biotech. biochem. 57, 1439–1444. Available: http://ci.nii.ac.jp/els/110002676529.pdf?id=ART0002948600&type=pdf&lang=en&host=cinii&order_no=&ppv_type=0&lang_sw=&no=1338944039&cp=.

[164] Dev A, Mohan JC, Sreeja V, Tamura H, Patzke GR, Hussain F, Weyeneth S, Nair SV, Jayakumar R (2010) Novel carboxymethyl chitin nanoparticles for cancer drug delivery applications. Carbohydr. polym. 79:1073–1079. ISSN: 01448617. DOI: 10.1016/j.carbpol.2009.10.038.

[165] Janes KA, Calvo P, Alonso MJ (2001) Polysaccharide colloidal particles as delivery systems for macromolecules. Adv. drug deliv. rev. 47: 83–97. ISSN: 0169409X. DOI: 10.1016/S0169-409X(00)00123-X.

[166] Haidar ZS, Hamdy RC, Tabrizian M (2008) Protein release kinetics for core-shell hybrid nanoparticles based on the layer-by-layer assembly of alginate and chitosan on liposomes. Biomaterials. 29: 1207–1215. ISSN: 01429612. DOI: 10.1016/j.biomaterials.2007.11.012.

[167] Calvo P, Remuñán-López C, Vila-Jato JL, Alonso M J (1997) Novel hydrophilic chitosan-polyethylene oxide nanoparticles as protein carriers. J. appl. polym. sci. 63, 125–132. ISSN: 00218995. DOI: 10.1002/(SICI)1097-4628(19970103)63:1<125::AID-APP13>3.0.CO;2-4.

[168] Bodmeier R, Chen H, Paeratakul O (1989) A Novel Approach to the Oral Delivery of Micro- or Nanoparticles. Pharm. Res. 6: 413–417. PMID:2748533.

[169] Bourtoom T (2008) Factor Affecting the Properties of Edible Film Prepared from Mung Bean Proteins, International Food Research Journal 15: 167-180. ISSN: 19854668.

[170] Guilbert S, Gontard N and Cuq B (1995) Technology and applications of edible protective films. Packag. technol. sci., 8: 339–346. ISSN: 08943214. doi: 10.1002/pts.2770080607.

[171] Kester J J and Fennema OR (1986) Edible films and coatings: A review in Food Technol. 40 : 47-59

[172] Muzzarelli R A A (1996) Chitosan-based dietary foods. Carbohydrate Polymer 29: 309-316. ISSN: 01448617. DOI: 10.1016/S0144-8617(96)00033-1

[173] Miranda SP, Garnica O, Lara-Sagahon V and Cárdenas G (2004) Water Vapor Permeability and Mechanical Properties of Chitosan Films. J. chil.chem. soc. 49:173-178. ISSN: 07179324. doi: 10.4067/S0717-97072004000200013

[174] Salvador L, Miranda S P, Aragón N and Lara V (1999) Recubrimiento de quitosán en aguacate. Journal of the Mexican Chemical Society 43: 18-23. Available : http://redalyc.uaemex.mx/redalyc/src/inicio/ArtPdfRed.jsp?iCve=47543204. ISSN 1870-249X.

[175] Trejo V, Aragón N, Miranda P. (2001) Estimación de la permeabilidad al vapor de agua en películas a base de quitosán. Journal of the Mexican Chemical Society .45:1-5. ISSN 1870-249X . Available: http://redalyc.uaemex.mx/redalyc/src/inicio/ArtPdfRed.jsp?iCve=47545101>

[176] Cardenas G and Miranda S P (2004) FTIR and TGA studies of chitosan composite films. J. Chil. Chem. Soc. 49:291-295 .Available: <http://www.scielo.cl/scielo.php?script=sci_arttext&pid=S0717-97072004000400005 &lng=es&nrm=iso>. ISSN 0717-9707. doi: 10.4067/S0717-97072004000400005.

[177] Sandford P A)1989) Chitosan: commercial uses and potential application. In Skjak-Braek, G., Anthosen, T. and Sandford, P. (Eds.). Chitin and Chitosan: Source, Chemistry, Biochemistry, Physical Properties, and Application, p. 51-69. New York: Elsevier Applied Science. ISBN 1851663959, 9781851663958.

[178] Xu J, McCarthy SP, Gross RA, Kaplan DL (1996) Chitosan film acylation and effects on biodegradability Macromolecules. 29: 3436-3440. DOI: 10.1021/ma951638b.

[179] Park IK, Lee YK, Kim MJ, Kim SD (2002) Effect of surface treatment with chitooligosaccharide on shelf-life of baguette. J. chitin chitosan. 7:214–8. ISSN 1229-4160.

[180] Butler BL, Vergano PJ, Testin RF, Bunn JM, Wiles JL (1996) Mechanical and barrier properties of edible chitosan films as affected by composition and storage. J. food sci 61:953–5, 961. ISSN 0022-1147. DOI: 10.1111/j.1365-2621.1996.tb10909.x.

[181] Nadarajah K, Prinyawiwatkul W, Hong KN, Sathivel S, Xu Z (2006) Sorption behavior of crawfish chitosan films as affected by chitosan extraction processes and solvent types .Journal of Food Science, 71 (2) ISSN: 00221147. DOI: 10.1111/j.1365-2621.2006.tb08894.x.

[182] Bhale S, No HK, Prinyawiwatkul W, Farr AJ, Nadarajah K, Meyers SP (2003) Chitosan coating improves shelf life of eggs. J. food s.ci 68:2378–83. ISSN: 00221147. DOI: 10.1111/j.1365-2621.2003.tb05776.x.

[183] De Reu K, Grijspeerdt K,Messens W, Heyndrickx M, Uyttendaele M, Debevere J, Herman L (2006) Eggshell factors influencing eggshell penetration and whole egg contamination by different bacteria, including Salmonella enteritidis. Int. j food microbial. 112:253–60. ISSN: 01681605. DOI: 10.1016/j.ijfoodmicro.2006.04.011 .

[184] Murray MW and Rutherford PP (1963) The relationship between the loss of water and carbon dioxide from eggs and the effect upon albumen quality. Poult. sci 42: 505–8. doi: 10.3382/ps.0420505.

[185] Stadelman WJ (1986) The preservation of quality in shell eggs. In: Stadelman WJ, Cotterill OJ, editors. Egg science and technology. Westport, Conn.: AVI Publishing.p 63–73. ISBN 1560220031, 9781560220039.

[186] Lee SH (1996) Effect of chitosan on emulsifying capacity of egg yolk. J. korean soc food nutr. 25:118–22. Available: http://210.101.116.28/W_kiss10/79900052_pv.pdf.

[187] El Ghaouth A, Ponnampalam R, Castaigne F, Arul J (1992b) Chitosan coating to extend the storage life of tomatoes. Hort. Science. 27:1016–1018. Available: http://hortsci.ashspublications.org/content/27/9/1016.full.pdf

[188] Park SI, Stan SD, Daeschel MA, Zhao Y (2005) Antifungal coatings on fresh strawberries (Fragaria × ananassa) to control mold growth during cold storage. J. food sci. 70:M202–M207. DOI: 10.1111/j.1365-2621.2005.tb07189.x.

[189] Bai RK, Huang MY, Jiang YY (1988) Selective permeabilities of chitosan-acetic acid complex membrane and chitosan-polymer complex membranes for oxygen and carbon dioxide. Polym. bull. 20:83–88. DOI: 10.1007/BF00262253 .

[190] Soto-Peralta NV, Mller H, Knorr D (1989) Effects of chitosan treatments on the clarity and color of apple juice. J. food sci 54:495–496. DOI: 10.1111/j.1365-2621.1989.tb03119.x.

[191] Imeri AG and Knorr D (1988) Effect of chitosan on yield and compositional data of carrot and apple juice. J. food sci 53:1707–1709. DOI: 10.1111/j.1365-2621.1988.tb07821.x.

[192] Knorr D (1985) Utilization of chitinous polymers in food processing and biomass recovery. In: Colwell RR, Pariser ER, Sinskey AJ, editors. Biotechnology of marine polysaccharides. Washington, D.C.: Hemisphere Publishing Corp. p 313–332. ISBN 0891164332, 9780891164333.

[193] No HK and Meyers SP (2000) Application of chitosan for treatment of wastewaters. Rev. environ. contam. toxicol. 163:1–27. DOI: Version: za2963e q8za2 q8zb9 q8zc0 q8zdf q8ze3 q8zfe q8zg3.

[194] Kim JW and Hur JW. 2002. Improvement of functional properties of mayonnaise with egg-shell calcium and chitosan. Food eng. prog. 6:195–200.

[195] Del Blanco LF, Rodríguez MS, Schulz PC, Agulló E(1999) Influence of the deacetylation degree on chitosan emulsification properties. Colloid. polym sci. 277:1087–1092. ISSN: 0303402X. DOI: 10.1007/s003960050495.

[196] Filar LJ, Wirick MG (1978) Bulk and solution properties of chitosan. In: Muzzarelli RAA, Pariser ER, editors. Proceedings of the first international conference on chitin/chitosan. Mass.: MIT Sea Grant Program. p 169–81

[197] Kamil JYVA, Jeon YJ, Shahidi F (2002) Antioxidative activity of chitosans of different viscosity in cooked comminuted flesh of herring (Clupea harengus). Food chem. 79:69–77. http://dx.doi.org/10.1016/S0308-8146(02)00180-2.

[198] Darmadji P, Izumimoto M (1994) Effect of chitosan in meat preservation. Meat sci. 38:243–54. http://dx.doi.org/10.1016/0309-1740(94)90114-7.

[199] Sagoo S, Board R, Roller S (2002) Chitosan inhibits growth of spoilage microorganisms in chilled pork products. Food microbio.l 19:175–82. http://dx.doi.org/10.1006/fmic.2001.0474.

[200] Ha TJ and Lee SH (2001) Utilization of chitosan to improve the quality of processed milk. J. korean soc. food sci. nut. 30:630–634.

[201] Park SM, Youn SK, Kim HJ, Ahn DH (1999) Studies on the improvement of storage property in meat sausage using chitosan-I. J. Korean soc. food sci. nutr. 28:167–71.

[202] Youn SK, Park SM, Kim YJ, Ahn DH (1999) Effect on storage property and quality in meat sausage by added chitosan. J Chitin Chitosan 4:189–95.

[203] Decker EA and Hultin HO (1992) Lipid oxidation in muscle foods via redox iron. In: Angelo AJ, editor. Lipid oxidation in food.Washington, D.C.: American Chemical Society. P 33–54. ISBN13: 9780841224612eISBN: 9780841213562.

[204] Jeon YJ, Kamil JYVA, Shahidi F (2002) Chitosan as an edible invisible film for quality preservation of herring and Atlantic cod. J. agric. food chem. 50:5167–78. ISSN: 00218561. DOI: 10.1021/jf011693l.

[205] Kim KW, Thomas RL (2007) Antioxidative activity of chitosans with varying molecular weights. Food chem. 101:308–13. ISSN: 03088146. DOI: 10.1016/jfood chem.2006.01.038.

[206] Photchanacha S, Singkaew J and Thamthong J (2006) Effects of chitosan seed treatment on Colletotrichum sp. and seedling growth of chili cv. 'Jinda'. Acta Horticulturae. 712:585-590. ISSN: 05677572. ISBN: 9066055391;978-906605539-1

[207] Guan YJ, Hu J, Wang XJ and Shao CX (2009) Seed priming with chitosan improves maize germination and seedling growth in relation to physiological changes under low temperature stress. J. Zhejiang univ. sci. B, 10, 6, pp. 427–433. DOI: 10.1631/jzus.B0820373.

[208] Lizárraga-Paulín EG, Torres-Pacheco I, Moreno-Martínez E and Miranda-Castro S P (2011a) Chitosan application in maize (Zea mays) to counteract the effects of abiotic stress at seedling level. Afr. j. biotechnol. 10:6439-6446. Available online at http://www.academicjournals.org/AJB. ISSN 1684–5315.

[209] Uthairatanakij A, Teixeira JA and Obsuwan K (2007) Chitosan for Improving Orchid Production and Quality. Science. 1: 1-5. Available: http://www.globalsciencebooks.info/JournalsSup/images/Sample/OSB_1(1)1-5.pdf.

[210] Zhou YG, Yang YD, Qi YG, Zhang ZM, Wang XJ and Hu XJ (2002) Effects of chitosan on some physiological activity in germinating seed of peanut. J. Peanut sci. 31: 22–25. DOI: cnki:ISSN:1002-4093.0.2002-01-004.

[211] Reddy MV, Arul J, Angers P and Couture L (1999) Chitosan treatment of wheat seeds induces resistance to Fusarium graminearun and improves seed quality. J. agric. food chem. 47: 1208–1216. DOI: 10.1021/jf981225k.

[212] Ruan SL and Xue QZ (2002) Effects of chitosan coating on seed germination and salt-tolerance of seedlings in hybrid rice (Oryza sativa L.). Acta agron. sinica, 28: 803–808. DOI: cnki:ISSN:0496-3490.0.2002-06-014.

[213] Shao CX, Hu J, Song WJ and Hu WM (2005) Effects of seed priming with chitosan solutions of different acidity on seed germination and physiological characteristics of maize seedling. J. Zhejiang univ. agric. life sci. 31: 705–708. DOI: CNKI:SUN:ZJNY.0.2005-06-007.

[214] Lizárraga-Paulín EG, Torres-Pacheco I, Moreno-Martínez E y Miranda-Castro S P. (2011b) Protección contra estrés biótico inducida por quitosán en plántulas de maíz (Zea mays). Rev. mex. cienc. agríc. 2: 813-827. ISSN 2007-0934. Available: http://www.inifap.gob.mx/revistas/ciencia_agricola/vol2_num6_2011.pdf.

[215] Devlieghere F, Vermeulen A and Debevere J (2004) Chitosan: antimicrobial activity, interactions with food components and applicability as a coating on fruit and vegetables. Food microbial. 21, 6, pp. 703-71 . http://dx.doi.org/10.1016/j.fm.2004.02.008.

[216] Bittelli M, Flury M, Campbell GS and Nichols EJ (2001) Reduction of transpiration through foliar application of chitosan. Agric. forest meteorol. 107: 167–175. http://dx.doi.org/10.1016/S0168-1923(00)00242-2.

[217] Iriti M, Picchi V, Rossoni M, Gomarasca S, Ludwig N, Garganoand M and Faoro F (2009) Chitosan antitranspirant activity is due to abscisic acid-dependent stomatal closure. Env. exp. bot. 66: 493 500. Doi: 10.1016/j envepbox.2009.01.004

[218] Rabea EI, El Badawy MT, Stevens CV, Smagghe G and Steurbaut W (2003) Chitosan as antimicrobial agent: Applications and mode of action. Biomacromolecules, 4: 57–1465. DOI: 10.1021/bm034130m.

[219] El Hadrami A, Lorne RA, El Hadrami I, Fouad D (2010) Chitosan in plant protection. Marine drugs. 8:968-87. doi: 10.3390/md8040968 .

[220] Utsunomiya N, Kinai H, Matsui Y, Takebayashi T (1998) The Effects of Chitosan Oligosaccharides Soil Conditioner and Nitrogen Fertilizer on the Flowering and Fruit Growth of Purple Passionfruit (Passiflora edulis Sims var. edulis). Journal of the Japanese Society for Horticultural Science. 67: 567-571. ISSN:0013-7626.

[221] Wu L and Liu M (2008) Preparation and properties of chitosan-coated NPK compound fertilizer with controlled-release and water-retention. Carbohyd. polym. 72: 240–247. http://dx.doi.org/10.1016/j.biortech.2006.12.027.

[222] Benhamou N, Lafontaine PJ and Nicole M (1994) Induction of systemic resistance to Fusarium crown and root rot in tomato plants by seed treatment with chitosan. Phytopathology 84: 1432–1444. Available: http://www.apsnet.org/publications/phytopathology/backissues/Documents/1994Articles/Phyto84n12_1432.pdf.

[223] Hirano S (1989) Production and application of Chitin and chitosan in Japan. In Chitin and chitosan, (Skjak-Braek G, Anthonsesn, T., Sandford P (Eds) pp 51-69, Elsevier Applied Science, London, UK. ISBN 1851663959, 9781851663958.

[224] R.A.A. Muzzarelli (2008) Aspects of chitin chemistry and enzymology, In Paoletti M and Musumeci S (Eds.). Binomium chitin-chitinase: emerging issues, auppauge. NY, Nova Science. ISBN: 978-1-60692-339-9.

[225] Hahn M (1996) Microbial elicitors and their receptors in plants. Annu. Rev. phytopathol. 34: 87-412. DOI: 10.1146/annurev.phyto.34.1.387.

[226] Radman R, Saez T, Bucke C and Keshavarz T (2003) Elicitation of plants and microbial cell systems. Biotechnol. Appl. Bioc. 37: 91-102. DOI: 10.1042/BA20020118.

[227] Angelova Z, Georgiev S and Ross W (2006) Elicitation of plants. Biotechnol. Biotechnol. 20: 72-83. Available: http://www.diagnosisp.com/dp/journals/view_pdf.php?journal_id=1&archive=1&issue_id=11&article_id=298.

[228] Utsunomiya N and Kinai H (1994) Effect of chitosan-oligosaccharides soil conditioner on the growth of passionfruit. Journal of the Japanese Society for Horticultural Science. 64: 176-177. ISSN:0013-7626.

[229] Wanichpongpan P, Suriyachan K and Chandrkrachang S (2001) Effect of chitosan on the growth of Gerbera flower plant (Gerbera jamesonii). Chitin and chitosan: Chitin and Chitosan in Life Science, Yamaguchi, Japan, pp 198-201. ISBN 906605087X, 9789066050877.

[230] Hadwiger L A, Klosterman S J and Choi J J (2002) The mode of action of chitosan and its oligomers in inducing pant promoters and developing disease resistance in plants. Advances in Chitin Science. 5: 452-457

[231] Asghari-Zakaria R, Maleki-Zanjani B, Sedghi E (2009) Effect of in vitro chitosan application on growth and minituber yield of Solanum tuberosum L. Plant soil environ. 55: 252-256. Available: http://www.agriculturejournals.cz/publicFiles/08422.pdf.

[232] Bell A A, Hubbard JC, Liu L, Davis R M and Subbarao K V (1998) Effects of chitin and chitosan on the incidence and severity of Fusarium yellows in celery. Plant dis. 82: 322–328. http://apsjournals.apsnet.org/doi/pdf/10.1094/PDIS.1998.82.3.322.

[233] Murphy J G, Rafferty S M and Cassells A C (2000) Stimulation of wild strawberry (Fragaria vesca) arbuscular mycorrhizas by addition of shellfish waste to the growth substrate: interaction between mycorrhization, substrate amendment and susceptibility to red core (Phytophthora fragariae). Appl. soil ecol .15: 153–158. http://dx.doi.org/10.1016/S0929-1393(00)00091-3.

[234] Daayf F, El Bellaj M, El Hassni M, J'aiti F and El Hadrami I (2003) Elicitation of soluble phenolics in date palm (Phoenix dactylifera L.) callus by Fusarium oxysporum f. sp. albedinis culture medium. Env. exp. bot. 49: 41–47. http://dx.doi.org/10.1016/S0098-8472(02)00048-5.

[235] Uppal A K, El Hadrami A, Adam LR, Tenuta M and Daayf F (2008) Biological control of potato Verticillium wilt under controlled and field conditions using selected bacterial antagonists and plant extracts. Biol. Control. 44: 90–100. http://dx.doi.org/10.1016/j.biocontrol.2007.10.020.

[236] Zhang M, Tang T, Yuan H and Rui C (2003) Insecticidal and Fungicidal Activities of Chitosan and Oligo-chitosan. J bioact. compat. pol. 18: 391-400. doi: 10.1177/0883911503039019.

[237] Rabea EI, Badawy MEI, Rogge TM, Stevens CV, Steurbaut W, Hofte M, Smagghe G (2006) Enhancement of fungicidal and insecticidal activity by reductive alkylation of chitosan. Pest manag. sci. 62: 890-897. DOI: 10.1002/ps.1263.

[238] Rabea EI, Badawy MEI, Rogge TM, Stevens CV, Hofte M, Steurbaut W, Smagghe G (2005) Insecticidal and fungicidal activity of new synthesized chitosan derivatives. Pest manag. sci. 61: 951-960. DOI: 10.1002/ps.1085.

[239] Ramírez MA, Rodríguez AT, Alfonso L and Peniche C (2010) Chitin and its derivatives as biopolymers with potential agricultural applications. Biotecnol. apl. 27: 270-276. ISSN: 08644551.

[240] Gohel V, Singh A, Vimal M, Ashwini P and Chhatpar HS (2006) Bioprospecting and antifungal potential of chitinolytic microorganisms. African j. biotechnol. 5: 54-72. ISSN: 16845315. Available: https://tspace.library.utoronto.ca/bitstream/1807/6640/1/jb06009.pdf.

[241] Gomes RC, Semêdo LT, Soares RM, Alviano CS, Linhares LF, Coelho RR (2000) Chitinolytic activity of actinomycetes from a cerrado soil and their potential in biocontrol. Lett, appl. microbiol. 30: 146-50. ISSN: 02668254. PubMed ID: 10736018.

[242] Rodríguez-Kabana R, Morgan-Jones G and Ownley-Gintis B (1984) Effects of chitin amendments to soil on Hetedodera glycines, microbial population and colonization of cyst by fungi. Nematropica, 14: 9-25.

[243] Bélair G and Tremblay N (1995) The influence of chitin-urea amendments applied to an organic soil on Meloidogyne hapla population and growth of green house tomatoes. Phytoprotection. 76: 75-80. ISSN: 0031-9511. Available: http://www.erudit.org/revue/phyto/1995/v76/n2/706087ar.pdf.

Chapter 3

Capsular Polysaccharides Produced by the Bacterial Pathogen *Burkholderia pseudomallei*

Shauna L. Reckseidler-Zenteno

Additional information is available at the end of the chapter

1. Introduction

Polysaccharide capsules are structures found on the cell surface of a broad range of bacterial species. The polysaccharide capsule often constitutes the outermost layer of the cell, and therefore is often involved in mediating direct interactions between the bacteria and its environment. It is due to these interactions that polysaccharide capsules have been implicated as important virulence factors for many bacterial pathogens.

Bacterial extracellular polysaccharides (EPS) may be classified as either capsular polysaccharides (CPS), where the polysaccharide is intimately associated with the cell surface, or as slime polysaccharides, where the polysaccharide is loosely associated with the cell [1]. Differentiation between these forms is difficult since CPS may be released from the cell, giving the appearance of a slime polysaccharide [1]. In turn, distinguishing between CPS and other cell surface polysaccharides, such as O-antigenic moieties of lipopolysaccharide (LPS), may also be difficult, since CPS may be found associated with LPS [1].

Capsular polysaccharides are highly hydrated molecules that are over 95% water [2]. They are often linked to the cell surface of the bacterium via covalent attachments to either phospholipid or lipid-A molecules, although some CPS may be associated with the cell in the absence of a membrane anchor [1, 3]. Capsular polysaccharides can be either homo- or heteropolymers composed of repeating monosaccharides joined by glycosidic linkages [4]. The multiple hydroxyl groups present within each monosaccharide may be involved in the formation of the glycosidic bond, therefore, any two monosaccharides may be joined in a number of configurations, which leads to large structural diversity among CPS types. In the case of human pathogens, a large number of different capsule serotypes have been identified, and certain CPS or K-antigens have been associated with specific infections [4]. For example, the *Escherichia coli* K1 antigen, a homopolymer of α2,8-linked N-acetylneuraminic acid (NeuNAc), is the major cause of neonatal meningitis [5]. While bacterial species may demonstrate great structural diversity in synthesizing capsules, chemically identical capsular polysaccharides may also be synthesized by different bacterial species. The *Neisseria meningitidis* group B capsular polysaccharide is identical to the K1 polymer of *E. coli,* and the *E. coli* K18, K22, and K100 antigens have the same constituents and structure as the *Haemophilus influenza* serotype b capsule [6, 7]. The conservation of CPS types between bacterial species raises interesting questions regarding the evolution of capsules and the transmission and acquisition of capsule biosynthesis genes [4].

The genetic loci necessary for the production of bacterial capsules are primarily clustered at a single chromosomal locus, which allows for the coordinate regulation of a large number of genes that may be involved in both the biosynthesis and export of capsular polysaccharides [4, 1]. In most bacterial species, the capsule gene clusters demonstrate conserved genetic organization. The capsules of *E. coli* have been classified into four groups based on genetic organization and biosynthetic criteria and capsules of other bacteria may resemble these prototypes. Group 1 capsules include the *E. coli* K30 capsule and the capsules of *Klebsiella* sp. and *Erwinia* sp. [8]. Group 2 capsules include the *E. coli* K1 and K5 capsules, as well as the capsules produced by *Neisseria* sp. and *Haemophilus* sp. Group 3 capsules include the *E. coli* K10 capsule. Group 4 capsules include the capsules of *E. coli* K40 and 0111. Both group 2 and group 3 capsule gene clusters are organized into 3 regions. Regions 1 and 3 are involved in the export and modification of the capsular polysaccharides and are conserved between members of the group, while region 2 contains the genes responsible for the biosynthesis of the capsule and is usually serotype specific [4, 8]. Generally these regions are organized into one transcriptional unit and the regions within the capsule locus are divergently transcribed [4, 8]. In addition, some genes within a region may be translationally coupled, such as the *kpsU* and *kpsC* genes in group 2 *E. coli* capsules, and the *kpsM* and *kpsT* genes in group 3 *E. coli* capsules, which allows for balanced expression of two proteins [4].

The A+T composition of capsule gene clusters is often significantly higher than the rest of the chromosome, suggesting a common ancestry of capsule genes in gram-negative bacteria [9]. It is likely that these A+T rich regions have been horizontally transferred between bacterial species. In addition, the A+T ratio of region 2 DNA of group 2 *E. coli* capsule gene clusters compared to regions 1 and 3 confirms that capsule diversity has been achieved in part through the acquisition of different region 2 sequences [4].

The production of a polysaccharide capsule is widespread in pathogenic bacteria. A number of functions have been assigned to bacterial capsules including: prevention of dessication, adherence, resistance to non-specific host immunity, resistance to specific host immunity, and mediating the diffusion of molecules through to the cell surface [4, 1].

Capsules may form a hydrated gel around the surface of the bacterial cell, which may protect the bacteria from the harmful effects of dessication [10]. This may increase the survival of encapsulated bacteria outside of the host, promoting the transmission of pathogenic bacteria from one host to another [4]. Mucoid isolates of *E. coli, Acinetobacter calcoaceticus,* and *Erwinia stewartii* are more resistant to drying than isogenic nonmucoid strains [11]. Studies with *E. coli* have shown that the expression of genes encoding for the colanic acid capsule is increased by dessication [11]. In addition, alginate production by *Pseudomonas aeruginosa* is triggered by high osmolarity, which may be a consequence of dessication [12].

Capsular polysaccharides may promote adherence of bacteria to both surfaces and other bacterial cells, which may facilitate colonization of a particular niche and may lead to the formation of biofilms [13]. Cell-surface polysaccharides have been shown to mediate the attachment of bacterial cells to one another, leading to biofilm formation and persistence of the organisms during colonization [1, 14].

Capsular polysaccharides are one of the components responsible for resistance to the non-specific immunity of the host. The presence of a capsule is thought to confer resistance to non-specific host defense mechanisms such as complement and complement-mediated opsonophagocytosis [4]. Bacterial capsules may resist complement-mediated killing by providing a permeability barrier to complement components, which masks the underlying cell surface structures that activate complement [15]. The capsule may also act in concert with O-antigens to confer resistance to complement-mediated killing [16]. As a result, a combination of cell surface structures is responsible for conferring resistance to killing by the complement cascade [4]. Finally, capsules are responsible for resistance to complement-mediated opsonophagocytosis. This resistance may be due to steric effects, which results in the capsule acting as a barrier between the C3b deposited on the bacterial surface and the C3b receptors present on phagocytes [4]. Alternatively, the resistance to opsonophagocytosis may be due to the net negative charge of the polysaccharide capsule [17].

Capsules may also confer resistance to the specific immune response of the host. Although most capsular polysaccharides can elicit an immune response, some capsules are poorly immunogenic [4]. Examples of such capsules include those containing NeuNAc, such as the *E. coli* K1 capsule or the capsule of *Neisseria meningitidis* serogroup B, and the *E. coli* K5 capsule, which is similar to desulfoheparin [18, 19]. Because these capsules are structurally similar to polysaccharides encountered on host tissue, these capsules are poorly immunogenic, and elicit a poor antibody response in the host [20].

Burkholderia pseudomallei, the causative agent of melioidosis, is a gram-negative, facultatively anaerobic, motile bacillus that is commonly found in the soil and stagnant waters in a number of regions around the world, particularly in areas that fall between 20° north and 20° south of the equator [21, 22]. Infection by *B. pseudomallei* is often due to either direct inoculation into wounds and skin abrasions or to inhalation of contaminated material [22, 23, 24,]. This would explain the prevalence of the disease among rice farmers as well as helicopter pilots in the Vietnam War who developed melioidosis due to inhalation of contaminated dust [24]. Melioidosis may present as an acute pneumonia or an acute septicemia, which is the most severe form of the disease. The disease may also manifest as a chronic infection involving long-lasting suppurative abscesses in numerous sites in the body. Infection with *B. pseudomallei* may even result in a subclinical infection and remain undetected for a number of years. Both the chronic and subclinical forms generally remain undiagnosed until activated by a traumatic event or a decrease in immunocompetence [25]. *B. pseudomallei* is inherently resistant to a number of antibiotics, and even with aggressive antibiotic therapy, the mortality rate remains high, and the incidence of relapse is common [26, 27].

At the time our studies were initiated some cell-associated antigens had been identified and characterized in *B. pseudomallei*. Cell-associated antigens include exopolysaccharide (EPS) and lipopolysaccharide (LPS) [28, 29, 30]. The EPS produced by *B. pseudomallei* was determined to be an unbranched polymer of repeating tetrasaccharide units with the structure -3)-2-*O*-acetyl-β-D-Gal*p*-(1-4)-α-D-Gal*p*-(1-3)-β-D-Gal*p*-(1-5)-β-D-KDO*p*-(2- [31, 32]. The role of EPS in virulence was not known, but sera from patients with melioidosis had been shown to contain antibodies against EPS [30]. Two other EPS structures were also identified; a branched 1,4-linked glucan polymer ((CP-1a) and a triple-branched heptasaccharide repeating unit composed of rhamnose, mannose, galactose, glucose, and glucuronic acid (CP-2) [33]. The genes involved in the synthesis of these capsules, and the role of these capsules in virulence had not been identified. The LPS of *B. pseudomallei* was structurally characterized and reported to contain two types of O-polysaccharide moieties termed type I O-PS and type II O-PS [34, 35]. Type II O-PS was found to be an unbranched heteropolymer with repeating D-glucose and L-talose residues with the structure -3)-β-D-glucopyranose-(1-3)-6-deoxy-α-L-talopyranose-(1-. Type II O-PS had been shown to be involved in serum resistance [36]. Type II O-PS mutants also demonstrated reduced virulence in three animal models of *B. pseudomallei* infection [36]. Type I O-PS was determined to be an unbranched homopolymer with the structure -3)-2-*O*-acetyl-6-deoxy-β-D-manno-heptopyranose-(1-, however, the role for this polysaccharide in infection had not been defined, nor the genes responsible for its biosynthesis been identified.

B. thailandensis is a nonpathogenic soil organism originally isolated in Thailand [37]. Based on biochemical, immunological, and genetic data, *B. pseudomallei* and *B. thailandensis* are closely related species. However, these two organisms differ in a number of ways and have been classified into two different species [38]. The rRNA sequence of *B. thailandensis* differs from that of *B. pseudomallei* by 15 nucleotides, and there are significant differences in genomic macrorestriction patterns between these organisms [39]. The biochemical profiles of these two species differ in that *B. thailandensis* can utilize L-arabinose whereas *B. pseudomallei* does not [38, 40]. The most distinct difference between these two species, however, is their relative virulence. The 50% lethal dose (LD_{50}) for *B. pseudomallei* in the Syrian hamster model of acute melioidosis is <10 organisms, whereas the LD_{50} for *B. thailandensis* is approximately 10^6 organisms [38]. It has also been shown that the two species can be differentiated based on their propensity to cause disease in humans. Environmental strains isolated in Thailand that are able to assimilate L-arabinose are not associated with human infection, whereas clinical isolates are not able to utilize L-arabinose [41].

2. Subtractive hybridization to identify Capsular Polysaccharide 1 (CPS I)

To identify the genetic determinants that confer enhanced virulence in *B. pseudomallei*, a method combining subtractive hybridization, insertional mutagenesis, and animal virulence studies was developed [42]. Subtractive hybridization was carried out between the virulent *B. pseudomallei* and the weakly virulent *B. thailandensis* in order to isolate DNA sequences encoding for virulence determinants unique to *B. pseudomallei*. The genomic DNA sample from *B. pseudomallei* containing the sequences of interest was known as the tester DNA, and genomic DNA from *B. thailandensis*, the reference sample, was called the driver DNA. Tester and driver DNAs were digested and subjected to two rounds of hybridization. The remaining unhybridized sequences were considered tester-specific sequences. To enrich for tester-specific sequences, excess driver DNA was added in the hybridizations. The tester-specific sequences were then amplified by PCR and cloned into the plasmids pPCR or pZErO-2.1. Screening of the subtraction library revealed a number of DNA sequences unique to *B. pseudomallei*. Fifteen distinct plasmid inserts from the library were sequenced. The DNA inserts ranged from 100 to 800 bp in length and were found to contain an average G+C content of approximately 44 to 52%, which is considerably lower than the 68% G+C content of the *B. pseudomallei* chromosome. The DNA sequences were analyzed using the NCBI BLASTX program. One of the plasmid inserts, pDD1015, was found to share limited homology with WbpX, a glycosyltransferase, from *Pseudomonas aeruginosa* [43].

2.1. Demonstration that CPS I is important for the virulence of *B. pseudomallei*

The 373-bp DNA insert from pDD1015 was cloned into a mobilizable suicide vector, pSKM11 [44]. The resulting plasmid, pSR1015, was mobilized into wildtype *B. pseudomallei* 1026b to create the mutant strain SR1015. Since the insert from pDD1015 was found to demonstrate homology to a glycosyltransferase from *P. aeruginosa*, it was postulated that it might encode a protein involved in carbohydrate synthesis. Since three carbohydrate structures had been previously purified and characterized, antibodies to each of these polysaccharides were available. To define the phenotype of SR1015, an ELISA was performed with the EPS-specific monoclonal antibody 3015, and *B. pseudomallei* 1026b and SR1015 were both found to contain EPS [45]. SR1015 was also shown to contain type II O-PS and to be serum resistant. Immunogold electron microscopy studies using rabbit polyclonal sera specific for a type I O-PS–flagellin conjugate was performed on the parent strain, 1026b, and SR1015 (Figure 1). *B. pseudomallei* 1026b reacted with antibodies to both flagellin and type I O-PS, as was evident by the distribution of gold particles around the bacterial surface and extending out along the flagella (Figure 1A). Unlike *B. pseudomallei* 1026b, SR1015 reacted only with the antibodies to flagellin, as the gold particles were found associated only with the flagella (Figure 1B). *B. thailandensis*, the negative control, did not react with the antibodies either to flagellin or to type I O-PS (Figure 1C). *B. stabilis* LMG7000 was also shown to react to the antibodies to type I O-PS, indicating this organism may produce a similar capsule (Figure 1D). Western blot analysis of proteinase K-digested whole cells from *B. pseudomallei* 1026b, *B. thailandensis* E264, and *B. pseudomallei* SR1015 using rabbit polyclonal sera raised to O-PS–flagellin protein conjugate confirmed the lack of type I O-PS in SR1015. Type I and type II O-PS were stained in *B. pseudomallei* 1026b, while only type II O-PS was stained in the lanes corresponding to *B. pseudomallei* SR1015 and *B. thailandensis*. These results indicated that we had identified and insertionally inactivated a gene involved in the synthesis of the type I O-PS of *B. pseudomallei* [42]. SR1015 was tested for virulence in the Syrian hamster model of acute septicemic melioidosis. The LD_{50} for SR1015 after 48 h was 3.5 x 10^5 CFU, while the LD_{50} of the parent strain, 1026b, was <10 CFU. The LD_{50} for SR1015 was similar to that for the weakly virulent *B. thailandensis* (6.8×10^5 CFU) [42]. This demonstrated that SR1015 is severely attenuated for virulence in this animal model of melioidosis and that type I O-PS is a major virulence determinant of *B. pseudomallei*. We later determined that the type I O-PS was a capsular polysaccharide (CPS I), not an O-PS moiety, which will be discussed below.

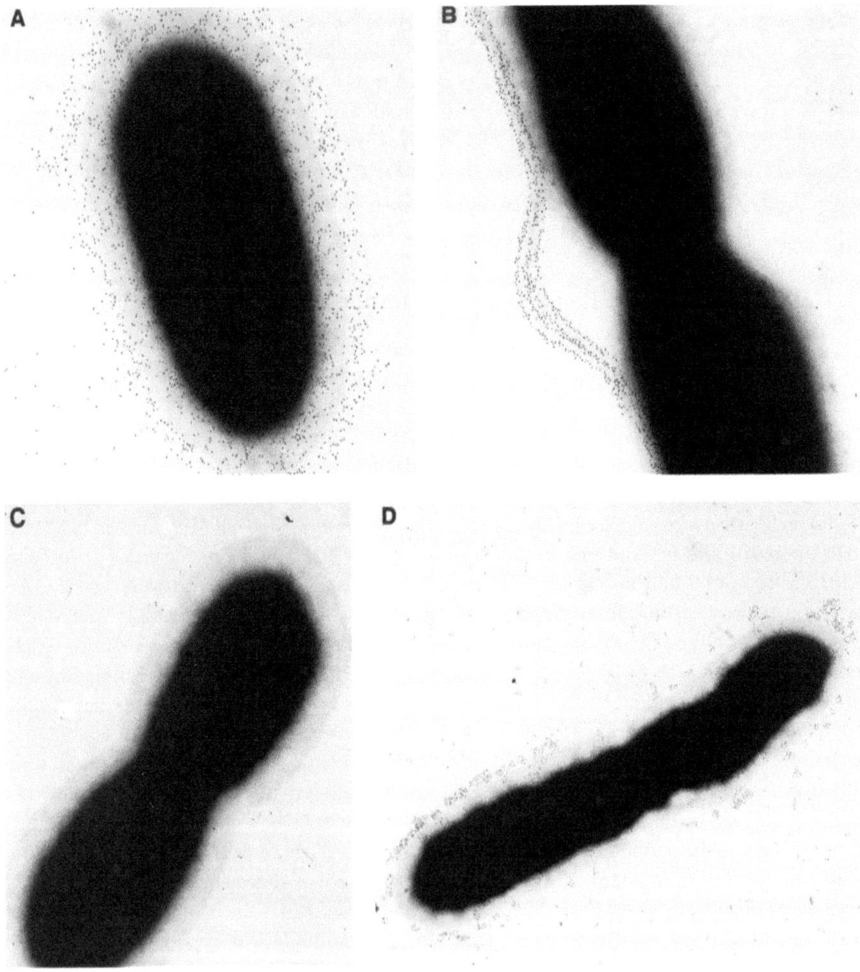

Figure 1. Immunogold electron microscopy of *B. pseudomallei* 1026b (A) and SR1015 (B), *B. thailandensis* E264 (C), and *B. stabilis* LMG7000 (D). Bacteria were reacted with polyclonal rabbit antiserum directed against an O-PS–flagellin protein conjugate absorbed with *B. thailandensis* E264 to remove the antibodies directed against type II O-PS, washed, and reacted with a goat anti-rabbit IgG-gold (5 nm) conjugate. Original magnification x330,000.

2.2. Cloning and sequencing of the genetic loci required for CPS I production and export

Two methods were used to clone the genes involved in the production and export of type I O-PS. The DNA flanking the insertion of pSR1015 was cloned from SR1015 and sequenced. We also used transposon mutagenesis to clone the genes involved in production of the polysaccharide; this was done to obtain any unlinked genes that may be involved in polysaccharide production. Approximately 1,300 transposon mutants were screened for loss of type I O-PS by ELISA. Six mutants were identified, and the DNA flanking the transposon insertion was cloned and sequenced. The Tn5-OT182 mutants SLR5, SLR8, SLR13, SLR18, and SLR19 mapped to the same region of the chromosome. Sequence analysis of the cloned fragments revealed the presence of 26 potential open reading frames involved in the synthesis and export of type I O-PS [42]. The open reading frames that predicted proteins involved in polysaccharide biosynthesis were found to demonstrate homology to proteins involved in the synthesis of a polysaccharide structure composed primarily of mannose. The other reading frames in the locus predicted proteins involved in the transport of capsular polysaccharides in a variety of bacteria, particularly those that produce group 2 and group 3 capsular polysaccharides [8]. The genes responsible for the production of type I O-PS were found to be similar to other loci encoding for capsular polysaccharides in that they are divergently transcribed [4]. The gene cluster involved in the production of this polysaccharide is also similar to group 3 capsule gene clusters in that there are no genes encoding KpsF and KpsU, which are present in group 2 capsule gene clusters [8]. However, the organization of the *B. pseudomallei* type I O-PS gene cluster differs in that it does not contain two export regions flanking a single biosynthetic region as seen in other group 3 capsule polysaccharide clusters [46]. The biosynthetic genes identified are not organized into one continuous transcriptional unit; instead, *wcbB, manC,* and *wcbP* are separated from the rest of the biosynthetic genes. The overall G+C content of this region is about 58%, lower than the G+C content of the rest of the chromosome (68%). The low G+C content in these clusters suggests that polysaccharide genes have a common origin and may have been transferred horizontally between species [9]. The genes involved in the production of this polysaccharide were named according to the bacterial polysaccharide gene nomenclature scheme [47]. The gene products associated with this cluster are shown in Figure 4. Mutations constructed in a number of these genes confirmed their role in the production of this polysaccharide [42].

2.3. CPS I was originally identified as a Type I O-polysaccharride moiety

The polysaccharide with the structure -3)-2-*O*-acetyl-6-deoxy-β-D-manno-heptopyranose-(1- was originally isolated and characterized as an O-PS component of LPS in *B. pseudomallei* and was designated type I O-PS [35]. However, our results suggested that this polysaccharide was a capsule rather than an O-PS moiety. The genes involved in the production of this capsule demonstrated strong homology to the genes involved in the production of capsular polysaccharides in many organisms, including *N. meningitidis, H. influenzae,* and *E. coli.* In addition, the export genes associated with this cluster are not associated with the previously characterized O-PS gene cluster [36]. Western blot analysis of proteinase K cell extracts and silver staining demonstrated that this polysaccharide has a high molecular mass (200 kDa) and lacks the banding pattern seen with O-PS moieties. Studies by our laboratory indicated that mutants in the production of the core oligosaccharide of the LPS are still capable of producing this polysaccharide [48]. Based on the above criteria and the genetic similarity to group 3 capsules, we proposed that this polysaccharide is a group 3 capsule and designated this capsule CPS I. This conclusion was further supported by Isshiki *et al* who separated this polysaccharide from a smooth lipopolysaccharide preparation of *B. pseudomallei* [49].

3. Role of the *B. pseudomallei* capsule in virulence

The role of CPS I in the pathogenicity of *B. pseudomallei* was investigated by performing further animal studies, serum bactericidal assays, complement protein C3b deposition assays, and radio-labelled phagocytic assays [50]. These experiments were facilitated by constructing a deletion strain harbouring a mutation in one of the CPS I genes and by complementation of this strain. An in-frame deletion was constructed in *wcbB*, a gene which encodes a glycosyltransferase, resulting in the capsule-minus strain SZ210. To confirm the role of *wcbB* in the biosynthesis of capsule, SZ210 was complemented by the introduction of a wild-type copy of the *wcbB* gene cloned into the mobilizable broad-host-range plasmid pBHR1 (MoBiTec). Western blot analysis of proteinase K-digested whole cells was performed using mouse monoclonal antibody directed to *B. pseudomallei* capsule to assess capsule production by these strains. Similar to the capsule minus strain SR1015 and *B. thailandensis* E264, which is known to lack this capsule, SZ210 was found to be negative for CPS I production, as indicated by the absence of a 200 kDa band that is present for wild-type 1026b. Complementation of SZ210 by providing the wild type *wcbB* gene in *trans* restored capsule production. Whole-cell extracts from the complemented strain SZ210(pSZ219) reacted to the capsule antibody producing the 200-kDa band corresponding to the *B. pseudomallei* capsule.

3.1. Production of CPS I correlates with clinical infection

To establish a correlation between CPS I production and clinical infection a number of strains of *B. pseudomallei* isolated from a variety of clinical specimens were tested for capsule production by western blot analysis with polyclonal rabbit antisera to *B. pseudomallei* CPS I. Out of the 55 clinical strains tested for capsule production, 52 were found to produce this capsule. Three strains, 420a, 415c, and 375a were found to be negative for capsule production, similar to *B. thailandensis* E264. However, one of the capsule genes, *wzt2*, was successfully amplified from these three strains and following inoculation in the animal model, all three of these strains were found to produce capsule by western blot analysis. This indicated that CPS I production may be regulated in some strains and its expression may be induced *in vivo*. Therefore all of the 55 clinical strains of *B. pseudomallei* tested were found to produce capsule, establishing a 100% correlation between CPS I production and clinical infection [51].

3.2. CPS I promotes survival of *B. pseudomallei in vivo*

Syrian golden hamsters were inoculated intraperitoneally with 10^1 to 10^5 cells of either wild type *B. pseudomallei* 1026b, capsule mutants SR1015 and SZ210, or the complemented strain SZ210(pSZ219). One group of animals inoculated with SR1015 also received 100 μg of purified *B. pseudomallei* capsule. After 48 h, the LD_{50} values were calculated, and the blood of the infected animals was diluted and plated for bacterial quantitation. The addition of purified capsule significantly increased the virulence of the capsule mutant strain SR1015. The LD_{50} value was calculated to be 34 CFU, similar to the LD_{50} value of wild-type *B. pseudomallei* 1026b (<10 CFU). In contrast, the LD_{50} value for SR1015 without the addition of purified capsule was calculated to be 3.5×10^5 CFU, 10,000- fold higher than when capsule was added to the inoculum. In addition, purified capsule enhanced the survival of SR1015 in the blood. Bacteria could not be detected in the blood of hamsters inoculated with SR1015 alone. However, the number of SR1015 CFU recovered from the blood of infected animals was 9.0×10^2 CFU/ml when capsule was added to the inoculum, an almost-1,000-fold increase. This number was comparable to the number of wild-type *B. pseudomallei* 1026b bacteria recovered from the blood. The addition of capsule was not toxic to the hamsters, as hamsters inoculated with 100 μg of purified capsule alone survived for the duration of the experiment without any ill effects. The LD_{50} value for the capsule mutant strain SZ210 containing an in frame deletion of the *wcbB* gene was calculated to be 9.6×10^4 CFU, and the number of bacteria in the blood was determined to be 10 CFU/ml. Complementation of this strain restored virulence in the animal model, resulting in an LD_{50} value of 12 CFU, comparable to that of wild type *B. pseudomallei* 1026b. Furthermore, the number of bacteria in the blood of animals infected with the complemented strain, SZ210(pSZ219), was determined to be 4.9×10^5 CFU/ml, similar to the number of bacteria recovered from animals infected with 1026b [50].

To further demonstrate the role of the capsule in infection by *B. pseudomallei*, an experiment was designed to investigate differences in tissue distribution between the capsule mutant strain and the wild type in infected hamsters. Animals were inoculated with 10^2 CFU of either wild-type *B. pseudomallei* 1026b or the capsule mutant SR1015. At different time points, the animals were sacrificed, and the numbers of bacteria in the blood, liver, lungs, and spleen of each animal were determined. As seen in Figure 2, the numbers of *B. pseudomallei* 1026b and SR1015 bacteria were nearly undetectable at 12 h (Figure 2A). By 24 h, the numbers of 1026b bacteria recovered from the blood, lung, liver, and spleen increased, while SR1015 was detected only in the spleen (Figure 2B). By 48 h, very high numbers of 1026b bacteria were recovered from all of the organs taken, representing a dramatic increase compared to the inoculum (Figure 2C). In contrast, all of the organs taken from hamsters infected with SR1015 contained fewer bacteria (Figure 2C). Of particular interest was the fact that the number of SR1015 bacteria recovered from the blood at 48 h was lower than in the inoculum, suggesting that the capsule mutant was cleared from the blood more effectively than the wild type. The number of SR1015 bacteria recovered from the spleen was higher than the number of SR1015 bacteria in the blood, suggesting that SR1015 was being cleared from the blood and sequestered in the spleen. The difference in virulence between the two strains can be attributed to capsule production, since the CPS I mutant strain was found to have a growth rate similar to that of the wild-type strain 1026b [50].

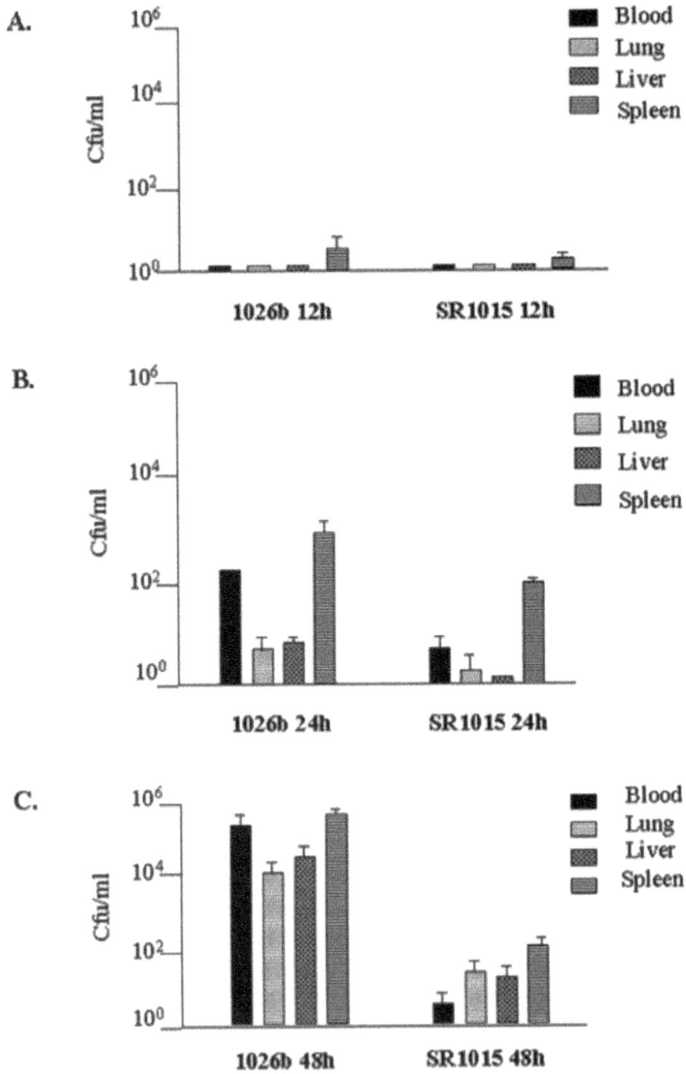

Figure 2. Differences in tissue distribution between *B. pseudomallei* strains 1026b and SR1015 in the Syrian hamster model of acute melioidosis. Female Syrian hamsters (three per group) were inoculated intraperitoneally with 10^2 CFU of either strain, and at 12, 24, and 48 h, two groups of animals were sacrificed and bacterial quantitation of the tissues was determined. The data represent the average number of bacteria found in each tissue and the standard deviation for a given time point.

3.3. CPS I production by *B. pseudomallei* is responsible for persistence in the blood by inhibiting complement

To define the role of the capsule for persistence in the blood, serum bactericidal assays were performed with the addition of purified capsule to determine if capsule had an effect on the survival of serum-sensitive strains of *B. pseudomallei*. For these experiments, we utilized a double mutant that we constructed in the laboratory, SLR5, which lacks both capsule and O-polysaccharide, since the capsule mutant SR1015 was previously found to be serum resistant [42]. The survival of SLR5 was extremely poor when incubated in the presence of 30% normal human serum (NHS). However, the addition of purified capsule increased the survival of SLR5 in NHS. The addition of 50 μg of capsule to the reaction increased the numbers of SLR5 to 5.9 x 10^1 CFU/ml, and the addition of 100 μg of capsule increased the survival of SLR5 by nearly 1,000-fold to 1.9 x 10^3 CFU/ml. Furthermore, pre-incubation of 30% NHS with 100 μg of capsule (PI-CPS) before the addition of bacteria increased the survival of SLR5 100,000-fold to 4.4 x 10^6 CFU/ml. This was similar to the survival of SLR5 when incubated with serum that was heat-inactivated (HI-NHS). These effects were found to be specific to capsule, since the addition of 50 or 100 μg of purified *B. pseudomallei* O-PS or preincubation of the serum with O-PS did not increase the survival of serum-sensitive SLR5 [50].

Since capsule mutants of *B. pseudomallei* are serum resistant in that they are not susceptible to lysis by the membrane attack complex (MAC) because they still produce O-PS, we postulated that the ability of the capsule to enhance survival in the blood could be due to its ability to inhibit C3b deposition and opsonization. To investigate the effect of capsule on C3b deposition, the amount of C3b deposited on the surfaces of wild-type *B. pseudomallei* 1026b and the capsule mutant, SR1015, in the presence of serum was determined by Western blot analysis using a mouse monoclonal antibody specific to human complement factor C3b. The deposition of C3b was found to be more pronounced in the capsule mutant SR1015 than in the wild type in both 10 and 30% NHS. Similar results were observed with the capsule mutant SZ210, a strain containing an in-frame deletion of the *wcbB* gene. More C3b was detected when SZ210 was incubated in both 10 and 30% NHS than with 1026b. Optical densitometry measurements were performed in order to quantitate the difference in C3b deposition between the strains. The average amount of C3b deposited on the surfaces of SR1015 and SZ210 bacteria was 3.5-fold higher than for 1026b in 10% NHS and 2.5-fold higher in 30% NHS. In addition, there was a shift in the molecular mass of C3b, which normally runs at 185 kDa, indicating a covalent attachment of the molecule to the bacterial surface. The nature of this attachment was not investigated; however, C3b is thought to covalently attach to the bacterial surface through an ester or amide linkage [50, 52].

Immunofluorescence microscopy analysis was also performed to demonstrate the difference in C3b deposition between the capsule mutant and the wild type. The same experiment described above was performed, and samples were reacted with the mouse monoclonal antibody to human complement factor C3b, except that the samples were reacted with a secondary antibody conjugated to Cy3 and stained with DAPI for visualization of bacterial cells. As shown in Figure 3, the *B. pseudomallei* capsule mutant SR1015 demonstrated more reactivity to the antibody to human C3b in the presence of serum than the wild-type 1026b.

This is evident from the red fluorescence that corresponds to the C3b bound to the bacterial surface surrounding the blue DAPI-stained cells seen when the capsule mutant was incubated in the presence of 10% NHS (Figure 3D to F). In contrast, the amount of red fluorescence surrounding the DAPI-stained wild-type cells was minimal in the presence of 10% NHS (Figure 3A to C). There was a dramatic difference in the amount of C3b deposited on the surface of the capsule mutant compared to the wild type, which was detectable after only 15 min of incubation of the bacteria with human serum (Figure 3B and E). By 60 min, there was some C3b deposition on wild-type *B. pseudomallei*; however, there was still more C3b deposited on the surface of the capsule mutant (Figure 3C and F) [50]. This experiment was not performed with 30% NHS due to excessive clumping of the samples during the fixation process, which resulted in inconsistent and poor staining of the cells. Western blot analysis was also performed to determine the amount of complement factor C3b deposition on the surface of *B. thailandensis* E264, a related nonpathogenic organism. The amount of C3b deposition in *B. thailandensis* E264 was more pronounced than with *B. pseudomallei* 1026b and was similar to the amount of C3b deposited on the surface of the capsule mutant, *B. pseudomallei* SR1015, in the presence of human serum. The amount of C3b deposition that occurred on the surface of *B. thailandensis* was expected, since the organism is known to lack this capsule [42, 50].

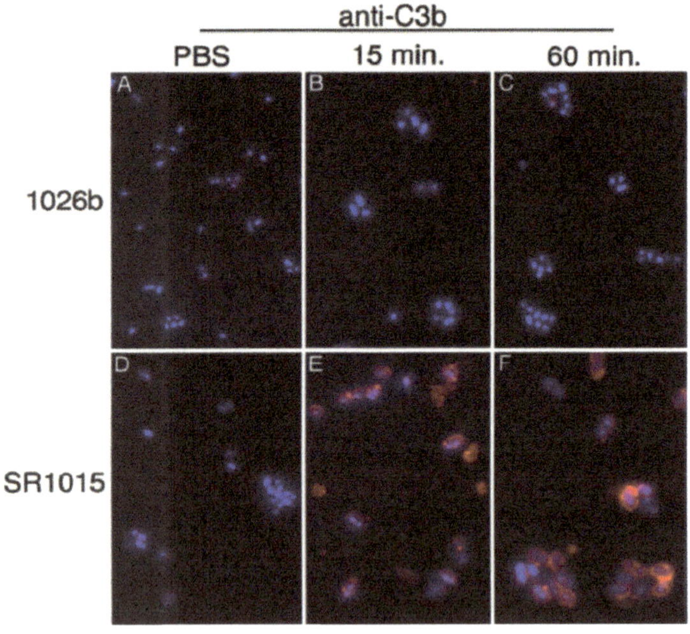

Figure 3. Immunofluorescence microscopy analysis of decreased complement factor C3b deposition in 10% normal human serum by *B. pseudomallei* capsule. *B. pseudomallei* 1026b and SR1015 were incubated in 10% normal human serum (NHS), reacted with a mouse monoclonal antibody to human complement factor C3b, reacted with a rabbit anti-mouse IgG conjugated to Cy3 (Jackson Laboratories), and stained

with DAPI for visualization of whole bacterial cells (Sigma). (A) *B. pseudomallei* 1026b incubated in PBS; (B) 1026b incubated in 10% NHS for 15 min; (C) 1026b incubated in 10% NHS for 60 min; (D) *B. pseudomallei* SR1015 incubated in PBS; (E) SR1015 incubated in 10% NHS for 15 min; (F) SR1015 incubated in 10% NHS for 60 min. The blue fluorescence indicates the DAPI stained bacteria, and the red fluorescence indicates the binding of complement factor C3b to the bacterial surface.

3.4. The *B. pseudomallei* capsule CPS I reduces phagocytosis

The capsule mutant SR1015 was phagocytosed more significantly by PMNL than the wild-type strain. The proportion of wild-type *B. pseudomallei* 1026b phagocytosed in the presence of 10% NHS was 35.9%, while the proportion of the capsule mutant SR1015 phagocytosed was 51.7% ($P < 0.001$). When each strain was incubated in the presence of 30% NHS, 59.3% of the wild-type strain 1026b was phagocytosed by the PMNL after 30 min compared to 82.3% for the capsule mutant ($P < 0.001$) [50].

3.5. *B. pseudomallei* CPS I expression is elevated in the presence of 30% normal human serum

The *lux* reporter strain *B. pseudomallei* SZ211 was constructed by cloning an internal fragment of the *wcbB* gene into pGSV3-*lux*, a suicide vector containing the *lux* operon from *Photorhabdus luminescens* [53]. Absorbance (OD540) and luminescence (in relative light units) measurements were taken every 2 h. Capsule expression was higher in the presence of M9 plus 1% glucose plus 30% normal human serum (NHS) and M9 plus 1% glucose plus 30% heat-inactivated serum (HI-NHS) than in M9 plus 1% glucose alone. The increase in light production of SZ211 in the presence of serum supports the requirement for capsule for survival in serum. The strain *B. pseudomallei* SZ213 was constructed by cloning an internal region of the *wbiA* gene, which encodes an *O*-acetyltransferase required for *O*-acetylation of the O-PS component of *B. pseudomallei* LPS [36]. Light production of this strain was measured under the same conditions to determine whether LPS expression was induced in the presence of serum. Similar to the capsule, LPS expression was elevated in the presence of both 30% NHS and 30% HI-NHS. The levels of expression of both capsule and LPS were not significantly different in NHS and HI-NHS, suggesting that the environment of the serum may be required for induction of gene expression rather than complement [50].

4. Other capsules produced by *B.pseudomallei*

Sequence analysis of the completed genome of *B. pseudomallei* revealed four operons with the predicted function of capsular polysaccharide biosynthesis and export [54]. One of these operons, with the gene identifiers BPSL2786-2810, corresponds to the previously characterized mannoheptose capsule designated CPS I [42, 50]. Three other operons were identified. These three capsule operons in the genome of *B. pseudomallei* were further analyzed using the BLAST program and Artemis. The operons are illustrated in Figure 4. The operon consisting of the genes BPSS0417-0429, was designated CPS II (Figure 4B). Another operon, BPSS1825-1835, was designated CPS III and the predicted homologues were investigated further (Figure 4C). A fourth operon, CPS IV, was found to contain genes that may be involved in the synthesis of a capsule with the gene identifiers BPSL2769-2785 (Figure 4D) [54, 55].

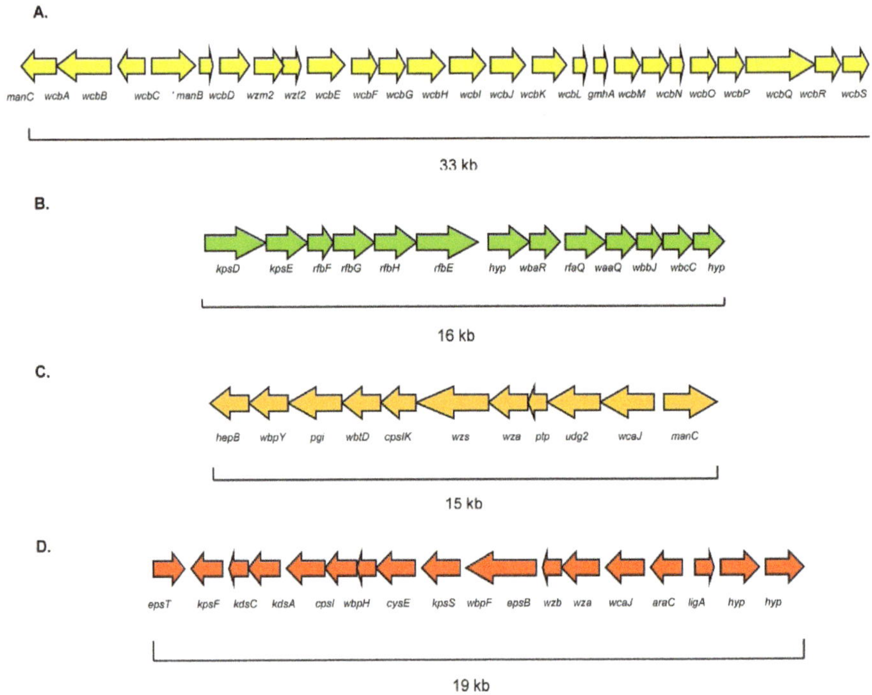

Figure 4. Organization of the chromosomal regions containing the genes comprising the *B. pseudomallei* capsule operons. The direction of transcription is represented by arrows, and the gene names demonstrating the highest degree of homology to the *B. pseudomallei* open reading frames are indicated. The relative sizes of each locus are indicated. (A) *B. pseudomallei* capsule cluster I (CPS I). (B) *B. pseudomallei* capsule cluster II (CPS II). (C) *B. pseudomallei* capsule cluster III (CPS III). (D) *B. pseudomallei* capsule cluster IV (CPS IV).

4.1. Distribution of capsule loci between three *Burkholderia* species

Comparative analysis of the genomes of three *Burkholderia* species, *B. pseudomallei*, *B. mallei* and *B. thailandensis*, was performed to determine whether all of the predicted *B. pseudomallei* capsule operons were present in *B. mallei* and *B. thailandensis* as well. CPS II, III, and IV were found to be present in *B. pseudomallei* and *B. thailandensis*, but not *B. mallei*. This is in contrast to CPS I, which is present in *B. pseudomallei* and *B. mallei*, but not *B. thailandensis* [42, 50, 54, 55, 56, 57]. CPS II was found to be identical between *B. pseudomallei* and *B. thailandensis*, but *B. thailandensis* was found to contain two flanking hypothetical genes not present in *B. pseudomallei*. The CPS II genes were found to be deleted entirely from *B. mallei*. A large chromosomal region ranging from open reading frames BPSS0404 to BPSS0491 including CPS II was shown to be deleted in *B. mallei* compared to *B. pseudomallei* and replaced with a large chromosomal region containing open reading frames BMAA0555 (IS407A *orfB*) to BMAA1784, a unique hypothetical protein not found in *B. pseudomallei*. The two genomes align with the presence of the alkyl hydroperoxidase reductase genes *ahpC* and *ahpF*, but these are organized in the opposite orientation in *B. mallei* compared to *B. pseudomallei*. The entire CPS III operon and flanking genes were shown to be the same in both *B. pseudomallei* and *B. thailandensis*. In contrast, the majority of the CPS III cluster was deleted from *B. mallei* with the exception of the *wcaJ* and *manC* genes, as well as the two flanking hypothetical genes on one side, and another hypothetical gene on the other side of the deleted region. The deletion of the CPS III genes in *B. mallei* was found to be replaced with the IS407A *orfA* and *orfB* genes. The entire CPS IV region was found to be replaced in *B. mallei* and flanked by two IS407A elements. The open reading frame BPSL2785, which encodes a hypothetical protein, is present in *B. mallei* (BMA2284.1) ATCC 23344 as well as a number of other *B. mallei* strains, but is organized in the opposite orientation. The genomes of *B. pseudomallei*, *B. thailandensis*, and *B. mallei* all diverge upstream of the CPS IV region, but all three organisms were found to align at the location of the *ompA* and hypothetical genes [55].

4.2. CPS III does not contribute to the virulence of *B. pseudomallei*

In order to assess the role of CPS III in virulence a mutant in the CPS III operon was tested for virulence compared to wild type *B. pseudomallei* in the Syrian hamster model of melioidosis. Syrian golden hamsters were inoculated intraperitoneally with 10^1 to 10^3 cells of either wild-type *B. pseudomallei* 1026b or the capsule mutant SZ1829. After 48 h, the LD_{50} values were determined. SZ1829 had LD_{50} values of <10 CFU, identical to that of wild type *B. pseudomallei*, indicating that this capsule is not required for virulence. In addition, the bacterial load in the blood of the infected hamsters was similar to that of wild-type and significantly higher than that of the non-pathogenic *B. thailandensis* E264 and the CPS I mutant, *B. pseudomallei* SR1015, both of which are incapable of establishing bacteremia [37, 50]. This indicates that CPS III does not contribute to persistence in the blood. Similar results were obtained for CPS II and CPS IV mutants, but we went on to further characterize CPS III.

4.3. Expression of CPS III in host and environmental conditions

A *lux* reporter strain was constructed in the CPS III operon by cloning an internal fragment of one of the genes into pGSV3-*lux*, a suicide vector containing a promoterless *lux* operon from *Photorhabdus luminescens* [53]. Regulation of this capsule in an environment similar to that encountered in the host was determined by growing the *lux* reporter strain, SZ1829, in the presence of M9 plus 1% glucose versus M9 plus 1% glucose plus 30% normal human serum (NHS). Absorbance (OD_{540}) and luminescence (in relative light units) measurements were taken every hour. The expression of CPS III (SZ1829) was higher in M9 plus 1% glucose alone compared to M9 plus 1% glucose plus 30% NHS. The expression of SZ1829 was 3-4 fold lower in 30% NHS. This was in contrast to the expression of CPS I (SZ211) (see section 3.5), which was significantly more highly expressed in 30% NHS at a level of 3-4 fold compared to growth in M9 plus 1% glucose alone [55]. Although CPS III demonstrated higher expression initially in 30% NHS, this may have been due to the fact that the addition of NHS caused precipitation in the media which affected the optical density of the cultures.

The expression of the *lux* operon in reporter strains SZ211 (CPS I$^-$) and SZ1829 (CPS III$^-$) was also measured in water to determine whether CPS III was induced in this environment. Overnight cultures of SZ211 and SZ1829 were inoculated into sterile water and incubated at 37°C without shaking. Capsule expression was determined as described above, but the luminescence/absorbance calculations for water were compared to the values for these strains when grown in LB. CPS III was found to be induced in water compared to LB. The expression of SZ1829 was found to be significant with an increase of 2-3 fold over the course of the experiment. The expression of CPS I was found to be greater than 4 fold higher in LB compared to water [55].

Microarray analysis of capsule expression was performed using a low-density DNA microarray. RNA was isolated from the livers and lungs of hamsters infected with *B. pseudomallei* and from *B. pseudomallei* grown in LB. The results of the microarray experiment are shown in Table 1 [55]. The level of gene expression, or fold change, is represented as the ratio of gene expression in the hamster compared to growth in LB. As shown in Table 1, CPS III genes were not found to be significantly expressed *in vivo* since most of the fold changes were determined to be less than 2-fold. Many of the genes had negative fold change values, indicating that these genes are suppressed in the host environment. The highest fold change result was 2.337866 for BPSS1827, a predicted glucose-6-phosphate isomerase, which is still much lower than the fold changes observed for CPS I genes, which were significantly higher [58].

Gene ID	**Predicted function**	**Fold Change (*in vivo* vs. *in vitro*)**
BPSS1825	Glycosyltransferase	1.757155
BPSS1826	Glycosyltransferase	0.093565
BPSS1827	Glucose-6-phosphate isomerase	2.337866
BPSS1828	Glycosyltransferase	-0.66607
BPSS1829	Glycosyltransferase	-0.18061
BPSS1830	Capsule export, tyrosine-protein kinase	-0.30655
BPSS 1831	Capsule export, outer membrane protein	0.725849
BPSS1832	Transport, tyrosine-protein phosphatase	1.52665
BPSS1833	UDP-glucose-6-dehydrogenase	-0.71411
BPSS1834	Sugar transferase	-0.07963
BPSS 1835	Mannose-1-phosphate guanyltransferase	-0.75075

*Note: Gene expression representative of the liver and lungs of infected hamsters.

Table 1. Microarray analysis of *B. pseudomallei* CPS III expression following intraperitoneal inoculation in the hamster model of melioidosis.

4.4. Carbohydrate composition of CPS III

Glycosyl composition analysis was performed on the purified capsule by combined gas chromatography/mass spectrometry (GC/MS). GC/MS results indicated that CPS III is composed of galactose, glucose, mannose, xylose, and rhamnose residues, with the highest proportion of carbohydrate being galactose and glucose. Glycosyl linkage analysis was also performed [55]. The predominant glycosyl residue detected was a terminally-linked heptopyranosyl (t-Hep) at a percentage of 23.2. Other residues detected were a terminally-linked and a 4-linked glucopyranosyl (t-Glcp) (4-Glcp) at percentages of 14.6 and 10.8, respectively [55].

5. Conclusion

Although significant advances have been made in the field, melioidosis continues to be a public health concern in many regions of the world [59]. Completion of the sequencing of the *B. pseudomallei* genome has revealed potential virulence determinants and comparative genomics between the genomes of *B. pseudomallei*, *B. thailandensis*, and *B. mallei* species has contributed to a better understanding of the organism. Further studies are ongoing to define the pathogenesis of *B. pseudomallei* and to identify effective vaccine candidates and diagnostic targets [54, 59].

To obtain virulence determinants unique to *B. pseudomallei*, we used subtractive hybridization between this organism and a related nonpathogenic organism, *B. thailandensis*. Analysis of the subtractive hybridization library revealed that *B. pseudomallei* contains a number of DNA sequences that are not found in *B. thailandensis*. One of the subtraction clones, pDD1015, demonstrated weak homology to a glycosyltransferase, WbpX, from *P. aeruginosa* [43]. The insert from pDD1015 was cloned into a mobilizable suicide vector for insertional inactivation of the glycosyltransferase gene in wild-type *B. pseudomallei*. The resulting strain, SR1015, was markedly less virulent than the parent strain in an animal model. We determined that SR1015 harbored a mutation in a glycosyltransferase gene involved in the production of a capsular polysaccharide which we subsequently designated as CPS I. We then identified the operon involved in the biosynthesis and transport of this capsular polysaccharide (CPS I) [42]. The genes identified encode for proteins that are similar to proteins involved in the biosynthesis and export of capsular polysaccharides, particularly those involved in the production of group 3 capsular polysaccharides. Group 3 capsules include the *E. coli* K10 capsule and may also include the *H. influenzae* group b capsule and the capsule produced by *N. meningitidis* serogroup B [8]. Group 3 capsules are always coexpressed with O serogroups, are not thermoregulated, are transported by an ABC-2 exporter system, and do not contain the *kpsU* and *kpsF* genes, and usually the gene clusters map near the *serA* locus [8]. Thus far, no *serA* locus that is associated with the type I O-PS cluster was identified, but this polysaccharide is coexpressed with O antigen and lacks the *kpsU* and *kpsF* genes, and genes encoding for a putative ABC-2 transporter have been identified. The genes involved in the production of group 3 capsules are organized into regions and are divergently transcribed. Regions 1 and 3 are generally conserved and contain genes involved in export of the polysaccharide. These regions flank region 2, which contains the biosynthetic genes and is not conserved between serotypes [4]. The genetic organization of the CPS I is also similar to that of other capsule gene clusters in that the genes are organized into more than one transcriptional unit and appear to be divergently transcribed [42].

The polysaccharide with the structure -3)-2-*O*-acetyl-6-deoxy-β-D-manno-heptopyranose-(1- was originally isolated and characterized as an O-PS component of LPS in *B. pseudomallei* and was designated type I O-PS [35]. However, our results suggested that this polysaccharide is a capsule rather than an O-PS moiety. The genes involved in the production of this capsule demonstrated strong homology to the genes involved in the production of capsular polysaccharides in many organisms, including *N. meningitidis, H. influenzae,* and *E. coli.* In addition, the export genes associated with this cluster are not associated with the previously characterized O-PS gene cluster [36]. Western blot analysis of proteinase K cell extracts and silver staining showed that this polysaccharide has a high molecular mass (200 kDa) and lacks the banding pattern seen with O-PS moieties. This conclusion was further supported by another group of researchers that demonstrated this polysaccharide is a capsule rather than an O-PS component of LPS because it lacks a lipid A moiety and was not capable of macrophage activation [49]. Studies by our laboratory have indicated that mutants in the production of the core oligosaccharide of the LPS are still capable of producing this polysaccharide [48]. Based on the above criteria and the genetic similarity to group 3 capsules, we proposed that this polysaccharide is a capsule.

Virulence genes of a number of pathogenic bacteria are located on pathogenicity islands (PAIs), regions on the bacterial chromosome that are present in the genome of pathogenic strains but rarely present in those of nonpathogenic strains. The PAIs may range in size from about 30 kb to 200 kb and often differ in G+C content from the remaining bacterial genome; the PAIs are often associated with the carriage of many virulence genes. These genetic units are often flanked by direct repeats and may be associated with tRNA genes or insertion sequence (IS) elements at their boundaries. They may also be associated with the presence of mobility genes, such as IS elements, integrases, transposases, and origins of plasmid replication. These DNA regions are considered to be unstable in that they may be subject to deletion with high frequency or undergo duplications and amplifications [7]. A number of PAIs have been described for both gram-positive and gram-negative bacteria, and the application of subtraction hybridization has been used to successfully identify such genetic elements [7]. The subtractive hybridization that was carried out between *B. pseudomallei* and *B. thailandensis* led to the identification of a number of sequences that were found to be A-T rich compared to the rest of the *B. pseudomallei* chromosome. This, combined with the fact that insertional mutagenesis of the glycosyltransferase gene identified by this method resulted in an avirulent strain, suggests that we may have identified DNA sequences from a putative PAI and that the capsular polysaccharide gene cluster may be located on this island. It is possible that *B. pseudomallei, B. mallei,* and *B. stabilis* acquired DNA encoding for capsule as well as other potential, yet unidentified virulence factors by horizontal transfer recently in evolution. *B. pseudomallei* and *B. mallei* are known to contain IS elements that are present in *B. cepacia* but not in *B. thailandensis* [56, 60].

Capsule production has been correlated with virulence in many bacteria, particularly those causing serious invasive infections of humans [61]. Our studies demonstrated that CPS I is critical for the virulence of *B. pseudomallei* [42, 50]. A number of functions have been suggested for polysaccharide capsules: prevention of desiccation for transmission and survival, adherence for colonization, resistance to complement-mediated phagocytosis and complement-mediated killing, and resistance to specific host immunity due to a poor antibody response to the capsule [4].

To establish a correlation between capsule production and clinical infection a number of *B. pseudomallei* strains isolated from clinical specimens were tested for CPS I production. All 55 strains tested were found to produce CPS I by western blot analysis [51]. In addition 10 strains of *B. thailandensis* were tested and found negative for CPS I production, confirming the importance of CPS I in virulence as well as clinical infection.

CPS I production by *B. pseudomallei* was shown to contribute to the persistence of the organism in the blood of the host. All CPS I mutants tested in the animal model could not be isolated from the blood following infection. The addition of purified capsule was shown to increase the virulence of the CPS I mutant strains SR1015 and SZ210 in the animal model. Differences in tissue distribution between wild type *B. pseudomallei* and SR1015 in infected hamsters indicated that SR1015 was cleared from the blood because the numbers of SR1015 in the blood of infected hamsters was 10,000-fold lower than that of wild type 1026b and lower than the initial inoculum of 100 cfu/ml [50].

CPS I production was shown to be responsible for persistence in the blood by evasion of the complement cascade and the mechanism for this was determined to be through the reduction of C3b deposition and opsonophagocytosis. The addition of purified CPS I to serum bactericidal assays showed that the capsule contributes to increased resistance of serum sensitive strains lacking the O-polysaccharide moiety (O-PS) of LPS to the bactericidal effects of normal human serum. However, CPS I mutants themselves were not found to be serum sensitive because they still produced O-PS, which was previously shown to be responsible for serum resistance, because it prevents lysis by the MAC complex [36]. This led us to postulate that CPS I was affecting the complement cascade through some other mechanism and it was found that this mechanism was through the reduction of C3b deposition and opsonization [50]. Both Western blot analysis and immunofluorescence microscopy experiments using a mouse monoclonal antibody to human C3b demonstrated the inhibition of C3b deposition by CPS I. In both experiments C3b deposition was more pronounced on the surface of the CPS I mutant compared to wild type. Also evident was that some C3b deposition occurred in the wild type, but this was expected since bacterial capsules are known to allow the diffusion of some C3b to the bacterial surface and *B. pseudomallei* is capable of activating the alternative pathway of complement culminating in the formation of the MAC complex [36, 62]. The accumulation of C3b affects the amplification step of the complement cascade and therefore, the less C3b deposited the less C5a is generated for phagocyte recruitment [63]. This explains the increased clearance of CPS I mutants from the blood. This conclusion was supported by the fact that *B. thailandensis*, the non-pathogenic organism which lacks CPS I, has been shown to be serum resistant, but is not capable of establishing a bacteremia in the Syrian hamster model of acute melioidosis [36, 37, 42]. Effective opsonization of invading bacteria results in enhanced phagocytosis and clearance of organisms form the blood of an infected host [52]. Quantitative radiolabelled phagocytic assays were also performed to establish a correlation between opsonization of the bacteria and phagocytosis by polymorphonuclear leukocytes. In the presence of serum, the CPS I mutant was more readily phagocytosed than wild type [50].

The expression of CPS I in the presence of normal human serum was found to be significantly elevated, also confirming that this capsule contributes to survival in the host. The presence of CPS I enables *B. pseudomallei* to survive in the blood through the inhibition of complement factor C3b deposition and phagocytosis [50]. The presence of this capsule facilitates survival as well as spreading to other organs, which can explain the overwhelming septicemia that is common in culture-positive melioidosis patients [64]. Therefore CPS I production is critical to the virulence of *B. pseudomallei* and further research will enhance the development of preventative strategies for melioidosis since this polysaccharide is one of the components of a *B. pseudomallei* subunit vaccine [28, 65].

Sequence analysis of the genome of *B. pseudomallei* revealed the presence of four operons possibly involved in polysaccharide capsule biosynthesis. One of these operons, (CPS I), corresponded to the previously identified and characterized mannoheptose capsule that was shown to be responsible for virulence and comprises one of the currently proposed melioidosis and glanders subunit conjugate vaccine [28, 66, 42, 50, 67]. The CPS I capsule cluster is present in the genome of *B. mallei* as well, but the complete cluster is not found in the genome of *B. thailandensis* [56, 57, 68]. This correlates with previous studies that have shown that this capsule is produced by *B. mallei*, but not by *B. thailandensis* [36, 37, 42, 56].

Three other putative capsule operons were identified by sequence analysis and all of these operons were found to be present in *B. pseudomallei* and *B. thailandensis*, but not *B. mallei*. Since these capsules are found in *B. thailandensis* and *B. pseudomallei*, they may be required for either survival in the host or in the environment; however, further studies are required to determine the roles of CPS II and CPS IV.

CPS III, located on chromosome 2, was found to contain 11 genes involved in the biosynthesis of a polysaccharide and was shown to be present in the genomes of *B. pseudomallei* and *B. thailandensis*, but not *B. mallei*. A mutation in the CPS III cluster did not affect production of CPS I and so it can be concluded that this operon encodes for gene products responsible for the biosynthesis of a separate capsule. CPS III was not found to contribute to the pathogenesis of *B. pseudomallei*. This capsule was not shown to be highly expressed *in vivo* by microarray analysis and was not required for virulence in the animal model. The CPS III mutant, SZ1829, which contains a mutation in the BPSS1829 gene as a result of insertional inactivation, was found to be as virulent in the animal model as wild type *B. pseudomallei*. The expression of this capsule was shown to be elevated when incubated in water, but suppressed in the presence of normal human serum [55]. The presence of the CPS III cluster in *B. pseudomallei* and *B. thailandensis*, both of which can survive for long periods in the environment compared to *B. mallei*, the increased expression of this capsule in water, and the low level of expression of this capsule *in vivo*, suggests that this capsule may contribute to the survival of *B. pseudomallei* in the environment [69].

Previous studies have demonstrated that *B. pseudomallei* produces three other capsular polysaccharides in addition to CPS I and these have been structurally characterized. One is an acidic polysaccharide with the structure, -3)-2-*O*-acetyl-β-D-Gal*p*-(1-4)-α-D-Gal*p*-(1-3)-β-D-Gal*p*-(1-5)-β-D-KDO*p*-(2-, which is recognized by patient sera [32]. The other two are: a branched 1,4-linked glucan polymer ((CP-1a) and a triple-branched heptasaccharide repeating unit composed of rhamnose, mannose, galatose, glucose, and glucoronic acid (CP-2) [49]. Combined GC/MS analysis of CPS III revealed that the composition of this capsule demonstrates some similarity to the composition of the previously described capsule CP-2 composed of rhamnose, mannose, galactose, glucose, and glucoronic acid; however, the proportions of carbohydrate residues were not similar, and the CPS III capsule was also found to contain xylose and not glucoronic acid. In addition, CPS III was determined to be composed primarily of heptose [55]. Therefore it is evident that the capsule identified in this study is not one of the previously described capsule structures. Some of the previously characterized capsules produced by *B. pseudomallei* have been shown to be produced under unique conditions [32, 49]. Strain variation, differences in expression of the capsules, and discrepancies between purification strategies may also explain why a number of capsules have been shown to be produced by this organism. Nevertheless, the genes BPSS 1825-1835 appear to be involved in the biosynthesis of a capsule with this composition. Further analysis by 2D NMR would be required to definitively establish a connection between CPS III and one of the other published structures.

Studies by another laboratory have also focused on the presence of these capsule clusters in *B. pseudomallei*. Sarkar-Tyson *et al.* identified two polysaccharide clusters, one of which corresponds to the CPS III presented in this paper, but the authors identified this cluster as type IV O-PS (2007). The type IV O-PS was found to be involved in virulence in a mouse model [70]. However, a mutant in this polysaccharide did not demonstrate any difference in hydrophobicity compared to wild-type, indicating that this polysaccharide does not contribute to making the cell surface more hydrophobic, which is an advantageous characteristic for some pathogenic bacteria. The differences in virulence compared to the current work can be attributed to the use of different animal models; however, all other data seem to indicate that this capsule is not required for virulence.

A study was recently published which outlines the identification of another capsule produced by *B. pseudomallei* [71]. This capsule was determined to be composed of 1,3-linked α-D-mannose residues. This capsular polysaccharide was also found to be produced by *B. mallei*. The genes involved in the synthesis of this polysaccharide have not yet been identified and work is also underway to determine the role for this novel capsule in the pathogenesis of melioidosis and glanders.

B. pseudomallei is an environmental saprophyte often found in soil and stagnant water and incidence of the disease is high in rice farmers in Southeast Asia [22, 69]. This organism harbors a large genome which explains its ability to survive for long periods of time in the environment as well exist as a significant pathogen in both humans and animals. The presence of multiple polysaccharide clusters in the genome and the production of multiple capsule structures under differing conditions may contribute to the ability of this organism to adapt to a variety of conditions. As demonstrated in this study, capsule expression is dependent on the particular environment, which indicates that *B. pseudomallei* produces these capsules to promote a survival advantage either in the host or in the environment. Further studies aimed at characterizing the capsules of *B. pseudomallei* will be beneficial to understand the pathogenesis of this organism and to advance further vaccine development.

Author details

Shauna L. Reckseidler-Zenteno
Athabasca University, Athabasca, Alberta, Canada

Acknowledgement

The author would like to thank Dr. Donald Woods and all colleagues in the *Burkholderia pseudomallei* research community for their support and discussions over the years. The work described in this chapter was funded by the following sources: Department of Defense contract DAMD 17-98-C-8003, the Medical Research Council of Canada, the Canadian Bacterial Diseases Network of Centers of Excellence, Canadian Institutes for Health Research MOP 36343, and a Research Incentive Grant from Athabasca University. Carbohydrate analysis was conducted by the Complex Carbohydrate Research Center at the University of Georgia and this was supported in part by the Department of Energy-funded (DE-FG09-93ER-20097) Center for Plant and Microbial Complex Carbohydrates. Microarray analysis was performed at the Southern Alberta Microarray Facility.

6. References

[1] Whitfield C (1988) Bacterial Extracellular Polysaccharides. Can. J. Microbiol. 34: 415-420.
[2] Costerton J W, Irvin R T, Cheng K-J (1981) The Bacterial Glycocalyx in Nature and Disease. Annu. Rev. Microbiol. 35: 299-324.
[3] Whitfield C, and Valvano M (1993) Biosynthesis and Expression of Cell-surface Polysaccharides in Gram-negative Bacteria. Adv. Microbiol. Phys. 35: 135-146.
[4] Roberts I S (1996) The Biochemistry and Genetics of Capsular Polysaccharide Production in Bacteria. Ann. Rev. Microbiol. 50: 285-315.
[5] Robbins J B, McCracken G H, Gotschlich E C, Ørskov I, Hanson L A (1974). *Escherichia coli* K1 Capsular Polysaccharide Associated with Neonatal Meningitis. N. Engl. J. Med. 90: 267-271.
[6] Grados O, Ewing V M (1970) Antigenic Relationships between *Escherichia coli* and *Neisseria meningitidis* group B. J. Infect. Dis. 122: 100-103.

[7] Hacker J, Blum-Oehler G, Mühldorfer I, Tschäpe H (1997) Pathogenicity Islands of Virulent Bacteria: Structure, Function and Impact on Microbial Evolution. Mol. Microbiol. 23: 1089-1097.

[8] Whitfield C, Roberts I S (1999) Structure, Assembly, and Regulation of Expression of Capsules in *Escherichia coli*. Mol. Microbiol. 31: 1307-1319.

[9] Frosch M, Müller D, Bousset K, Müller A (1992) Conserved Outer Membrane Protein of *Neisseria meningitidis* Involved in Capsule Expression. Infect. Immun. 60: 798-803.

[10] Roberson E, Firestone M (1992) Relationship between Dessication and Exopolysaccharide Production in Soil *Pseudomonas* sp. Appl. Environ. Microbiol.58: 1284-1291.

[11] Ophir T, Gutnick D (1994) A Role for Exopolysaccharide in the Protection of Microorganisms from Dessication. Appl. Environ. Microbiol. 60: 740-745.

[12] Berry A, DeVault J, Chakrabarty A (1989) High Osmolarity is a Signal for Enhanced *algD* Transcription in Mucoid and Nonmucoid *Pseudomonas aeruginosa* strains. J. Bacteriol. 171: 2312-2317.

[13] Costerton J W, Cheng K-J, Geesy G G, Ladd T I, Nickel J C et al. (1987) Bacterial Biofilms in Nature and Disease. Annu. Rev. Microbiol. 41: 435-464.

[14] Jenkinson F F (1994) Adherence and Accumulation of Oral Streptococci. Trends. Microbiol. 2: 209-212.

[15] Howard C J, Glynn A A (1971) The Virulence for Mice of Strains of *Escherichia coli* Related to the Effects of K antigens on their Resistance to Phagocytosis and Killing by Complement. Immunology 20: 767-777.

[16] Kim K, Kang, J, Cross A (1986) The Role of Capsular Antigens in Serum Resistance and In Vivo Virulence of *Escherichia coli*. FEMS Microbiol. Lett. 35: 275-278.

[17] Moxon E, Kroll J (1990) The Role of Bacterial Polysaccharide Capsules as Virulence Factors. Curr. Top. Microbiol. Immunol. 150: 65-85.

[18] Bhattacharjee A, Jennings H, Kenny C, Martin A, Smith I (1975) Structural Determination of the Sialic Acid Polysaccharide Antigens of *Neisseria meningitidis* Serogroups B and C with Carbon 13 Nuclear Magnetic Resonance. J. Biol. Chem. 250: 1926-1932.

[19] Vann W F, Schmidt M A, Jann B, Jann K (1981) The Structure of the Capsular Polysaccharide (K5 antigen) of Urinary Tract Infective Escherichia coli O10:K5:H4. A Polymer Similar to Desulfo-heparin. Eur. J. Biochem. 116: 359-364.

[20] Wyle F, Artenstein M, Brandt B L, Tramont E C, Kapser D L et al. (1972) Immunologic Response of Man to Group B Meningococcal Polysaccharide Vaccines. J. Infect. Dis. 126: 514-521.

[21] Dance, D A B (1991) Melioidosis: the Tip of the Iceberg? Clin. Microbiol. Rev. 4: 52–60.

[22] Leelarasamee A, Bovornkitti S (1989). Melioidosis: review and update. Rev. Infect. Dis. 11: 413–425.

[23] Chaowagul, W, White N J, Dance D A, Wattanagoon Y, Naigowit P, Davis T M, Looareesuwan S, Pitakwatchara N (1989). Melioidosis: a Major Cause of Community Acquired Septicemia in northeastern Thailand. J. Infect. Dis. 159: 890–899.

[24] Howe C, Sampath A, Spotnitz M (1971) The Pseudomallei Group: a Review. J. Infect. Dis. 124: 598–606.

[25] Ip, M, Osterberg L G, Chau P Y, and Raffin T A (1995) Pulmonary Melioidosis. Chest 108: 1420–1424.

[26] Currie B J, Fisher D A, Anstey N M, Jacups S P (2000) Melioidosis: Acute and Chronic Disease, Relapse and Re-activation. Trans. R. Soc. Trop. Med. Hyg. 94: 301-304.

[27] White N J, Dance D A B, Chaowagul, W, Wattanagoon Y, Wuthiekanun V, Pitakwatchara N (1989) Halving of Mortality of Severe Melioidosis by Ceftazadime. Lancet 2: 697-701.

[28] Brett P J, Woods D E (1996) Structural and Immunologic Characterization of *Burkholderia pseudomallei* O-polysaccharide–flagellin Protein Conjugates. Infect. Immun. 64: 2824–2828.

[29] Bryan L E, Wong S, Woods D E, Dance D A, Chaowagul W (1994). Passive protection of diabetic rats with antisera specific for the polysaccharide portion of the lipopolysaccharide isolated from *Pseudomonas pseudomallei.* Can. J. Infect. Dis. 5: 170–178.

[30] Steinmetz I, Rohde M, Brenneke B (1995) Purification and Characterization of an Exopolysaccharide of *Burkholderia* (*Pseudomonas*) *pseudomallei.* Infect. Immun. 63: 3959–3965.

[31] Masoud, H, Ho M, Schollaardt T, Perry M B (1997) Characterization of the Capsular Polysaccharide of *Burkholderia pseudomallei* 304b. J. Bacteriol. 179: 5663–5669.

[32] Nimtz M, Wray V, Domke T, Brenneke B, Haussler S, Steinmetz I (1997). Structure of an Acidic Exopolysaccharide of *Burkholderia pseudomallei.* Eur. J. Biochem. 250: 608–616.

[33] Karawara K, Dejsirilert S, Ezaki T (1998) Characterization of Three Capsular Polysacharides Produced by *Burkholderia pseudomallei. FEMS Microbiol Lett* 169: 283-287.

[34] Knirel Y A, Paramonov N A, Shashkov A S, Kochetkov N K, Yarullin R G, Farber S M, Efremenko V I (1992) Structure of the Polysaccharide Chains of *Pseudomonas pseudomallei* Lipopolysaccharides. Carbohydr. Res. 233: 185–193.

[35] Perry M B, MacLean L L, Schollaardt T, Bryan L E, Ho M (1995) Structural Characterization of the Lipopolysaccharide O antigens of *Burkholderia pseudomallei.* Infect. Immun. 63: 3348–3352.

[36] DeShazer D, Brett P J, Woods D E (1998). The Type II O-antigenic Polysaccharide Moiety of *Burkholderia pseudomallei* Lipopolysaccharide is Required for Serum Resistance and Virulence. Mol. Microbiol. 30: 1081–1100.

[37] Brett P J, DeShazer D, Woods D E (1997) Characterization of *Burkholderia pseudomallei* and *Burkholderia pseudomallei*-like strains. Epidemiol. Infect. 118:137–148.

[38] Brett P J, DeShazer D, Woods D E (1998) *Burkholderia thailandensis* sp. nov., description of a *Burkholderia pseudomallei*-like species. Int. J. Syst. Bacteriol. 48: 317–320.

[39] Chaiyaroj S C, Kotrnon K, Koonpaew S, Anantagool N, White N J, Sirisinha S (1999) Differences in Genomic Macrorestriction Patterns of Arabinose-positive (*Burkholderia thailandensis*) and Arabinose-negative *Burkholderia pseudomallei.* Microbiol. Immunol. 43: 625–630.

[40] Wuthiekanun V, Smith M D, Dance D A, Walsh A L, Pitt T L, White N J (1996) Biochemical Characteristics of Clinical and Environmental Isolates of *Burkholderia pseudomallei.* J. Med. Microbiol. 45: 408–412.

[41] Trakulsomboon S, Dance D A B, Smith M D, White N J, Pitt T L (1997) Ribotype Differences between Clinical and Environmental Isolates of *Burkholderia pseudomallei.* J. Med. Microbiol. 46: 565–570.

[42] Reckseidler S L, DeShazer D, Sokol P A, Woods D E (2001) Detection of Bacterial Virulence Genes by Subtractive Hybridization: Identification of Capsular Polysaccharide of *Burkholderia pseudomallei* as a Major Virulence Determinant. Infect. Immun. 69: 34–44.

[43] Rocchetta H L, Burrows L L, Pacan J C, Lam J S (1998) Three Rhamnosyltransferases Responsible for Assembly of the A-band D-rhamnan Polysaccharide in *Pseudomonas aeruginosa*: a Fourth Transferase, WbpL, is Required for the Initiation of both A band and B-band Lipopolysaccharide Synthesis. Mol. Microbiol. 28: 1103–1119.

[44] Mongkolsuk S, Rabibhadana S, Vattanaviboon P, Loprasert S (1994) Generalized and Mobilizable Positive-selective Cloning Vectors. Gene 143: 145–146.

[45] Steinmetz I, Reganzerowski A, Brenneke B, Haussler S, Simpson A, White N J (1999) Rapid Identification of *Burkholderia pseudomallei* by Latex Agglutination Based on an Exopolysaccharide-specific Monoclonal Antibody. J. Clin. Microbiol. 37: 225–228.

[46] Clarke B R, Pearce R, Roberts I S (1999) Genetic Organization of the *Escherichia coli* K10 Capsule Gene Cluster: Identification and Characterization of Two Conserved Regions in Group III Capsule Gene Clusters Encoding PolysaccharideTransport Functions. J. Bacteriol. 181: 2279–2285.

[47] Reeves P R, Hobbs M, Valvano M A, Skurnik M, Whitfield C, Coplin D, Kido N, Klena J, Maskell D, Raetz C R, Rick P D (1996) Bacterial Polysaccharide Synthesis and Gene Nomenclature. Trends Microbiol. 498: 495–503.

[48] Burtnick M N, Woods D E (1999) Isolation of Polymyxin B-susceptible Mutants of *Burkholderia pseudomallei* and Molecular Characterization of Genetic Loci Involved in Polymyxin B Resistance. Antimicrob. Agents Chemother. 43: 2648–2656.

[49] Isshiki Y, Matsuura M, Dejsirilert S, Ezaki T, Kawahara K (2001) Separation of 6-deoxy-heptan from a Smooth-type Lipopolysaccharide Preparation of *Burkholderia pseudomallei*. FEMS Microbiol. Lett. 199: 21-25.

[50] Reckseidler-Zenteno S L, DeVinney R, Woods D E (2005). The Capsular Polysaccharide of *Burkholderia pseudomallei* Contributes to Survival in Serum by Reducing Complement Factor C3b Deposition. Infect. Immun. 73: 1106-1115.

[51] Reckseidler-Zenteno S L (2002) Role of the *Burkholderia pseudomallei* Capsule. University of Calgary. 228 p.

[52] Joiner K A, Brown E J, Frank M M (1984) Complement and Bacteria: Chemistry and Biology in Host Defense. Annu. Rev. Microbiol. 2: 461–491.

[53] Moore R A, Reckseidler-Zenteno S, Kim H, Nierman W, Yu Y, Tuanyok A, Warawa J, DeShazer D, Woods D E (2004) The Contribution of Gene Loss to the Pathogenic Evolution of *Burkholderia pseudomallei* and *Burkholderia mallei*. Infect. Immun. 72: 4172–4187.

[54] Holden M T, Titball R W, Peacock S J, Cerdeño-Tárraga A M, Atkins T, Crossman L C, Pitt T, Churcher C, Mungall K, et al. (2004) Genomic Plasticity of the Causative Agent of Melioidosis, *Burkholderia pseudomallei*. Proc. Natl. Acad. Sci. USA 101: 14240-14245.

[55] Reckseidler-Zenteno S L, Viteri D F, Moore R, Wong E, Tuanyok A, Woods D E (2010) Characterization of the Type III Capsular Polysaccharide Produced by *Burkholderia pseudomallei*. J. Med. Microbiol. 59(12): 1403-1414.

[56] DeShazer D, Waag D M, Fritz D L, Woods D E (2001) Identification of a *Burkholderia mallei* Polysaccharide Gene Cluster by Subtractive Hybridization and Demonstration that the Encoded Capsule is an Essential Virulence Determinant. Microb. Pathog. 30: 253-269.

[57] Yu Y, Kim H S, Chua H H, Lin C H, Sim S H, Lin D, Derr A, Engels R, DeShazer D et al. (2006) Genomic Patterns of Pathogen Evolution Revealed by Comparison of *Burkholderia pseudomallei*, the Causative Agent of Melioidosis, to Avirulent *Burkholderia thailandensis*. BMC Microbiol. 6: 46.

[58] Tuanyok A, Tom M, Dunbar J, Woods D E (2006) Genome-wide Expression Analysis of *Burkholderia pseudomallei* Infection in a Hamster Model of Acute Melioidosis. Infect. Immun. 74: 5465-5476.

[59] Cheng A C, Currie B J (2005) Melioidosis: Epidemiology, Pathophysiology, and Management. Clin. Microbio.l Rev. 18: 383-416.

[60] Mack K, Titball R W (1998) The Detection of Insertion Sequences within the Human Pathogen *Burkholderia pseudomallei* which have been Identified Previously in *Burkholderia cepacia*. FEMS Microbiol. Lett. 162: 69–74.

[61] Boulnois G J, Roberts I S (1990) Genetics of Capsular Polysaccharide Production in Bacteria. Curr. Top. Microbiol. Immunol. 150: 1–18.

[62] Egan, A M, Gordon, D L (1996) *Burkholderia pseudomallei* Activates Complement and is Ingested but not Killed by Polymorphonuclear Leukocytes. Infect. Immun. 64: 4952-4959.

[63] Densen P (1995) Complement. In: Mandell G L, Bennett J E, Dolin R, editors. Principles and Practice of Infectious Diseases. New York: Churchill Livingstone. pp. 58-78.

[64] Supputamongkol Y, Rajchanuwong A, Chaowagul W, Dance D A B, Smith M D, Wuthiekanun V, Walsh A L, Pukrittayakamee S, White N J (1994) Ceftazadime vs. Amoxicillin/Clavulanate in the Treatment of Severe Melioidosis. Clin. Infect. Dis. 19: 846-853.

[65] Patel N, Conejero L, De Reynal M, Easton A, Bancroft GJ, Titball RW (2011) Development of Vaccines against *Burkholderia pseudomallei*. Front. Microbiol. 2: 198.

[66] Brett P J, Woods D E (2000) Pathogenesis of and Immunity to Melioidosis. Acta. Trop. 74: 201-210.

[67] Warawa J, Woods D E (2002) Melioidosis Vaccines. Expert Rev. Vaccines 1: 477-482.

[68] Nierman W C, DeShazer D, Kim H S, Tettelin H, Nelson K E, Feldblyum T, Ulrich R L, Ronning C M, Brinkac L M, et al (2004) Structural Flexibility in the *Burkholderia mallei* Genome. Proc. Natl. Acad. Sci. USA 101: 14246-14251.

[69] Aldous P (2005) Tropical Medicine: Melioidosis? Never heard of it… Nature 434: 692-693.

[70] Sarkar-Tyson M, Thwaite J E, Harding S V, Smither S J, Oyston P C F, Atkins T P, Titball, R W (2007) Polysaccharides and Virulence of *Burkholderia pseudomallei*. J. Med. Microbiol. 56: 1005-1010.

[71] Heiss C, Burtnick M N, Wang Z, Azadi P, Brett P J (2012) Structural Analysis of Polysaccharides Expressed by *Burkholderia mallei* and *Burkholderia pseudomallei*. Carbohydrate Res. 349: 90-94.

Chapter 4

Mechanisms of O-Antigen Structural Variation of Bacterial Lipopolysaccharide (LPS)

Rosa Eugenia Reyes, Carolina Romo González,
Rafael Coria Jiménez, Maribel Ortiz Herrera and Alejandra Aquino Andrade

Additional information is available at the end of the chapter

1. Introduction

One of the most complex topics within bacterial anatomy and physiology is that of exopolysaccharides. These molecules have various structures and functions and also provide different types of advantages to their producing microorganisms, including surface variability, resistance to innate and acquired immunity mechanisms, the ability to adhere to different surface and cell types and resistance to antibiotic activity.

These bacterial systems are closely linked to the different genera and species that contain them. The organisation and expression of the genes that code for these external structures, genetic expression mechanisms and biosynthesis routes are extremely heterogeneous.

Although numerous classic microbial physiology and biochemical studies have focused on analysis of the external structures of microorganisms, not until recently has the study of exopolysaccharides become important, due to the role of exopolysaccharides in bacterial pathogenicity and the ecology of microbial populations and their possible role in the colonisation, residence and adaptation mechanisms in various ecosystems. Among the most important exopolysaccharides are those described below.

2. The glycocalyx

Although it can not be considered a bacterial structure, the glycocalyx is a heterogeneous set of exopolymers that have diverse biochemical compositions [1]. The exopolymers are located immediately adjacent to the microorganism wall and are present as hydrophobic gels that are weakly associated with the external bacterial structures.

The production and presence of these exopolymers provide bacteria with a high degree of surface hydrophobicity that enables interaction between cellular and inert surfaces and subsequent bacterial colonisation through the development of microcolonies and biofilms.

Furthermore, the presence of certain biopolymers is related to the ability to resist antibiotic action by capturing these compounds through periplasmic glucans. An example of this is resistance to tobramycin, which is captured by cyclic-β (1,3)-glucan [2].

3. Capsules

In contrast to the glycocalyx, a bacterial capsule is a well-defined external structure with a characteristic composition for each bacterial genus and species that provides a number of advantages, which are primarily related virulence, to the producing microorganism.

One of the classic examples of the importance of the capsule is provided by *Streptococcus pneumoniae,* which is a gram-positive diplococci agent that causes pneumonia, meningitis and septicaemia. The virulent strains of this species produce a capsule that inhibits opsonisation and phagocytosis. The composition of the capsule is dependent on the producing bacteria. In the case of *Escherichia coli* K1, the capsule is formed by polysialic acid, in *S. pyogenes,* by hyaluronic acid and in *Streptococcus* group B by sialic acid. The biochemical composition of the capsules may be extremely diverse, which gives rise to a great amount of antigenic variability and presents a problem for the immunological mechanisms of the host in recognising these organisms [2].

It has been documented that the capsule participates in bacterial adhesion mechanisms and that its synthesis is stimulates by low stress conditions, such as the presence of serum, low Fe^{++} concentration and high CO_2 tensions. Although in certain microorganisms, the presence of a capsule is discreet, in others, such as *Cryptococcus neoformans,* the capsule plays an essential role in the mechanisms for aggression against the host.

4. Biofilms

Because of the production and exportation of bacterial exopolymers, the strains increase their degree of surface hydrophobicity, which facilitates interaction, adsorption and residence on a wide range of surfaces that, in principle, hinder bacterial colonisation. Regarding the production of biofilms, it has been documented that each bacterial genus and species responds to different signals from the environment and the host, as is the case of induction by tobramycin and the response capacity of the *quorum sensing* mechanisms. One of the most studied examples of the formation of bacterial biofilms is the case of *Pseudomonas aeruginosa,* which is an opportunistic pathogen that is associated with infections in immunocompromised hosts, such as cystic fibrosis (CF) patients. In CF patients colonised by *P. aeruginosa,* the bacteria exhibits two colonial phenotypes. The first phenotype is associated with the production of alginate (a polymer of mannuronic acid and glucuronic acid that forms a viscous gel around the bacteria), and the second phenotype is rough and is related to a lack of alginate. The production of alginate is a marker of virulence in which the producing strains are more aggressive than the non-producing strains. Additionally, alginate provides bacteria with the ability to form microcolonies and biofilms [3].

In the relationship between *P. aeruginosa* and the infected patient, the host exerts different types of selective pressure that favour the persistence of the mucoid strains with elevated alginate production. This biosynthesis is regulated by different bacterial signalling systems (quorum sensing) that detect changes in the host environment and modulate the metabolism to adapt. To date, it is not known precisely which signals in hosts with CF favour and allow for a predominance of mucoid strains compared to non-mucoid strains. *In vitro* studies, it has been demonstrated that both bacterial phenotypes exhibit similar behaviour and that successive passages in culture media inhibit the expression of the non-mucoid phenotype [4,5].

The physiology of bacteria that are found in biofilms is heterogeneous and depends on the specific site that the microorganism occupies in the microcolony. Nutrient gradients occur from the surface of the biofilm to the most internal parts, thereby influencing the bacterial physiology and consequently modifying the speed of growth, the generation time, the susceptibility to antibiotics (due to factors such as the presence of a diffusion barrier to antibiotics), the antigenic variability of individuals, the susceptibility to opsonisation and phagocytosis and even the alginate functions as a negative immunomodulator for the host [4].

Another example of the formation of biofilms is that produced by bacteria of the genus *Staphylococcus*, predominantly the coagulase-negative forms in which the production and excretion of biopolymers is capable of increasing cell surface hydrophobicity and, through hydrophobic interactions, adhering to surfaces. These mechanisms enable the bacteria to enter hosts, who subsequently require invasive procedures or therapeutic procedures, such as catheters. In the case of the bacterial populations that colonise the oral cavity, heterogeneous populations participate in the biofilms, of which *Streptococcus mutans* has a fundamental role in adherence because of its capacity to produce biofilms, thereby providing a high degree of surface hydrophobicity. This hydrophobicity allows the bacteria to adhere to dental enamel and begin the process of colonisation and the subsequent adherence to an increasingly heterogeneous bacterial population, which may ultimately cause harm and generate a cariogenic process.

5. Gram-negative lipopolysaccharide (LPS): Structure and function

LPS is essential in the structure and function of the external membrane of gram-negative bacterial cell walls. LPS intervenes in the transportation of hydrophobic molecules to the interior of bacterial cells and are an essential factor in host-microorganism interactions.

LPS is an amphipathic glycoconjugate that constitutes 10% to 15% of the total molecules in the external membrane and represents 75% of the total of bacterial surface [1]. There are three different LPS domains: a) Lipid A, which is the domain that is anchored to the membrane and the hydrophobic and endotoxic portions of the structure; b) The core oligosaccharide, which is the domain that connects lipid A to antigen O and is divided into the inner core and the outer core. The inner core is joined to lipid A and consists of unusual monosaccharides, including 2-keto-3-deoxy-octanoate (Kdo) and L-glycero-D-mannoheptose.

The outer core is joined to the O antigen and is made of common sugars such as hexoses and hexosamines [2]; and c) The O polysaccharide, which is the hydrophilic and immunodominant domain of LPS and is an oligosaccharide of repeated units that is projected from the core toward the exterior of the bacterial surface.

The O antigen has a polysaccharide chain that varies in length with up to 40 repeated units of dideoxyhexoses. At least 20 different sugar molecules may compose the O antigen, including molecules that are rarely found in nature, such as abequose, colitose, paratose and tyvelose. These components are strain-specific. The O antigen displays a large degree of inter-species and intra-species variation, which is related to the nature, order and union of the different sugars (**Figure 1**) [3].

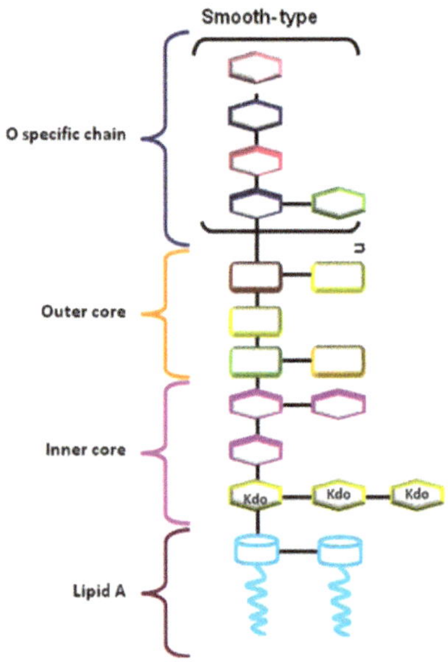

Figure 1. Schematic representation of lipopolysaccharide structure. Smooth-type (left), Semi-Rough type with only one O-chain subunit (center) and rough-type (right).

The O antigen is the immunodominant part of LPS and therefore is the easiest target for the humoral response of the host. For this reason, the O antigen is the basis for the serological classification of gram-negative bacteria. The O antigen is recognised by the innate immune response and participates in complement activation and in the inhibition of the formulation of the complex that attacks the membrane [6,7].

6. Biosynthesis of LPS

LPS is the primary component in the surface of gram-negative bacteria. The synthesis of LPS structures, which consist of lipid A, the core and antigen O, begins in the cytoplasm, where these structures are assembled. The structures are translocated to compartments such as the periplasm until the final destination is reached, which is the surface of the external membrane. The synthesis process has been widely studied in *E. coli* and *Salmonella.* The biosynthesis and exportation pathways of LPS are common among the majority of gram-negative bacteria, as is explained below (**Figure 2**). However, unique characteristics may exist in certain bacteria with respect to the types of enzymes and particular pathways.

The formation of lipid A is carried out in the internal face of the cytoplasmic membrane, and nine enzymes participate: LpxA, LpxC, LpxD, LpxH, LpxB, LpxK, KtdA, LpxL and LpxM. The biosynthesis of LPS begins with the formation of uridine diphosphate-diacyl- *D*-glucosamine (UDP-diacyl-GlcN) from uridine diphosphate *N*-acetylglucosamine (UDP-GlcNAc). This reaction is catalysed by the enzymes LpxA, LpxC and LpxD and results in two 3-OH fatty acid chains in the 2 and 3 position of UDP-glucosamine (UDP-GlcNAc) to form UDP-2,3-diacyl- glucosamine (UDP-diacyl-GlcN). Subsequently, this molecule is hydrolysed by LpxH to form lipid X, the enzyme LpxB condenses the lipid X and its precursor UDP-diacyl-GlcN to form disaccharide-1-P and the enzyme LpxK phosphorylates this molecule at the 4 position of the disaccharide-1-P molecule to form the lipid IV_A. Subsequently, KdtA incorporates the waste from 3-deoxy-D-manno-octulosonic acid (Kdo) in the 6′ position of lipid IV_A using the nucleotide cytidine monophosphate-3-deoxy-D-manno-octulosonic acid (CMP-Kdo) as a donor to produce Kdo_2 – lipid IV_A, which is exposed to more reactions catalysed by LpxL and LpxM to form Kdo_2 – lipid A. The enzyme LpxL adds a second lauryl, and LpxM adds a residue of myristoyl to the distal glucosamine unit.

Notably, the acyltransferases – LpxA, LpxD, LpxL and LpxM – selectively catalyse the different substrates and employ different acyl donors. For the first steps of the synthesis pathway of lipid A, the enzymes LpxA, LpxB and LpxD are required, with 3R-hydroxyacyl - Acyl Carrier Protein (3R-hydroxyacyl-ACP) serving as a donor. This compound is dehydrated by FabZ to form trans-2-acyl-ACP, which is also used as a donor of fatty acids in the biosynthesis of phospholipids. The synthesis of other LPSs in bacteria, such as *Neisseria meningitidis*, also involve trans-2-acyl-ACP[8]. Importantly, the structure of lipid A is the most conserved compared to the structure of the core oligosaccharides and antigen O [9,10].

6.1. The core oligosaccharides

The assembly of lipid A from the core oligosaccharides (Kdo_2 – lipid A) is the next step in the synthesis of LPS. This step is performed on the cytoplasmic surface of the internal membrane by glycosyltransferases, which are associated with the membrane and with nucleotide sugars as donors.

The core oligosaccharides normally contains 10 to 15 monosaccharides and may be divided into two structural regions, which are the inner core and the outer core, which are ultimately connected to lipid A and antigen O, respectively, in the final structure of LPS. The inner core contains residues of Kdo and Hep (L-glycero-D-manno-heptose). Kdo is the most conserved component in the nuclear region of the LPS. In contrast, the outer core is more variable, depend on the strain. However, the vertebral column of the oligosaccharide is typically composed of six units, and upon joining with other units, the column forms structures. The sugars commonly found in the core oligosaccharides are D-glucose, D-galactose, Kdo and Hep [9,10].

6.2. The O antigen

The majority of O antigens are heteropolymers, although a portion of O antigens may be composed of a single monosaccharide. The synthesis of the O antigen is performed in the same location as the core oligosaccharides, and this synthesis also uses nucleotide sugars as donors. In the majority of bacteria, a cluster of genes known as *rfb* codes the enzymes necessary for 1) the biosynthesis of the nucleotide sugars of antigen O, 2) the transfer of the sugars to form the polysaccharide chain (glycosyltransferases) and 3) the assembly and transfer of antigen O toward the periplasm. The synthesis routes of the nucleotide sugars are grouped according to the nucleotide that bonds to the sugar, which may be CDP, UDP, dTDP or GDP. Antigen O may be a homopolymer or a heteropolymers, and the sugars may be formed linearly or in a ramification. Glycosyltransferases may be grouped according to their function, and they carry undecaprenyl phosphate, which is also used for the synthesis of capsular polysaccharides and peptidoglycans.

The following hypotheses have been proposed regarding the assembly and transfer process of antigen O: a) a pathway dependent on Wzy, which is the prototype system; b) a pathway dependent on ABC transporters, which are typically used by linear polysaccharide structures; c) a pathway dependent on synthase, which involves glycosyltransferases capable of synthesising within a single polypeptide and is an uncommon pathway and finally d) seroconversion reactions, in which the addition of acetyl residues or glucose residues modifies antigen O. Within the prototype pathway dependent on Wzy, in bacteria such as *Salmonella enterica* and *E. coli*, a multi-step process occurs. When the lipid A-core and the O antigen are synthesised, they are transported to the periplasm. The protein MsbA, which displays homology with MDR (multi-drug resistant) eukaryote proteins, transports the lipid A-core, and Wzx transports the O antigen, which was previously polymerised by the proteins Wzy and Wzz. With the help of WaaL, the structures of the lipid A-core and the O antigen are assembled, finally producing the LPS [10].

6.3. LPS and its transportation to the external membrane

When an LPS is formed, it must pass through the periplasmic space to reach the external membrane [9,10], and this process is facilitated by protein LptA (periplasmic), LptB (cytosolic), LptC, LptF, LptG (internal membrane) and LptD and LptE (external membrane). Several of these proteins act in complexes. For example, in the case of the transporter ABC, LptBFG and LptA and LptC translocate the LPS to the internal side of the external membrane such that the proteins LptD and LptE place it on the surface of the membrane. It has been observed an absence of LptA or LptB or both causes the accumulation of LPS in the periplasm [11–17].

In the majority of bacteria, the genes that code for the enzymes involved in the biosynthesis of the O antigen are found in clusters. However, in the case of *Helicobacter pylori,* which is a pathogenic bacteria of the human stomach, these enzymes are found distributed throughout its chromosome, which probably contributes to the fact that the assembly pathway of its LPS has not been completely characterised. However, several enzymes that participate in the synthesis of *H. pylori* LPS has been identified and characterised, including several glycosyltransferases [18,19]. The glycosyltransferase WecA and the ligase WaaL also participate in the biosynthesis of *H. pylori* LPS. However, translocases are typically not involved in the translocation of the O antigen to the periplasm for its assembly with the lipid A-core structure [20]. Only the participation of a translocase named Wzk, which directs N-glycosylation in other bacteria, has been observed, and this fact suggests that the translocase Wzk of *H. pylori* could indicate an evolutionary connection between the biosynthesis pathways of LPS and glycoproteins [18]. Recently, it has been observed that there is an analogy and homology between the biosynthesis of LPS and the biosynthesis of glycoproteins in other bacteria, such as *S. enterica, P. aeruginosa* PAO1, *Neisseria* spp., *Paenibacillus alvei, Campylobacter jejuni and E. coli O8.* This homology could have enormous biotechnological potential. However, further studies are required to confirm this fact [21].

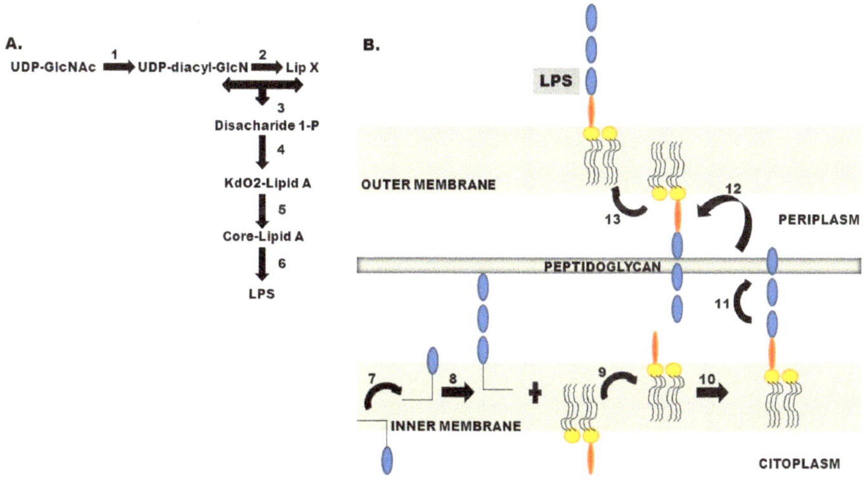

Figure 2. The biosynthetic pathway and transport of lipopolysaccharide. **A.** A representation of the biosynthetic pathway of the structures of LPS in *E. coli.* **B.** The assembly and transport of the LPS; the antigen O is assembled to the structure core-lipid A in the periplasm and is later transported toward the outer membrane. The names of the enzymes involved in these processes are: 1. LpxA, LpxC, LpxD; 2. LpxH; 3. LpxB; 4. LpxK, KdtA, LpxL, LpxM; 5. Glycosyltransferases; 6 and 10. WaaL; 7. Wzx; 8. Wzy, Wzz; 9. MsbA; 11. Lpt B C F G; 12. LptA; 13. Lpt D E. The lipid A (Yellow circles), the core (Red ovals) and antigen O (Blue ovals)

7. Regulation of the expression of LPS

The biosynthesis of LPS is performed through two separate pathways. One pathway involves the formation of lipid A and the core, and the other pathway involves the formation of the O antigen. In the synthesis of LPS, a large number of genes participate, many of which are part of clusters located in different regions of the bacterial chromosome and, in some organisms, in plasmids [22,23].

7.1. Regulation of the biosynthesis of lipid A

Lipid A and the core oligosaccharide are formed in a continuous process, which is separate from the synthesis of the O antigen. In the majority of *Enterobacteriaceae*, the genes involved in this synthesis are found in a single copy and share several characteristics with pathogenicity islands (PAI) [10,24].

The genes involved in the first steps of the biosynthesis of lipid A in *E. coli, S. enterica* serovar Typhimurium, *Yersiia enterocolitica, Haemophilus influenzae* and *Rickettsia rickettsii* are grouped into the *lpx*D-*fabZ*-*lpx*A-*lpx*B cluster[8]. The genes *lpx*A and *lpx*D code for N-acyltransferases, which add fatty acids to the glucosamine disaccharide. Both enzymes contain a conserved repeated structure, which is the hexapeptide $[(I,V,L)GXXXX]_n$. The gene *lpx*B is a co-transcript with *lpx*A and codes for the disaccharide synthase of lipid A, which catalyses the formation of the disaccharide of lipid A from UDP-2.3-diacylglucosamine and 2,3-diacylglucosamine-1-phosphate. The gene *fab*Z codes the enzyme that catalyses the dehydration of (3R)-hydroxyacyl-ACP to trans-2-acyl-ACP, which is used as a donor of fatty acid in the biosynthesis of phospholipids [10].

The proteins involved in the biosynthetic pathways of UDP-GlcNAc, UDP-Glc and UDP-Gal are coded in constitutive genes.

L-glycero-D-manno-heptose is added to its derivative ADP, which is synthesised from sedoheptulose 7-phosphate in four steps. The genes *gmhA* and *gmhD* code for enzymes in the first and last steps. The G+C content of the *gmhA, gmhD* and *waaE* genes are 51%, 51% and 52.7%, respectively.

Kdo is transferred from CMP-Kdo and synthesised from arabinose-5P and PEP by a three-stage pathway[9]. Two of the genes in this process, namely, *kdsA* and *kdsB,* are well-characterised and have G+C contents of 51.6% and 52.7%, respectively.

The gene *waaA,* which is located in the *waa* cluster, codes for a bifunctional Kdo transferase that adds residues from 2 Kdo. These genes have a G+C content of between 51% and 54%. Several of the modifications of lipid A are regulated by the concentration of Mg^{2+} through the regulon *phoP-phoQ* [25].

PhoP-PhoQ is a two-component system that regulates virulence through adaptation to limited magnesium environments and regulates numerous cellular activities in gram-negative bacteria. This regulon consists of an external membrane sensor, PhoQ, and a cytoplasmic regulator, PhoP, and is activated by the acidic pH and by certain antimicrobial peptides (APs). PhoP-PhoQ is repressed by millimolar concentrations of magnesium and calcium. PhoQ senses the concentration of magnesium and of APs throughout a periplasmic domain, which undergoes a conformational change when it is joined to these compounds and results in autophosphorylation. The activation of the PhoP-PhoQ system may allow for the activation or repression of 40 genes [26].

The regulon PrmAB of *Salmonella* is also a two-component system coded in the operon *pmrCAB*, which has protein products that include a phosphoethanolamine phosphotransferase (PmrC), a response regulator (PmrA) and a kinase sensor (PmrB). It has been confirmed that PmrAB regulates more than 20 genes in *Salmonella*. However, several studies suggest that more than 100 genes show activity[27]. One of the primary roles of the activation of PmrAB is the modification of LPS, such as the addition of Ara4N to lipid A, which, as explained below, impacts the susceptibility to some antimicrobial agents and the addition of phosphoethanolamine (pEtN) in the core of the LPS [26].

7.2. Regulation of the biosynthesis of the core

In *E. coli*, *Salmonella* and *K. pneumoniae*, the genes involved in the biosynthesis of the core are grouped on the chromosome. These loci code for the activities required in the assembly of the outer core and also code for the transferases necessary for the synthesis of the inner core. In *E. coli* and *Salmonella*, the *waa* locus (also called *rfa*) is formed by three operons and is located between the genes *cysE* and *pyrE*. The operons are defined by the name of the first gene in each transcriptional unit, such as *gmhD*, *waaQ* and *waaA* (**Figure 3**) [28].

Genes in the *waa* operon code for all of the transferases that assemble the core, including the gene that codes for the enzyme for the last step of the synthesis of ADP-L-Glycero-D-manno-heptose and the gene of the ligase of the O antigen, *waaL*. In addition, the *waa* operon contains the gene *waaA*, which codes for a bifunctional transferase Kdo.

The operon *gmhD*, which is located on the extreme 5′ end of the *waa* cluster, contains *gmhD*, *waaF*, *waaC* and *waaL*. The genes *gmhD-waaFC* are required for the biosynthesis and transfer of L, D-heptose. GmhD catalyses the last reaction of the synthesis of ADP-L-Glycero-D-manno-heptose. WaaC is the transferase for Hepl, and WaaF is a putative transferase of Hepll. WaaL is the ligase that bonds the O polysaccharide to the lipid A and core.

The transcription of the operon *gmhD* in *E. coli* K12 is regulated by heat shock promoters, indicating a requirement for the heptose domain of LPSs for growth at high temperatures [23,28,29].

The central operon *waaQ* contains 10 genes that are necessary for the biosynthesis of the outer core and for the modification of the core. WaaQ is the transferase for HepIII. The proteins WaaG, WaaO and WaaR are transferases for Glcl, Glcll and GlcIII, respectively, and WaaB is the transferase for the GaL residue. WaaP and WaaY are involved in the phosphorylation of the residue of heptose, whereas the functions of WaaU, WaaS and WaaZ are still not clear. In isolates of *E. coli* with the R1 and R4 cores, this operon contains the structural gene of ligase *waaL*, which must be produced for the union of the O polysaccharide with the complete core.

The *waaQ* operon is preceded by a JUMPStart sequence (Just Upstream of Many Polysaccharide-associated gene Starts), which includes a conserved region of 8 bp that is known as *ops* (operon polarity suppressor). The expression of the cluster for the biosynthesis of the core is regulated by the protein RfaH and also in response to thermal shock. RfaH is homologous to the factor NusG, which regulates the expression of the operon of hemolysin, genes of the polysaccharides of the capsule and genes for the transfer of the F plasmid. Regulation by RfaH occurs at the level of the polymerisation of the mRNA and depends on *ops* sequences that act in *cis*, as in the case of the operon *waaQGPSBIJYZK*, which includes 10 genes of the cluster of the core of LPS [22,30].

The transcript *waaA* contains the structural gene *waaA* (formally called *kdtA*), which codes for the transferase Kdo and a "non LPS" gene that codes for the adenylyltransferase pantetheine (*coaD*, formally *kdtB*) [31].

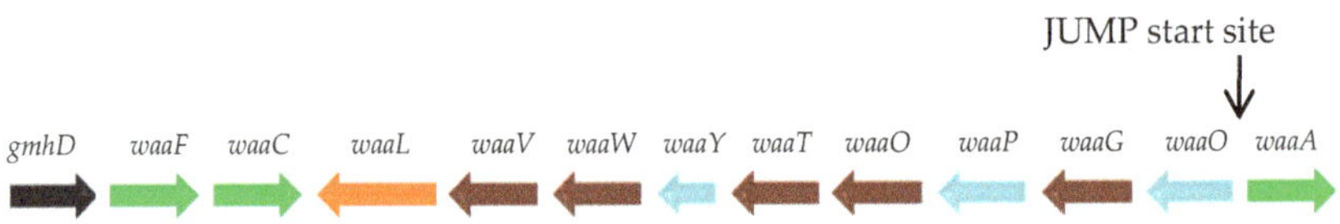

Figure 3. Structure of the genes involved in the synthesis of the core of *E. coli* R1. The genes of the glycosyltransferases that form the bonds of the inner core are shown in green. Those genes of the enzymes that modify the structure are in red, and the glycosyltransferases of the outer core are in blue. In orange are genes that modify the ligase enzyme.

7.3. Regulation of the biosynthesis of the O polysaccharide

The genes involved in the biosynthesis of the O antigen are generally found in the chromosome in the cluster of the O antigen or *rfb*. These genes have a lower GC content than the average for genomes, between 30% and 40%, which provides evidence that these genes have been acquired through lateral inter-species transference [24].

The genes that code for the proteins that participate in the synthesis of the O antigen form three main groups: a) proteins involved in the biosynthesis of the precursors of nucleotide sugars of the O antigen; b) protein glycosyltransferases, which sequentially transfer various precursor sugars to form an oligosaccharide of a lipid carrier, undecaprenyl phosphate (UndP), which is located in the cytoplasmic face of the internal membrane and c) genes for the processing of the O antigen, which are involved in the translocation through the membrane and polymerisation (**Figure 4**) [32,33]. A fraction of O antigens includes acetyl-O groups, and others include residues; therefore, in the corresponding clusters, the transferases for them are coded. The differences among the many forms of the O antigen are due to the genetic variation in the cluster of the O antigen. The genes for the initial steps, which are also involved in conserved functions, do not duplicate in the cluster of the O antigen [24].

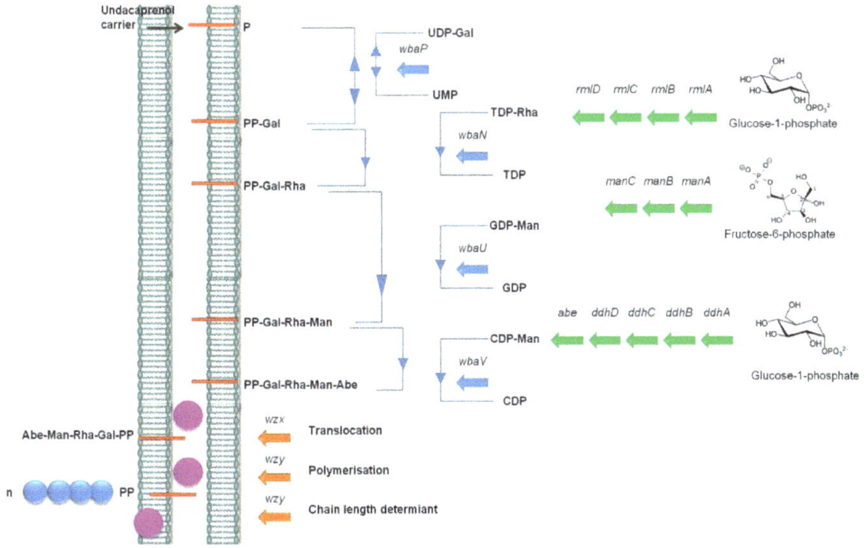

Figure 4. The synthesis of the O antigen involves three types of genes: a) genes related to the biosynthesis of precursors of sugar (green boxes), whose protein products perform their function in the cytoplasm; b) genes for glycosyltransferases (blue boxes), whose corresponding proteins transfer nucleotide sugars in the UndP lipid to form the O unit. This process occurs on the cytoplasmic side of the internal membrane and c) genes for the assembly of the O antigen and its exportation (orange box). Modified from [38].

In *E. coli* and *S. enterica,* the genes involved in the synthesis of the O antigen are typically found among the constituent genes *galF* and *gnd*. In *P. aeruginosa,* these genes are found among *himD* and *tyrB* [15]). In *V. cholerae,* O antigen synthesis genes are found among *gmhD* and *rjg* [34], and in *Yersinia* spp., these genes are found among *hemH* and *gsk.* Some exceptions, such as the polysaccharide O54 of *S. enterica*, which is a cluster that is coded in a plasmid [35] (**Figure 5A**).

The genes of the biosynthetic pathway of the three precursors of the nucleotides of sugar are grouped within the genetic cluster of the O antigen. *manB and manC* code for enzymes that convert mannose-6-P into GDP-mannose. The operon *rmlABCD* codes for enzymes that form TDP-rhamnose from glucose-1-P. *ddhABCD* and *abe* code for enzymes to make CDP-abequose from glucose 1-P. UDP-Gal is used in other pathways and is synthesised by conserved enzymes. The transferase galactose, which is coded by *wbaP*, initiates the synthesis of the O units by transferring galactose phosphate from UDP-Gal to UndP.

The transferases coded by *wbaZ*, *wbaW* and *wbaQ* are positioned over two residues of mannose and one of rhamnose before the residues of mannose, which is acetylated by WbaL. The residue of abequose is bonded to acetyl rhamnose by WbaR to form a complete O unit. The flippase of the bond of the O antigen is coded by *wzx* and is responsible for moving the O unit bonded to UndP through the internal membrane before the polymerase Wzy exerts its action over the long-chain O antigen.

The majority of the O antigen operons are constitutively expressed and are preceded by a sequence of 39 bp, known as JUMPStart **(Figure 5B)**.This sequence includes two elements known as *ops*, which are involved in the recruiting of elongation factors during the transcription. The JUMPStart region is controlled by RfaH, which acts as a positive regulator for the genes involved in the outer core of LPS and also increases the expression of the locus *rfb* [1,18,19.]

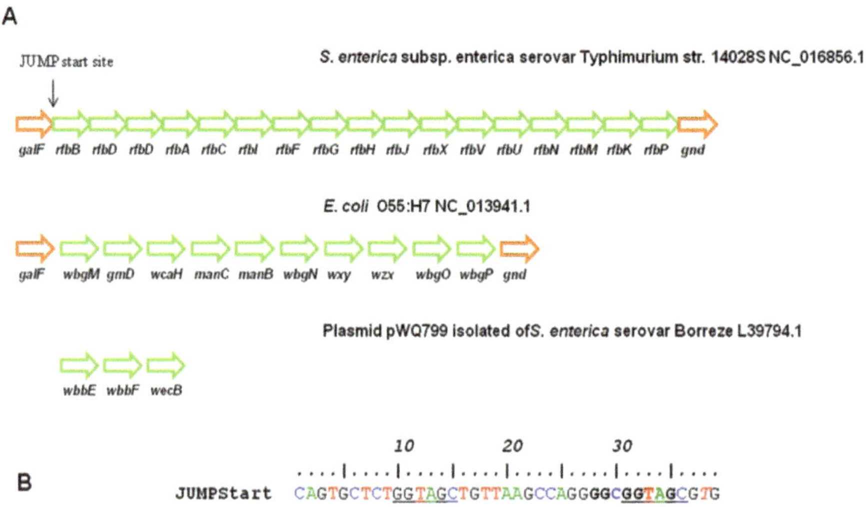

Figure 5. A. The location of the genes involved in the synthesis of the O antigen in several species of *Enterobacteriaceae*. **B.** The JUMPStart sequence of *S. flexneri* strain IB1833. The *ops* sequence is indicated in bold for eight base pairs. A residue of thymine that is highly conserved in the far 3′ is indicated by cursive text, and repeated sequences are underlined. Accession number: U118168.1.

8. Variability of the LPS

In bacterial pathogens, the most variable structures are those that are expressed on the cell surface. Intra-species genetic individuality has inheritable traits that may be observed in a sample of individuals. The genetic variability is derived from the combination of a certain number of genes, which each exists as a family of alleles that differ in structure and function. An example is the family located in the biosynthetic locus of the O antigen, and an example of this phenomenon at the population level is the hypervariability in the structure of this biomolecule [39]. Among the components of the surface of the gram-negative bacteria, LPS is the primary constituents of the external membrane, which is a heterogeneous surface that is significantly involved in the process of the microorganism adapting to its environment.

LPS primarily consists of conserved segments, such as lipid A and the core, and secondarily consists of a hypervariable segment, which is the O antigen. The conserved domains of the LPS are shared regions among bacterial species, which intervene in the development and in the survival of the bacteria. The O antigen may display modifications such alterations in the length of the oligosaccharide chain and changes in the surface composition and in the chemical configuration, due to the addition of glycosyl or fucosyl groups or even non-hydrocarbonated substitutes, such as acetyl or methyl groups, which could affect the cellular structure [40].

8.1. Intra-species variability in the composition of the O antigen

The heterogeneity in the expression of LPS may provide a medium to discriminate among the bacterial species. This heterogeneity is responsible for the well-known ladder profile that can be detected in silver-stained SDS-PAGE gels (**Figure 6).** This method is used to determine the number and repeated units of oligosaccharides that constitute the O antigen, which has been useful in epidemiological studies [41]. The smooth strains contain the entire LPS, whereas the semi-rough strains have a subunit of the O antigen, and the rough strains lose the subunits of the O antigen.

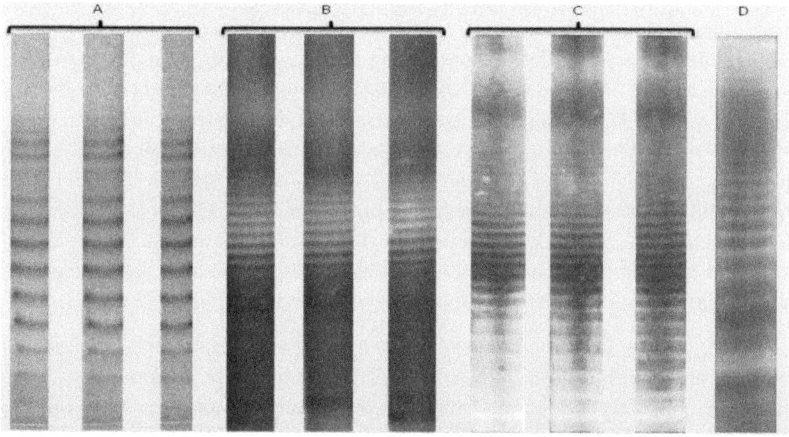

Figure 6. LPS profiles in SDS-PAGE. The smooth forms of LPS display a ladder profile that is strain-specific. **A)** *C. freundii* E9750; **B)** *S. senftenberg* 74210; **C)** *E. coli* O157:H7; **D)** *S. marcescens* biotype TC.

In **Table 1**, examples are presented of various sequences from different genera and species, which mark the variability of intra-species and inter-species LPS that impedes the immune response control of bacterial infections, among other issues.

Strain	Antigenic unit
E. coli O157 *S. enterica* O30 *C. freundii* F90	[D-PerNAc $\xrightarrow[1,3]{\alpha}$ L-Fuc $\xrightarrow[1,4]{\alpha}$ D-Glc $\xrightarrow[1,3]{\beta}$ D-Gal2NAc] $\xrightarrow[1,2]{\alpha}$
E. coli O55 *S. enterica* O50	[D-Gal $\xrightarrow[1,3]{\alpha}$ D-Gal2NAc $\xrightarrow[1,6]{\beta}$ D-Glc2NAc] $\xrightarrow[1,3]{}$ (D-Glc2NAc ↑ β 1,3 from D-Gal; Col $\xrightarrow[1,2]{\alpha}$ D-Gal)
E. coli O111	$\xrightarrow[4]{}$ [D-Glc $\xrightarrow[1,4]{\alpha}$ D-Gal $\xrightarrow[1,3]{\alpha}$ D-GlcNAc] $\xrightarrow[1]{\beta}$ (Col ↓ α 1,3 to D-Glc; Col ↑ α 1,6 to D-Glc)
E. coli 113	$\xrightarrow[4]{}$ [D-GalA $\xrightarrow[1,3]{\alpha}$ D-Gal $\xrightarrow[1,3]{\alpha}$ D-GlcNAc $\xrightarrow[1,4]{\beta}$ D-GalNAc] $\xrightarrow[1]{\beta}$ (D-GalNAc ↑ β 1,3 from D-Gal)
S. enterica E1	$\xrightarrow[6]{}$ [L-Man $\xrightarrow[1,4]{\beta}$ L-Rha $\xrightarrow[1,3]{\alpha}$ L-Gal] $\xrightarrow[1]{\alpha}$

D-Col, D-colitose; L-Fuc, L-fucose; D-Gal, D-galactose; D-Gal2NAc, 2-*N*-acetylgalactosamine; D-Glc, D-glucose; D-Glc2NAc, D-2-*N*-acetylglucosamine; D-Per, D-perosamine; D-Per4NAc, D-4-*N*-acetylperosamine.

Table 1. Examples of different structures of the bacterial O antigen.

An increase in temperature during microbial growth causes changes in the concentration of carbohydrates in LPS, which modifies their composition [42]. In a study on *in vitro* passages of the *C. freundii* E9750 strain cultivated under different temperatures, variability was observed in the ladder profiles of the O antigen, in the concentration of carbohydrates and in the agglutination reactions with the anti-O serum of *C. freundii* E9750 and *Salmonella senftenberg* 74210 [43]. According to the ladder profiles of the isolates of *C. freundii* E9750, six profiles were distinguished: A (control strain), B, C, D, E (immunoreactive isolates) and F (isolates that had lost the immunoreactive chain). These profiles were generated according to their similarity. Each LPS profile displayed a typical ladder profiles (**Figure 7).**

The different profiles from the isolates obtained from the sub-cultures of *C. freundii* E9750 were out of phase with respect to the control. This result may be explained by variation in the number of oligosaccharide units present in the O antigen. However, when the length of the chain is increased, the differences are more marked, and the bands of the ladder profiles are out of phase, as Lawson et al., report with strains of *S. enterica* serovar Typhimurium [41]. There are other reports in the literature on the variability of the length of the O antigen chain as a response to change in temperature, as in strains of *P. aeruginosa* and enteric bacteria [44,45].

Several isolates of *C. freundii* E9750 displayed cross-reactivity in agglutination tests with the anti-O serum of *S. senftenberg* 74210, which suggests changes in the conformation of the epitopes that may be associated with the addition of glucosyl groups or residues derived from N-acetyl [46].

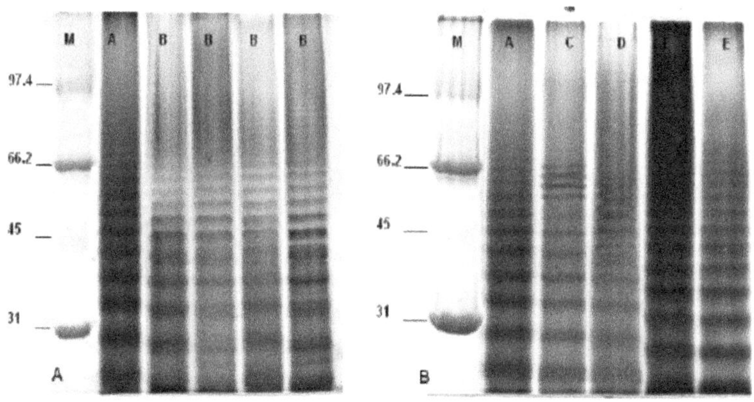

Figure 7. LPS profiles in SDS-PAGE stained with silver. Profiles of colonies of *C. freundii* E9750. A) isolates at a temperature of 42°C; B) isolates at a temperature of 37°C. M= weight markers.

The isolates of *C. freundii* E9750 that display variability suggest heterogeneity in the composition or conformation of the epitopes of the O antigen. The cross-reactivity of the isolates of *C. freundii* and *S. senftenberg* 74210 is associated with the specificity of the epitopes of the O polysaccharide that these species share. The addition at O antigen of glucosyl groups or derived N-acetyl residue could be involved in these cross-reactions.

8.2. Heterogeneity of populations and phase variation

Phase variation is used by various bacterial species to generate diversity within a population. Phase variation is a process of change in the expression of the epitopes of the cellular surface of the bacteria [19]. Bacterial cells may phenotypically vary even within a clone population, which allows them to adapt to their environment or even to evade the immune response of the host. Phase variation is a phenomenon that generates phenotypic heterogeneity within a population by means of gene regulation, which changes genes from a state of expression in which they are "turned on" to a state of non-expression in which they are "turned off". The state of expression is inheritable, reversible and affects the same phenotype.

Antigenic variation is referred to as the expression of an alternative form of an antigen of the cellular surface, such as polysaccharides, lipoproteins and type IV pili, which at the molecular level, share characteristics with the phase variation mechanisms. During this adaptation process, the bacteria display reversible phenotypic changes as a result of genetic changes or epigenetic alterations in a specific locus. The mechanisms which allow for phase variation are: genetics (Slipped-strand mispairing, recombination) and epigenetic (DNA methylation) [47,48].

a. Genetic mechanisms

Numerous studies have been performed to reveal the genetic basis of the variation of the O antigen. In certain cases, the variability in the expression of the genes is regulated by elements in *cis*, which cause changes in the composition of the structure of the antigens of the bacterial surface. Certain pathogens change the structure of the O antigen through the acquisition of phage genetic material followed by recombination processes [7].

i. Slipped-strand mispairing

One of the mechanisms that regulate phase variation at the molecular level is the slipping of one of the DNA strands, which causes a mispairing between the daughter strand and the parent strand during the replication of the DNA. This process is known as slipped-strand mispairing (SSM). The genomic sequences susceptible to SSM are those which contain short repetitions, microsatellites or a variable number of in tandem repetitions, which may cause a change in the expression of genes at the level of the transcription processes or translation according to the location of the repeated sequence in relation to the promoter and the codifying sequence [49]. At a transcriptional SSM, this may lead to the activation or deactivation of the promoter region of the target gene, as occurs in *H. influenzae (hif*A/B). At a translational level, SSM may affect the codifying region, as for example, with the genes involved in the biosynthesis of the LPS of *H. influenzae* and *Neisseria* spp. [50].

Within the genome of *H. pylori*, certain loci have been identified with repeated sequences of a single nucleotide or a pair of nucleotides. Several of these repetitions are within the open reading frames (ORFs) (**Figure 8**). In the transcription process, the mispairing between nucleotides, when one of the strands of DNA slips over another chain, causes the "gain" or "loss" of a unit in the reading frame, which leads to the loss of the start codon or mutations in the proteins. Therefore, SSM increases the genetic variability of *H. pylori*. Similar repetitive sequences have been found in other microorganisms, such as *H. influenzae* [51].

One group of genes that generate phase variation are those that code for enzymes that intervene in the biosynthesis of LPS, which may cause variants of the gene product in the same bacterial population. The LPS of the majority of *H. pylori* strains contains complex carbohydrates known as the Lewis antigen. Type 1 (Le^a,Le^b) and type 2 (Le^x,Le^y) Lewis antigens are epitopes of fucosylated oligosaccharides. At least 80% of the strains of *H. pylori* express type 2. Some research on the antigenic determinants involved in the biosynthesis of the Lewis antigen have allowed for the identification of the fucosyltransferases (FucTs) that are involved in the formation of these antigens [19].

The genes that code for the FucTs have elements in *cis* that are differentiated by containing poliA and poliC sequences of different lengths that mediate SSM. The size of these sequences regulates the activation and deactivation of the genes of the FucTs. However, in some cases, such as in *H. pylori* UA948, the inhibition of the expression of *futB* is due to mutations outside of the hypervariable region (the elimination of 80 nucleotides in the promoter region) [52].

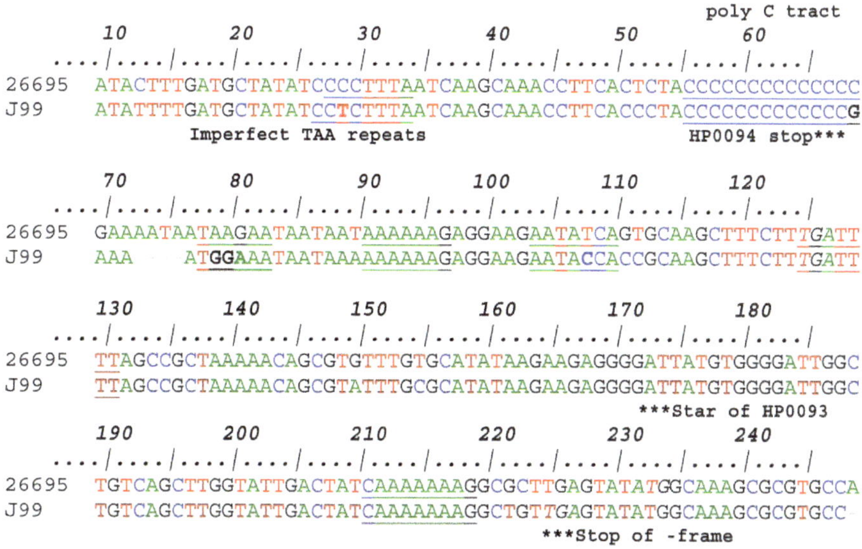

Figure 8. The nucleotide sequence of the central region of the Hp *fuc*T2 gene. The sequences show the characteristics (simple repetitions) responsible for the ORF in the *H. pylori* J99 strain and the 26695 type variant. Due to the number of different repetitions of the residues of poli C, start sequences of an ORF of *fucT2* of the 26695 strain are found in the TGA stop codon (marked with asterisks) shortly after reading 1 (HP0093), which is the same as the marker of the *fuc*T J99 reading. The three supposed motifs X XXY YYZ are highlighted in bold and are underlined.

The α 1,2-fucosyltransferase (FutC) catalyses the addition of fucose in the conversion process of Le^X to Le^Y. Sanabria-Valentín et al., through *in vitro* and *in vivo* studies, confirmed the main function of *futC* slipped-strand mispairing in the variation of the Le antigen [53].

The *futC* gene includes an internal Shine-Dalgarno type sequence and a heptamer (AAAAAAG) followed by a loop structure. During translation, when the ribosomes are in the heptameric sequence of the mRNA, a phase shift occurs in the reading frame. The presence of Shine-Dalgarno type sequences and the loop structure accelerate the translation process by an interaction with the ribosome components [19].

ii. Phage recombination: seroconversion

The O antigen is a determinant of the virulence necessary for the pathogenicity of *S. flexneri*. The O antigen of *S. flexneri* is called the Y serotype and consists of repeated units of a tetrasaccharide of N-acetylglucosamine-rhamnose I-rhamnose II, rhamnose III, which forms the structure of the vertebral column of the polysaccharide unit of all the serotypes of *S. flexneri*, except the 6 and 6a serotypes. There are 13 serotypes, which are differentiated by the addition of glucosyl groups or acetyl residue to the different sugar molecules in the tetrasaccharide unit [54,55].

The temperate phages of *S. flexneri* play an important role in the processes of seroconversion (antigenic variation). The bacteriophages SfV, SfII, SfX, the cryptic prophages SfI and SfIV code for the *gtr* genes, which are the proteins involved in the glycosylation of the O antigen. When these phagic elements lysogenise, a conversion of Y serotype strains into 5a, 2a, X, 1a and 4a serotypes, respectively, occurs (**Figure 9**).

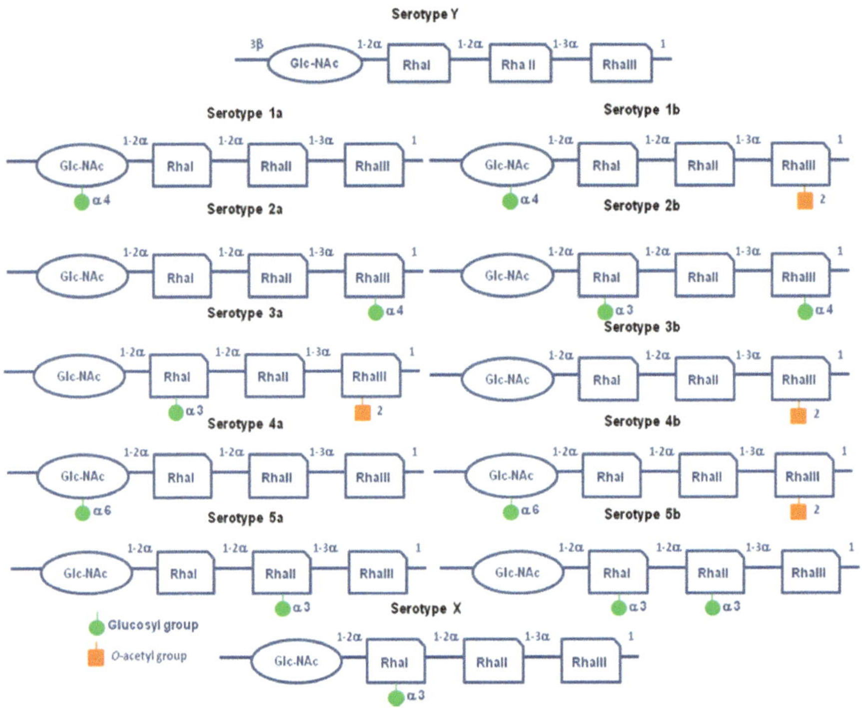

Figure 9. The chemical composition of the different serotypes of *S. flexneri*. The serotype Y formed by repeated units of tetrasaccharide N-acetylglucosamine-rhamnose I–rhamnose, II–rhamnose III. The serotypes are differentiated by the bonding of the glucosyl or the acetyl group. Adapted from [54].

The bacteriophage codes for an acetyltransferase and produces a conversion to the 3b serotype. The lysogenisation of the SfV bacteriophage produces modifications of the type V O antigen, which involves the addition of a glycosyl group through a bond of α 1,3 to the rhamnose II of the repeated tetrasaccharide unit. Similar to other phages that intervene in the glycosylation process, the genes involved in the conversion of serotypes are located immediate downstream from the *attP* locus, which is preceded by the genes *int* and *xis*. These phages are inserted into the *thrW* locus of the host [56].

The *gtr* genes of the temperate phages that code the glycosyltransferases are located in the genome of the phage, downstream from the *attP* locus. These genes are found in a cluster of three genes: *gtr*A, *gtr*B and *gtr* (type), which are cotranscribed. *grt*A and *grt*B are homologous and are interchangeable among the serotypes of *S. flexneri*. The Gtr (type) protein is specific for the formation of the glucosyl bond in a particular sugar molecule of the O antigen [54].

It has been suggested that GtrB catalyses the transfer of glucose from UDP-glucose to bactoprenol phosphate to form UndP-β-glucose in the cytoplasm. This molecule is subsequently translocated by GtrA in the periplasm before the glucosyl residue is joined by Gtr(type) for the growth of the O antigen unit [57].

The genes *gtr*V and *grt*X code for the glycosyltransferases GtrV and GtrX, respectively, which are membrane proteins that catalyse the transference of glucosyl residues through the bonding of the 1,3 rhamnose II and rhamnose I of the O antigen unit. This intervenes in the conversion of the serotype of *S. flexneri* from Y to the 5a serotype and the X serotype, respectively. GtrlV adds glucosyl residues to N-acetylglucosamine of the repeated unit of the O antigen through an α 1,6 bond, converting the Y serotype into the 4a serotype [57,58].

b. Epigenetic mechanisms: DNA methylation

The term epigenetic is defined as "inheritable changes in genetic expression that occur without alterations in the DNA nucleotide sequence". Thus, an epigenetic mechanism may be understood as a complex system to use the genetic information selectively by activating and deactivating various functional genes. Epigenetic modifications may imply methylation of cytosine residue in the DNA. DNA methylation has been observed in various bacterial species. In bacteria, methylation is part of a defence mechanism to reduce the amount of horizontal genetic transference among species. DNA methylation constitutes an epigenetic marker that identifies the template strand during the replication of the DNA. Generally, the methylation of the regulatory elements of genes, such as promoters, enhancers, insulators and repressors, suppresses this function [59].

The modifications of the O antigen that may affect the serotype are related to those that contain the operon that code for the glycosyltransferases (*gtr*).

Within a clone population of *S. enterica* serovar Typhimurium, the lysogenic phage P^{22} may lead to variability of the O antigen. The phase variation of the gtr (glycosylation of the O antigen) indirectly contributes to the diversity of the serotypes of *Salmonella*. The cluster that codes for the glycosyltransferases consists of three genes: *gtrA* codes for a membrane protein, *gtrB* codes for a glycosyl translocase and *gtrC* codes for the glycosyltransferase, which mediates the bonding of glucose to the O antigen [58].

Through studies based on the analysis of gene expression, the presence of mutations, the level of DNA methylation and the *in vitro* interaction of DNA-proteins, Broadbent et al., demonstrated that the Dam methyltransferase proteins together with OxyR regulate phase variation at the level of the *gtr* $_{P22}$ promoter in *S.enterica* serovar Typhimurium. OxyR is an activator or a repressor of the *gtr* system, which depends on the alternative side (GATC sequences) to which OxyR bonds in the *gtr* $_{P22}$ regulatory region (**Figure 10**). The bonding of OxyR is inhibited by the methylation of the Dam target sequence, and the state of expression of the system is inheritable [60].

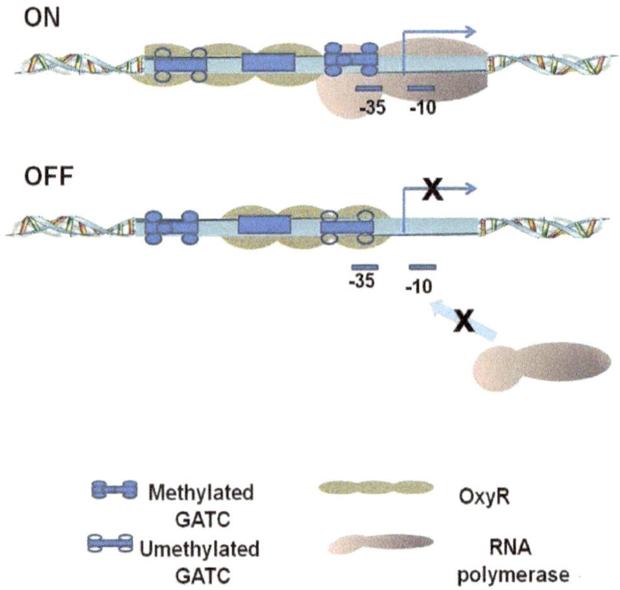

Figure 10. Model of the phase variation of the regulatory region of the *gtr* operon. Illustration of the interaction of DNA proteins in the regulatory region of the *grt* P22 operon, which consists of methylation and demethylation in the GATC sequence in the activated and deactivated phases. Adapted from [60].

Understanding the variation of the LPS structure is important because the composition and the length of the O antigen chain may be an indicator of the virulence, and this characteristic often differs within a single bacterial strain [7].

9. The importance of the variability of the O antigen of LPS

The modifications that are present in the O antigen and that cause its variability play an important role in infections by gram-negative bacteria, given that the modifications may influence adherence, colonisation and the ability to evade the host's defence mechanisms.

9.1. The role of the variation of LPS in the immune response.

LPS activates not only the innate immune response but also the adaptive response. The first contact that LPS has with the immune system is with lipid A, which is recognised by the receptors involved in the innate immune response, while the structure of the O antigen participates in the adaptive response (synthesis of antibodies). LPS is a potent stimulator of the cells of the immune system, given that it induces the production of pro-inflammatory cytokines, such as tumor necrosis factor α (TNF-α), interleukin 6 (IL-6), interleukin 1β (IL-1β) and acute phase proteins.

Although the variability is greater in the O antigen within LPS, there is also variability within lipid A. This variability is due to the length of the chains and the saturation of its acyl groups. Because these groups are strong immunostimulants, both the changes in the number of their chains and the presence of phosphorylations within the structure of the lipid A may influence its biological activity [61].

Lipid A is the structure of LPS that is recognised by the TLR4 receptors, which are part of the so-called Toll Like Receptors (TLRs), which are expressed by cells of the innate immune system and are stimulated by pathogen-associated molecular patterns (PAMPs). The stimulation of the LPS in certain cells of the monocyte-macrophage lineage, lymphoid cells and even cells that are not part of the immune system, such as epithelial cells, endothelial cells and vascular cells, occurs with the participation of other molecules, such as LPS binding protein (LBP), CD14 and MD-2. The transduction signals of TLR4 are divided into MyD88-dependent and MyD88-independent (also called TRI-dependent) groups. These signals may be regulated at various levels. For example, the RP105 and SIGIRR (Single immunoglobulin IL-IR related molecule) molecules inhibit the start of the signalling cascade [62].

Recognition through TLR4 is crucial for the control of infection, but changes in the signalling pathways may cause sepsis or evasion of the pathogen. The importance of signalling via TLR4-MD2 in response to gram-negative pathogens make this pathway an alternative to search for therapeutic targets not only for infectious diseases but also for other diseases with inflammatory aetiology, such as cancer, atherosclerosis, asthma and autoimmune conditions.

Antagonists of TLR4-MD2 have been identified, and several of these are based on the lipid A structures and other inhibitor molecules [63–65]. The intention is to use this type of antagonist therapy to treat septic shock. Additionally, several TLR4 antagonists primarily those that activate the TRAF (TNF receptor- associated factor) or TRIM (Tripartite motif) pathways have been proposed as adjuvants.

However, certain pathogens have the ability to modify the structure of lipid A and its detection by the host. For example, some isolates of *P. aeruginosa* are capable of modifying the structure of lipid A into a penta-acylated moiety, which does not activate the TLR4 and allows the immune response to be evaded. Other isolates of *P. aeruginosa* colonise respiratory pathways of patients with CF during its adaptation, producing hexa-acylated structures that are highly pro-inflammatory [61].

The large variability that the O antigen displays allows for the existence of various clones within a single species, which offers a selective advantage in the niche occupied by this clone and is precisely the interaction between the O antigen and the immune system that permits this advantage.

Many pathogens have the capability of varying the antigens that are attached to their surface and therefore can vary their antigenic composition. This variation is typically mediated by the regulation of the expression of genes. By varying their antigenicity, the pathogens have a greater ability to evade the immune response of the host, and this variability makes it more difficult to design vaccines for these pathogens [66].

The O antigen is considered to be highly immunogenic and induces the production of antibodies that may activate the complement pathway, either through the classic pathway or an alternate pathway, which leads to cellular death or phagocytosis. Certain modifications in the oligosaccharide chain of the O antigen may alter the interaction of the complement pathway. Several O antigens of pathogens are similar to host molecules and facilitate invasion through mimicking in the host; for example, O antigens of the LPS of *H. influenzae* and of *N. gonorrhoeae* mimic epitopes of glycosphingolipids [67].

The mimicking property may also serve to evade the immune system, as is the case of *H. pylori*. The chains of the O antigen that contain the surface of the LPS of *H. pylori* express Lewis antigens, mainly Le^x and Le^y, although some isolates may contain other antigens, such as Le^a, Le^b, $Le^{c,}$ Sialyl-Le^x and H-1, in addition to type A and B blood groups [68].

The expression of Lewis antigens and their fucosylation have biological effects in the pathogenesis of this bacterium. The O antigen of *H. pylori* exhibits molecular mimicking with the Lewis antigens of the host within the gastric epithelium. The expression of the Lewis antigens is subject to phase variation, given that the regulation of the glycosyltransferase genes is regulated by SSM, which in the O antigen structure promotes variations among the strains. The antigenic mimicking is essentially involved in the evasion of the immune system and gastric adaptation. Several studies show that mimicking also plays a role in the colonisation and adhesion of Le^x of the bacteria with the galectin-3 of a gastric receptor.

Moreover, *H. pylori* is capable of evading the binding effect of surfactant protein D, which is expressed in the gastric mucosa and is a component of the innate immune response [69]. This microorganism impedes the bonding of the surfactant protein through the variation of its LPS. This phenomenon is associated with changes in the fucosylation of the O antigen chain. In addition, the expression of Lewis antigens affects both the inflammatory response and the polarisation of the T cells that are triggered after an infection. Because it is a chronic pathogen, several studies have shown that *H. pylori* may induce anti-Lewis auto-reactive antibodies, which enable the gastric mucosa to be recognised and contribute to the development of gastric atrophy [70].

9.2. Changes in the LPS related to resistance

The hydrophobic antibiotics that reach the interior of the cells due to the permeability of the external membrane are aminoglycosides, macrolides, rifamycins, novobiocin, fusidic acid and cationic peptides. The tetracycline and the quinolones use pathways that are mediated by lipids and porins. The central region of the LPS is important because it provides a barrier against hydrophobic antibiotics and other components; isolates that express a long LPS have intrinsic resistance to these factors [71].

The polymyxins, which include polymyxin B and colistin (polymyxin E), belong to a group of natural antimicrobials that are found in eukaryotic cells; this group is known as the cationic antimicrobial peptides. The polymyxins are active against gram-negative pathogens, such as *P. aeruginosa, Acinetobacter baumanii, Klebsiella* spp., *E. coli* and other *Enterobacteriaceae* [71–73].

The LPS has a negative charge and provides integrity and stability to the external membrane of the bacteria. Polymyxin has a positive charge, displacing the Mg^{+2} or Ca^{+2} and bonding to LPS, which as a consequence, destabilises and destroys the internal and external membrane [74].

Gram-negative bacteria may develop resistance to colistin and polymyxin B. The most important mechanisms involve modifications in the external membrane through changes of the LPS. The modification of the LPS occurs with the addition of 4-amino-4-deoxy-L-arabinose (Lara4N) to a phosphate group in lipid A. This addition causes an decreases in the negative charge of the lipid A, which decreases the affinity of the positively charged polymyxins [68–70].

The biosynthesis of LAra4N is mediated by the regulatory systems PmrA/PmrB and PhoP/PhoQ [26]. One of the primary roles of the activation of PmrAB is the modification of the LPS. These modifications include additions of Ara4N and pEtN to the lipid A and of pEtN to the core of LPS. The modifications mask phosphate groups with positive charges, thereby affecting the electrostatic interaction with certain cationic compounds. The biosynthesis of LAra4N depends on the genes of the operon of resistance to polymyxin, which is known as *arn*. This operon includes the genes *pmrHFIJKLM* [71–73].

10. Conclusion

In the bacterial pathogens, the most variable structures are those expressed in the cell surface. LPS is one of the principal antigenic structures of cell surface of gram-negative bacteria. A great variability in LPS has been demonstrated and principally in O antigen of gram-negative bacteria. This variability is present not only in the longitude of the oligosaccharide chains but also in the composition and structure of LPS.

Many of the functions of O antigen are associated to the longitude of the chain and to the variability of its structural features. This variability could affect the function, physical and chemical properties as well as the target site of LPS and determines the changes in the virulence of the microorganism that favor its adaptation to fluctuating enviroment which in many occasions are hostile to the microorganism and permit its evasion of the immune response of the host. The variation in O antigen structure has demonstrated that its composition and the longitude of its chain could be biological markers of virulence and this characteristic could differ within the same bacterial strain. The variability of LPS could derive from adaptations that involve associated changes to the synthesis of this molecule. The antigenic variability could occur by means of genetic and epigenetic mechanisms. The lost or gain of genes associated to variability of LPS is due to the events of genetic material interchange produced by lateral transference of genes which leads to strain selection with new characteristics and the evolution of the bacteria by modification of this structure.

One of the most important aspects of LPS function is its participation as immunogenic molecule and its role in bacterial classification based on O antigen and its variability. In general, it is seen that the modifications of O antigen play an important role in the process of infection including the adherence, the colonization, and the ability to evade defensive mechanisms of the host especially the innate resistance.

The study of the events of variation of LPS and its effects on pathogenecity and virulence represents a field of study of great interest to understand bacterial physiology and its mechanisms of adaptation and evolution.

The immunogenicity and variability of O antigen confer to gram-negative bacteria an important characteristic for its serological typification. The 0-antigen is subject to an intense selection on the part of immune system, which could be the principal factor for the different forms in which it is presented. For this reason, the variability of O antigen has been an area of intense research.

Author details

Rosa Eugenia Reyes, Alejandra Aquino Andrade, Rafael Coria Jiménez,
Carolina Romo González and Maribel Ortiz Herrera
Departamento de Medicina Experimental, Instituto, Nacional de Pediatría, México D.F., México

Rosa Eugenia Reyes
Departamento de Sistemas Biológicos, División CBS, Universidad Autónoma Metropolitana Unidad, Xochimilco, Coyoacán, México D.F., México

11. References

[1] Erridge C, Bennett-Guerrero E, Poxton IR. (2002) Structure and function of lipopolysaccharides. Microbes Infect 4:837–51.

[2] Silhavy TJ, Kahne D, Walker S. (2010) The bacterial cell envelope. Cold Spring Harb Perspect Biol 2:a000414.

[3] Flemming H-C, Neu TR, Wozniak DJ. (2007) The EPS matrix: the «house of biofilm cells». J. Bacteriol 189:7945–7.

[4] Pier GB. (2007) *Pseudomonas aeruginosa* lipopolysaccharide: a major virulence factor, initiator of inflammation and target for effective immunity. Int. J. Med. Microbiol 297:277–95.

[5] Franklin MJ, Nivens DE, Weadge JT, Howell PL. (2011) Biosynthesis of the *Pseudomonas aeruginosa* extracellular polysaccharides, Alginate, Pel, and Psl. Front Microbiol 2:167.

[6] Erridge C, Stewart J, Bennett-Guerrero E, McIntosh TJ, Poxton IR. (2002) The biological activity of a liposomal complete core lipopolysaccharide vaccine. J. Endotoxin Res 8:39–46.

[7] Lerouge I, Vanderleyden J. (2002) O-antigen structural variation: mechanisms and possible roles in animal/plant-microbe interactions. FEMS Microbiol. Rev 26:17–47.

[8] Steeghs L, Jennings MP, Poolman JT, van der Ley P. (1997) Isolation and characterization of the *Neisseria meningitidis lpx*D-*fab*Z-*lpx*A gene cluster involved in lipid A biosynthesis. Gene 190: 263–70.

[9] Raetz CRH, Whitfield C. (2002) Lipopolysaccharide endotoxins. Annu. Rev. Biochem 71: 635–700.

[10] Wang X, Quinn PJ. (2010) Lipopolysaccharide: Biosynthetic pathway and structure modification. Prog. Lipid Res 49:97–107.

[11] Sperandeo P, Cescutti R, Villa R, Di Benedetto C, Candia D, Dehò G, Polissi A.(2007) Characterization of *lpt*A and *lpt*B, two essential genes implicated in lipopolysaccharide transport to the outer membrane of *Escherichia coli*. J. Bacteriol 189:244–53.

[12] Sperandeo P, Lau FK, Carpentieri A, De Castro C, Molinaro A, Dehò G, Silhavy TJ, Polissi A. (2008) Functional analysis of the protein machinery required for transport of lipopolysaccharide to the outer membrane of *Escherichia coli*. J. Bacteriol 190:4460–9.

[13] Sperandeo P, Dehò G, Polissi A. (2009) The lipopolysaccharide transport system of Gram-negative bacteria. Biochim. Biophys. Acta 1791:594–602.

[14] Ruiz N, Gronenberg LS, Kahne D, Silhavy TJ. (2008) Identification of two inner-membrane proteins required for the transport of lipopolysaccharide to the outer membrane of *Escherichia coli*. Proc. Natl. Acad. Sci. U.S.A. 105:5537–42.

[15] Ma B, Reynolds CM, Raetz CRH. (2008) Periplasmic orientation of nascent lipid A in the inner membrane of an *Escherichia coli* LptA mutant. Proc. Natl. Acad. Sci. U.S.A. 105:13823–8.

[16] Wu T, McCandlish AC, Gronenberg LS, Chng S-S, Silhavy TJ, Kahne D. (2006) Identification of a protein complex that assembles lipopolysaccharide in the outer membrane of *Escherichia coli*. Proc. Natl. Acad. Sci. U.S.A. 103:11754–9.

[17] Bos MP, Tefsen B, Geurtsen J, Tommassen J. (2004) Identification of an outer membrane protein required for the transport of lipopolysaccharide to the bacterial cell surface. Proc. Natl. Acad. Sci. U.S.A. 101:9417–22.

[18] Hug I, Couturier MR, Rooker MM, Taylor DE, Stein M, Feldman MF. (2010) *Helicobacter pylori* lipopolysaccharide is synthesized via a novel pathway with an evolutionary connection to protein N-glycosylation. PLoS Pathog 6:e1000819.

[19] Wang G, Ge Z, Rasko DA, Taylor DE. (2000) Lewis antigens in *Helicobacter pylori*: biosynthesis and phase variation. Mol. Microbiol 36:1187–96.

[20] Raetz CRH, Reynolds CM, Trent MS, Bishop RE. (2007) Lipid A modification systems in gram-negative bacteria. Annu. Rev. Biochem.76:295–329.

[21] Hug I, Feldman MF. (2011) Analogies and homologies in lipopolysaccharide and glycoprotein biosynthesis in bacteria. Glycobiology 21:138–51.

[22] Marolda CL, Valvano MA. (1998) Promoter region of the *Escherichia coli* O7-specific lipopolysaccharide gene cluster: structural and functional characterization of an upstream untranslated mRNA sequence. J. Bacteriol. 180:3070–9.

[23] Schnaitman CA, Klena JD. (1993) Genetics of lipopolysaccharide biosynthesis in enteric bacteria. Microbiol. Rev. 57:655–82.

[24] Reeves PP, Wang L. (2002) Genomic organization of LPS-specific loci. Curr. Top. Microbiol. Immunol. 264:109–35.

[25] Guo L, Lim KB, Gunn JS, Bainbridge B, Darveau RP, Hackett M, Miller SI. (1997) Regulation of lipid A modifications by *Salmonella typhimurium* virulence genes phoP-phoQ. Science 276:250–3.

[26] Gunn JS. (2008) The *Salmonella* PmrAB regulon: lipopolysaccharide modifications, antimicrobial peptide resistance and more. Trends Microbiol. 16:284–90.

[27] Marchal K, De Keersmaecker S, Monsieurs P, van Boxel N, Lemmens K, Thijs G, Vanderleyden J, De Moor B. (2004) In silico identification and experimental validation of PmrAB targets in *Salmonella typhimurium* by regulatory motif detection. Genome Biol. 5:R9.

[28] Heinrichs DE, Yethon JA, Whitfield C. (1998) Molecular basis for structural diversity in the core regions of the lipopolysaccharides of *Escherichia coli* and *Salmonella enterica*. Mol. Microbiol. 30:221–32.

[29] Rietschel ET, Kirikae T, Schade FU, Mamat U, Schmidt G, Loppnow H, Ulmer AJ, Zähringer U, Seydel U, Di Padova F. (1994) Bacterial endotoxin: molecular relationships of structure to activity and function. FASEB J. 8:217–25.

[30] Bailey MJ, Hughes C, Koronakis V. (1997) RfaH and the ops element, components of a novel system controlling bacterial transcription elongation. Mol. Microbiol. 26:845–51.

[31] Geerlof A, Lewendon A, Shaw WV. (1999) Purification and characterization of phosphopantetheine adenylyltransferase from *Escherichia coli*. J. Biol. Chem. 274:27105–11.

[32] Feng L, Han W, Wang Q, Bastin DA, Wang L. (2005) Characterization of *Escherichia coli* O86 O-antigen gene cluster and identification of O86-specific genes. Vet. Microbiol. 106:241–8.

[33] Clark CG, Kropinski AM, Parolis H, Grant CCR, Trout-Yakel KM, Franklin K, Ng L-K, Paramonov NA, Parolis LAS, Rahn K, Tabor H. (2009) *Escherichia coli* O123 O antigen genes and polysaccharide structure are conserved in some *Salmonella enterica* serogroups. J. Med. Microbiol. 58:884–94.

[34] Sozhamannan S, Deng YK, Li M, Sulakvelidze A, Kaper JB, Johnson JA, Nair GB, Morris JG Jr. (1999) Cloning and sequencing of the genes downstream of the *wbf* gene cluster of *Vibrio cholerae* serogroup O139 and analysis of the junction genes in other serogroups. Infect. Immun. 67:5033–40.

[35] Keenleyside WJ, Whitfield C. (1996) A Novel Pathway for O-Polysaccharide Biosynthesis in *Salmonella enterica* Serovar Borreze. J. Biol. Chem. 271:28581–92.

[36] Wang L, Jensen S, Hallman R, Reeves PR. (1998) Expression of the O antigen gene cluster is regulated by RfaH through the JUMPstart sequence. FEMS Microbiol. Lett. 165:201–6.

[37] Liu B, Knirel YA, Feng L, Perepelov AV, Senchenkova SN, Wang Q, Reeves PR, Wang L. (2008) Structure and genetics of *Shigella* O antigens. FEMS Microbiol. Rev. 32:627–53.

[38] Samuel G, Reeves P. (2003) Biosynthesis of O-antigens: genes and pathways involved in nucleotide sugar precursor synthesis and O-antigen assembly. Carbohydr. Res. 338:2503–19.

[39] Olson MV, Kas A, Bubb K, Qui R, Smith EE, Raymond CK, Kaul R. (2004) Hypervariability, suppressed recombination and the genetics of individuality. Philos. Trans. R. Soc. Lond., B, Biol. Sci .359:129–40.

[40] Dixon DR, Darveau RP. (2005) Lipopolysaccharide heterogeneity: innate host responses to bacterial modification of lipid a structure. J. Dent. Res. 84:584–95.

[41] Lawson AJ, Chart H, Dassama MU, Threlfall EJ. (2002) Heterogeneity in expression of lipopolysaccharide by strains of *Salmonella enterica* serotype Typhimurium definitive phage type 104 and related phage types. Lett. Appl. Microbiol. 34:428–32.

[42] Tso MD, Dooley JS. (1995) Temperature-dependent protein and lipopolysaccharide expression in clinical *Aeromonas* isolates. J. Med. Microbiol. 42:32–8.

[43] Reyes RE, Ramírez Saad H, Pérez J, Meléndez E, Navarro A, Murata Ch, Cravioto A. (2010) Intra-strain variation of lipopolysaccharides from *Citrobacter freundii* E9750 by *in vitro* passages. Afr. J. Microbiol. Res. 4:1324–31.
[44] McGroarty EJ, Rivera M. (1990) Growth-dependent alterations in production of serotype-specific and common antigen lipopolysaccharides in *Pseudomonas aeruginosa* PAO1. Infect. Immun. 58:1030–7.
[45] Guard-Petter J, Lakshmi B, Carlson R, Ingram K. (1995) Characterization of lipopolysaccharide heterogeneity in *Salmonella enteritidis* by an improved gel electrophoresis method. Appl. Environ. Microbiol. 61:2845–51.
[46] Nishiuchi Y, Doe M, Hotta H, Kobayashi K. (2000) Structure and serologic properties of O-specific polysaccharide from *Citrobacter freundii* possessing cross-reactivity with *Escherichia coli* O157:H7. FEMS Immunol. Med. Microbiol 28:163–71.
[47] van der Woude MW, Bäumler AJ. (2004) Phase and antigenic variation in bacteria. Clin. Microbiol. Rev. 17:581–611.
[48] Wisniewski-Dyé F, Vial L. (2008) Phase and antigenic variation mediated by genome modifications. Antonie Van Leeuwenhoek 94:493–515.
[49] Torres-Cruz J, van der Woude MW. (2003) Slipped-strand mispairing can function as a phase variation mechanism in *Escherichia coli*. J. Bacteriol. 185:6990–4.
[50] van Belkum A, Scherer S, van Leeuwen W, Willemse D, van Alphen L, Verbrugh H. (1997) Variable number of tandem repeats in clinical strains of *Haemophilus influenzae*. Infect. Immun. 65:5017–27.
[51] Olivares D, Gisbert JP. (2006) Factors involved in the pathogenesis of *Helicobacter pylori* infection. Rev Esp Enferm Dig 98:374–86.
[52] Wang G, Rasko DA, Sherburne R, Taylor DE. (1999) Molecular genetic basis for the variable expression of Lewis Y antigen in *Helicobacter pylori*: analysis of the alpha (1,2) fucosyltransferase gene. Mol. Microbiol. 31:1265–74.
[53] Sanabria-Valentín E, Colbert M-TC, Blaser MJ. (2007) Role of *fut*C slipped strand mispairing in *Helicobacter pylori* Lewisy phase variation. Microbes Infect. 9:1553–60.
[54] Allison GE, Verma NK. (2000) Serotype-converting bacteriophages and O-antigen modification in *Shigella flexneri*. Trends Microbiol. 8:17–23.
[55] Perepelov AV, L'vov VL, Liu B, Senchenkova SN, Shekht ME, Shashkov AS, Feng L, Aparin PG, Wang L, Knirel YA. (2009) A similarity in the O-acetylation pattern of the O-antigens of *Shigella flexneri* types 1a, 1b, and 2a. Carbohydr. Res. 344:687–92.
[56] Allison GE, Angeles D, Tran-Dinh N, Verma NK. (2002) Complete genomic sequence of SfV, a serotype-converting temperate bacteriophage of *Shigella flexneri*. J. Bacteriol. 184:1974–87.
[57] Nair A, Korres H, Verma NK. (2011) Topological characterisation and identification of critical domains within glucosyltransferase IV (GtrIV) of Shigella flexneri. BMC Biochem. 12:67.
[58] Korres H, Verma NK. (2006) Identification of essential loops and residues of glucosyltransferase V (GtrV) of Shigella flexneri. Mol. Membr. Biol. 23:407–19.
[59] Rodríguez-Dorantes M, Téllez-Ascencio N, Cerbón MA, López M, Cervantes A. (2004) DNA methylation: an epigenetic process of medical importance. Rev. Invest. Clin. 2004;56:56–71.

[60] Broadbent SE, Davies MR, van der Woude MW. (2010) Phase variation controls expression of *Salmonella* lipopolysaccharide modification genes by a DNA methylation-dependent mechanism. Mol. Microbiol. 77:337–53.
[61] Bryant CE, Spring DR, Gangloff M, Gay NJ. (2010) The molecular basis of the host response to lipopolysaccharide. Nat. Rev. Microbiol. 8:8–14.
[62] Lu Y-C, Yeh W-C, Ohashi PS. (2008) LPS/TLR4 signal transduction pathway. Cytokine 42:145–51.
[63] Kawamoto T, Ii M, Kitazaki T, Iizawa Y, Kimura H. (2008) TAK-242 selectively suppresses Toll-like receptor 4-signaling mediated by the intracellular domain. Eur. J. Pharmacol. 584:40–8.
[64] Toshchakov VY, Fenton MJ, Vogel SN. (2007) Cutting Edge: Differential inhibition of TLR signaling pathways by cell-permeable peptides representing BB loops of TLRs. J. Immunol. 178:2655–60.
[65] Lee H-K, Brown SJ, Rosen H, Tobias PS. (2007) Application of beta-lactamase enzyme complementation to the high-throughput screening of toll-like receptor signaling inhibitors. Mol. Pharmacol. 72:868–75.
[66] Maskell D, Frankel G, Dougan G. (1993) Phase and antigenic variation the impact on strategies for bacterial vaccine design. Trends Biotechnol. 11:506–10.
[67] Reeves P. (1995) Role of O-antigen variation in the immune response. Trends Microbiol. 3:381–6.
[68] Nilsson C, Skoglund A, Moran AP, Annuk H, Engstrand L, Normark S. (2008) Lipopolysaccharide diversity evolving in *Helicobacter pylori* communities through genetic modifications in fucosyltransferases. PLoS ONE 3:e3811.
[69] Khamri W, Moran AP, Worku ML, Karim QN, Walker MM, Annuk H, Ferris JA, Appelmelk BJ, Eggleton P, Reid KBM, Thursz MR. (2005) Variations in *Helicobacter pylori* lipopolysaccharide to evade the innate immune component surfactant protein D. Infect. Immun. 73:7677–86.
[70] Moran AP. (2008) Relevance of fucosylation and Lewis antigen expression in the bacterial gastroduodenal pathogen *Helicobacter pylori*. Carbohydr. Res. 343:1952–65.
[71] Delcour AH. (2009) Outer membrane permeability and antibiotic resistance. Biochim. Biophys. Acta 1794:808–16.
[72] Gatzeva-Topalova PZ, May AP, Sousa MC. (2005) Structure and mechanism of ArnA: conformational change implies ordered dehydrogenase mechanism in key enzyme for polymyxin resistance. Structure 13:929–42.
[73] Falagas ME, Rafailidis PI, Matthaiou DK. (2010) Resistance to polymyxins: Mechanisms, frequency and treatment options. Drug Resist. Updat. 13:132–8.
[74] Clausell A, Garcia-Subirats M, Pujol M, Busquets MA, Rabanal F, Cajal Y. (2007) Gram-negative outer and inner membrane models: insertion of cyclic cationic lipopeptides. J Phys Chem B 111:551–63.

Chapter 5

Exopolysaccharide Biosynthesis in *Rhizobium leguminosarum*: From Genes to Functions

Tanya V. Ivashina and Vladimir N. Ksenzenko

Additional information is available at the end of the chapter

1. Introduction

Gram-negative soil α-proteobacteria belonging to the genera *Allorhizobium, Azorhizobium, Bradyrhizobium, Mesorhizobium, Rhizobium,* and *Sinorhizobium* (*Ensifer*) are able to infect the roots of their legume hosts in a host-specific way and induce the formation of specialized new plant organs - nodules, in which endosymbiotic bacteria reduce atmospheric nitrogen to ammonia. *Rhizobium leguminosarum* comprises two biovars, namely *trifolii* (nodulating *Trifolium*) and *viciae* (nodulating *Pisum, Vicia, Lathyrus,* and *Lens*) [1]. Closely related to the *R. leguminosarum* is *R. etli* (formely *R. leguminosarum* bv. *phaseoli*) nodulating *Phaseolus* beans. To avoid confusions, we will use former names for *R. leguminosarum* bv. *phaseoli* strains as they were described in original papers.

The *R. leguminosarum* strains investigated so far synthesize different types of polysaccharides, including acidic exopolysaccharide (EPS), capsule polysaccharide (CPS), gel-forming polysaccharide (GPS), cellulose fibrils, galactomannan, lipopolysaccharide, and cyclic glycans [2]. The cyclic neutral β-(1,2)-glucans are predominantly accumulated within the periplasmic space and play an essential role during hypoosmic adaptation as well as during plant infection [3]. Glucomannan was shown to be important for lectin-mediated polar attachment to *Vicia sativa* and *Pisum sativum* root hairs and competitive nodulation [4,5]. The CPS is tightly associated with the cell surface of bacteria forming a polysaccharide matrix surrounding the bacteria [6]. Differences in noncarbohydrate substitutions, such as *O*-acetyl, pyruvate, and 3-hydroxybutyrate, may distinguish anionic capsule-bound polysaccharides from secreted EPS. In late-stationary-phase cultures, CPS was replaced by a polysaccharide with strong gel-forming properties having an unknown function [7]. The LPS present at the outer leaflet of the outer membrane and consists of three parts: lipid A, the core polysaccharide and the O-antigen polysaccharide [8]. More and more data appear indicating that LPS plays a specific role in the later stages of establishment of symbiosis (reviewed by [9]). EPS refers to extracellular polysaccharides with slight or no cell association [9-11].

Among the above mentioned polysaccharides, acidic EPSs attract attention due to their diverse functions both in free-living rhizobia and during the establishment of nitrogen-fixing symbiosis with host legume plants. EPS forms a biofilm layer on the cell surface which is thought to contribute to the following processes: cellular protection against environmental stresses, attachment to surfaces, nutrient gathering, and the preferential absorption of flavonoids secreted by plants along the membrane surface [9]. In addition, EPS biosynthesis is required for the effective nodulation of legumes such as *Medicago, Pisum, Trifolium, Leucaena,* and *Vicia* spp., which form indeterminate-type nodules (otherwise called meristematic or cylindrical) [12]. EPS-deficient mutants of *Rhizobium leguminosarum* bv. *viciae* (hereafter *Rlv*), *R. leguminosarum* bv. *trifolii* (*Rlt*), and *Sinorhizobium meliloti* induce symbiotically defective phenotypes which include delayed root hair curling, nodules devoid of bacteria due to infection threads that abort within peripheral cells of the developing nodule, and small, partially infected, non-nitrogen-fixing nodules [12]. The precise function(s) of the EPS molecules in these associations is still unclear; however, recent studies suggest that these extracellular polymers may function as symbiotic signaling molecules which regulate plant responses in the infection process [13-17].

The current knowledge of EPS biosynthesis in rhizobia was based almost exclusively on the research with succinoglycan (EPS I) produced by *S. meliloti* [18]. More than 26 genes needed for the synthesis, modification, polymerization, export, and processing of *S. meliloti* EPS I were identified, and a pathway for the polysaccharide biosynthesis was proposed [19-21]. Recent discoveries on succinoglycan biosynthesis and its functioning in symbiosis are summarized in the exhaustive reviews [22,23].

The EPSs produced by *R. leguminosarum* strains investigated to date structurally differ from those of *S. meliloti.* At present there is fragmentary information about the *R. leguminosarum* EPS assembly, modification, and processing, as well as the role of the individual gene products in these processes. These data were reviewed in [9-11,24]. Here we report some recent and complementary data obtained in our group. On the basis of these data as well as the analysis of the sequenced *R. leguminosarum* and *R. etli* genomes we tried to outline main regularities in the EPS synthesis.

2. Structural characteristics of exopolysaccharides

Structures of known *R. leguminosarum* and *R. etli* (*Re*) EPSs are presented in Table 1. It should be noted that EPSs secreted by the most of the *Rlv* and *Rlt* strains were shown to have repeating units of identical structure [25-35]. These EPSs are branched heteropolysaccharides of the octasaccharide repeating unit consisting of a backbone of two glucose and two glucuronic acid residues, a side chain of three glucose and one galactose residues. The backbone contains β1-4 and α1-4 linkages only while the branch point is linked β1-6. The side chain contains β1-4 and β1-3 linkages. In general, the octasaccharide is modified by two pyruvyl, one or two non-stoichiometric O-acetyl, and one non-stoichiometric hydroxybutanoyl groups. The distribution pattern of O-acetyl and 3-hydroxybutanoyl groups may vary for some *R. leguminosarum* strains and was shown to be dependent on the growth phase of bacteria and culture medium [27,32].

	Repeating unit structure	Strains	References
Backbone chain	→4)-α-Glc-(1→4)-α-GlcA-(1→4)-β-GlcA-(1→4)-β-Glc-(1 6 ↑	all strains	[25-41]
Side chains	β ↑ β-Gal-(1→3)-β-Glc-(1→4)-β-Glc-(1→4)-β-Glc 6 4 6 4 Pyr Pyr	*Rlv*: VF39, 3841, LPR1, 128c53, 128c63; *Rlt*: TA1, LPR5, ANU843, NA30, 0403; *Re* CFN42; *Rlp*: LPR49, 127K36; *Rhizobium* sp. GRH2	[25-35]
	β ↑ β-Glc-(1→3)-β-Glc-(1→4)-β-Glc 6 4 Pyr	*Rlt* 4S	[37]
	β ↑ β-Gal-(1→3)-β-Glc-(1→3)-β-Glc-(1→3)-β-GlcA-(1→4)-β-Glc 6 4 6 4 Pyr Pyr	*Rlv* 248	[38]
	β ↑ α-Gal-(1→4)-β-GlcA-(1→3)-β-Glc-(1→4)-β-Glc-(1→4)-β-Glc 6 4 Pyr	*Rlp* 127K44	[39]
	β ↑ β-Glc-(1→6)-α-Gal-(1→4)-β-GlcA-(1→4)-β-Glc-(1→4)-β-Glc-(1→4)-β-Glc 6 4 Pyr	*Rlp* 127K38	[40]
	β ↑ β-Gal-(1→6)-β-Gal-(1→6)-β-Glc-(1→6)-α-Gal-(1→4)-β-GlcA-(1→4)-β-Glc-(1→4)-β-Glc 6 4 Pyr	*Rlp* 127K87	[41]

Table 1. Structure of *R. leguminosarum* and *R. etli* EPS repeating units. Abbreviations: Glc, glucose; GlcA, glucuronic acid; Gal, galactose and Pyr, ketal pyruvate group.

However, several *R. leguminosarum* strains produce EPS with divergent side chains though with identical backbones and the same β1-6-linked glucosyl residue branching the side chain. Side chains of these EPS may consist of three to seven sugar residues. Up to three Gal residues can be present as in the *R. leguminosarum* bv. *phaseoli* (*Rlp*) 127K87 EPS, or terminal Gal residue can be absent as in the *Rlt* 4S EPS. In addition, in some cases side chains of EPS contain GlcA residues (*Rlp* 127K44, *Rlp* 127K38, *Rlp* 127K87, *Rlv* 248). Besides β1-4 and β1-3 linkages, sugar residues can be attached by α1-6 or β1-6 glycosidic bonds (*Rlp* 127K38, *Rlp* 127K87).

The acidic nature of EPS is explained by the presence of uronic acids and negatively charged pyruvyl groups. Similar to other representatives of *Rhizobiaceae*, the *R. leguminosarum* strains synthesize EPS in high-molecular-weight (HMW) and low-molecular-weight (LMW) forms [13,42]. The latter were proposed to act as signaling factors during the development of symbiosis [14,16-17,42].

3. Organization of exopolysaccharide biosynthesis genes

According to the modern conception, the synthesis of heteropolysaccharides requires a complex pathway starting with the synthesis of sugar nucleotide precursors as well as of the non-carbohydrate donors followed by sequential assembly of the repeating unit on polyprenyl lipid carries, their modification, polymerization, and export outside of the cell [20,21,43].

We started the study of the genetic control of the biosynthesis of acidic exopolysaccharide with isolation of non-mucoid Tn*5*-derived mutants in *Rlv* VF39. As a result, five non-slimy mutants (GL1-5) were obtained. The mutations were mapped within four separate chromosomal loci. The open reading frames (*orfs*) interrupted by insertion of the Tn*5* transposon were named as *pss* (**p**oly**s**accharide **s**ynthesis) according to Borthakur and co-workers [44]. The Tn*5* insertion in the GL4 mutant was localized within the *pssA* gene [45], the ortholog of which was previously identified in *Rlp* 8002 [44]. The mutations in GL2 and GL6 were mapped within the *pssE* and *pssD* genes, respectively. Their orthologs were found earlier in *Rlt* LPR5 [46]. Chromosomal walking around these genes in *Rlv* VF39 allowed us to identify a 15.5–kb multi-cistronic operon which included a core set of genes needed for the assembly of the repeating unit of the EPS (*pssEDCFGHIJS*), its modification (*pssKMR*), polymerization (*pssL*) and processing (*pssW*) (Fig. 1) [47]. It should be mentioned that the *pssV-E* operon was found in all *R. leguminosarum* and *R. etli* genomes, whose complete or partial sequences are available now. Moreover, nine out of the fifteen genes from this operon have orthologs in all these genomes. At the same time, certain *Rlv* VF39 genes are absent in some other genomes, certain genes are substituted for non-orthologous genes, and some additional genes are also present (Fig. 1). We will discuss the functioning of all these genes bellow. Here we would like only to consider the problem with their names.

All fifteen genes from the *pssV-E* operon were named as *pss* genes. In addition, the same gene name abbreviation was assigned to six genes (*pssA*, *pssB*, *pssN*, *pssO*, *pssP* and *pssT*) localized in other operons. It is easy to count up that only five letters of the alphabet left that can be used with the *"pss"* body in the names of new genes involved in EPS biosynthesis. Meanwhile, in our opinion even at present new names for eight genes from *Rlt* WSM2304, *Re* CFN42, *Re* CNPAF512 and *Re* CIAT 652 have to be assigned. Therefore, we propose (i) to retain the existing names for all orthologous genes, and (ii) to introduce a new set of genes with the body name *"psa"* (**p**oly**s**accharide repeating unit **a**ssembly). Our propositions concerning new names to be assigned for certain genes involved in EPS biosynthesis are summarized in Table 2.

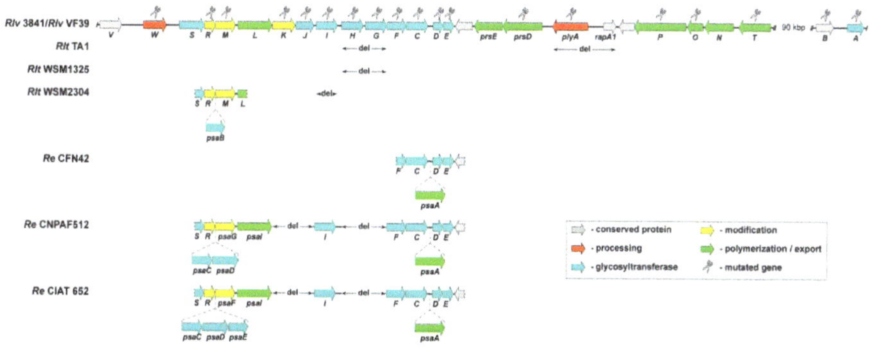

Figure 1. Gene arrangements in PssI-cluster of *R. leguminosarum* and *R. etli* strains.

Proposed gene name	Existing gene names in *R. leguminosarum* and *R. etli* genomes*				Predicted function
	Rlt WSM2304	*Re* CFN42	*Re* CNPAF512	*Re* CIAT 652	
psaA	Rleg2_0274	RHE_CH03224	RHECNPAF_4 30080	RHECIAT_CH 0003464	O-antigen ligase
psaB	Rleg2_2965	none	none	none	GT, family 2
psaC	none	none	RHECNPAF_4 30069	RHECIAT_CH 0003456	GT, family 2
psaD	none	none	RHECNPAF_4 30070	RHECIAT_CH 0003457	GT, family 1
psaE	none	none	none	RHECIAT_CH 0003458	GT, family 2
psaF	none	none	none	RHECIAT_CH 0003459	Polysaccharide pyruvyl transferase
psaG	none	none	RHECNPAF_4 30071	none	Polysaccharide pyruvyl transferase
psaI	none	none	RHECNPAF_4 30072	RHECIAT_CH 0003459	Polysaccharide biosynthesis protein

* EMBL/GenBank/DDBJ accession numbers: *Rlt* WSM2304 (CP001191), *Re* CFN42 (CP000133), *Re* CNPAF512 (AEYZ01000266) and *Re* CIAT 652 (CP001074)

Table 2. Proposed names for genes controlling EPS biosynthesis in some *R. leguminosarum* and *R. etli* strains.

The *pssV-E* operon is neighboring with the region comprising several operons containing genes for the Type I secretion system (*prsED*), EPS processing (*plyA*), and EPS polymerization/export (*pssTNOP*). This whole chromosomal region is known now as the Pss-I gene cluster [48].

The *pssA* gene controlling the first step in the repeating unit assembly is localized approximately at 90 kb from the Pss-I cluster. The gene was shown to be transcribed as a monocistronic mRNA [45]. Upstream of the *pssA* is the *pssB* gene encoding inositol monophosphatase [49-51]. Effects of mutations within the *pssB* gene on the synthesis of EPS and symbiotic behavior have been analyzed in the *Rlv* VF39 and *Rlt* TA1 backgrounds and have been shown to be contradictory. In *Rlv* VF39 the *pssB* mutants retained the ability to produce EPS in amounts equal to those of the wild-type strain. In *Rlt* TA1 the *pssB* inactivation displayed an increased overall production of EPS versus the wild-type strain, and alterations in the LPS PAGE-banding pattern and the O-antigen sugar composition [51,52]. Nevertheless, *pssB* mutants of both strains elicited non-effective nodules on the *Vicia faba* or *Trifolium pratense* roots, respectively [49,52].

In the case of GL1 and GL3 mutants we have not carried out the extended chromosomal walking around the Tn*5* insertions, but only short genome sequences flanking Tn*5* were determined. We localized the Tn*5* insertion in the GL3 mutant within some small *orf* (213 bp only). The ortholog of this *orf* (RL2260) was found in the chromosome sequence of *Rlv* 3841 located far away from the Pss-I cluster. It encodes a 7.2 kDa positively charged protein (pI 10.8), which is conserved in numerous bacteria. As for the GL1 mutation, we have not been able to map it up to now. Probably, this mutation targets a gene located in one of the *Rlv* VF39 strain-specific plasmids.

Taking together these data, one can conclude that the core set of EPS synthesizing genes are clustered in the chromosomal Pss-I region. Clustering of genes involved in EPS biosynthesis is not unique for *R. leguminosarum*, but is widespread in different polysaccharide producing bacteria [53,54]. Such a type of genes organization could reflect their coordinated expression and tightly regulated control.

Evidently, EPS biosynthesis is linked with other metabolic pathways in the cell. Therefore, localization of mutations affecting EPS production distant from the Pss-I region can reflect this linkage. For example, the *pssA* gene is located in another chromosomal region probably due to involvement in the initiation of not only EPS synthesis, but the synthesis of the other polysaccharides. Recently, several regulatory genes influencing EPS production (*psiA*, *psrA*, *exoR*, *expR*, *rosR*, *praR*) were found to be localized either in different regions of the chromosome or at the endogenious plasmids (reviewed by [24]). We can not exclude that mutations in GL3 and GL1 also target the regulatory genes.

4. Synthesis of nucleotide-sugar precursors

According to the sugar composition of acidic EPS, biosynthesis of its repeating units requires nucleotide sugars: UDP-glucose, UDP-glucuronic acid and UDP-galactose that are formed by central carbon metabolism. Several genes involved in synthesis of nucleotide-sugar precursors were identified in *R. leguminosarum* genomes. The *exoB* encodes UDP-glucose 4-epimerase, responsible for UDP-galactose production [55]. Mutations in this gene have a pleiotropic effect and influence the synthesis of different classes of galactose containing polysaccharides; namely acidic EPS, glucomannan, lipopolysaccharide, and probably the galactose-rich gel-forming polysaccharide [5,55,56].

The *exo5* gene encodes a UDP-glucose dehydrogenase responsible for oxidation of UDP-glucose to UDP-glucuronic acid. A mutation in *exo5* affects the production of extracellular acidic polysaccharide and capsular polysaccharide both of which contain glucuronic acid residues in the backbone chain [57,58].

5. Assembly of the EPS repeating unit

As mentioned above, heteropolysaccharides are polymers consisting of identical repeating units which can vary only by the distribution of modifying groups per monomer. Obviously, their assembly has to be stringently controlled. The biosynthetic pathways for the number of heteropolysaccharides have been elucidated [54,59]. It is evident from the obtained data that the unique structure of the repeating unit is governed by the specificity of nonprocessive glycosyltransferase (GT) catalyzing a certain step of the biosynthesis. This specificity is likely based on the ability of GT to recognize the sugar residue to be transferred, the acceptor, and the linkage to be formed. At present the complete pathway of EPS biosynthesis has not been determined for any of *R. leguminosarum* biovars, and only some individual steps have been characterized. Nevertheless, a detailed consideration even of these fragmentary data together with a comparative analysis of the available *R. leguminosarum* genome sequences allowed us to predict the genetic control of all steps in the repeating unit assembly at least for *Rlv* VF39 and closely related strains.

It was shown previously that assembly of a repeating unit of *R. leguminosarum* EPS starts with the addition of a glucose residue to the lipid carrier [60]. Biochemical studies and complementation analysis provided strong evidence that this reaction is conducted by the *pssA* gene product [36,61]. The *pssA* gene encodes the integral inner membrane protein UDP-glucose:polyprenyl-phosphate glucosephosphotransferase, belonging to a family of diverse bacterial sugar transferases. Members of this family catalyze the formation of a phosphodiester bond between polyprenol phosphate and hexoso-1-phosphate, which is donated by nucleotide sugars. The *pssA* gene is highly conserved in *R. leguminosarum* biovars and *R. etli* [44-46,50]. The *pssA* mutants do not produce EPS and as a consequence impaired the normal development of the nitrogen-fixing nodules on the appropriate plant hosts and the formation of biofilms both *in vitro* and on root hairs [5,45,46,50]. Some contradictory results exist concerning the influence of *pssA* mutations on the synthesis of CPS which displays similar structure to EPS in terms of glycosyl composition [5,30,62]. In the *Rlt* 5599 genetic background, *pssA* mutants still produce CPS at the level similar to that of a wild-type strain [46]. In contrast, both EPS and CPS are absent in the *pssA* mutant of *Rlv* 3841 [5], which indicates that the PssA protein might be involved also in the initiation of the CPS synthesis. It has been shown that expression of *pssA* depends on the environmental factors such as phosphate and ammonium concentrations and also on root exudates [63].

Bossio and co-workers have demonstrated that subsequent to addition of glucose to isoprenylpyrophosphate, two glucuronic acid residues are attached [64]. The attachment of the first GlcA is catalyzed by the *pssE* and *pssD* gene products [47,61]. This conclusion is based on the results of *in vivo* reciprocal complementation between the *pssED* and *spsK* genes of the *Sphingomonas* strain S88 and the *in vitro* sugar incorporation studies. Recently we confirmed directly the function of PssED. The corresponding mRNAs were translated in a wheat-germ cell-free system in the presence of liposomes obtained from *Rlv* VF39 phospholipids. The resulting proteoliposomes were used as an enzyme source in experiments on PssDE specificity (Ivashina et al., unpublished).

Both, PssD and PssE display similarity to GT family 28 (CAZy database, http://www.cazy.org/). Notably, the amino acid sequences of PssD and PssE are similar to the N-terminal and C-terminal halves of the glucuronosyl-(ß1,4)-glucosyltransferase SpsK, respectively. This observation has led us to the conclusion that PssD and PssE represent two subunits of the same enzyme. The proposed catalytic domain is localized in the peripheric inner membrane of the *pssE* subunit, in contrast to PssD, which was shown to be an integral inner membrane protein. We have also found that the integration of PssE into a membrane or liposomes strongly depended on the presence of PssD and *vice versa* (Ivashina et al, unpublished). It is interesting to note that the same was observed in the case of yeast proteins Alg13 and Alg14. It was found that Alg14 was needed for the correct positioning of Alg13 on the cytosolic face of the endoplasmic reticulum membrane mediating the formation of the active UDP-N-acetylglucosamine transferase complex [65,66]. Mutations in the *pssE* and *pssD* genes fully abolished EPS production and as a consequence resulted in defects of nodule infection [67-69].

PssC belongs to the GT family 2 containing a variety of inverting glycosyltransferases (enzymes that form glycosidic bonds with stereochemistry opposite to that of the glycosyl donor) that utilize a diverse range of nucleotide-sugar donors and participate in the synthesis of various types of polysaccharides [70]. This GT was assigned by Pollock and co-workers to glucuronosyl-(β-1,4)-glucuronosyltransferase, which catalyzed the attachment of the third sugar residue (GlcA) to the disaccharide (GlcA-β-1,4-Glc) lipid-linked intermediate with the formation of the β,1-4 glycosidic bond [61]. This conclusion was based on the comparative data on the genetic control of the first three steps of *R. leguminosarum* EPS and the *Sphingomonas* strain S88 sphingan assembly. Obviously, these data can not be considered as direct evidence, and conclusion on the PssC assignment needs additional experimental proofs.

Several *pssC* mutations have been characterized in various *R. leguminosarum* biovars backgrounds to date [46,47]. All of them were mapped at the N-terminus of PssC and resulted in the decreased amount (27-38%) of EPS in culture supernatants. However, structural analysis of EPSs secreted by these mutants showed them to be identical to that of the wild-type strains [46,47]. We proposed that the initiation of translation of *pssC* could be realized from the second potential start codon GTG located downstream from the spots of mutations and therefore leading to the synthesis of protein retaining enzymatic activity. In fact, Western blot analysis with antibodies against PssC demonstrated the synthesis of a truncated protein in the *pssC* mutant (Ivashina et al., unpublished). Attempts to introduce mutations into the central part of *pssC* in the *Rlv* VF39 background were unsuccessful. However, it was easy to homogenote *pssC* in the same strain carrying a mutation in the *pssD* gene, which failed to produce EPS. These results pointed to the detrimental effect of such mutations most probably due to the accumulation of lipid-linked intermediates in the cytoplasmic membrane and as a result to inability of the lipid carrier to be released for other essential cellular functions. The data obtained with the use of different genetic systems led to the same conclusions [61].

The PssJ protein is the last glycosyltransferase for which its biochemical function was ascertained experimentally. The *Rlt* RBL5515 strain carrying mutation in the *pssJ* gene (known as *exo*344::Tn*5*), synthesizes residual amounts of EPS, the repeating unit of which lacks the terminal galactose of the side chain. On the basis of the structural features of the polysaccharides synthesized and the results of an analysis of the enzyme activities involved, it was hypothesized that the galactosyltransferase catalyzing formation of the β1-3 linkage between sub-terminal (Glc) and terminal (Gal) sugar residues in the octasaccharide unit is affected in this strain [6]. PssJ did not reveal any homologs in protein databases and therefore it could be referred to a family of "not-classified glycosyltransferases" (CAZy database).

It can be seen from the EPS repeating unit structure that the third (GlcA) and the fourth (Glc) sugar residues in the backbone chain are linked by the α1-4 glycosidic bond (Table 1). PssS is the only enzyme which can be responsible for this reaction. According to homology search data, the PssS was referred to the GT family 1 (CAZy database), which integrates the retaining glycosyltransferases forming glycosidic bonds with stereochemistry identical to that of the glycosyl donor. These enzymes were shown to be involved in exopolysaccharide, lipopolysachharide, and slime polysaccharide colanic acid biosynthesis. We were unable to disrupt the *pssS* gene in the wild-type strain *Rlv* VF39, but easily inactivated it in the *pssD* mutant (Eps^-). It is likely that in this case the inactivation of *pssS* also leads to the accumulation of toxic lipid-linked intermediates as it was proposed for *pssC* mutants.

Summarizing, specific GTs were assigned to the assembly of the backbone chain of the octasaccharide unit, as well as in the attachment of the terminal Gal in the side chain. It should be emphasized that in all known *R. leguminosarum* EPS structures the backbone chains are identical (Table 1). At the same time, in genomes of *R. leguminosarum* strains, which synthesize similar EPSs the orthologs of *pssAEDCS* genes are present (Fig. 1). Therefore, we can conclude that the prediction has been made correctly. The same statement is true for PssJ catalyzing the attachment of the terminal Gal residue: in the *R. leguminosarum* genomes, where *pssJ* has not been found, the *pssK*, which has been predicted to modify this Gal residue (see below), is also missing.

As mentioned above, *R. leguminosarum* and *R. etli* can produce EPSs with side chains varying in their length, sugar composition and type of glycosidic linkages. It should be noted, that the set of GT genes in Pss-I clusters can also vary. Studies on the genetic control of EPS biosynthesis are impeded to a considerable degree due to the fact that when genome sequences are available, nothing is known about the structure of synthesized EPS and *vice versa*. The *Rlv* VF39 and *Rlt* TA1 represent the only pair when (i) the structure of EPSs and sequences of the Pss-I clusters are determined; (ii) both strains produce structurally identical EPSs but differ in the sets of GT genes; (iii) data on mutational analysis of *Rlv* VF39 GT genes are obtained. Taking into account all these considerations, we have picked the *Rlv* VF39/*Rlt* TA1 pair for prediction of the pathway of side chain biosynthesis. In this case the question arises, which glycosyltransferase initiates branching by attachment of the Glc residue via the β1-6 bond, and which GT(s) is (are) responsible for the attachment of two subsequent Glc residues by formation of the β1-4 glycosidic linkage. It is obvious, that in *Rlt* TA1 only two GTs (PssF and PssI) can perform these functions. In addition to PssF and PssI, two other GTs (PssH and PssG) can participate in the side chain assembly in the *Rlv* VF39. We introduced mutations into all four genes (*pssFGHI*) in *Rlv* VF39 and found that the structures of EPSs of mutant strains were identical to that of the parental strain, and only the level of acidic EPS production decreased. Based on these results, we can conclude that the action of each GT considered in this system can be interchangeable.

In our opinion, PssF is the best candidate to play the role of GT which catalyzes the attachment of the Glc residue by formation of the β1-6 bond. Firstly, in all *Rhizobium leguminosarum* strains the EPS side chain starts with the Glc residue attached to the backbone chain via the β1-6 bond. At the same time, PssF is present in all PssI-clusters sequenced up to now. Secondly, PssI, PssH and PssG reveal a rather high level of similarity with each other especially in the N-terminal parts of their amino acid sequences where catalytic domains are located. In contrast, PssF is practically non-homologous to that of three GTs but shows although weak but yet reliable homology of its N-terminal half with GTs attaching the Glc residue via the β1-6 bond (e.g. ExoO from *S. meliloti* [20]).

If our prediction on the PssF function is correct, the attachment of two subsequent Glc residues could be achieved by single GT (PssI) in the *Rlt* TA1, and as many as three GTs (PssI, PssH and PssG) could participate in this process in the *Rlv* VF39. Apparently, PssI in the *Rlt* TA1 strain is to a certain extent tolerant to the acceptor structure and the identity of EPS repeating units probably is attained at the expense of high specificity of PssJ, which catalyzes the last step of the EPS assembly.

It seems that in the *Rlv* VF39 the subsequent attachment of Glc residues is achieved by two separate GTs, namely PssI and PssH. This assumption is based on a comparison of the amino acid sequences of these homologous GTs. A rather low level of similarity of their C-terminal parts, containing the putative acceptor recognition domain was observed. In contrast, PssG reveals a very high level of homology to PssI over its entire amino acid sequence (more than 80% similarity). It is plausible to assume that PssI and PssG are isoenzymes, which handle the same step in the EPS assembly. Thereby, genetic control of the repeating unit biosynthesis in the *Rlv* VF39 resembles that of *S. meliloti*, where the attachment of the sugar residue at each step of biosynthesis is catalyzed by specific GT, and even two GTs can participate in catalysis at some steps of the pathway.

A presumptive circuit of the EPS repeating unit assembly in *Rlt* TA1 and *Rlv* VF39 is presented in Figure 2. It should be noted that PssA, PssDE, PssC, PssS, PssF and PssI/PssG apparently comprise the basic set of GTs involved in biosynthesis of all EPS under consideration. Taking into account the mentioned variability of side chains of these EPS, one can expect that the basic set of GTs in corresponding *R. leguminosarum* strains is augmented. Indeed, PssJ is present in *Rlv* 3841, *Rlv* VF39, *Rlt* TA1, *Rlt* WSM2304, *Re* CFN42, but is absent in *Re* CNPAF512 and *Re* CIAT 652. At the same time, genes for family 2 GT PsaC and family 1 GT PsaD were found in the PssI-clusters of the latter strains. In addition, genes encoding family 2 GTs PsaB and PsaE, amino acid sequences of which are similar to GTs attaching the Glc residue via the β1-6 bond, were localized in *Rlt* WSM2304 and *Re* CIAT 652, respectively.

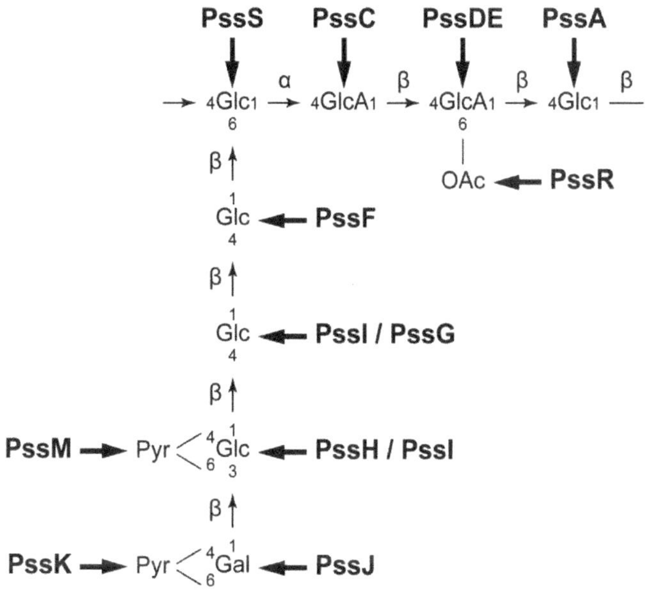

Figure 2. Model for the EPS repeating unit assembly in *Rlv* VF39 and *Rlt* TA1 strains. Abbreviations: Glc, glucose; GlcA, glucuronic acid; Gal, galactose and Pyr, ketal pyruvate group.

6. Genes involved in modification of EPS

Three genes, *pssR*, *pssM*, and *pssK*, were identified within the Pss-I cluster that may be involved in the modification of EPS in all *R. leguminosarum* strains as well as in *R. etli* CFN42 (Fig. 1). A homology search for the *pssR* gene product revealed similarity of its C-terminal region (amino acids 87-136) with the corresponding parts of a large number of acetyltransferases, including well characterized rhizobial NodL [71-73], *E. coli* LacA and CysE [74,75], *S. aureus* Cap1G [76]. The hexapeptide repeat [LIV]-G-x(4) (IPR001451, bacterial transferase hexapeptide repeat) is present in the homologous regions of all these proteins including the PssR. It should be noted that PssR shares similarity over nearly its entire length with conservative putative acetyltransferases from different representatives of *Bacillaceae*. Proteins from this family use acetyl coenzyme A as the acetyl donor and acetylate different substrates including capsular and extracellular polysaccharides, lipooligosaccharides, chitin fragments, N-acetylglucosamine and antibiotics [64,77,78].

Notably *pssR* orthologs were found in all PssI-clusters (Fig. 1). This observation is in agreement with the data concerning the major site of O-acetylation localized at the second GlcA residue in the backbone chains which are identical in all EPS with known structure.

Insertional inactivation of *pssR* in the *Rlv* VF39 genome does not result in a complete absence of acetyl groups in EPS. This suggests the existence of other gene(s) elsewhere in the *Rlv* VF39 genome needed for the EPS acetylation. Decreasing of the level of acetylation has no effect on nodule development and nitrogen fixation. Similar data were obtained for *S. meliloti* ExoZ mutants, which failed to acetylate succinoglycan. It was shown that the acetyl decoration of succinoglycan is not absolutely required for a nodule formation; however it increased the efficiency of infection threads initiation [79,80].

The amino acid sequence of PssM shares homology with several known and putative ketal pyruvate transferases, including ExoV from *S. meliloti* and GumL from *Xanthomonas campestris*. Knock-out of the *pssM* gene does not result in the loss of ability to produce HMW EPS, but leads to the absence of the pyruvic acid ketal group at subterminal glucose in the repeating unit of EPS as it was shown by ^{13}C and ^{1}H NMR analyses. Complementation *in trans* restored the EPS modification in the *pssM* mutant [81]. Disruption of the *pssM* gene led to essential disturbances in symbiosis. Thus, the *pssM* mutation resulted in the formation of aberrant non-nitrogen-fixing nodules on peas. Ultrastructural studies of mutant nodules indicated that the infection thread formation, release of bacteria into the plant cell cytoplasm and early steps of differentiation of bacteroids were not affected. However, further stages in the symbiosome development and maintenance were arrested. We proposed that the induction of early senescence of symbiosomes depends on the failure in recognition mechanisms and, what is essential, that recognition of a micro-symbiont by the host plant is important not only at early stages of symbiosis, but also during its intracellular period of life [81]. Moreover, an accumulation of very large starch granules observed in infected and non-infected cells, suggests that the plant-derived photosynthates, which serve as an energy source for nitrogen fixation [82] are not fully consumed in *pssM* induced nodules. The mechanisms which modify the "symbiotic" nodule to starch accumulation may include alteration in the starch phosphorylase activity and (or) its expression [83].

Our finding that mutation in *pssM* abolishes pyruvylation of only one of the two sugar residues in *Rlv* VF39 EPS permits to propose that pyruvylation of the terminal galactose may be controlled by the *pssK* gene localized within the Pss-I cluster. The PssK amino acid sequence was similar to proteins containing the pyruvyltransferase domain IPR007345, including Pvg1p from *Shizosaccharomyces pombe*, YveS, YvfF and YxaB from *Bacillus subtilis* [84], and EpsL from *Streptococcus thermophilus* [85]. It was shown that Pvg1 catalysed the transfer of the pyruvyl group to Galβ1,3-residues in N-linked galactomannan chains [86]. Interestingly, no sequence homology was observed between the PssM and PssK proteins that can reflect different substrate specificities of these enzymes. No direct evidence for the *pssK* function was obtained in any of the *R. leguminosarum* strains. Our preliminary data indicate that knock-out of *pssK* abolishes the EPS synthesis and results in a non-slimy phenotype of colonies (Ivashina et al., unpublished). It is possible that pyruvyl modification of the terminal sugar residue may be necessary for the efficient polymerization or export of EPS as it was proposed for *S. meliloti exoV* which is involved in pyruvylation of succinoglycan [19,21].

As seen from Figure 1, in the *Re* CNPAF512 and *Re* CIAT 652 the *pssK* gene is absent and *pssM* is replaced by non-orthologous *psaG* and *psaF* genes, respectively. The latter genes presumably can also encode ketal pyruvate transferases since IPROO7345 domain (Polysacch_pyruvyl_Trfase) was found in their amino acid sequences. Unfortunately, EPS structure of both *R. elti* strains remains unknown, but one can suppose that at least in their side chains it differs from that of *Rlv* VF39.

This assumption is based on the observation that the sets of GTs in their Pss-I clusters differ from that of *Rlv* VF39. One can see that genes of ketal pyruvyl transferases are different also. Therefore, this finding additionally argues towards high substrate specificity of these modifying enzymes.

7. Polymerization and secretion of EPS

At present three pathways are known for the export of carbohydrate polymers in bacteria: (i) Wzx/Wzy-dependent; (iii) ATP-binding cassette (ABC) transporter-dependent; and (iii) synthase-dependent (reviewed in detail by [87,88). In the Wzx/Wzy dependent mode individual undecaprenol diphosphate-linked polysaccharide repeating units are assembled and translocated across the cytoplasmic membrane by a transport process requiring a Wzx protein (putative translocase or "flippase") followed by their polymerization at the periplasmic space by the Wzy protein [87,89]. Further export of polysaccharides from the periplasm to the cell surface has been shown to be dependent upon additional protein(s) assigned to the polysaccharide co-polymerase (PCP) and the outer membrane polysaccharide export (OPX; formerly OMA) families [87,89]. The best characterized member of the OPX family is an *E. coli* K30 outer membrane lipoprotein Wza which forms a multimeric 'secretin-like' structure mediating translocation of the group 1 capsular polysaccharide across the outer membrane. The high-resolution crystal structure of Wza has been determined and this shed light on the CPS traffic across the outer membrane [90]. It has been postulated that the Wza protein together with co-polymerase Wzc form a molecular scaffold that spans the cell envelope and promotes the export of CPS (reviewed by [87]).

Current data indicate that polymerization and secretion of acidic EPS in *R. leguminosarum* biovars might be realized in a Wzx/Wzy-dependent manner. This supposition is based on the structural similarity of *R. leguminosarum* proteins PssTNOP and PssL to enzymes involved in CPC/EPS biosynthesis. The main data concerning the elucidation of the role of the mentioned proteins in EPS biosynthesis were obtained in the group of A. Skorupska.

The precise function of PssL has not been determined due to the inability to knock-out the *pssL* gene in the *Rlt* TA1 strain. However, the amino acid sequence similarity and hypothetical protein secondary structure allow placing the PssL protein within Wzx-like translocases that belong to the polysaccharide specific transport (PST) family. The predicted secondary structure of the *Rlt* TA1 PssL inner membrane protein has been supported experimentally with a series of PssL-PhoA and PssL-LacZ translational fusions. The obtained results clearly show that PssL displays characteristic features of members of the PST protein family comprising transporters with 12 membrane spanning segments, a large cytoplasmic domain, located between the sixth and seventh transmembrane segments, and amino and carboxyl termini located in the cytoplasm [91].

In addition to *pssL*, four closely linked *pssTNOP* genes were identified in the Pss-I cluster of various representatives of *R. leguminosarum* (Fig. 1) and assigned to be involved in polymerization and export of EPS [42,92,93]. The PssT protein has been predicted to be a Wzy-like protein that together with PssL might be responsible for Wzx/Wzy-like-dependent EPS polymerization and translocation. This conclusion is based on structural homology of PssT with inner membrane proteins belonging to the PST family of proteins that are involved in transport of complex polysaccharides [42]. The PssT consists of 12 transmembrane helices, a large periplasmic loop between the ninth and tenth transmembrane segments, and cytoplasmic N- and C-termini. The predicted topology of PssT has been confirmed with the use of a series of PssT-PhoA fusion proteins and a complementary set of PssT-LacZ fusions. The role of PssT in EPS biosynthesis has been investigated further by plasmid integration mutagenesis. The *Rlt* TA1 *pssT* mutant lacking the C-terminal part of PssT (starting after the 363-rd amino acid located in the periplasmic loop) produced increased amounts of total EPS with an altered distribution of high- and low-molecular-weight forms in comparison to the wild-type strain [42]. The PssT was structurally and functionally homologous to *S. meliloti* ExoT, which together with ExoP and ExoQ proteins is involved in the final stages of succinoglycan biosynthesis [21,94].

The PssP protein displays significant structural features with members of the copolymerase (PCP2a) family that are involved in the synthesis of high-molecular-weight CPS/EPS including the well characterized ExoP protein from *S. meliloti* and a Wzc protein from *E. coli* [95-98]. Membrane topology of the PssP protein resembles that of ExoP. Both proteins consist of a periplasmic hydrophilic N-terminal domain flanked by two potential transmembrane helices and a cytoplasmic C-terminal domain. The C-terminus contains the conserved Walker motifs A and B for ATP binding. Coiled-coil regions characteristic of PCP2a members were found both in periplasmic and cytoplasmic C-terminal domains of PssP [97]. ExoP has been shown to be an autophosphorylating protein tyrosine kinase. Site-directed mutagenesis of specific tyrosine residues in the cytoplasmic domain of ExoP has been demonstrated to result in an altered ratio of LMW succinoglycan to HMW succinoglycan [98]. It has been hypothesized that the phosphorylation state of ExoP might regulate the degree of succinoglycan polymerization by controlling polymerization activities of other proteins, e.g., ExoQ and ExoT [94]. A putative site for tyrosine phosphorylation has been found in the PssP protein, however, the functional significance of this site for phosphorylation of PssP is still unknown. Unlike ExoP, no tyrosine-rich region is found at the C-terminus of PssP.

Several mutations have been introduced into the *Rlt* TA1 *pssP* gene and shown to display different effects. The *Rlt* TA1 mutant with the deletion of the entire coding region of *pssP* is deficient in EPS production. A mutant that synthesizes a functional N-terminal periplasmic domain but lacks the C-terminal part of PssP produces significantly reduced amounts of EPS with a slightly changed low- to high-molecular form ratio. A *pssP* mutant with the disrupted 5'-end of the gene synthesizes exclusively low-molecular-weight EPS suggesting the importance of the functional N-terminal domain in the degree of polymerization [99].

The *pssN* gene encodes a protein which is homologous to the outer membrane polysaccharide export OPX protein family involved in CPC/EPS export [96,99]. Like other members of the OPX family, PssN contains a conserved signal peptidase II cleavage site in the lipobox. With the use of *pssN-phoA* and *pssN-lacZ* gene fusions and *in vivo* acylation with [^{3}H]-palmitate it has been shown that PssN is a lipoprotein associated with the outer membrane and with the N-terminal signal sequence directed to the periplasm. Several experimental approaches (indirect immunofluorescence with anti-PssN and fluorescein isothiocyanate-conjugated antibodies and protease digestion of spheroplasts and intact cells of *Rlt* TA1) indicated that PssN is not exposed to the surface, but oriented towards the periplasmic space. Investigation of the secondary structure of the purified PssN-His$_6$ protein by Fourier transform infrared spectroscopy (FTIR) revealed the predominant presence of beta-structure; however, alpha-helices, which could be involved in association with murein and/or other proteins, were also detected. Similar to OPX proteins, PssN has been shown to exist in a homo-oligomeric form of at least two monomers suggesting that together with PssP it might be involved in the formation of efflux channels for EPS export. No *pssN* mutants have been obtained so far. However, the increased amount of the PssN protein in *Rlt* TA1 correlated with a moderate enhancement of EPS production [92,99].

It was hypothesized by Mazur and co-workers [42], that the PssT protein, acting in complex with PssP and PssN, could be involved in controlling the rate of polymerization of repeating units and export of EPS to the cell surface. The PssN could interact with the periplasmic loop of the PssP protein, whereas the transmembrane regions of PssP could associate with the corresponding PST transporter, facilitating polymer export across the bilayer sructure.

The *pssO* gene product reveals no homology with known bacterial proteins. However, its participation in EPS biosynthesis has been confirmed by mutagenesis analysis: deletion of *pssO* in *Rlt* TA1 abolished EPS production and overproduction of PssO increased EPS secretion. Subcellular fractionation, *pssO–phoA and pssO-LacZ* translational fusion analyses and immunolocalisation of PssO on the *Rlt* TA1 cell surface by electron microscopy demonstrated that PssO is secreted to the extracellular medium and remains attached to the cell. The secondary structure of PssO–His$_6$, as determined by FTIR spectroscopy, is rich in α-helices (32%) [100]. It was speculated by Marczak and co-workers, that PssO may function as a periplasmic "chaperon" coating the EPS polymer and protecting it from the action of glycanases and/or be co-transported with the polysaccharide through a channel formed in the outer membrane. However, the authors can not exclude that PssO forms some kind of a cell surface structure essential for the assembly of the EPS transporter complex and its stability [100].

Using plasmid-borne transcriptional fusions of promoters of *pss* genes with the reporter gene *lacZ,* the effect of root exudate, phosphate, and ammonia on expression of *pssT, pssN, pssO,* and *pssP* genes in the wild-type *Rlt* TA1 background was examined. A stimulating effect of these environmental factors on *pssO* and *pssP* was observed. Interestingly, within the putative *pssO* promoter the divergent *nod*-box element was found. The *pssO* promoter was slightly inducible in a flavonoid-dependent manner in wild-type strains *Rlt* TA1 and *Rlt* 843 and very weakly in a mutant of *Rlt* 843 that lacks the regulatory *nodD* gene. The regulation of EPS production by NodD might be an important finding that connects EPS synthesis to the symbiosis of *R. leguminosarum* with clovers [101].

pssTNOP genes from *Rlt* TA1 have corresponding orthologs in genomes of *R. leguminosarum* and *R. etli* (Fig. 1) suggesting that there is a common mechanism of their action at least in these strains. The *pssL* gene is not so conserved: in *Re* CNPAF512 and *Re* CIAT 652 it was replaced by the non-homologous gene designated as *psaI.* However, the PsaI protein can be assigned to the same family IPROO2797 (Polysacc_synth.) as PssL. As mentioned above, these are just the strains for which EPS side chains were predicted to have different structure. We suggest that namely PssL/PsaI can be specific for the structure of EPS to be translocated across the inner membrane to the periplasmic space. Probably the main function of PssL-like proteins consists in the stringent control of the identity of repeating units which are further polymerized by the action of PssT.

Noteworthy, another gene designated as *psaA* was found in the *pssV-E* operons of *R. etli* strains described here. It encodes the protein, which can be assigned to the O-antigen ligase-like protein family PF13425. The function of this protein in EPS biosynthesis still remains unclear, however its participation in translocation or polymerization of the EPS repeating units can not be excluded. It should be noted that *psaA* homologs were found in all *R. leguminosarum* genomes under consideration, but they localize in different chromosomal regions far from the Pss-I cluster.

8. Processing of EPS

It was well documented that several rhizobial species (e.g., *S. meliloti, R. leguminosarum, Rhizobium sp.* NGR234, *Bradyrhizobium*) produce two distinct EPS classes that differ in size: HMW forms symbiotically active LMW forms [13-17,27,102]. LMW forms of succinoglycan can be produced in *S. meliloti* either by direct export of the low-polymerized octasaccharide repeating units [95,103] or by depolymerization of HMW polysaccharide by specific glycanases [104,105]. These observations are also true for *R. leguminosarum* model. As discussed above, mutations in the genes controlling the polymerization and transport of EPS in the *Rlt* TA1 strain could contribute to the production of LMW EPS [93].

Moreover, three gene products have been shown to participate in degradation of HMW EPS in *R. leguminosarum*: PlyA, PlyB, and PssW (formely PssT in *Rlv* VF39 and PssT1 in *Re* CFN42). PlyA and PlyB, which are similar to each other display homology to bacterial and fungal polysaccharide lyases [106,107]. Ten copies of a novel heptapeptide repeat motif were found in the sequences of these proteins which may constitute a fold similar to that found in the family of extracellular pectate lyases. These proteins are secreted via the PrsDE system of the Type I secretion system which is conserved in different *Rhizobium* species [108]. PlyA appears to remain attached to the cells, while the PlyB diffuses beyond the edge of the colony [109]. It has been proposed that the presence of extra 50 amino acids near the C-terminal domain of PlyA could be responsible for maintaining the protein attachment to the cell surface. Both proteins were inactive in the EPS-defective mutants and did not degrade mature EPS. They may be only active in association with the rhizobial cell surface suggesting the activation of PlyB by an EPS-related (nascent EPS or an intermediate in EPS biosynthesis) component [109]. The PlyA and PlyB glycanases are not specific for EPS but can also degrade carboxymethyl cellulose (CMC). In cultured bacteria the *plyA* gene is expressed at a very low level, while a *plyB* mutant has a very large reduction in degradation of EPS and CMC [106]. Cultures of *plyB* mutants contained an increased ratio of EPS repeating units to the reducing ends indicating that EPS was present in a longer-chain form, and this correlated with a significant increase in the culture viscosity. A double *plyAB* mutant retained residual CMC degradation, indicating the existence of additional activities in the cell. Recently, a third gene named *plyC* has been found in the genome of *Rlv* 3841. PlyC is secreted via the PrsDE secretion system and displays common structural features with PlyA and PlyB proteins [107], indicating that it may perform a similar function. The analysis of the symbiotic properties of a *plyAB* double mutant revealed that genes involved are not required for symbiotic nitrogen fixation and that nodulation was not significantly affected [109]. In *Rlt* TA1 the *plyA* gene is missing.

Recently we have characterized another glycosylhydrolase encoded by the *pssW* gene and provided experimental evidence for its participation in the EPS processing (Kanapina et al, unpublished). The *pssW* gene has its counterparts in *R. leguminosarum* and *R. etli* genomes, and is located within the *pssV-E* operon. The PssW protein was referred to the family 10 of glycosylhydrolases of the GH-A clan, which are retaining glycoside hydrolases displaying endo-1,3-β-xylanase and endo-1,4-β-xylanase activities (CAZy database). We have shown that PssW is synthesized as a precursor of 42.4 kDa followed by its translocation across the cytoplasmic membrane and by cleavage of the 46 amino acid signal peptide. The periplasmic localization of the PssW indicates that it might be active toward the nascent EPS before its secretion outside the cell. The deletion of the *pssW* gene resulted in approximately a 3-fold increase in the ratio of HMW to LMW EPS and as a result in the increase of viscosity of the culture supernatant. Complementation of the *pssW* mutation restored the wild-type phenotype, and even increased the level of secreted LMW EPS. The PssW purified from the periplasmic space did not degrade CMC and succinoglycan, and revealed a 2-fold decrease in hydrolysis of EPS in the *Rlv* VF39 *pssM* mutant which lacks one of the two pyruvyl groups. The latter suggests the importance of pyruvyl modification on the degradation activity of the PssW. The knock-out of *pssW* did not significantly affect nodulation of peas probably due to the incomplete block in the LMW EPS synthesis.

It can be concluded that *R. leguminosarum* strains like *S. meliloti* were able to realize different strategies for production of LMW forms of polysaccharide: regulation of the degree of polymerization of EPS and hydrolase-mediating cleaving of EPS. Complex mechanisms directing the synthesis of LMW EPS can reflect the evolutionary benefit of rhizobia possessing different pathways of LMW EPS production that can be considered as important molecules for cell to cell communications during the development of nitrogen-fixing nodules.

9. EPS-deficient mutants as a possible model for studying bacterial gene expression in symbiosis

A decade ago in collaboration with the group of Prof. B. Rolfe we demonstrated that mutations in the *pssA* gene in *Rlv* VF39 and *Rlt* ANU794 strains led not only to abolishing the capacity of these strains to synthesize EPS but also to induction or up-regulation of at minimum 22 proteins. The differences identified in the *pssE* and *pssD* mutants of the *Rlv* VF39 strain were a distinct subset of the same protein synthesis changes that occurred in the *pssA* mutant (9 out of 22 changes) [110]. Genetic complementation of *pssA* restored wild-type protein synthesis levels. We concluded that the observed alterations in protein synthesis are caused either by dysfunction of the PssA protein itself or are a response to the absence of EPS.

The N-terminal sequence analysis of 15 members of the *pssA* mutant stimulon was performed and the unique amino acid sequences were determined for 11 proteins [110]. However, our attempts to identify these proteins were unsuccessful, and it was clear why. None of *R. leguminosarum* and *R. etli* genomes was sequenced at that time. We have repeated our attempts now and assigned 9 proteins to their known orthologs in *Rhizobiaceae* species. The results of this analysis are summarized in Table 3. The functions of all of these orthologs are predicted. Five proteins belong to ABC transporter systems; two proteins are assigned as NADH-dependent FMN reductases, one as taurine dioxygenase, and the last as disulfide isomerase. Most genes encoding the proteins were mapped on indigenous plasmids.

It is interesting why mutation in a single gene resulted in induction or up-regulation of genes encoding proteins with such different functions. We propose here the following hypothesis. It is based on the data that the expression of the *pssA* gene was not observed in bacteroids [63,111]. In our opinion, members of the *pssA* mutant stimulon might be just those proteins which are expressed at the symbiotic state of bacteria. In this case, PssA, directly or indirectly, could play the role of a negative regulator of their gene expression in free-living bacteria. EPS itself could play the same role.

Obviously, the hypothesis can be easily verified. It is sufficient to examine whether the genes listed in Table 3 are expressed in bacteroids. If our assumption turns out to be correct, EPS-deficient mutants of *R. leguminosarum* could be considered as an attracting model for studying bacterial gene expression during symbiosis.

Table 3. Identification of proteins from the *pssA* mutant stimulon in *Rhizobiaceae* species.

Rlt ANU437 mutant-responsive proteins			Characteristics of *Rlt* ANU437 orthologs from the other *Rhizobiaceae* species					
Spot/status*	M_r/pI*	N-terminal sequence (position in the ortholog)**	Protein (accession number)/organism	M_r/pI***	Signal peptide position	Domain/motif	Predicted function	Gene location
n2/up-reg. (+)	37.436/6.31	**QTDAPLLSKV** 27 36	PRL100398 (YP_770674)/*Rlv* 3841	34,376/5.65	1-26	IPRO15168, NMT1/TH15-like	ABC transporter protein	pRL10
n4/induced n6/induced	22,147/5.90 22,147/6.23	**SNDLIVGFSG** 2 11 **SNDLIVGFSGNL** 2 13	ATCR1_07864 (EHH06861)/*Agrobacterium tumefaciens* CCNWWGS0386	19,791/5.35	-	IPRO05025, FMN_red	NADH-dependent FMN reductase	Not known
n5/induced n7/induced	22,880/6.09 22,880/6.38	**SAPRLVGLAGSF** 2 13 **SAPRLVGLAGSF** 2 13	PRL90304 (YP_765590)/*Rlv* 3841	20,211/5.46	-	IPRO05025, FMN_red	NADH-dependent FMN reductase	pRL9
n8/up-reg. (++)	34,000/7.50	**SNPVLVNQIIPESRVV**(G/T)**LT**(A/D)SV 2 23	Rleg2_5751 (YP_002278025)/*Rlt* WSM2304	33,910/6.48	-	IPRO03819, Taurine_dOase	Taurine dioxygenase	pRLG202
n15/up-reg. (++) n21/up-reg. (+++)	34,321/5.15 34,197/5.02	**ADKKVVVAYQTDALP** 21 35 **ADKKVVVAYQTDALP** 21 35	PRL100239 (YP_770519)/*Rlv* 3841	34,400/4.94	1-20	IPRO15168, NMT1/TH15-like	ABC transporter protein	pRL10
n16/induced	30,566/5.11	LXX**QQKKEFGEFIKQYL** 26 42	RL2091 (YP_767689)/*Rlv* 3841	24,820/5.04	1-25	IPRO01853, DSB-like_Thioredoxin_dom	Disulfide isomerase	chromosome
n17/up-reg. (+) n22/up-reg. (+++)	30,456/4.97 114,862/4.99	**EDKSIKVGIMAGEEE** 40 55 **EDKSIKVGIMAGEEEDIXRVVASEAAK** 40 66	PRL90303 (YP_765589)/*Rlv* 3841	26,417/4.86	1-39	IPRO04872, Lipoprotein NlpA	Lipoprotein, component of ATP-driven transport system	pRL9
n18/up-reg. (+++)	45,109/4.78	**DGLSGAPAPFDKGGV** 25 39	PRL90245 (YP_765531)/*Rlv* 3841	34,271/4.69	1-24	PF13407, Peripla_BP_4	Periplasmic solute-binding component of ABC transporter	pRL9
n19/up-reg. (++)	34,633/4.94	**TDTVKLRYLASQGNL** 25 39	Rleg_6388 (YP_002979383)/*Rlt* WSM1325	32,191/4.98	1-25	IPRO15168, NMT1/TH15-like	ABC transporter protein	pR132505

* Spot number: arbitrary numbers were assigned to proteins which exhibited a change in their relative synthesis levels in response to a Tn*5* insertion in the *pssA* gene of *Rlt* ANU794. Status: induced, when protein is absent in the wt strain and synthesized in the mutant strain; up-reg (+, ++, or +++), when 2-, 3- or 5-fold (or more) level of individual protein synthesis was observed in the mutant strain. Observed M_r and pI values of the protein based on the migration of proteins in gels. Data were taken from Table 2 [110].

** Data on N-terminal amino acid sequences were taken from Table 3 [110]. Identical amino acids in the matched sequences in the ortholog proteins are shown in bold letters, and similar amino acids are underlined. Positions of these sequences in the orthologs are indicated bellow the sequences.

*** M_r and pI of the ortholog proteins were calculated from their amino acid sequences.

10. Conclusion

The EPSs described in this review have common structure in that their repeating units possess identical backbone chains and the same β1-6-linked glucosyl residue starting the side chain. The diversity of EPSs is specified by the structure of their side chains. Presumably, structural information contained in side chains determines participation of EPS in symbiosis as signaling factors. This assumption follows from at least the observation that the absence of a single pyruvyl group in the side chain dramatically disturbed symbiotic properties of bacteria.

Most of the *R. leguminosarum* and *R. etli* genes involved in EPS biosynthesis are localized within the single cluster Pss-I. Gene arrangement in this cluster is similar. This implies that Pss-I clusters of these bacterial species have been evolved from a common ancestor. The differences mostly include genetic rearrangements of GTs genes, genes for modification of certain glycosyl residues in the side chain, and genes *pssL/psaI* thought to be involved in controlling translocation of the repeating units. It seems that each variant of Pss-I cluster originated as a result of co-evolution of all these genes.

Author details

Tanya V. Ivashina*
Skryabin Institute of Biochemistry and Physiology of Microorganisms, Russian Academy of Sciences, Pushchino, Moscow Region, Russia

Vladimir N. Ksenzenko
Institute of Protein Research, Russian Academy of Sciences, Pushchino, Moscow Region, Russia

Acknowledgement

We are thankful to Yu.A. Knirel for providing an information dealing with rhizobia EPS structures and A.S. Glukhov, T.B. Kuvshinkina and L.A. Kulakov for the help in the manuscript preparation. This work was supported by grant 11-04-02120a from the Russian Foundation for Basic Research.

11. References

[1] Weir BS (2011) The Current Taxonomy of Rhizobia. New Zealand Rhizobia Website. Available: http://www.rhizobia.co.nz/taxonomy/rhizobia.html.
[2] Zevenhuizen LP, Scholten-Koerselman HJ (1979) Surface Carbohydrates of *Rhizobium*. I. Beta-1, 2-glucans. Antonie Van Leeuwenhoek. 45: 165-175.
[3] Breedveld M, Miller KJ (1994) Cyclic β-Glucans of Members of the Family *Rhizobiaceae*. Microbiol. Rev. 58: 145-161.

* Corresponding Author

[4] Laus MC, Logman TJ, Lamers GE, Van Brussel AA, Carlson RW, Kijne JW (2006) A Novel Polar Surface Polysaccharide from *Rhizobium leguminosarum* Binds Host Plant Lectin. Mol. Microbiol. 59: 1704-1713.

[5] Williams A, Wilkinson A, Krehenbrink M, Russo DM, Zorreguieta A, Downie JA (2008) Glucomannan-Mediated Attachment of *Rhizobium leguminosarum* to Pea Root Hairs is Required for Competitive Nodule Infection. J. Bacteriol. 190: 4706-4715.

[6] Breedveld MW, Cremers HC, Batley M, Posthumus MA, Zevenhuizen LP, Wijffelman CA, Zehnder AJ (1993) Polysaccharide Synthesis in Relation to Nodulation Behavior of *Rhizobium leguminosarum*. J. Bacteriol. 175: 750-757.

[7] Zevenhuizen LPTM, van Neerven ARW (1983) Surface Carbohydrates of *Rhizobium*. IV. Gel-Forming Capsular Polysaccharide of *Rhizobium leguminosarum* and *Rhizobium trifolii*. Carbohydr. Res. 124: 166-171.

[8] Campbell GR, Reuhs BL, Walker GC (2002) Chronic Intracellular infection of Alfalfa Nodules by *Sinorhizobium meliloti* Requires Correct Lipopolysaccharide Core. Proc. Natl. Acad. Sci. USA. 99: 3938-3943.

[9] Fraysse N, Couderc F, Poinsot V (2003) Surface Polysaccharide Involvement in Establishing the *Rhizobium*-Legume Symbiosis. Eur. J. Biochem. 270: 1365-1380.

[10] Lepek VC, D'Antuono AL (2005) Bacterial Surface Polysaccharides and Their Role in the Rhizobia-Legume Association. Lotus Newsletter. 35: 93-105.

[11] Skorupska A, Janczarek M, Marczak M, Mazur A, Król J (2006) Rhizobial Exopolysaccharides: Genetic Control and Symbiotic Functions. Microb. Cell Factories. 5:

[12] Gage DJ (2004) Infection and Invasion of Roots by Symbiotic, Nitrogen-Fixing Rhizobia During Nodulation of Temperate Legumes. Microbiol. Mol. Biol. Rev. 68: 280-300.

[13] Djordjevic SP, Chen H, Batley, M, Redmond JW, Rolfe BG (1987) Nitrogen Fixation Ability of Exopolysaccharide Synthesis Mutants of *Rhizobium* sp. Strain NGR234 and *Rhizobium trifolii* is Restored by the Edition of Homologous Exopolysaccharides. J. Bacteriol. 169: 53-60.

[14] Battisti L, Lara JC, Leigh JA (1992) Specific Oligosaccharide Form of the *Rhizobium meliloti* Exopolysaccharide Promotes Nodule Invasion in Alfalfa. Proc. Natl. Acad. Sci. USA. 89: 5625-5629.

[15] Urzainqui A, Walker GC (1992) Exogenous Suppression of the Symbiotic Deficiencies of *Rhizobium meliloti exo* Mutants. J. Bacteriol. 174: 3403-3406.

[16] Wang L-X, Wang Y, Pellock B, Walker GC (1999) Structural Characterization of the Symbiotically Important Low-molecular Weight Succinoglycan of *Sinorhizobium meliloti*. J. Bacteriol. 181: 6788-6796.

[17] Staehelin C, Forsberg LS, D'Haeze W, Gao MY, Carlson RW, Xie ZP, Pellock BJ, Jones KM, Walker GC, Streit WR, Broughton WJ (2006) Exo-oligosaccharides of *Rhizobium* sp. Strain NGR234 are Required for Symbiosis with Various Legumes. J. Bacteriol. 188: 6168-6178.

[18] Becker A, Puhler A (1998) Production of Exopolysaccharides. In: Spaink HP, Kondorosi A, Hooykaas JJ, editors. The *Rhizobiaceae*. Kluwer Academic Publishers, Dordrecht, the Netherlands. pp. 97–118.

[19] Becker A, Kleickmann A, Kuster H, Keller M, Arnold W, Pühler A (1993) Analysis of the *Rhizobium meliloti* Genes *exoU, exoV, exoW, exoT* and *exoI* Involved in Exopolysaccharide Biosynthesis and Nodule Invasion: *exoU* and *exoW* Probably Encode Glucosyltransferases. Mol. Plant-Microbe Interact. 6: 735-744.
[20] Glucksmann MA, Reuber TL, Walker GC (1993) Family of Glycosyl Transferases needed for the Synthesis of Succinoglycan by *Rhizobium meliloti*. J. Bacteriol. 175: 7033-7044.
[21] Glucksmann MA, Reuber TL, Walker GC (1993) Genes Needed for the Modification, Polymerization, Export, and Processing of Succinoglycan by *Rhizobium meliloti*: a Model for Succinoglycan Biosynthesis. J. Bacteriol. 175: 7045-7055.
[22] Jones K.M., Kabayashi H, Davies BW, Taga ME, Walker GC (2007) How Rhizobial Symbionts Invade Plants: the *Sinorhizobium-Medicago* Model. Nat. Rev. Microbiol. 5: 619–633.
[23] Long SR (2001) Genes and Signals in the *Rhizobium*-Legume Symbiosis. Plant Physiol. 125: 69-72.
[24] Janczarek M (2011) Environmental Signals and Regulatory Pathways that Influence Exopolysaccharide Production in Rhizobia. Int. J. Mol. Sci. 12: 7898-7933.
[25] Robertsen BK, Aman P, Darvill AG, McNeil M, Albersheim P (1981) Host-Symbiont Interactions. V. The Structure of Acidic Extracellular Polysaccharides Secreted by *Rhizobium leguminosarum* and *Rhizobium trifolii*. Plant Physiol. 67: 389-400.
[26] Dudman WF, Franzén L-E, McNeil M, Darvill AG, Albersheim P (1983) The Structure of the Acidic Polysaccharide Secreted by *Rhizobium phaseoli* strain 127 K87. Carbohydr. Res. 117: 169–183.
[27] McNeil M, Darvill J, Darvill A, Albersheim P, van Veen R, Hooykaas P, Schilperoort R, Dell A (1986) The Discernible Structural Features of the Acidic Exopolysaccharides Secreted by Different *Rhizobium* Species are the Same. Carbohydr. Res. 146: 307-326.
[28] Philip-Hollingsworth S, Hollingsworth RI, Dazzo FB (1989) Host-Range Related Structural Features of the Acidic Extracellular Polysaccharides of *Rhizobium trifolii* and *Rhizobium leguminosarum*. J. Biol. Chem. 264: 1461-1466.
[29] Philip-Hollingsworth S, Hollingsworth RI, Dazzo FB, Djordjevic MA, Rolfe BG (1989) The Effect of Interspecies Transfer of *Rhizobium* Host-Specific Nodulation Genes on Acidic Polysaccharide Structure and *In Situ* Binding by Host Lectin. J. Biol. Chem. 264: 5710-5714.
[30] Carlson RW, Lee RP (1983) A Comparison of the Surface Polysaccharides from *Rhizobium leguminosarum* 128C53 Sm^r Rif^r with the Surface Polysaccharides from Its exo^{-1} Mutant. Plant Physiol. 71: 223-228.
[31] Orgambide G, Philip-Hollingsworth S, Cargill L, Dazzo F (1992) Evaluation of Acidic Heteropolysaccharide Structures in *Rhizobium leguminosarum* Biovars Altered in Nodulation Genes and Host range. Mol. Plant-Microbe Interact. 5: 484-488.
[32] Canter-Cremers HC, Batley M, Redmond JW, Wijfjes AH, Lugtenberg BJ, Wijffelman CA (1991) Distribution of O-acetyl Groups in the Exopolysaccharide Synthesized by *Rhizobium leguminosarum* Strains is not Determined by the Sym Plasmid. J. Biol. Chem. 266: 9556-9564.

[33] O'Neill MA, Darvill AG, Albersheim P (1991) The Degree of Esterification and Point of Substitution by *O*-acetyl and *O*-(3-hydroxybutanoyl) Groups in the Acidic Extracellular Polysaccharides Secreted by *Rhizobium leguminosarum* Biovars *viciae, trifolii,* and *phaseoli* are not Related to Host-Range. J. Biol. Chem. 266: 9549-9555.
[34] Gil-Serrano A, Sanchez del Junco A, Tejero-Mateo P, Megias M, Caviedes MA (1990) Structure of the Extracellular Polysaccharide Secreted by *Rhizobium leguminosarum* var. *phaseoli* CIAT 899. Carbohydr Res. 204:103-107.
[35] Lopez-Lara IM, Orgambide G, Dazzo FB, Olivares J, Toro N (1993) Characterization and Symbiotic Importance of Acidic Extracellular Polysaccharides of *Rhizobium* sp. Strain GRH2 Isolated from Acacia Nodules. J. Bacteriol. 175: 2826-2832.
[36] Ksenzenko VN, Ivashina TV, Dubeikovskaya ZA, Ivanov SG, Nanazashvili MB, Druzhinina TN, Kalinchuk, NA, Shibaev VN (2007) The *pssA* Gene Encodes UDP-glucose:polyprenyl Phosphate-glucosyl Phosphotransferase Initiating Biosynthesis of *Rhizobium leguminosarum* Exopolysaccharide. Rus. J. Bioorg. Chem. 33: 150-155.
[37] Amemura A, Tokuya H, Abe M, Shiro H (1983) Structural Studies on the Extracellular Acidic Polysaccharide from *Rhizobium trifolii* 4S. Carbohydr. Res. 115: 165–174
[38] Canter Cremers HC, Stevens K, Lugtenberg BJ, Wijffelman CA, Batley M, Redmond JW, Breedveld MW, Zevenhuizen LP (1991) Unusual Structure of the Exopolysaccharide of *Rhizobium leguminosarum* bv. *viciae* Strain 248. Carbohydr. Res. 218: 185-200.
[39] Franzén L-E, Dudman WF, McNeil M, Darvill AG, Albersheim P (1983) The Structure of the Acidic Polysaccharide Secreted by *Rhizobium phaseoli* Strain 127 K44. Carbohydr. Res. 117: 157.
[40] Aman P, Franzen L-E, Darvill JE, McNeil M, Darvill AG, Albersheim P (1982) The Structure of the Acidic Polysaccharide Secreted by *Rhizobium phaseoli* strain 127 K38. Carbohydr. Res. 103: 77-100.
[41] Ziegler T, Eckhardt E, Keller D (1997) Rhizobial Saccharides.3. Unexpected Alpha-Selective Glycosylation During Synthesis of Pyruvated Tetrasaccharide Fragments Related to Exopolysaccharides of *Rhizobium leguminosarum.* J. Carbohydr. Chem. 16: 719-738
[42] Mazur A, Król JE, Marczak M, Skorupska A (2003) Membrane Topology of PssT, the Transmembrane Protein Component of the Type I Exopolysaccharide Transport System in *Rhizobium leguminosarum* bv. *trifolii* strain TA1. J. Bacteriol. 185: 2503-2511.
[43] Whitfield C (2006) Biosynthesis and Assembly of Capsular Polysaccharides in *Escherichia coli.* Ann. Rev. Biochem. 75: 39-68.
[44] Borthakur D, Barker RF, Latchford JW, Rossen L, Johnston AW (1988) Analysis of *pss* Genes of *Rhizobium leguminosarum* Required for Exopolysaccharide Synthesis and Nodulation of Peas: Their Primary Structure and Their Interaction with *psi* and Other Nodulation Genes. Mol. Gen. Genet. 213: 155-162.
[45] Ivashina TV, Khmelnitsky MA, Shlyapnikov MG, Kanapin AA, Ksenzenko VN (1994) The *pssA* Gene from *Rhizobium leguminosarum* bv. *viciae* VF39: Cloning, Sequence and the Possible Role in Polysaccharide Production and Nodule Formation. Gene 150: 111-116.

[46] Van Workum WAT, Canter Cremers HCJ, Wijfjes AHM, van der Kolk C, Wijffelman CA, Kijne JW (1997) Cloning and Characterization of Four Genes of *Rhizobium leguminosarum* bv. *trifolii* Involved in Exopolysaccharide Production and Nodulation. Mol. Plant-Microbe Interact. 10: 290–301.

[47] Sadykov MR, Ivashina TV, Kanapin AA, Shlyapnikov MG, Ksenzenko VN (1998) Structure-functional Organization of Exopolysaccharide Biosynthetic Genes in *Rhizobium leguminosarum* bv. *viciae* VF39. Mol. Biol. (Mosk). 32: 665-671.

[48] Król J, Mazur A, Marczak M, Skorupska A (2007) Syntenic Arrangements of the Surface Polysaccharide Biosynthesis Genes in *Rhizobium leguminosarum*. Genomics. 89: 237-247.

[49] Ivashina TV, Sadykov MR, Chatuev BM, Kanapin AA, Shlyapnikov MG, Ksenzenko VN (1996) The *Rhizobium leguminosarum bv. viciae* VF39 *pssB* Gene Product Participates in Symbiotic Fixation of Molecular Nitrogen. Dokl. Akad. Nauk (Russian). 350: 712-715.

[50] Janczarek M, Krol J, Skorupska A (1999) The *pssB* Gene Product of *Rhizobium leguminosarum* bv. *trifolii* is Homologous to a Family of Inositol Monophosphatases. FEMS Microbiol. Lett. 173: 319–325.

[51] Janczarek M, Skorupska A (2001) The *Rhizobium leguminosarum* bv. *trifolii pssB* Gene Product is an Inositol Monophosphatase that Influences Exopolysaccharide Synthesis. Arch. Microbiol. 175: 143-151.

[52] Kutkowska J, Janczarek M, Kopcinska J, Urbanik-Sypniewska T, Skorupska A (2007) Effects of *pssB* Mutation on Surface Polysaccharides and Symbiotic Phenotype of *Rhizobium leguminosarum* bv. *trifolii.* Acta Biologica Cracoviensia/Botanica. 49: 81–89.

[53] Bernhard F, Coplin DL, Geider K (1993) A Gene Cluster for Amylovoran Synthesis in *Erwinia amylovora*: Characterization and Relationship to *cps* Genes in *Erwinia stewartii.* Mol. Gen. Genet. 239: 158-168.

[54] Franklin M, Nivens DE, Weadge JT, Howell PL (2011) Biosynthesis of the *Pseudomonas aeruginosa* Extracellular Polysaccharides, Alginate, Pel, and Psl. Front. Microbiol. 2: 167.

[55] Canter Cremers HCJ, Batley M, Redmond JW, Eydems L, Breedveld MW, Zevenhuizen LPTM, Pees E, Wijffelman CA, Lugtenberg BJJ (1990) *Rhizobium leguminosarum exoB* Mutants are Deficient in the Synthesis of UDP-glucose 4′-epimerase. J. Biol. Chem. 265: 21122-21127.

[56] Sanchez-Andujar B, Coronado C, Philip-Hollingsworth S, Dazzo FB, Palomares AJ (1997) Structure and the Role in Symbiosis of the *exoB* Gene of *Rhizobium leguminosarum* bv. *trifolii.* Mol. Gen. Genet. 255: 131-140.

[57] Laus MC, Logman TJ, van Brussel AAN, Carlson RW, Azadi P, Gao MY, Kijne JW (2004) Involvement of *exo5* in Production of Surface Polysaccharides in *Rhizobium leguminosarum* and its Role in Nodulation of *Vicia sativa* subsp. Nigra. J. Bacteriol. 186: 6617–6625.

[58] Muszyński A, Laus M, Kijne JW, Carlson RW (2011) Structures of the Lipopolysaccharides from *Rhizobium leguminosarum* RBL5523 and its UDP-Glucose Dehydrogenase Mutant (*exo5*). Glycobiology. 21: 55-68.

[59] Reuber TL, Walker GC (1993) Biosynthesis of Succinoglycan, a Symbiotically Important Exopolysaccharide of *Rhizobium meliloti*. Cell. 74: 269-280.

[60] Bossio JC, De Lannino NI, Dankert MA (1986) *In vitro* Synthesis of a Lipid-linked Acetylated and Pyruvylated Oligosaccharides in *Rhizobium trifolii*. Biochem. Biophys. Res. Commun. 134: 205-21.

[61] Pollock TJ, van Workum WA, Thorne L, Mikolajczak MJ, Yamazaki M, Kijne JW, Armentrout RW (1998) Assignment of Biochemical Functions to Glycosyl Transferase Genes which are Essential for Biosynthesis of Exopolysaccharides in *Sphingomonas* Strain S88 and *Rhizobium leguminosarum*. J. Bacteriol. 180: 586-593.

[62] Abe M, Sherwood JE, Hollingsworth RI, Dazzo FB (1984) Stimulation of Clover Root Hair Infection by Lectin-Binding Oligosaccharides from the Capsular and Extracellular Polysaccharides of *Rhizobium trifolii*. J. Bacteriol. 160: 517-520.

[63] Janczarek M, Skorupska A (2004) Regulation of *pssA* and *pssB* Gene Expression in *Rhizobium leguminosarum bv. trifolii* in Response to Environmental Factors. Antonie Van Leewenhoek. 85: 217-227.

[64] Bossio JC, Semino CE, De Lannino NI, Dankert MA (1996) The *in vitro* Biosynthesis of the Exopolysaccharide Produced by *Rhizobium leguminosarum* bv. *trifolii* strain NA30. Cell. Mol. Biol. 42: 737-758.

[65] Gao XD, Moriyama S, Miura N, Dean N, Nishimura S (2008) Interaction Between the C termini of Alg13 and Alg14 Mediates Formation of the Active UDP-N-acetylglucosamine Transferase Complex. J. Biol. Chem. 283: 32534-32541.

[66] Lu J, Takahashi T, Ohoka A, Nakajima K-I, Hashimoto R, Miura N, Tachikawa H, Gao X-D (2012) Alg14 Organizes the Formation of a Multiglycosyltransferase Complex Involved in Initiation of Lipid-linked Oligosaccharide Biosynthesis. Glycobiology. 22: 504-516.

[67] van Workum WAT, van Slageren S, van Brussel AAN, Kijne JW (1998) Role of Exopolysaccharides of *Rhizobium leguminosarum* bv. *viciae* as Host Plant-Specific Molecules Required for Infection Thread Formation During Nodulation of *Vicia sativa*. Mol. Plant-Microbe Interact. 11: 1233-1241.

[68] van Workum WAT, van Slageren S, van Brussel AAN, Kijne JW (1998) Role of Exopolysaccharides of *Rhizobium leguminosarum* bv. *viciae* as Host Plant-Specific Molecules Required for Infection Thread Formation During Nodulation of *Vicia sativa*. Mol. Plant-Microbe Interact. 11: 1233-1241.

[69] Król J, Wielbo J, Mazur A, Kopcinska J, Lotocka B, Golinowski W, Skorupska A (1998) Molecular Characterization of *pssCDE* Genes of *Rhizobium leguminosarum* bv. *trifolii* Strain TA1: *pssD* Mutant is Affected in Exopolysaccharide Synthesis and Endocytosis of Bacteria. Mol. Plant-Microbe Interact. 11: 1142-1148.

[70] Saxena IM, Brown RMJr, Fevre M, Geremia RA, Henrissat B (1995) Multidomain Architecture of β-Glycosyl Transferases: Implications for Mechanism of Action. J. Bacteriol. 177: 1419–1424.

[71] Downie A (1989) The *nod*L Gene from *Rhizobium leguminosarum* is Homologous to the Acetyl Transferases Encoded by *lacA* and *cysE*. Mol. Microbiol. 3: 1649-1651.

[72] Bloemberg GV, Thomas-Oates JE, Lugtenberg, BJJ Spaink HP (1994) Nodulation Protein NodL of *Rhizobium leguminosarum* O-Acetylates Lipo-Oligosaccharides, Chitin Fragments and N-Acetylglucosamine *in vitro*. Mol. Microbiol. 11: 793-804.

[73] Bloemberg GV, Lagas RM, van Leeuwen S, van der Marel GA, van Boom JH, Lugtenberg BJ, Spaink HP (1995) Substrate Specificity and Kinetic Studies of Nodulation Protein NodL of *Rhizobium leguminosarum*. Biochemistry. 34: 12712-12720.

[74] Pye VE, Tingey AP, Robson RL, Moody PC (2004) The Structure and Mechanism of Serine Acetyltransferase from *Escherichia coli*. J. Biol. Chem. 279: 40729-40736.

[75] Wang XG, Olsen LR, Roderick SL (2002) Structure of the *lac* Operon Galactoside Acetyltransferase. Structure. 10: 581-588.

[76] Luong TT, Ouyang S, Bush K, Lee CY (2002) Type 1 Capsule Genes of *Staphylococcus aureus* are Carried in a Staphylococcal Cassette Chromosome Genetic Element. J. Bacteriol. 184: 3623-3629.

[77] Claus H, Borrow R, Achtman M, Morelli G, Kantelberg C, Longworth E, Frosch M, Vogel U (2004) Genetics of Capsule O-Acetylation in Serogroup C, W-135 and Y Meningococci. Mol. Microbiol. 51: 227-239.

[78] Rende-Fournier R, Leclereq R, Galimand M, Duval J, Courvalin P (1993) Identification of the *sat*A Gene Encoding a Streptagramin A Acetyltransferase in *Enterococcus faecium* BM4145. Antimicrob. Agents Chemother. 10: 2119-2125.

[79] Buendia A M, Enenkel B, Koplin R, Niehaus K, Arnold W, Pühler A (1991) The *Rhizobium meliloti exoZ1 exoB* Fragment of Megaplasmid 2: ExoB Functions as a UDP-glucose 4-epimerase and ExoZ Shows Homology to NodX of *Rhizobium leguminosarum* Biovar *viciae* Strain TOM. Mol. Microbiol. 5: 1519-1530.

[80] Reuber TL, Walker GC (1993) The Acetyl Substituent of Succinoglycan is not Necessary for Alfalfa Nodule Invasion by *Rhizobium meliloti* Rm1021. J. Bacteriol. 175: 3653-3655.

[81] Ivashina TV, Fedorova EE, Ashina NP, Kalinchuk NA, Druzhinina TN, Shashkov AS, Shibaev VN, Ksenzenko VN (2010) Mutation in the *pssM* Gene Encoding Ketal Pyruvate Transferase Leads to Disruption of *Rhizobium leguminosarum* bv. *viciae-Pisum sativum* Symbiosis. J. Appl. Microbiol. 109: 731-742.

[82] Postma JG, Jager D, Jacobsen E, Feenstra WJ (1990) Studies on a Non-Fixing Mutant of Pea (*Pisum sativum* L.): Phenotypical Description and Bacteroid Activity. Plant Sci. 68: 151-161.

[83] Colebatch G, Desbrosses G, Ott T, Krusell L, Montanari O, Kloska S, Kopka J, Udvardi MK (2004) Global Changes in Transcription Orchestrate Metabolic Differentiation During Symbiotic Nitrogen Fixation in *Lotus japonicus*. Plant J. 39: 487–512.

[84] Fabret C, Quentin Y, Chapal N, Guiseppi A, Haiech J, Denizot F (1996) Integrated Mapping and Sequencing of a 115 kb DNA Fragment from *Bacillus subtilis*: Sequence Analysis of a 21 kb Segment Containing the *sigL* Locus. Microbiol. 142: 3089-3096.

[85] Stingele F, Neeser JR, Mollet B (1996) Identification and Characterization of the *eps* (Exopolysaccharide) Gene Cluster from *Streptococcus thermophilus* Sfi6. J. Bacteriol. 178: 1680-1690.

[86] Andreishcheva EN, Kunkel JP, Gemmill TR, Trimble RB (2004) Five Genes Involved in Biosynthesis of the Pyruvylated Galβ1,3-Epitope in *Schizosaccharomyces pombe* N-Linked Glycans. J. Biol. Chem. 279: 35644-35655.

[87] Cuthbertson L, Mainprize IL, Naismith JH, Whitfield C (2009) Pivotal Roles of the Outer Membrane Polysaccharide Export and Polysaccharide Copolymerase Protein Families

in Export of Extracellular Polysaccharides in Gram-Negative Bacteria. Microbiol. Mol. Biol. Rev. 73: 55-77.
[88] Cuthbertson L, Kos V, Whitfield C (2010) ABC Transporters Involved in Export of Cell Surface Glycoconjugates. Microbiol. Mol. Biol. Rev. 74: 341-362.
[89] Morona R, Purins L, Tocilj A, Matte A, Cygler M (2009) Sequence-Structure Relationships in Polysaccharide Co-polymerase (PCP) Proteins. Trends Biochem. Sci. 34: 78-84.
[90] Dong C, Beis K, Nesper J, Brunkan-LaMontagne AL, Clarke BR, Whitfield C, Naismith JH (2006) Wza the Translocon for *E. coli* Capsular Polysaccharides Defines a New Class of Membrane Protein. Nature. 444: 226-229.
[91] Mazur A, Marczak M, Król JE, Skorupska A (2005) Topological and Transcriptional Analysis of *pssL* Gene Product: a Putative Wzx-like Exopolysaccaride Translocase in *Rhizobium leguminosarum* bv. *trifolii* TA1. Arch. Microbiol. 184: 1-10.
[92] Mazur A, Król JE, Skorupska A (2001) Isolation and Sequencing of *Rhizobium leguminosarum* bv. *trifolii pssN, pssO* and *pssP* Genes Encoding the Proteins Involved in Polymerization and Translocation of Exopolysaccharide. DNA Seq. 12: 1-12.
[93] Mazur A, Król JE, Wielbo J, Urbanik-Sypniewska T, Skorupska A (2002) *Rhizobium leguminosarum* bv. *trifolii* PssP Protein is Required for Exopolysaccharide Biosynthesis and Polymerization. Mol. Plant-Microbe Interact. 15: 388-397.
[94] Gonzalez JE, Semino CE, Wang L-X, Castellano-Torres LE, Walker GC (1998) Biosynthetic Control of Molecular Weight in the Polymerization of the Octasaccharide Subunits of Succinoglycan, a Symbiotically Important Exopolysaccharide of *Rhizobium meliloti*. Proc. Natl. Acad. Sci. USA. 95: 13477-13482.
[95] Becker A, Niehaus K, Pühler A (1995) Low-molecular-weight Succinoglycan is Predominantly Produced by *Rhizobium meliloti* Strains Carrying a Mutated ExoP Protein Characterized by a Periplasmic N-terminal Domain and a Missing C-terminal Domain. Mol. Microbiol. 16: 191-203.
[96] Paulsen IT, Beness AM, Saier MH Jr (1997) Computer-based Analyses of the Protein Constituents of Transport Systems Catalysing Export of Complex Carbohydrates in Bacteria. Microbiol. 143: 2685-2699.
[97] Morona R, Van Den Bosch L, Daniels C (2000) Evaluation of Wzz/MPA1/MPA2 Proteins Based on the Presence of Coiled-coil Regions. Microbiol. 146: 1-4.
[98] Niemeyer D, Becker A (2001) The Molecular Weight Distribution of Succinoglycan Produced by *Sinorhizobium meliloti* is Influenced by Specific Tyrosine Phosphorylation and ATPase Activity of the Cytoplasmic Domain of the ExoP Protein. J. Bacteriol. 183: 5163-5170.
[99] Marczak M, Mazur A, Król JE, Gruszecki WI, Skorupska A (2006) Lipoprotein PssN of *Rhizobium leguminosarum* bv. *trifolii*: Subcellular Localization and Possible Involvement in Exopolysaccharide Export. J. Bacteriol. 188: 6943-6952.
[100] Marczak M, Mazur A, Gruszecki WI, Skorupska A (2008) PssO, a Unique Extracellular Protein Important for Exopolysaccharide Synthesis in *Rhizobium leguminosarum* bv. *trifolii*. Biochimie. 90: 1781–1790.

[101] Wielbo J, Mazur A, Król JE, Marczak M, Skorupska A (2004) Environmental Modulation of the *pssTNOP* Gene Expression in *Rhizobium leguminosarum bv. trifolii.* Can. J. Microbiol. 50: 201-211.
[102] Louch HA, Miller KJ (2001) Synthesis of a Low-molecular-weight Form of Exopolysaccharide by *Bradyrhizobium japonicum* USDA 110. Appl. Environm. Microbiol. 67: 1011-1014.
[103] Gonzalez JE, Semino CE, Wang L-X, Castellano-Torres LE, Walker GC (1998) Biosynthetic Control of Molecular Weight in the Polymerization of the Octasaccharide Subunits of Succinoglycan, a Symbiotically Important Exopolysaccharide of *Rhizobium meliloti.* Proc. Natl. Acad. Sci. USA. 95: 13477-13482.
[104] York GM, Walker GC (1997) The *Rhizobium meliloti exoK* Gene and *prsD/prsE/exsH* Genes are Components of Independent Degradative Pathways which Contribute to Production of Low-molecular-weight Succinoglycan. Mol. Microbiol. 25: 117-134.
[105] York GM, Walker GC (1998) The Succinyl and Acetyl Modifications of Succinoglycan Influence Susceptibility of Succinoglycan to Cleavage by the *Rhizobium meliloti* Glycanases ExoK and ExsH. J. Bacteriol. 180: 4184-4191.
[106] Finnie C, Zorreguieta A, Hartley NM, Downie JA (1998) Characterization of *Rhizobium leguminosarum* Exopolysaccharide Glycanases that are Secreted via a Type I Exporter and Have a Novel Heptapeptide Repeat Motif. J. Bacteriol. 180: 1691-1699.
[107] Krehenbrink M, Downie JA (2008) Identification of Protein Secretion Systems and Novel Secreted Proteins in *Rhizobium leguminosarum bv. viciae.* BMC Genomics. 9: 55.
[108] Finnie C, Hartley NM, Findlay KC, Downie JA (1997) The *Rhizobium leguminosarum prsDE* Genes are Required for Secretion of Several Proteins, Some of Which Influence Nodulation, Symbiotic Nitrogen Fixation and Exopolysaccharide Modification. Mol. Microbiol. 25: 135-46.
[109] Zorreguieta A, Finnie C, Downie JA (2000) Extracellular Glycanases of *Rhizobium leguminosarum* are Activated on the Cell Surface by an Exopolysaccharide-related Component. J. Bacteriol. 182: 1304-1312.
[110] Guerreirro N, Ksenzenko VN, Djordjevic MA, Ivashina TV, Rolfe BG (2000) Elevated Levels of Synthesis of Over 20 Proteins After Mutation of the *Rhizobium leguminosarum* Exopolysaccharide Synthesis Gene *pssA.* J. Bacteriol. 185: 4521-4532.
[111] Latchford JW, Borthakur D, Johnston AW (1991) The Products of *Rhizobium* Genes, *psi* and *pss,* which Affect Exopolysaccharide Production, are Associated with the Bacterial Cell Surface. Mol. Microbiol. 5: 2107-2114.

Chapter 6

Polysaccharides from Larch Biomass

Natalya Nikolaevna Trofimova, Elena Nikolaevna Medvedeva, Nadezhda Viktorovna Ivanova, Yuriy Alekseevich Malkov and Vasiliy Anatolievich Babkin

Additional information is available at the end of the chapter

1. Introduction

There are two species of the genus *Larix* Mill, *Larix sibirica* Ledeb. (Figure 1) and *Larix gmelinii* (Rupr.) Rupr., considered to be the most abundant trees in the Russian Federation: the total stock of their wood exceeds 26 billion m^3. Traditionally, the main economic value of larch wood consists of the manufacture of roundwood (timber), the value of which is connected with the high quality of lumber from this breed of tree. Larch wood can sometimes be used in insignificant quantities in pulp-and-paper manufacture to obtain pulp. About 40% of this valuable breed (as bark, sawdust) enters waste through existing lumber manufacturing processes. Such an irrational approach to the development of larch wood does not allow exploitation of the richest potential of the given renewed source. Meanwhile, biologically active compounds contained in larch biomass can be used for the manufacture of products for medical, food and agricultural purposes with maximal benefit. The development of complex technology for chemical processing of larch biomass and waste timber will considerably raise the economic value of this biological resource.

This chapter is devoted to polysaccharides contained in the wood and bark of larch, i.e. heteropolysaccharide arabinogalactan (AG) and pectin polysaccharides (pectin substances, PS). The chapter also deals with the development of technologies for the preparation of these polysaccharides and the study of their useful properties. Wood hemicelluloses and cellulose represent a potential source of valuable monosaccharides, namely glucose, which can be readily obtained by hydrolysis transformation of these polysaccharides.

Larch wood contains a high quantity of arabinogalactan, a considerable part of which can be found in the butt of a tree, which frequently enters waste. Arabinogalactan possesses a wide spectrum of biological activity. It shows immunomodulating activity, gastroprotective, membranotropic and prebiotic properties, and can be used in the medical, veterinary, food and cosmetic industries. Arabinogalactan is a perspective matrix for obtaining on its basis metal-, sulpho-, amino- and other derivatives due to the reactive hydroxyl and aldehyde groups contained in its molecule. The ability of arabinogalactan to form water-soluble stable substances with inorganic nanoparticles and low-molecular medical substances (MS) provides serious prospects for the development of materials with unique properties.

Figure 1. *Larix sibirica* Ledeb.

In the USA, arabinogalactan has been extracted from the wood of *L. occidentalis* Nutt. and *L. laricina* (Du Roi) by K. Koch for more than 40 years as a commercial product. Effective immunity-modulating and prebiotic biologically active food additives have been developed to improve quality of human life. Application of arabinogalactan in agriculture as fodder additive allows the greater efficiency of animal industries. Arabinogalactan is not currently manufactured in Russia.

The bark of larch does not however have industrial application. Annually, wood-processing industries and pulp-and-paper enterprises waste more than 30 million m^3 in volume. It has become a serious environmental problem because the bark is badly exposed to biodegradation. At the same time, the chemical compounds of the bark can be a source of valuable biologically active substances, including polysaccharides. The creation of medical, food and other useful products on the basis of polysaccharides is also possible.

Larch bark contains about 7–12% of pectin polysaccharides, based on the weight of absolutely dry raw material. Pectin is acid polysaccharide–glycogalacturonane, and is contained in practically all plants. It is obtained from diverse sources that differ in their chemical structure due to distinctions in the qualitative structure of carbohydrates and their quantitative parities. Pectin substances promote digestive processes, and help organisms to resist many diseases such as atherosclerosis, diabetes, cancer, etc. There are two basic sources of pectin production in Germany and Denmark. However, there is no industrial production of pectin in Russia.

As one of the possible sources of polysaccharides, cellulose–lignin residue can be formed by the extraction of polyphenolic substances from larch wood in the scheme for complex processing of larch biomass. Its chemical processing allows carbohydrate products to be produced, mainly crystalline glucose. Glucose is contained, mainly in the bound form, in considerable amounts of natural products as a constituent of various glycosides and polysaccharides. Taking into consideration that about 40% of the dry substance of plant mass is accounted for by cellulose formed by photosynthesis at a rate of about 70 kg/day per each inhabitant of our the planet, glucose is the most abundant sugar in nature.

All developed technologies are environmentally friendly. They are focused on processing of timber and industrial wood-processing waste, and possess high technical and economic parameters.

2. Larch Arabinogalactan

2.1. Physicochemical and biological properties of Arabinogalactan

Larch wood is distinctive for its high content of water soluble polysaccharide arabinogalactan, reaching up to 35% by weight of dry wood [1]. This valuable substance has been studied since the 1950s [1-6]. The physicochemical and biological properties of AG from the wood of *L. occidentalis* Nutt. and Siberian larch species *L. sibirica* Ledeb. and *L. gmelinii* (Rupr.) Rupr. are the most explored. The wood of Siberian larches contains up to 10–15% of AG [2,7] and is a reliable source of industrial raw material for AG production.

The AG macromolecule from larch wood has a highly branched structure. Its main chain consists of β-(1→3) linked galactose residues (Figure 2). Approximately one half of the side chains (in *L. occidentalis* Nutt., *L. sibirica* Ledeb. and *L. gmelinii* (Rupr.) Rupr.) is formed of β-(1→6)-linked dimers of galactopyranose; galactopyranose monomers comprise about a quarter; and the remainder contains the major part of the polysaccharide's arabinose in aggregates of two or more monomers [1,8,9]. Arabinose fragments mainly occur as side chains consisting of 3-O-substituted β-L-arabinofuranose residues and terminal residues of β-L-arabinopyranose, β-D-arabinofuranose and α-L-arabinofuranose [8,9]. However, arabinose fragments have also been found in the main chain [1]. Glucuronic acid fragments in AG from wood of various larch species is low in content. As for AG from the above species, no glucuronic acid fragments have been detected in the purified samples [6].

Monosaccharide composition and molecular mass (MM) of AG macromolecules differ among the species and also varies within single species. It has been established that the composition of AG macromolecules is dependent on the conditions under which it is isolated from larch wood and on the purification procedure [9,10] as well as on molecular weight [1,11]. AG macromolecules have low molecular weights (13–20 kDa, according to HPLC data) and a narrow molecular weight distribution (degree of polydispersity 1.1–2.3) [10].

Biological activity of AG in higher plants is directly linked to their structural characteristics, such as length of galactan chain, structure of side chains, molecular weight and ability to form intermolecular associates [12,13]. Larch AG is characterized by low toxicity, showing neither acute poisoning for doses of 5 g/kg nor chronic poisoning for doses of 500 mg/kg per day [14]. Diverse biological activity of AG includes immunomodulatory, prebiotic, hypolipidemic, gastro- and hepatoprotective, mitogenic, antimutagenic and antiviral effects, etc. There are reports on the inhibitory and destructive action of AG against certain types of malignant tumours [6,15-22]. Moreover, it has a good solubility in cold water, uniquely low viscosity of concentrated aqueous solutions, an ability to bind fat and retain liquid and dispersive capacity, etc. All these benefits are in high demand in medicine and veterinary science as well as in food and cosmetic industries [5,6]. There are a number of biologically active food supplements which incorporate AG [23-25]. In medicine, an ophthalmic composition (eye drops/contact lens care solution) has been developed [26]. Membranotropicity caused by galactose fragments and realized through receptor mediated endocytosis makes AG a promising drug carrier to increase absorbability and selectivity of medical substances that are characterized by low bioavailability [27-36]. Applications of AG in photodynamic diagnostics, in oncological disease therapy and in gene therapy (targeted delivery of functional genes) are currently being explored [37-39]. The unique properties of AG are prominent among the known polysaccharide carriers of medical substances [40,41].

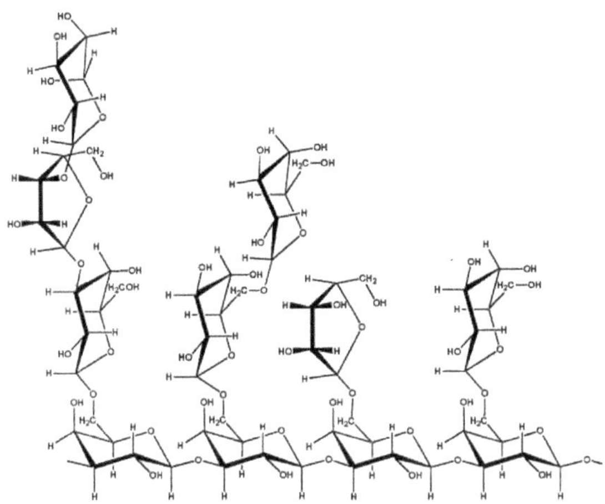

Figure 2. Structural fragment of AG macromolecule

Significant interest for medicine is raised by the products of AG modification [6]. Introducing diverse functional groups into the AG macromolecule makes it a manifold synthone to obtain a wide range of new biologically active substances. Oxidation is a promising way to functionalise AG. Methods of selective oxidation developed in classical carbohydrate chemistry [42] have been used to develop functionalized AG products [43,44]. Among others, oxidative destruction with simultaneous introduction of carboxylic groups into the macromolecule under the action of hydrogen peroxide in aqueous medium has been carried out [45]. It has been revealed that oligomeric products show anti-inflammatory and antiulcer activities. Reactions of AG and oxidized AG with some known MS give intermolecular complexes [46-48]. Such complexes of AG with 5-aminosalicylic acid show high antiulcer activity, while complexes incorporating 4-aminosalicylic or isonicotinic acid hydrazide demonstrate an antituberculosis effect. It has been established that conjugates of AG and products of its modification increase the physiological effect of MS and decrease their toxicity [28,46-48]; for instance, a conjugate of AG (9kDa) 9-β-D-arabinofuranosyladenine-5'-monophosphate was 25-fold more active than the parent compound 9-β-D-arabinofuranosyladenine (araA) in decreasing the amount of hepatitis B virus, and the toxicity of the complex preparation is much lower than araA [28]. To increase reactivity of AG with MS, the synthesis of conjugates proceeds by bromination, phosphorylation, amination, formation of hydrazides or reaction with $NaBH_4$ [27,33,34].

Mechanochemical activation is another promising method of AG modification, in which the target products are obtained in one stage without the use of solvents. Mechanochemical treatment can give rise to numerous physicochemical transformations of AG macromolecules, which are associated, first, with the breaking and formation of valence bonds and, second, with the disturbance and origination of weak intermolecular interactions (disordering, conformational rearrangements, etc.). As a result, the polysaccharide can change its biological activity, toxicity and pharmacological properties. HPLC and quantitative ^{13}C NMR spectroscopy has established [49] that mechanochemical treatment of AG isolated from Siberian larch wood changes its molecular weight distribution, monosaccharide composition and degree of branching due to partial destruction of the macromolecules and subsequent recombination of their fragments. The extent of these changes depends on the activation conditions. The IR and ^{13}C NMR spectra have not shown any functionalization of AG macromolecules in the conditions studied. Toxic-pharmacological study has established that a mechanochemically activated AG sample has the same LD_{50} (more 5000 mg/kg) than the starting AG. As for the effect of mechanochemically activated AG upon the central nervous system, it has demonstrated anxiolytic activity similar to sibason in a dose of 20 mg/kg tested in laboratory animals. Meanwhile, single intravenous injection of the substance, in a dose of 3.5 mg/kg, slightly but statistically significantly decreases arterial pressure (by 6%) in normotensive rats without affecting electrocardiogram parameters and heart rate. Thus, mechanochemically activated AG is a promising drug carrier [49].

Combined mechanochemical activation of MS with AG (pharmacon clathration) is even more effective in improving safety and bioavailability of drugs [50-53]. Essentially (up to 50 times) increased solubility of MS and dramatically decreased therapeutic doses of the same efficiency are reported for clathrates of poorly soluble anti-inflammatory, psychotropic and hypotensive drugs [51,52]; for instance, clathrates of AG with nifedipine containing a 10 times lower dose than the starting MS show pronounced hypotensive and antiarrhythmic effects [52]. Additionally, the side effects of MS in clathrates are decreased, for instance in the ulcerogenity of nonsteroidal anti-inflammatory drugs. HPLC, ^{13}C NMR and IR spectroscopic studies have shown the absence of any chemical reaction between AG and MS at pharmacon clathration [51,53]. The X-ray phase study and thermal analysis prove the destruction of the crystal structure of MS and its dispersion within the AG matrix. The polysaccharide macromolecules are cleaved similarly to the mechanochemical activation of AG alone [49].

L. sibirica Ledeb. and *L. gmelinii* (Rupr.) Rupr. contain much bioflavonoid dihydroquercetin (DHQ, taxifolin) and their diverse biological activity is well studied. It is an officinal drug with a wide range of therapeutic action and is also the basis for a number of efficient medical preparations and food supplements [54]. Complexes of AG and DHQ combining these unique properties are very promising. Such complexes have been obtained by pharmacon clathration [54] and have shown essentially improved solubility (up to 38 times) in comparison with starting DHQ and untreated AG/DHQ mixture. No chemical reaction between AG and DHQ takes place, as in clathrates [51]. According to HPLC data, AG in AG/DHQ clathrates has a narrower molecular weight distribution in comparison with pure mechanochemically activated AG, due to a decrease in both high- and low-molecular fractions, and thus DHQ stabilizes the polysaccharide macromolecules in mechanochemical treatment.

The most recent application of AG as a stabilizing polymer matrix in hybrid nanosized materials is based on iron oxides, cobalt, copper, nickel, ferrites and zero-valence metals such as silver, palladium and platinum [18,55-57]. Metal content in nanocomposite samples depends on synthetic conditions and on the type of metal ion used, varying in the range of 0.1–21.0%. In the case of metal oxide nanocomposites, AG shows properties of a nanostabilizing matrix, while in the composites of noble metals it reduces metal to a zero-valence state and stabilizes the metal nanoparticles formed. Nanocomposites based on AG retain high biological activity. Ferroarabinogalactans show synergy between the pronounced antianaemic activity of the ferric core and the unique membranotropic and immunomodulatory properties of AG. Parenteral administration of ferroarabinogalactan normalizes quantitative and qualitative characteristics of the erythrocyte system and iron depot level in animals (white rats) [18]. The original synthetic method of ferroarabinogalactan retains both membranotropic and immunomodulatory properties of AG. Studies of the natural effects of immunomodulators, together with the investigation of specific immunity to plague, have revealed that ferroarabinogalactan activates peritoneal macrophages in guinea pigs, in comparison with the animals' cells being immunized only with vital plague vaccine.

Antibiotic resistance of microorganisms has led to a new interest in silver preparations. The most efficient are preparations of ultradispersed silver. Highly dispersed (nanosized) particles increase bactericidal activity. It has been established that silver-containing nanocomposites with AG possess high antimicrobial activity against gram-negative enterobacteria (*Escherichia coli, Salmonella typhimurium, Candida albigans, Bacillus subtilis* and *Staphyllococcus aureus*) [57].

Thus, the method for synthesis of nanocomposites with available polysaccharide AG is an easy way to synthesize universal materials. The AG-based nanobiocomposites synergistically combine the properties of the stabilizing natural polysaccharide matrix and the nanocore materials. They are applicable as nanosized water-soluble enantioselective catalysts, magnet-controlled medical substances, materials for coherent and nonlinear optics, high-sensitivity optical markers, universal antimicrobial preparations, etc. The use of AG as a bioactive polysaccharide matrix participating in the processes of receptor-induced endocytosis leads to new approaches to therapy for metal deficiency states and to the development of new biomaterials of target action, which are in high demand in medicine and biology, both as controlled composite materials and as new water-soluble biodegradable metal-containing drugs.

At present, healthy diet is a question of public policy in all developed countries due to the undisputed role of food in public health, working capacity of people, adaptation, child growth and longevity. The increasing popularity of healthy diets has made manufacturers pay more and more attention to functional food, i.e. medically fortified food products. Food supplements and enriched products are becoming increasingly popular. Such products are functionalized by both natural substances (pectin, inulin, gum arabic, etc.) and semi-synthetic compounds (lactulose, polydextrose, resistant starches, chitosan, etc.).

Larch arabinogalactan adds nutrition and function to beverages, snack foods, nutrition bars, and more [6,58]. Not only does AG function as a prebiotic fibre and immunity enhancer, it also retains moisture, enhances mouthfeel and bulk, and improves shelf stability. Because of AG's low-viscosity profile and emulsification-enhancing properties, the most immediate applications include refrigerated and non-refrigerated beverages, and beverage mixes. Interest has also been expressed in snack foods, bars, ready-to-eat cereals, yogurt/dairy products and baked goods. Commercially available larch AG-containing products include beverages and nutrition bars.

Not only does AG add nutrition, it also provides functional benefits. An independent food laboratory confirmed that the inclusion of AG improved white pan bread make-up, external symmetry and internal grain scores. Fat-free flour tortillas with AG showed better handling, taste and aroma than the control. AG is a low-calorie additive for artificial sweeteners. It delivers mouthfeel, taste and bulking attributes that are most like sugar.

In confectionery and baked goods, AG lowers water activity and aids flavour and oil retention. AG can be used in browning compositions for uncooked foods, in seasoning powders to improve flow and reduce hygroscopicity, and in starch-containing foods to inhibit swelling.

Recent clinical investigations of AG have demonstrated not only benefits to gastrointestinal health and immunity, but also a significant reduction in serum cholesterol, glucose and insulin levels [58,59]. This opens the door to potential heart health label claims and provides an option for consumers looking for foods that are beneficial in terms of body weight, blood glucose or blood insulin control.

AG benefits, as determined by human and animal clinical trials, have been observed as low as 1.5 g/day, or more specifically 20 mg/kg of body weight. AG has been found to be totally safe as a food ingredient. On average, finished products containing a minimum of 60 mg/kg of body weight or about 4.5 g/day is recommended for foods [58]. Using arabinogalactan additives isolated from Siberian larch, the authors of one study [60] examined the soft wheat flour quality and quantity of gluten, physical properties of the dough, and quality of finished bread, depending on the quantity of the added polysaccharide. The addition of 1% of arabinogalactan to flour causes a significant improvement in the qualitative indices of bread. In this case, AG is totally consumed in the course of bread making because it is utilized by yeast. It is recommended that bread quality can be improved when the flour incorporates 1% mass of AG. When 2–3% of AG is added to flour, the AG content decreases. An excess of AG inhibits yeast growth, which leads to a decrease in bread quality.

The optimum compositions for AG-enriched bakery goods and pastry have been proposed [61,62]. It has been shown that when AG is added to flour in proportions of 1–5% by weight, bakery products are rich in dietary fibre of prebiotic and immunostimulating action, while their energy density is lower due to the decreased amount of sugar in the recipes. The AG-enriched bakery and pastry products have a medical effect [59]. Production technology for AG-enriched prebiotic cultured milk products has been developed [63]. The efficiency of AG in veterinary medicine has been proven [64-67].

2.2. Technology of AG production

The only regular production of AG is realized in the USA where the raw materials are derived from *L. occidentalis* Nutt. [5,6]. The known AG production techniques are all based on extraction of the polysaccharide from larch wood particles (chips, shavings, sawdust) and differ in pre-treatment of the raw material, conditions of extraction and methods of purifying the extracts and the final product [10]. The extraction process is highly dependent on the wood reduction range. Sawdust (specific surface of 164 cm^2/g) yields almost 90% of AG during the first 10 min of extraction while chips (specific surface of 17 cm^2/g) yield only 10% [68]. To intensify extraction, mechanochemical activation of larch wood and its further treatment with superheated water steam ("autohydrolysis explosion") is proposed [69]. The AG yields increase significantly (more than twice) if larch sawdust extraction is conducted with microwave or pulse ultrasound activation. Microwave or shock-and-acoustic processing of sawdust allows 90% of AG extraction to be completed in 30–60 seconds [70]. The products of larch wood water extraction, together with AG, are various phenolic compounds (low-molecular and oligomeric flavonoids, lignans, lignine substances and tannins), making production of high purity AG a complicated problem [71,72]. Purification of AG water extracts from those impurities is a current issue.

We propose a beneficial, economically and ecologically sound method to obtain dry 95–97% AG [17,73]. There are two main stages: first, extraction and purification of the extract, and second, dry product isolation. Every stage has been carefully studied and theoretically rationalized.

Study of extraction process kinetics is primarily aimed at determining duration of contact between the phases to give the target degree of isolation. Extraction kinetics data determine the geometry of the apparatus. AG water extraction proceeds in two stages, the first of which is fast and the second is slow (Figure 3).

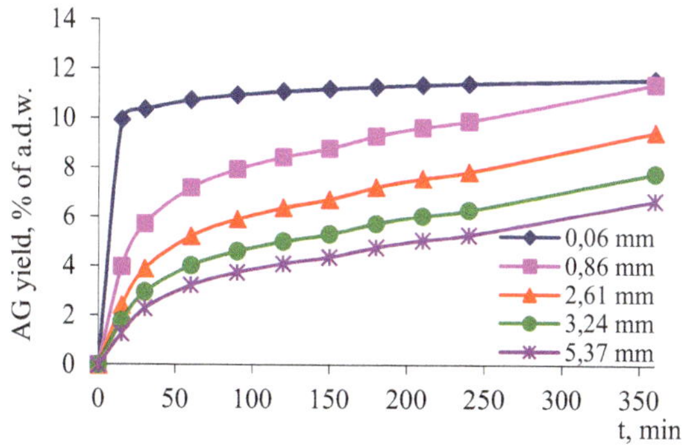

Figure 3. AG yield (% of absolutely dry wood, a.d.w.), showing dependence on particle size

The extraction process is determined by AG diffusion and penetration of extracting solution into wood pores as well as by hydrodynamic conditions. Thirty minutes is sufficient to isolate almost the total amount of AG from sawdust, while, in the same conditions, wood chips need several days to yield the major fraction of AG. The quality of AG and other water-soluble components is not affected by wood reduction range.

On the basis of the experimental data, we calculated diffusion constants, mass-transfer coefficients and the Biot diffusion criterion (Bi), characterizing the influence of hydrodynamic conditions on AG isolation rate. For the extract (Y_p), extraction was conducted until the equilibrium of the AG solution concentrations in wood pores reached equilibrium. As $\tau \to \infty$, Fourier diffusion criterion $Fo_g \to \infty$. Thus, the kinetic equation of the extraction process can be described by the equation [74]:

$$\frac{X - Y^*}{X_s - Y_s} = \sum_{n=1}^{\infty} A_n e^{-\mu_n^2 Fo_g} \quad (1)$$

where X is the average AG solution concentration in wood pores, at the moment of time τ, g/cm^3; Y^* is the equilibrium concentration of the substance isolated in the solution, g/cm^3; X_s is the initial AG solution concentration in the wood pores, g/cm^3; and Y_s - is the initial concentration of the substance isolated in the solution, g/cm^3;

$$A_n = \frac{6Bi^2}{\mu_n^2\left(\mu_n^2 + Bi^2 - Bi\right)} \quad (2)$$

where μ_n represents characteristic equation roots.

According to substance balance data,

$$Y^* - Y_i = b\left(X - Y_p\right) \quad (3)$$

where Y_i is the concentration of the substance isolated in the solution at the moment of time τ, g/cm³; and

$$b = \frac{G\varepsilon}{\rho V} \tag{4}$$

where G is the mass of the solid, g; V is the volume of the liquid phase, cm³; ρ is the density of the solid, g/cm³; ε is the specific volume of wood pores occupied by solution, cm³/cm³.

The right side of the equation (3) determines the fraction of the substance transferred from the solid into the solution during the time interval from the moment under consideration until the end of the experiment. The left side determines the increase of the solution concentration during the interval mentioned. Substituting (X – Y^*) into equation (1) according to (2) leads to:

$$\frac{Y^* - Y_i}{X_s - Y_s} = \sum_{n=1}^{\infty} B_n e^{-\mu_n^2 Fo_g} \tag{5}$$

where $B_n = bA_n$.

For a regular mode of extraction, the first member of the series in equation (5) is sufficient.

Figure 4 graphs the dependence of $\ln\frac{Y^* - Y_i}{X_s - Y_s}$ on τ according to equation (5), on the basis of experimental data.

Extrapolating the straight line (Figure 4) $f(\tau) = \ln\frac{Y^* - Y_i}{X_s - Y_s}$ to τ = 0 gives a diffusion constant, according to the equations:

$$\mathrm{tg}(\alpha) = -\mu_1^2 \frac{D\tau}{l^2} \tag{6}$$

and

$$tg\mu_1 = \frac{\mu_1}{b} \tag{7}$$

where α is the angle of inclination of the straight line to the time axis.

Solving the characteristic equation (6) regarding μ_1 gives, according to equation (7), the diffusion constant D. On the basis of the experimental data, the diffusion constant at temperatures of 20–25 °C is $1.55–2.67 \cdot 10^{-10}$ m²/s. For particles of average size of 5 mm, $Bi >> 1$, which demonstrates the insignificant affect of hydrodynamic conditions upon AG isolation rate.

These data on mass transfer were used to calculate the AG extraction process from larch wood particles in similar hydrodynamic conditions. The value of the diffusion constant D was used to calculate the processes conducted at a given temperature.

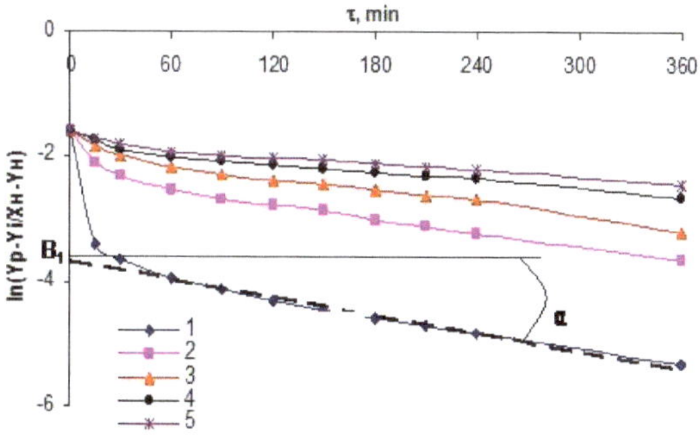

Figure 4. Graph of regular mode of AG extraction. Particle sizes, mm: 1 - 0.056; 2 - 0.86; 3 - 2.61; 4 - 3.24; 5 - 5.37

Based on the experimental data, a mathematical model was developed, material balance was calculated, and the optimal parameters of the extraction process were determined [75]. The model was used to optimize the technological process and production algorithm.

According to the method developed, firstly DHQ and other phenolic extractive substances were isolated from larch wood particles by organic solvent, and then were exposed to extraction by circulating water at 60–80 ºC for 2–3 h. The extract obtained was treated with a cationic flocculant solution to remove mechanical and colloidal impurities. Decolourized extract was ready for being concentrated and further purified by ultrafiltration.

Concentrations of AG water extracts were decolourized by flocculation, which was achieved by ultrafiltration using UAM-150P cellulose acetate membranes (Russia) [76]. Filtration rate decreased with time due to an increase in solution concentrations, viscosities and sedimentation of high-molecular particles on the membrane surface. Initial productivity decreased as initial AG solution concentration increased. However, initial productivity increased with increasing pressure gradient. At the final stage of ultrafiltration, when the process rate approaches a constant level, the higher the pressure gradient the lower the productivity, due to the higher rate of concentration at high pressure during equal time intervals.

Studies have been made of the influence of pressure upon the ultrafiltration process. The maximum degree of concentration is reached at the pressure gradient $\Delta P = 0.4$ MPa. However, the optimal ratio between productivity and degree of concentration is at $\Delta P = 0.2$ MPa.

Ultrafiltration results in simultaneous concentration of AG extracts and their purification by almost entirely filtering out low-molecular phenolic impurities. Purification efficiency depends on composition of the extract, membrane characteristics and conditions of filtration as well as a degree of concentration.

According to the IR spectroscopy and HPLC data, filtrate contains, together with phenolic substances, an oligomeric fraction of AG. Atomic absorption and X-ray fluorescence analyses have shown that ultrafiltration purifies AG from metal cations [76]. The total content of dry substances in filtrates is not more than 1–2.5%.

To increase productivity of the ultrafiltration module, we also tested the UAM-500P membrane (Russia). Ultrafiltration dynamics of decolourized AG extracts have shown that filtration rate is in an inverse ratio to initial extract concentration. The use of a macroporous membrane allows ultrafiltration without pre-treatment of AG extracts by a flocculating agent. It has been proven experimentally that productivity of this process is comparable to that of extract decolourized by flocculation. Thus, for the UAM-500P membrane, pore blocking at the initial stage is not a limiting factor, unlike the case for UAM-150P. The optimal conditions of ultrafiltration have been determined to make the technology profitable.

After ultrafiltration, the concentrate was dried in a drying unit. The known methods of dry product isolation, by precipitation in alcohol or acetone [6], are disadvantageous for industrial use from a technological, economical and ecological point of view. The filtrate, without additional treatment, was mixed with fresh water and reused for DHQ extraction.

The method proposed, as compared to known methods, enables the following improvements:

- AG extraction from larch wood is realized after the isolation of DHQ and resin substances, giving a rather high purity of the extract
- the process is simple, energy-efficient and economically viable
- no expensive sorbents are needed and no toxic or combustible organic solvents are involved
- concentrates of dry substance content of up to 40% can be produced
- the closed water cycle allows water consumption and the amount of waste water to be decreased.

Additional AG purification from high-molecular phenolic impurities was realized by treatment of the water extracts with ecologically harmless oxidant (hydrogen peroxide) [17]. The optimal conditions were found to oxidize impurities without affecting the polysaccharide macromolecule.

For final product isolation from the concentrate, spray drying, lyophilic drying or fluid bed drying can be used.

The experiments showed that spray drying is technically and economically optimal. In a manufacturing pilot, different modes of AG spray drying were tested by varying the starting concentration of AG solution, air temperature at the drier input, air temperature at the drier output and pressure of compressed air at spraying. The temperature of the drying gas (air) was the most technologically relevant. The required humidity of the final product (less than 7%) was reached at an air temperature higher than 100 ºC. The optimal process conditions produced AG of high quality.

On the basis of our study, a technological scheme for isolating high purity AG was developed, involving:

1. Flocculation of AG solution (**FR**)
2. Oxidation of impurities in the AG solution (**OR**)
3. Microfiltration of the AG solution (**MF**)
4. Concentration of the AG solution by ultrafiltration (**UF**)
5. Spray drying of the concentrate (**SD**).

The manufacturing pilot revealed drawbacks to the proposed scheme: low productivity and high time consumption per product unit. Thus, the scheme was optimized.

The most reasonable sequence of steps was determined using a decision tree [77-79] with limited operation sequence combinations: spray drying was always the last step and, thus, was excluded from the decision tree; microfiltration and oxidation always followed flocculation. The decision tree, taking into account the limitations mentioned, is shown in Figure 5.

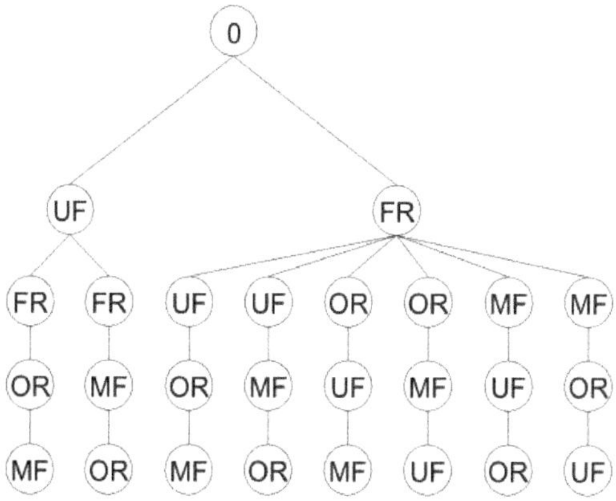

Figure 5. Decision tree

The variants were ranged from 1 to 7 according to these criteria:

- productivity
- material and reagent consumption
- steel intensity (investment)
- laboriousness (timetable)
- manufacturability.

The optimal variant is UF-FR-OR-MF (see Table 1), which has a four times higher productivity and a two times lower investment in comparison with the initial scheme.

An optimized technological scheme of arabinogalactan production was implemented at an experimental–industrial scale.

Criteria	Possible algorithms							
	FR-UF-OR-MF	FR-UF-MF-OR	FR-OR-UF-MF	FR-OR-MF-UF	FR-MF-UF-OR	FR-MF-OR-UF	UF-FR-OR-MF	UF-FR-MF-OR
Productivity	1	1	1	1	1	1	7	7
Material and reagent consumption	3	3	3	3	3	3	1	1
Steel intensity	1	1	1	1	1	1	7	7
Laboriousness	3	5	3	5	7	7	2	1
Manufacturability	2	1	2	5	7	6	6	4
Total	**10**	**11**	**10**	**15**	**19**	**18**	**23**	**20**

Table 1. Criteria of variant estimation

3. Larch bark pectins

3.1. Physicochemical and biological properties of pectins

Pectin substances are present in the majority of land and water plants, and in some freshwater algae [80]. Being an important component of cell walls, they are involved in ion exchange, water metabolism and cell wall structure formation. They stimulate seed germination and germ growth, provide turgor, etc.

The unique physicochemical properties of pectin make it indispensable in medical, food and cosmetic industries as a gelling agent, thickener, stabilizer and dietary fibre. Recently, it has become widely used as a matrix carrier for biologically active components in drugs. Pectins have physiological activities of their own (immunomodulating, hepatoprotective, anticarcinogenic, antimetastatic, etc.) making them applicable as medical preparations and biologically active food supplements.

Industrial demand for pectins in Russia is estimated at 2000 t/year, of which 10% is for the fragrance and cosmetic industries, 15% goes to medicine and pharmaceuticals, and 75% is for the food industry [81]. However, this demand is generally met by imported production. There are recent innovative Russian developments that are ecologically harmless and economically viable (there is no need to utilize aggressive acid media and to support treatment facilities), therefore having a low cost price. The raw material for pectin is the marc of citrus fruit, apple, sugar beet and sunflower head pith. There are proposals for using other plants as raw materials, such as amaranth, small mallow, duckweed, silene, coffee beans, etc. [82-86].

The bark of *L. sibirica* Ledeb. and *L. gmelinii* (Rupr.) Rupr., having a 12% pectin content, is a promising alternative raw material. At our laboratory, we are conducting systematic studies of the structure and properties of pectin from these larch species to determine a suitable technology for its industrial production.

3.2. Isolation of pectin substances from larch bark

There are a number of methods to isolate pectin polysaccharides from plant tissues, including hydrolysis extraction of dry raw material particles of certain sizes [87] using hot water, organic and inorganic acid solutions as well as salts, alkali or their mixtures as extracting solutions. Basic parameters of the pectin isolation process, such as raw material pre-processing, hydromodulus, temperature, extraction duration, medium pH and precipitator used, can all be varied depending on characteristics of the raw material [88]. We studied the influence of the following combinations of the basic parameters upon yield and product quality:

Experiment 1: 0.5% ammonium oxalate solution (hydromodulus 1:5)
Experiment 2: 0.5% oxalic acid solution (1:5)
Experiment 3: equimolar mixture of 0.5% oxalic acid and 0.5% ammonium oxalate solutions (1: 5)
Experiment 4: 0.25% sodium hydroxide solution (1:5)
Experiment 5: similar to experiment 3 (1:7)
Experiment 6: similar to experiment 3 (1:10)

In all the experiments, extraction process were the same (at 80 °C for 2 h of constant stirring).

The pectin samples obtained were white or light cream-coloured powders, tasteless and with no smell (Figure 6).

Figure 6. A laboratory sample of larch bark PS

Table 2 sets out the yield (% of weight of absolutely dry bark, a.d.b.) and composition data of pectin substances obtained in experiments 1–6.

The highest yields were observed with a weak alkali solution, but the ash content was too high (16.66%), which affected gelling ability [89]. The lowest ash content was found in the preparations isolated using an equimolar mixture of ammonium oxalate and oxalic acid (5–5.6%).

Experiment	Yield, % of a.d.b.	Medium pH	Composition, %		
			C	H	Ash
1	0.77	6.95	30.28	6.66	7.89
2	0.97	1.90	31.17	5.75	8.34
3	1.64	3.90	32.05	5.42	5.58
4	5.81	11.97	36.55	4.51	16.66
5	2.71	2.86	34.24	6.30	5.20
6	2.93	2.84	31.48	6.82	5.08

Table 2. Yield and elemental composition of pectin substances in larch bark

The optimal hydromodulus was observed in experiments 1–4 (hydromodulus 1:5) with yield increasing by 1.5 times. Hydromodulus provides insufficient penetration of extracting agent, lower than 1: 5. A 1:10 rise of hydromodulus (experiment 6) had no essential effect upon the yield and qualities of the product. Thus, the equimolar mixture of ammonium oxalate and oxalic acid used as an extracting solution at hydromodulus 1:7 (experiment 5) was the most effective in isolating pectin substances from larch bark, leading to a 2.7% yield of absolutely dry bark mass with ash content of 5.2%.

Raw material pre-processing by solvents of increasing polarity (hexane, ethyl acetate and water) resulted in both enzyme deactivation and elimination of the impurities, therefore increasing the extracting solution's ability to access the plant cell walls. Notably, the pre-extracted substances are valuable for medicine [90] and the leather industry [91].

We experimentally compared the yields of pectins isolated with and without raw material pre-processing in the conditions described above (see Figure 7). It was shown that prior elimination of impurities leads to higher yield of the product (about 1.5 times), clearly due to higher availability of pectin substances.

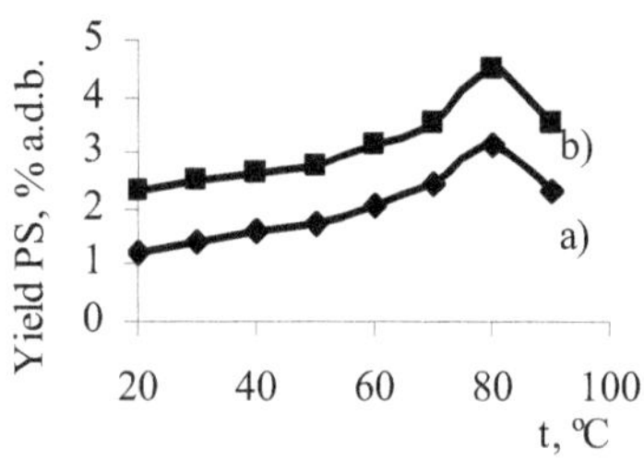

Figure 7. PS yield dependence on: a) raw material pre-processing, and b) extraction temperature

The PS yield increased as the extraction temperature rose, reaching a maximum at 80 °C. The data obtained were in good correspondence with the literature on classic pectin isolation [89]: raising temperature causes partial hydrolysis of protopectin. Thus, pectin yield increases while at temperatures higher than 80 °C the superstructure of pectin substances is broken. This is also confirmed by the dependence of the molecular weights of the resulting pectins on extraction temperatures (see Figure 8.).

Kinetic studies, particularly those concerning the pectin hydrolysis extraction process, have a particular interest. Pectin yields vs. extraction times are charted in Figure 9. A major part of PS is transferred into the extract within 1 h of extraction, after which there is no significant increase of yield.

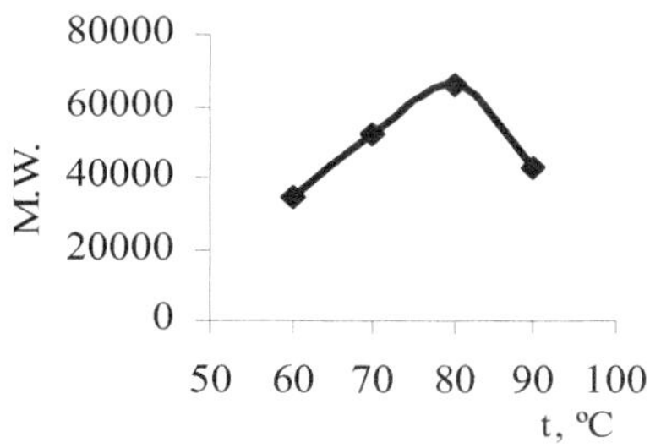

Figure 8. PS molecular weight (M.W.) dependence on extraction temperature

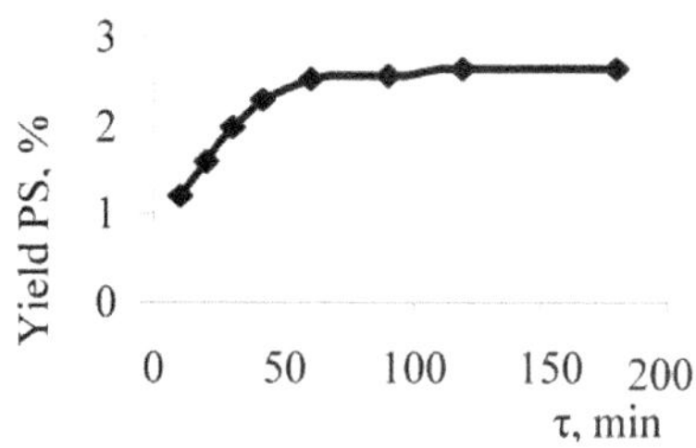

Figure 9. Dependence of PS yield on extraction time

We also studied the influence of type of precipitator used upon the yield and qualities of the product. For this purpose, pectin extract was prepared from larch bark by treating it with an equimolar mixture of 0.5% ammonium oxalate and 0.5% oxalic acid (hydromodulus 1:7) at 80 °C for 2 h. The extract was concentrated in a circulation vacuum evaporator until it reached one third of its original volume. One half of the concentrate was precipitated with acetone and the other half with ethanol. Precipitators were added in equal quantity, dropwise while continuous stirring was applied. The precipitate was vacuum-filtered, dissolved in 100 ml of distilled water by heating to 40–50 °C when necessary, and then again precipitated and filtered. The final precipitates were washed with the same precipitator and then with diethyl ether, dried in the air and then in a drier at 50 °C, cooled to room temperature in a desiccator and measured to determine yield. It is noteworthy that the dropwise addition of precipitator into the extract increased yield by 0.5% compared to the usual precipitation procedure. We established that acetone is less selective, and ethanol therefore gives a purer product. Purity of the pectin preparation obtained can also be estimated based on galacturonic acid content [92]: for larch bark pectins precipitated by acetone and ethanol, the result was 69.77 and 78.12%, respectively.

Thus, the optimal procedure for isolating pectin substances from larch bark involves pre-treatment by hexane, ethyl acetate and water, extraction by an equimolar mixture of 0.5% oxalic acid and ammonium oxalate solutions at hydromodulus 1:7 and an extraction temperature of 80 °C for 1 h, and precipitation by ethanol. The method has been patented [93] and used for preparing the samples for physicochemical and application studies.

3.3. Characterization of pectin substances

Pectinase enzyme hydrolysis of PS samples isolated from larch bark by the above method, and further analysis of hydrolysis products by paper chromatography (PC), have shown an essential destruction of PS with formation of free D-galacturonic acid.

Table 3 sets out the main maxima of absorption bands in the IR spectra of PS and their assignment, proving the PS pectin nature of the samples [94].

Frequency (ν, cm^{-1})	Assignment
3460	ν(OH), ν(H_2O)
3260	ν(NH)
2962, 2872	ν(CH_3)
2573	ν(OH),
1730	ν(C=O) в COOH
1640	$\delta(OH)_A$
1540	δ(NH)
1380–1450	δ(C-CH_3), ν(C-O) pyranose rings
1331	δ(OH) in pyranose rings
1265	ν(C-O) in esters
1150	ν(C-O-C)
1095	ν(C-C)
1027	ν(C-OH)
890	δ(C1-H) in glucopyranose ring
766, 629, 528	pulse vibrations of pyranose ring

Table 3. Absorption band maxima in IR spectra of PS and their assignments

Thus, enzyme hydrolysis and IR spectroscopy data prove that the polysaccharide isolated from larch bark refers to the pectin group.

The monosaccharide composition of PS was determined by total acid hydrolysis with trifluoroacetic acid (TFA). Monosaccharide identification of PS was performed using gas–liquid chromatography (GLC) and the sample was shown to consist of galacturonic acid, protein compounds and monosaccharides of arabinose, galactose, rhamnose, glucose, mannose and (in minor quantities) xylose. Dominant monosaccharides were galactose and arabinose, in a ratio of 2.7: 1.

The degree of homogeneity for PS was determined by ion exchange chromatography on DEAE cellulose with sodium chloride aqueous solutions. Four fractions were detected (Table 4). In the fractions PS-1 and PS-2, arabinose and galactose were predominant (18.26/52.96% and 11.65/30.83%, respectively); thus, they refer to acidic arabinogalactans. The acidic nature of PS was developed with D-galacturonic acid residues with the PS-1 proportion five times less than in PS-2, while in PS-3 and PS-4 it was a major monosaccharide, and thus they refer to pectins. The content of neutral monosaccharides in PS-4 was minimal compared to other fractions (3% mass). All the fractions contained protein compounds that were not eliminated by gel filtration. It seems likely that the protein and polysaccharide compounds were strongly aggregated, or that their molecular weights were close to each other.

Sample*	Yield, %	Content, %								
		Gal*p*A	Protein	Monosaccharides						
				Rha	Ara	Xyl	Man	Glu	Gal
PS-1	12.1	5.67	6.9	*traces*	18.26	1.54	2.53	5.95	52.92
PS-2	5.9	29.12	7.3	0.53	11.65	1.02	2.71	8.81	30.83
PS-3	17.0	65.93	5.7	1.91	4.45	0.75	1.18	1.12	9.42
PS-4	37.0	79.87	3.6	0.35	0.93	0.18	0.24	0.21	1.06

* PS-1 isolated with use of 0.01M NaCl solution, PS-2 – 0.1M NaCl solution, PS-3 and PS-4 – 0.2M NaCl solution

Table 4. Chemical characterization of PS sample after DEAE-cellulose fractioning

The amino acid composition of PS proteins was studied. The major components of PS were glutamic acid (6%) and aspartic acid (2.8%), while total content of amino acids with aliphatic side chains (glycine, alanine, valine, isoleucine, leucine) was 9% (Figure 10).

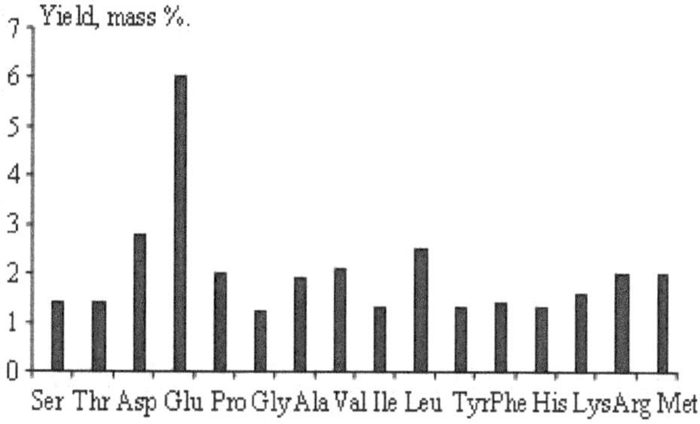

Figure 10. Amino acid composition of PS proteins

Thus, we isolated PS from the bark of *L. sibirica* Ledeb. and *L. gmelini* (Rupr.) Rupr., pre-treated with ethyl acetate and hot water. Enzyme hydrolysis with pectinase and IR spectroscopy were employed to prove the presence of pectin polysaccharides in the samples isolated. DEAE-cellulose chromatography revealed that PS includes four fractions, two of which are acidic arabinogalactans and the other two belong to the pectin group.

3.4. Structural study of main chain of larch bark pectin

Acid hydrolysis of PS by 2M TFA results in galacturonan PVG-1. The high value and positive sign of the rotation angle of +245.3° (*c* 0.1, H_2O) suggest α-D-configuration of D-galactopyranosyluronic acid residues.

Values of chemical shifts (CS) of carbon atoms in the ^{13}C NMR spectrum of PVG-1 (Table 5), compared to other data, [95] corresponded to those for carbon atoms in D-galacturonic acid residues in pyranose form which compose the linear fragment of pectin molecules (pectin core). The presence of an anomeric carbon atom signal at 101.9 ppm indicated both (1→4)-bonding between D-galacturonic acid residues and α-configuration of C-1 anomeric atoms. Signals at 176.2 ppm were assigned to the C-6 atom and indicated a free carboxyl group in D-galacturonic acid residue. Additionally, there were galacturonic acid residues esterified by methoxyl in the PVG-1 molecule, according to signals with CS at 172.2 ppm ($\underline{C}$-6-OCH_3) and 54.4 ppm (-$O\underline{C}H_3$). The ratio of the integrated signal intensity of carbon atoms observed in methoxyl and carboxyl groups suggests a high degree of galacturonan methoxylation. The ^{13}C NMR spectrum also showed signals at δ 76.1 and 74.9 referring to the C-3 carbon atom substitute in (…→4)-α-D-Gal*p*A-(1→…) galacturonic acid residue in the galacturonan molecule (the non-substituted atom has CS at 72.1 ppm).

Residue	C-1	C-2	C-3	C-4	C-5	C-6	C-6-(OCH_3)	-$O\underline{C}H_3$
→4)-α-D-Gal*p*A-(1→	100.9	68.9	70.1 76.1 74.9	79.2	73.4	176.2	172.2	54.4

Table 5. Chemical shifts of signals of galacturonic acid carbon atoms in ^{13}C NMR spectrum of PVG-1

Thus, according to spectral and chromatographic data, linear polysaccharide from larch bark has a structure of homogalacturonan consisting of (…→4)-α-D-Gal*p*A-(1→…)-linked fragments D-galacturonic acid has partially etherified by methoxyl groups with branching points at C-3 atom of galacturonopyranosyl residue.

3.5. Structural study of side branches of larch bark pectin

Partial acid hydrolysis of PS with 0.01M TFA for 3 h resulted in galacturonan PVG-2. According to ^{13}C NMR data, it was a pectin polysaccharide. The spectrum contained both typical signals of galacturonic acid residues, namely pronounced signals of anomeric carbon atoms at 100.4 and 104.4 ppm, and signals of carboxyl carbon atoms at 171.4, 166.5 and 53.7

ppm, the latter two being signals of carbon atoms in uronic acid residues methoxylated by the C-2 and/or C-3 atoms (Table 6).. Intensities and spectral positions of signals at 68.9, 70.8, 78.9 and 72.2 ppm corresponded to data in the literature for α-D-Gal$_p$A residues connected by 1→4 bonds. There is a ratio of 1:5 between integral signal intensities of carboxyl and methoxyl carbon atoms, which suggests a high degree of PS methoxylation.

Residue	C1	C2	C3	C4	C5	C6	-O<u>C</u>H$_3$ (<u>C</u>H$_3$-)
→4)-α-D-Gal$_p$A-(1→	100.4	68.9	70.8	78.9	72.2	171.4	-
2-MeO-α-D-Gal$_p$A-(1→	100.9	166.5	69.6	78.9	73.8	171.4	53.7
3-MeO-α-D-Gal$_p$A-(1→	100.9	68.9	166.5	78.9	73.8	171.4	53.7
β-D-Gal$_p$-(1→	104.64	71.7	74.1	69.6	76.1	62.0	-
→6)-β-D-Gal$_p$-(1→	104.38	71.7	73.8	69.6	74.3	70.8	-
→3,6)-β-D-Gal$_p$-(1→	104.64	71.7	82.5	69.57	74.1	71.4	-
α-L-Ara$_f$-(1→	108.6	80.7	78.9	84.9	62.0	-	-
β-L-Ara$_p$-(1→	101.1	69.6	-	-	-	-	-
→3,5)-α-L-Ara$_f$-(1→	108.6	80.7	84.9	83.2	67.8	-	-
→2,5)-α-L-Ara$_f$-(1→	108.0	84.9	77.6	83.2	67.8	-	-

Table 6. Chemical shifts in signals of carbon atoms in the ^{13}C NMR spectrum of PVG-2

In the ^{13}C NMR spectrum of PVG-2 samples there were upfield signals at 17.9 and 18.13 ppm belonging to C-6 atoms in terminal rhamnose residues and in polysaccharide chains, respectively. The integral intensities of these signals and those of C-2 and/or C-3 and C-6 carbon atoms for galacturonan residues at 166.5 and 171.4 ppm were found to have a ratio 1:5. The total integral intensity of signals for anomeric C-1 atoms for rhamnose and the total integral intensity of signals of anomeric atoms of galacturonan residues were equal to each other, *i.e.* they had the same ratio for rhamnose and galacturonan residue content in the chain. According to data in the literature, signals at δ 99.7, 77.6, 70.8, 82.5, 68.9 and 17.9 ppm are assigned to C-1, C-2, C-3, C-4, C-5 and <u>C</u>H$_3$ carbon atoms in →2,4)- α-L-Rha$_p$-(1 → residues.

Thus, according to ^{13}C NMR spectral data, linear fragments of pectin polysaccharide isolated from larch bark are rhamnogalacturonans where D-galacturonic acid residues in pyranose form with an α-configuration of their anomeric centre are connected 1–4 by glycosidic bonds. One fifth of galacturonan residues associated with the C-6 atom were esterified by methoxyl groups. The ratio between 2,4-substituted rhamnopyranosyl and galacturonosyl residues (1:5), thus, the main chain structure of the pectin polysaccharide was highly branched at the C-4 atoms of rhamnopyranosyl residues.

Further ^{13}C NMR spectrum analysis of the PVG-2 sample showed that arabinogalactan fragments are present in rhamnogalacturonan as side chains. Concerning signals of anomeric carbon atoms, the ^{13}C NMR spectrum of the PVG-2 sample showed that there are signals at 101.1, 104.38, 104.64 and 108.6 ppm, as well as signals of anomeric carbon atoms in galacturonopyranosyl residues of the galactan core. According to [9], intensities and values of CS can be assigned to signals of anomeric carbon atoms in β-L-Ara$_p$, α-L-Ara$_f$ and β-D-Gal$_p$ residues. The most upfield of the signals mentioned (δ 101.1 ppm) belong to terminal β-L-Ara$_p$ residues. Signals at 104.38 and 104.64 ppm belong to C-1 in β-D-Gal$_p$ residues while CS values of C-2, C-3, C-4, C-5 and C-6 atom signals are calculated according to the official data for β-D-galactopyranosyl residues. Bonding at the C-3 and C-6 positions of β-D-galactopyranose was proven by downfield shifts of these signals at 8.7 and 8.8-9.4 ppm, respectively, due to glycosylation of these atoms as compared to their positions in non-substituted 1→3,6 linked β-D-Gal$_p$ residues. Signals at δ 108.6 ppm, like those at 80.7, 78.9, 84.9 and 62.0 ppm, are terminal α-L-arabinofuranose. The anomeric atoms of arabinose and galactose are monosaccharides integrated at a ratio of 1:2.

Hence, according to spectral data for the PVG-2 fragment of the pectin polysaccharide from larch bark, highly branched arabinogalactan was detected as side chains consisting of linear chains with →3,6)-β-D-Gal$_p$-(1→ residues with branching at C-6 atoms. Side chains of arabinogalactan fragments contain terminal arabinose, both in pyranose and in furanose form, as well as →2,5)-α-L-Ara$_f$-(1→ and →3,5)-α-L-Ara$_f$-(1→ residues as intermediate fragments.

3.6. Larch bark pectin peculiarities and implementation fields

It has been determined that larch bark pectin substances possess immunomodulatory, antineoplastic, gastroprotective and antitoxic action [96-98]. In order to understand larch bark pectin's physiological and pharmacological action, we have started research focussed on examining its membrane-acting action. The vacuoles of isolated cell plants and their membranes were found to be an appropriate object for our research. The influence of pectin on membranes and the peculiarities of their barriers were estimated according to the change of destruction dynamics in isolated vacuoles in comparison with the control. The results are depicted in Figure 11. It has been established that implementation of pectin aqueous solutions leads to their protective action on vacuolar membranes, exceeding the control threefold. Thereby, the experiments proved that larch bark pectin possesses a membrane stabilizing activity.

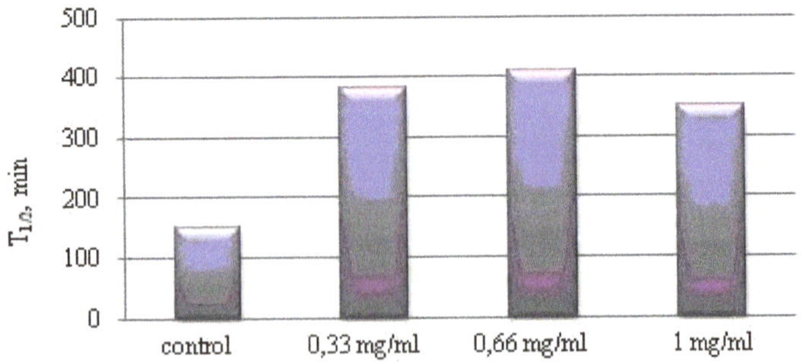

Figure 11. Influence of pectin upon isolated vacuole half-lives ($T_{1/2}$)

In order to broaden larch bark pectin implementation fields, we carried out research into its implementation as a reducer and stabilizer of noble metal particles in nanosized state. Supramolecular structure peculiarities, optical activity, carboxyl and abundance of hydroxyl groups, and polymeric pectin molecule stabilizing effect provided significant potential in nanobiocomposite formation processes in metals with a polysaccharide matrix ("pectin – metal (0)").

Synthesis of nanobiocomposites was carried out using the redox reactions of PS with silver nitrate. Nanobiocomposite samples 0.5 "pectin – Ag(0)" up to 72% content of silver were obtained in different reaction conditions. It was discovered that the effectiveness of the reaction to create a silver nanoparticle flow depends on medium spectrum pH. The spectra of the mixtures of pectin and silver nitrate water solutions versus time reaction are depicted in Figure 12a. It was determined that, with a reduction of pH to 3.5, the Ag(I) reaction proceeds very slowly. This is demonstrated by the appearance of a link in the absorption spectrum in the range of λ 280–470 nm only 24 h after the beginning of the reaction (Fig. 12a). The wide maximum low intensity link was indicated by the formation of silver metal primary centres. Despite this, the reaction speed of the reduction was so slow that even 96 h was not enough to create fully recovered Ag(0) centres. With pectin and Ag(I) interactions in reaction mixtures beginning at a pH of 7, a symmetric bond at λ_{max} 420 nm can be observed in the electron spectra at the start of reaction by proving the formation of Ag(0) nanoparticles (Fig.12b, line 2). Even so, it takes about 24 h for the full silver cation conversion which was experimentally evaluated according to the absorption bond intensity growth. Ag(I) reduction with pectin at pH 11–12 proceeded swiftly immediately after mixing of the components (Figure 12b, line 4 and Figure 12c line 1) and finished within 30 min. Reduction under these conditions was also accompanied by variations in the particle size of Ag(0), as shown by the shift in the Plasmon pick position into the short-wave region at 10 nm (Figure 12c).

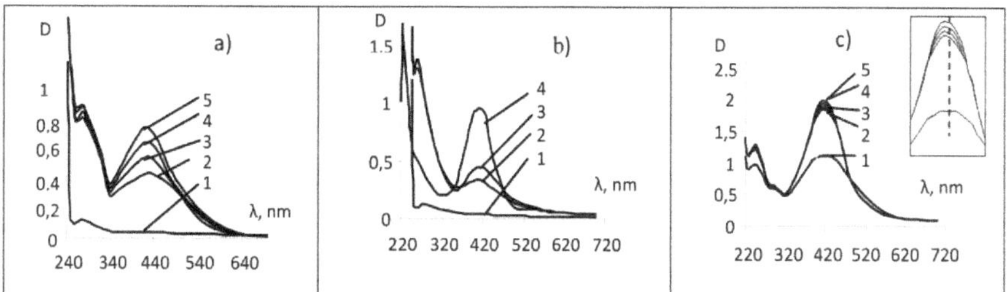

Figure 12. Absorption spectra of mixtures of aqueous solutions of pectin (0.5%) and silver nitrate (0.1%) in a ratio of 1:1 depending on: a) reaction duration: 1 min (1), 24 h (2), 48 h (3), 72 h (4), 96 h (5); b) medium pH: 3.5 (1), 7 (2), 9.7 (3), 11.5 (4); c) reaction duration at pH 11.5: 1 min (1), 30 min (2), 60 min (3), 180 min (4), 24 h (5)

Radiographic phase analysis of obtained nanobiocomposites of "pectin-Ag(0)" demonstrated it to be a mixture of radioamorphous and crystalline phases. There was a wide halo with maximum intensity at d ~ 0.46 nm in 2θ angle intervals from 8 to 60E (Figure13a) in a radioamorphous phase diffraction pattern typical of a pectin source. There were quite intensive but broadened lines typical of metallic silver (Figure 13b) during silver loading in the diffraction patterns of reaction products against a background of pectin reflection. The calculation of silver unit-cell parameters showed that in their quantity in the provided samples was lower than for massive silver and changed from 0.4036 to 0.4050 nm (∀ 0.0008 nm). Moreover, the average size of the coherent-scattering region (CSR) was calculated according to Selyakov–Sherarar's formula [99] to be in the range of 3 nm. The data obtained demonstrated that, in the samples of Ag (0), the persistence of nanosized particles was stabilized by an amorphous phase with pectin.

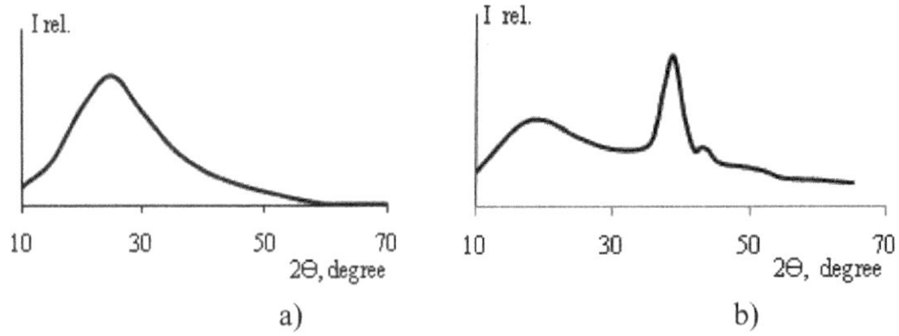

Figure 13. Diffraction patterns of pectin sample (a) and of the "pectin-Ag(0)" nanobiocomposite sample (b)

"Pectin – Ag (0)" nanobiocomposite scanning electron microscopy (Figure 14) showed that the analysed samples contain particles considerably smaller than 100 μm.

Figure 14. Electron microphotography of "pectin – Ag(0)" sample

Microphotography analysis of nanobiocomposites, obtained by the use of transmission electron microscopy, demonstrated that there are isolated silver particles of null valency in globular form (Figure 15a), of a size within the range from 4 to 17 nm (predominance (up to 80%) at 6–7 nm, Figure 15b).

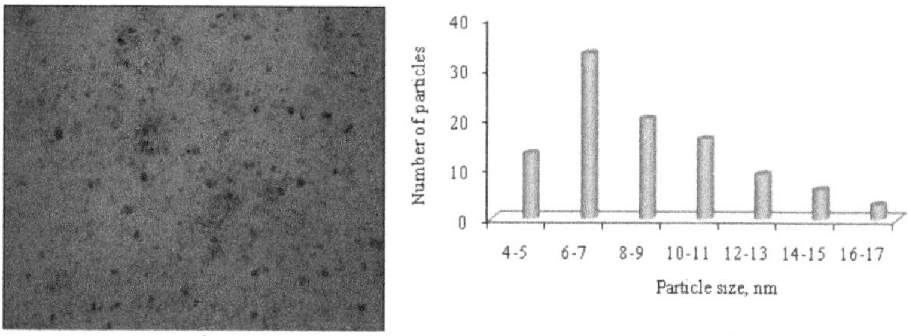

a) b)

Figure 15. Transmission electron microphotography, a), and size distribution graph of "pectin-Ag(0)" nanobiocomposite sample, b)

Thereby, "pectin-Ag(0)" nanobiocomposite formation takes place as a result of the interaction of pectin water solutions with Ag(I). Process speed increases significantly with variation within the alkaline pH range of the medium. The initial component proportion influences the results of the reaction: the more Ag(I) that falls per 1g of pectin, the less the quantity of Ag(0) particles that is created in a nanosized condition. Using pectin implements reduction and stabilizing functions, and also adjusts the sizes of obtained Ag(0) nanoparticles.

4. Cellolignin residue of larch wood as raw material for crystalline glucose production

The cellolignin residue is formed during the chemical processing of larch wood using the technology for obtaining dihydroquercetin and arabinogalactan [100]. CR represents larch wood chips initially extracted by ethyl acetate and hot water. The larch wood chip basically consists of cellulose, hemicellulose, and lignin. The polysaccharide content in larch wood chips is 65–75% of the mass of its absolutely dry wood (a.d.w.) [101], and the content of water-soluble substances is 10–16% (in some samples up to 30%) [102]. The content of holocellulose in CR (without water-soluble substances) is about 54% of its a.d.w. weight, whereas the content of holocellulose in the original larch wood is about 40% of its a.d.w. (also without water-soluble substances). The gain in the relative content of polysaccharides per mass of a.d.w. is about 13%, which permits one to consider the CR as a polysaccharide-"enriched" raw material from which it is possible to obtain sugar and other products by hydrolysis. Hemicellulose polysaccharides in the CR of larch wood are mainly represented by the water-soluble polysaccharide arabinogalactan and 4-O-methylglucuronoaraboxylans and galactoglucomannans [103], which are associated to differing degrees with cellulose. The yield of arabinogalactan from larch wood using dihydroquercetin production technology in which it is isolated as a by-product (without special optimization) accounts for 67% of its total content in the original raw material; therefore, the content of water-soluble substances in CR remains rather high, at 8.9%.

The data on the group composition of the components of CR were reported in [104]. The acid hydrolysis of hemicelluloses results in the formation of mono- and oligosaccharides, the presence of which substantially impairs the quality of hydrolysates and hinders the crystallization of glucose from these syrups. Therefore, it is necessary to purify CR from hemicelluloses. Hemicelluloses are commonly removed (to a particular limit) from the raw material, either by hot water extraction or by hydrolysis with diluted acid at elevated temperature.

In experiments with water hydrolysis of larch wood chips in laboratory autoclaves, the parameters of hydrolysis, i.e., temperature, hydromodulus, duration and the number of hydrolysis steps, were varied. The maximum total yield of water-soluble substances (18%) was attained based on four-step hydrolysis. However, it is economically more attractive to perform one-step hydrolysis, with the conditions specified for obtaining a maximum yield of sugars with this hydrolysis method being as follows: a gradual increase in temperature from 25 °C for 1 h and 160 °C for 1 h with a hydromodulus of 1:6. The yield of substances was 15%. As indicated by paper chromatography (PC) and thin layer chromatography (TLC), the pre-hydrolysate contained arabinose, galactose, xylose, mannose and trace amounts of glucose [105].

Mild hydrolysis of hemicelluloses of CR was performed using 1–5% sulphuric or hydrochloric acids at the boiling temperature of the solution (100–105 °C). By choosing the optimal hydrolysis conditions, a maximum yield of reducing substances (RS) in hydrolysate, 1.1%, was achieved with the use of 5% sulphuric acid, which corresponded to a 23% content of hemicelluloses or noncellulose polysaccharides in CR. The hexose content in the prehydrolysate was 33.9% of total sugars, as determined from the content of RS, and that of pentoses was 62.1%; arabinose, galactose, xylose, mannose, and glucose were identified qualitatively. Significantly less amounts of polysaccharides were hydrolyzed with the use of 2% hydrochloric acid under the same hydrolysis conditions - 9.6% .

Aqueous and acidic sugar solutions obtained during the pre-hydrolysis of CR (technological sugar solutions) contained greater amounts of fermentable sugars compared to other conifers and can be used in the production of feed for animals [106, 107] and other products.

To obtain pure glucose syrups, two alternative variants of acid hydrolysis of cellulose were considered: hydrolysis of CR and of cellulose itself after delignification of CR. Using the first variant, we studied the hydrolysis of CR by diluted (1.0–5.0%) hydrochloric acid at high temperatures (160–170 °C) and by concentrated (50–85%) sulphuric acid at room temperature to compare the crystallization properties of glucose syrups obtained by the two methods. A high temperature hydrolysis of the raw material by diluted acid was carried out in laboratory autoclaves in four steps. As the acid concentration was increased from 1 to 5%, a gain in the sugar content of the hydrolysates was observed. Thus, the mass portion of RS relative to the mass of a.d.w. during hydrolysis of sawdust was 5.4% using 1% H_2SO_4, 7.9% with 2% H_2SO_4, 9.2% with 3% H_2SO_4, and 11.6% with 5% H_2SO_4. However, simultaneously to an increase in the acid concentration, the quality of the hydrolysate was markedly impaired; an intensive dark colour appeared due to the formation of sugar degradation products, and the content of colloidal impurities in syrups at later stages caused severe problems. Conversely, the low temperature hydrolysis of CR (of the same raw material) by concentrated sulphuric acid made it possible to obtain a high yield of sugar in hydrolysates with a minimum content of degradation products. The maximum yield of RS was 64% using 70% sulphuric acid at the hydromodulus (1:5) and a hydrolysis time of 5 h, and 62.6% using 80% sulphuric acid at the hydromodulus (1:5) and hydrolysis time of 2 h. Hydrolysates contained no pentoses. These characteristics meet the requirements imposed upon hydrolysates from which crystalline glucose is isolated.

Thus, the use of concentrated sulphuric acid in the concentration range of 65–80% makes it possible to obtain hydrolysates with a maximum content of RS (up to 64%) [108]. After additional hydrolysis (the inversion stage), the hydrolysate has a pH value close to 2. On further evaporation of this hydrolysate, the acid is concentrated, which leads to further degradation of sugar; as a result, the yield of glucose decreases, and the syrup is contaminated with stained products.

Sulphuric acid was neutralized using barium acetate, sodium hydroxide and calcium hydroxide, and the contribution of each compound to formation of the mineral ash component in syrups was determined. The lowest ash content in the hydrolysate (0.2%) was achieved by applying barium acetate. However, because of the toxicity of barium and its salts, the use of barium was abandoned. Upon neutralization of sulphuric acid by calcium hydroxide the ash content in the hydrolysate was initially as high as 10%; nevertheless, preference was given to this compound alone since it is nontoxic, readily available, and convenient in operation. According to the "Glucose, crystalline hydrated" GOST 97588 Regulations [109], the ash content in the final product must not exceed 0.06–0.07%, calculated for dry substance; therefore, it was necessary to provide neutralization conditions to decrease the ash content in hydrolysates. For neutralizing sulphuric acid, pH value was brought to 4–4.5 at a temperature no higher than 80 °C. The amount of calcium hydroxide was calculated according to the sulphuric acid neutralization reaction so as to exclude the over-alkalization of the solution [110]. The resulting dihydrate gypsum crystals were filtered off. The ash content in hydrolysate decreased by up to 0.5%. In addition to glucose and mineral contaminations, the hydrolysate contained, dependent on hydrolysis conditions, the products of partial hydrolysis of the lignocarbohydrate complex of wood—mono-, di-, tri-, and oligosaccharides—as well as impurities belonging to different classes of organic compounds: acid-soluble lignin, furfurol, oxymethylfurfurol, a lignohumic complex, colloids, levulinic acid, and other organic acids [106]. At the next stage, we assessed the nature of substances by determining the colour of hydrolysates and selected how to remove them from sugar solutions [111]. It was possible to assign some impurities, that by their nature are associated with lignin, to substances based on their colouring. First of all, this was acid-soluble lignin. According to the published data, 2–3% of total lignins are dissolved in the hydrolysis of coniferous wood by a solution of 72% sulphuric acid [103].

In addition, coloured substances are formed during sugar degradation: hexosans form high molecular weight substances of brown colour, which partially precipitate from solution, and pentosans form furfurol, which imparts a yellow colour to hydrolysate. Under acidic conditions, lignin and sugar degradation products, i.e., furfurol and oxymethylfurfurol, form condensed products in small amounts, and insoluble humic compounds; and the products of incomplete hydrolysis of polysaccharides, oligosaccharides, can be partially adsorbed by acid-insoluble lignin [106]. Hydrolysates were clarified using activated carbon BAU (Russia). The UV spectrum of a neutralisate from CR shows an absorption band at 280 nm, which disappears after treatment of the neutralisate by activated carbon. Treating the hydrolysate with dichloroethane followed by IR analysis of the concentrated extract made it possible to identify it as acid-soluble lignin. Thus, the treatment of hydrolysates with activated carbon significantly reduces the content of acid-soluble lignin in hydrolysates (from 1.4 to 0.3%).

The scheme of the acid–hydrolytic transformation of cellulose to glucose with preliminary delignification of lignocellulose raw material by industrial methods [112] is the second variant, which also makes it possible to obtain high purity glucose syrups. Its main advantage is that it enables glucose syrups to be obtained with a factor of merit of no less than 85%, which are not contaminated with colouring impurities of ligno-carbohydrate origin and ash components. It is known that, during the low temperature hydrolysis of cellulose by concentrated acids, partial destruction of cellulose with the formation of water-soluble products occurs [113]. Under optimized conditions, concentrated sulphuric acid almost completely dissolves cellulose, and the cleavage of glycoside bonds proceeds in a homogeneous medium. As a result of hydrolysis, a mixture of products differing by the polymerization degree (PD) is formed: from comparatively high molecular weight cellulose and cellodextrins (PD from 7 to 50–60) to oligosaccharides (mainly di- and trisaccharides) and glucose. The composition of the mixture and the ratio of the products in the hydrolysate depend on the hydrolysis conditions. These products also vary in water solubility. Thus, cellodextrins, oligosaccharides, cellobioses and monosaccharides are water-soluble, and part of the cellulose itself, mainly its crystalline moiety, and hydrocellulose do not dissolve in water.

Thus, the hydrolysis of cellulose enables firstly the isolatation of intermediate water-soluble hydrolysis products with simultaneous removal of sulphuric acid without its chemical neutralization, which in turns prevents the entry of mineral impurities into syrups; and, secondly, obtaining of the required monomeric sugar, i.e., glucose, in one stage, by subsequent additional hydrolysis of the intermediate product. The hydrolysis of industrial cellulose was carried out by 72% sulphuric acid at room temperature for 1 h with regular stirring of the hydrolysate mass; in this case, cellulose had completely dissolved within the first 15 min. Increasing the hydrolysis duration up to 2 and 3 h did not significantly affect the final yield of the product, which was 80–90% of the weight of absolutely dry cellulose (a.d.c.) [114].

We arbitrarily called this product the inverted polysaccharide (IPS) since this name reflects its position in the technological scheme. Dried IPS is a white or pale cream powder. The product is partially soluble in water (the insoluble fraction accounts for 43% of the weight of IPS), is soluble in aqueous alkaline solutions, and exhibits a lower PD than the starting cellulose (150; PD for starting cellulose, 573). The content of cellulose in an aqueous IPS solution was estimated, using HPLC, to be 2% of the a.d.c. A comparative analysis of IR spectra of the starting cellulose and IPS indicated that IPS is cellulose with a high degree of amorphism [94]. In particular, this is evidenced by strong changes in the IR spectrum of IPS in the region of 600–1500 cm^{-1}, which accompanies changes in the polysaccharide hypomolecular structure, and smoothing of the intensity of so-called crystallinity bands at 1100, 1140, 1190, 1250, 1360, and 1420 cm^{-1} [115]. The IR spectrum of the product contained no absorption bands at 1112 and 1162 cm^{-1}, which are typical of the spectra of a highly ordered cellulose structure. The residual sulphuric acid content in an aqueous IPS solution was 0.2%, indicating that 98% of the acid taken for hydrolysis is removed simultaneously with the isolation of IPS (without chemical neutralization).

In order to convert IPS to the monomeric form of sugar (glucose), an inversion was carried out at high temperature using a diluted acid. We studied the kinetics of the IPS inversion using diluted (0.075–1.5%) hydrochloric acid [116]. The choice of this acid was primarily dictated by the fact that sodium chloride formed during the neutralization of the acid by NaOH is a part of the complex composite (CC) of glucose with the formula $(C_6H_{12}O_6)_2{\cdot}NaCl{\cdot}H_2O$, the decomposition of which results in the release of crystalline glucose. Considering that the potential yield of glucose, on inversion by 5% sulphuric acid at 100 °C for 5 h, is 1.5% in the hydrolysate, which corresponds to a glucose yield equal to 82% of the weight of a.d.c., the acid concentration of 0.125% and temperature of 170 °C represent the optimum inversion conditions under which the yield of RS in the invertion is at its maximum; the time taken for attaining the maximum yield in these conditions is minimal. Thus, during the hydrolysis of cellulose and the subsequent inversion of IPS, the main glucose content in the inverted solution is about 70% of the mass of RS; i.e., the real yield of glucose is 35–45% of a.d.c.

It should be noted that these glucose syrups are transparent, of a light yellow colour, and are distinguished by a high factor of merit (85–90% and more) (Figure 16). In comparison, the yield of glucose from CR (the first variant of hydrolysis) is 23–25%. CG is isolated from glucose solutions either by direct or salt crystallization [117]. We studied the crystallization properties of glucose syrups obtained by hydrolysis of CR and cellulose using both the direct and salt methods. As mentioned above, the factor of merit of starting syrups must be no less than 85% for the successful crystallization of glucose using the direct method. Glucose syrups obtained by hydrolysis of cellulose completely meet this requirement. The application of activated carbon increased the quality of the syrup since direct crystallization occurred only in clarified syrups.

In order to perform direct crystallization, cellulose hydrolysis was used to obtain a hydrolysate with a RS content of 1%, pH 4.4, and a factor of merit of 94%, which was allowed to stand at room temperature for spontaneous crystallization. After two weeks, the onset of crystallization was visually observed. Crystallization by itself, without the creation of special temperature conditions, progresses slowly (taking a month and more). The method of salt crystallization of glucose has some advantages over direct crystallization. It does not require a deep purification of hydrolysates, the crystallization process is shorter and simpler (there is no need for a multiple recrystallization), and the yield of glucose increases.

Figure 16. Glucose syrup from larch wood CR

We studied the crystallization conditions of glucose using its CC with sodium chloride $(C_6H_{12}O_6)_2 \cdot NaCl \cdot H_2O$ and examined CC crystals obtained from model mixtures of glucose, sodium chloride and water, as well as from experimental glucose syrups obtained by the hydrolysis of CR and cellulose [118].

The crystallization of CC was studied using artificial mixtures in which the NaCl content varied from 15% to a twofold excess relative to the glucose content. In all cases, a crystalline phase formed. The composition of the crystalline phase was determined by elemental analysis. We determined that the range of NaCl:glucose ratios from 0.2:1 to 0.7:1 (parts by weight) is optimal for the formation of CC. Similarly, we determined the crystallization conditions for CC in hydrolysates of CR and cellulose. CC crystals were isolated from hydrolysates; the artificial mixtures were colourless and transparent and had well-defined facets. According to the chemical analysis data, they have a composition close to being stoichiometric: C – 32.5%, H – 5.9% and Cl – 9.7%. Theoretically, CC with the general formula $(C_6H_{12}O_6)_2 \cdot NaCl \cdot H_2O$ contains C – 33%, H – 6%, and O – 18.13%.

According to X-ray phase analysis, CC monocrystals synthesized from pure solutions have the unit cell parameters a = b = 16.8, c = 17.0 Å and β = 120°, and represent a hexagonal prism. Based on the symmetry of lauegrams and weissenbergograms, they belong to the diffraction class $P\bar{3}m$ with regular extinctions at $1 \neq 3n$. Therefore, the spatial group of CC crystals was determined as $P3_1 12$ (151) and $P3_2 12$ (153).

The set of diffraction maxima obtained by X-ray phase analysis of CC crystals and the reference indicates that, under the experimental conditions used, glucose in the presence of sodium chloride crystallizes as CC with the formula $(C_6H_{12}O_6)_2 \cdot NaCl \cdot H_2O$.

Thus, when studying the crystallization properties of glucose syrups produced by acid hydrolysis of the crystalline glucose of larch wood, we obtained CC glucose crystals with sodium chloride, upon decomposition of which D-glucose is released in crystalline form. In addition, glucose can be directly crystallized from glucose syrups produced by the hydrolysis of cellulose with a high factor of merit (more than 85%).

The ways in which the efficiency of hydrolysis of polysaccharides from wood CR can be increased are of a great interest. One method is the steam explosion hydrolysis of cellulose containing raw material, which makes possible the efficient and completely ecologically safe decomposition of lignocellulose material into its constituents: lignin, hemicellulose, and cellulose.

We studied the steam explosion hydrolysis of CR from larch wood, and showed that this method can be used for effective prehydrolysis processing of larch wood CR [119].

Hence, a laboratory scheme has been developed which will be used as the technological basis for obtaining crystalline glucose from the CR in larch wood [120].

5. Experimental section

Arabinogalactan was extracted using technology from *L. sibirica* Ledeb. [76] at the experimental–industrial plant and purified with methods described in [73, 17]. Molecular masses of arabinogalactan were defined with a high-performance liquid chromatography (HPLC) method, assisted by the Agilent Technologies 1260 Infinity chromatographic system using 0.1 M sample solutions of $LiNO_3$ on PL aquagel-OH-40.8 mm, 300x7.5 mm column, with a PL aquagel-OH Guard 8 mm 50x7.5 mm precolumn, standardized according to dextrans with 25, 12 and 5 kDa molecular mass standard solutions and monosaccharides.

IR spectra were registered in KBr tablets on a "Specord 75IR" spectrophotometer with a 500–4000 cm^{-1} interval. UV spectra were registered with a "Specord UV-vis" spectrophotometer (10 mm layer thickness). NMR ^{13}C spectra of AG samples were registered with a "Varian VXR 500S" spectrometer with a 125.1 Hz operating frequency; D_2O was used as solvent. Deuteroaceton was used as an internal standard. The correlation of galactose and arabinose chains, composed of AG macromolecules, were calculated according to the correlation between integral intensities of carbon galactose anomeral atomic signals and arabinose. The ratio of galactose to arabinose units in AG macromolecules was calculated from the ratio of the intensities of signals from anomeric carbon atoms of glactose and arabinose [10].

Pectin polysaccharides (PS) were extracted from the bark of *L. sibirica* Ledeb. and from *L. gmelini* (Rupr.) Rupr. according to the scheme depicted in Figure 17. Larch air-dried bark (500 g), which was initially ground and treated with ethyl acetate, was extracted using distilled water at 70 °C over 3 h. Raw material residue was poured with a mixture (1:1, v/v) 0.5% of ammonium oxalate water solution and 0.5% oxalic acid water solution and heated at 80 °C for 2 h. The extract was concentrated. Polysaccharides were precipitated with a triple volume of ethyl alcohol or acetone and dried with lyophilization. As a result, we obtained PS. The PS was dissolved in water and aminoacids were detected on AAA339M automatic analyser.

PS (50 mg) was dissolved in 20 ml of water. Pektinaza (2 mg; Sigma, USA) water solution was added. The mixture was temperature-controlled at 37 ° C for 3 h. Then, a reaction mixture was heated for 5 min in a water bath at 100 °C. Coagulated protein was separated by centrifugation. The obtained supernatant was concentrated and up to 5 ml 96% ethyl alcohol was added (4 volumes). Deposition was separated with centrifugation. Alcoholic supernatant was concentrated and analysed with the help of PC.

Galacturonic acid content in PS was defined according to the reaction with 3,5-dimethyl phenol in the presence of concentrated H_2SO_4, protein using the Lowry method [121] and based on the calibrating schedule for a bovine serum albumin 80000 Da. Paper chromatography was carried out on "Filtrak FN-13" paper with a descending method in a n-butanol-pyridine-water system (volume correlations 6:4:3, respectively). To define carbohydrates, aniline phthalate was poured on the paper and heated at 105 °C. Gas–liquid chromatography was carried out using a Hewlett-Packard 4890A (USA) chromatograph equipped with a flame-ionization detector, RTX-1 (0.25 mm x 30 m) capillary column, argon carrier gas, and 1:60 dumping. Temperature rate: 175 °C (1 min)–250 °C (2 min), Δ 3°/min.

A full acid hydrolysis PS (5 mg) was carried out for the implementation of 2M trifluoroacetic acid (2 ml) which contained *myo*-inositol (1 mg/ml). The mixture was heated in a soldered ampoule for 5 h at 100 °C, and the acid was removed with a repeated dry evaporation with added methanol. As a result, we obtained PVG-1.

Ion-exchange chromatography PS (100 mg) was carried out on a DEAE-cellulose (25x2 cm) column. NaCl solutions were used as an eluent with increasing concentrations (0.01M–1M, 60 ml/h elution speed, fractions selection by 12 ml). Pick correspondent fractions at the output bents were combined, dialysed and lyophilized. As a result, we obtained PS 1-4 fractions. The monosaccharide composition of each fraction was defined with GLC in acetate polyol after preliminary hydrolysis.

In order to obtain acetate polyol, each PS 1-4 fraction was dissolved in a 1M ammonia solution (1 ml) and 5 mg of $NaBH_4$ was added. The mixture was kept for one day at a room temperature. Then, $NaBH_4$ was eliminated by adding 2–3 drops of concentrated acetic acid; 0.2 ml of dry pyridine and acetic anhydride were added to the dry residue. The mixture was acetilized at 100 °C for 1 h. The solution was dry-evaporated until pyridine and acetic anhydride were removed, first by adding 1 ml of toluene and then 1 ml of methanol. The obtained acetate mixture of PS 1-4 polyol fractions was dissolved in 0.2 ml of dry chloroform and moved quantitatively to Appendorf tubes, concentrated up to 0.1–0.2 ml and analysed with the GLC method.

PS (5 mg) partial acidic hydrolysis was carried out using 0.01M TFA (2 ml), which contained *myo*-inositol (1 mg/ml). The mixture was heated in a soldered ampoule at 100 °C for 3 h. The acid was removed using a repeated dry evaporation with added methanol. As a result, we obtained PVG-2.

After extraction of dihydroquercetin, arabinogalactan and resin, the larch chip presented as a cellolignin residue. The chip had the following dimensions: 25x15x5mm, and sawdust fraction, 1x2x2 mm.

Bleached pulp from Baikal Pulp Mill was used for hydrolysis: polymerization degree 573, ash 1.1%, humidity 3%. Cellulose hydrolysis was carried out using 72% sulphuric acid and water in a ratio of 1:3 at room temperature for 1 h. Hydrolysis products—inverted polysaccharides—were precipitated with a five-fold ethanol volume. The precipitate was filtered and washed with alcohol the last washed portion achieved a neutral reaction. The product was dried in the air at up to 6% humidity. Acid content of inverted polysaccharides was defined using 1N HCl titration. Inversion of IPS was carried out in 0.75–1.50% solutions, hydromodulus 1:30, at 100–170 °C, for 0.25–3.0 h inversion duration. The potential content of reducing substances in hydrolyzates was defined by inversion of water-soluble polysaccharides with 5% sulphuric acid. 20% NaOH was used to neutralize the hydrochloric acid. Glucose quantitative content in neutralisate (pH 4–5) was defined by HPLC methodology.

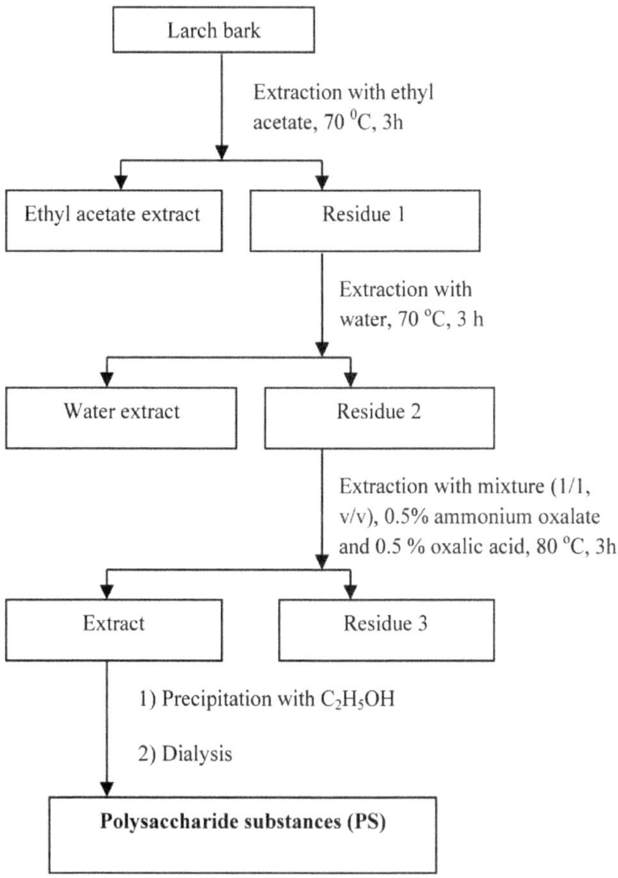

Figure 17. Extraction scheme of pectin substances from larch bark

Larch cellolignin timber residue with particles having dimensions of 25x15x5 mm was used for explosive autohydrolysis. Autoexplosive hydrolysis was carried out in a special 200 ml capacity autoclave, which allowed us to conduct a quick decompression of the reactor (steam explosion). Hydrolysis conditions were: 200 and 220 °C, duration 2 and 5min.

The laboratory scheme for obtaining crystalline glucose is depicted in Figure 18.

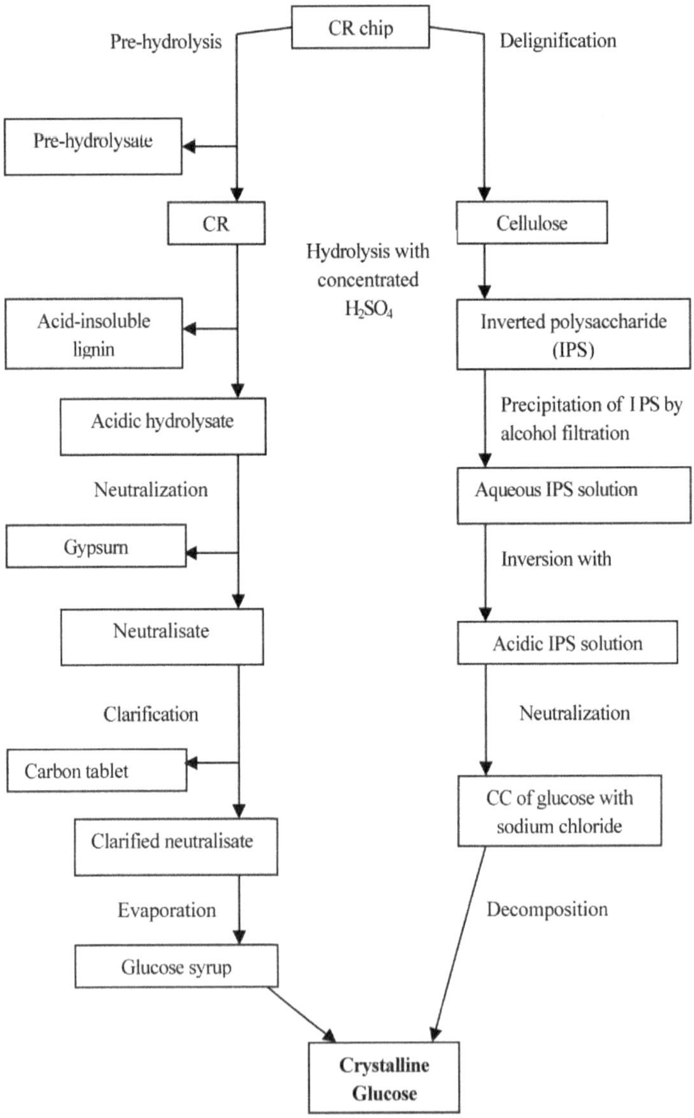

Figure 18. Laboratory scheme for obtaining crystalline glucose

6. Conclusion

This chapter therefore summarizes studies on polysaccharides in the context of the development of technology for 100% processing of larch wood and bark as forestry waste in order to provide new medicines, veterinary drugs, dietary supplements and valuable materials for the cosmetics and agricultural industries. There are data on larch wood and bark extraction by the two-phase solvent system, namely the kinetic study of extraction processes, diffusion constants, mass-transfer coefficients, mechanisms and physicochemical characterization of the transfer process, its mathematical modelling and structural characteristics of the samples isolated. This work aims to support the development of economically and ecologically viable production technology for high-demand products on the basis of renewable raw materials with a 15–20% increase of forestry efficiency due to waste processing. The technology will provide new medicines and food supplements, as well as cheaper, by 40–50%, analogues, to those currently known.

Author details

Natalya Nikolaevna Trofimova, Elena Nikolaevna Medvedeva, Nadezhda Viktorovna Ivanova, Yuriy Alekseevich Malkov and Vasiliy Anatolievich Babkin
Laboratory of Wood Chemistry, Federal Research Budget Institution, A.E. Favorsky Institute of Chemistry, Siberian Branch, Russian Academy of Sciences, Irkutsk, Russia

Acknowledgement

This project has been supported by a State Contract with the Department of Industry, Science and Technology of the Russian Federation №43.044.1.1.26.38(2002–2004); by a State Contract with the Agricultural Department of the Government of Irkutsk Oblast №02-66 (2010–2011); and by RAS and SB RAS Grants (2005–2012). The authors are grateful to all their colleagues at the Laboratory of Wood Chemistry of A.E. Favorsky Irkutsk Institute of Chemistry who have contributed to this project.

7. References

[1] Clarke A, Anderson R, Stone B (1979) Form and function of arabinogalactans and arabinogalactan-proteins. Phytochemistry. 18(4): 521-540.

[2] Antonova GF, Tukavkina NA (1983) Water soluble substances of larch and their possible applications. Khimia Drevesiny. 2: 89-96.

[3] Adams MF, Douglas C (1963) Arabinogalactan—a review of literature. TAPPI. 46: 544-548.

[4] Aspinall GO (1964) Some recent developments in the chemistry of arabinogalactans. In: Chimie et Biochimie de la Lignine, de la Cellulose et des Hemicelluloses. Grenoble: Actes du Symposium International de Grenoble. pp89-97.

[5] Medvedeva SA, Aleksandrova GP (2003) Modification strategy and biopotential of natural polysaccharide arabinogalactan. In: Panorama Sovremennoi Khimii Rossii [Polymer Synthesis and Modification]. Moscow: Khimia. pp328-356.

[6] Medvedeva EN, Babkin VA, Ostroukhova LA (2011) Larch arabinogalactan: properties and application prospects. Khimia rastitel'nogo syr'ia. 1: 27-37.
[7] Chochieva MM, Tsvetayeva IP, Yurieva MK, Zaytseva AF (1958) Arabinogalactan distribution in the Dahurian larch wood. Works of the USSR AS Forest Institute. 45: 31-49.
[8] Ponder G, Richards G (1997) Arabinogalactan from Western larch. Part III: Alkaline degradation revisited, with novel conclusions on molecular structure. Carbohydrate Polymers. 34: 251-261.
[9] Karacsonyi S, Kovacik V, Alfoldi J, Kubackova M (1984) Chemical and ^{13}C NMR studies of arabinogalactan from *Larix sibirica* L. Carbohydrate Research.134: 265-274.
[10] Medvedeva EN, Fedorova TE, Vanina AS, Rokhin AV, Es'kova LA, Babkin VA (2006) Arabinogalactan structure and properties vs. methods of its isolation from larch wood and purification techniques. Khimia rastitel'nogo syr'ia. 1: 25-32.
[11] Prescott J, Groman E, Gyongyi G (1997) New molecular weight forms of arabinogalactan from *Larix occidentalis.* Carbohydrate Research. 301: 89-93.
[12] Ovodov YS (1998) Polysaccharides of phanerogams: their structure and physiological activity (review article). Russian Journal of Bioorganic Chemistry (Bioorganicheskaya Khimiya). 24: 483-501.
[13] Arifkhodzhaev AO (2000) Galactans and galactan-containing polysaccharides of higher plants. Khimiya prirodnykh soedinenii. 3: 185-197.
[14] Groman E, Enriquez P, Jung Chu, Josephson L (1994) Arabinogalactan for hepatic drug delivery. Bioconjugate Chemistry. 5: 547-556.
[15] Grieshop CM, Flickinger EA, Fahey GC Jr (2002) Oral administration of arabinogalactan affects immune status and fecal microbial populations in dogs. The Journal of Nutrition. 132: 478-482.
[16] Robinson RR, Feirtag J, Slavin JL (2001) Effects of dietary arabinogalactan on gastrointestinal and blood parameters in healthy human subjects. Journal of the American College of Nutrition. 20: 279-285.
[17] Medvedeva EN, Babkin VA, Makarenko OA, Nikolaev SM, Khobrakova VB, Shulunova AM, Federova TE, Es'kova LA (2004) Preparation of high purity larch arabinogalactan and study of its immunomodulating properties. Khimiya rastitel'nogo syr'ia. 4: 17-23.
[18] Dubrovina VI, Medvedeva SA, Vityazeva SA, Kolesnikova OB, Aleksandrova GP, Gutsol LO, Grishenko LA, Chetverikova TD (2007) Structure and Immunomodulating Action of Arabinogalactan of Siberian Larch and its Metal Derivatives. Irkutsk: Asprint. 145 p.
[19] Robinson RR, Causey J, Slavin JL (2001) Nutritional benefits of larch arabinogalactan. In: McCleary BV, Prosky L. (eds.) Advanced Dietary Fiber Technology. Oxford: Blackwell Science Ltd. p443-451.
[20] Causey J, Robinson R, Feirtag J, Fulcher R, Slavin J (1999) Long term supplementation of arabinogalactan in humans and its effects on blood lipids, glucose and insulin. The FASEB Journal. Part I:3.
[21] Zarubaev VV, Ostrouhova LA, Medvedeva EN, Babkin VA, Kiselyov OI (2010) Antiviral preparations based on larch wood biologically active substances. Bulletin of Eastern-Siberian Science Centre SB RAMS. 1(71): 76-81.

[22] European Patent Office. Arabinogalactan as Immune Enhancer, EP 2 327 411 A1, Bulletin 2011/22. European Patent Office; 2011.
[23] Lonza. Nutrition. http://www.lonza.com/products-services/nutrition.
[24] Gaines Nutrition. Ara-Plus (90 capsules) (Mfr: D'Adamo Personalized Nutrition/North American Pharmacal) (Professional quality product). http://www.gaines.com/store/NorthAmerPharmacal/EM-ARAPLinfo.html (accessed Nov.1994).
[25] Kiselev OI, Babkin VA, Tikhonov VP (2010) Drug for Prophylaxis of Influenza and Upper Respiratory Tract Infections. RF Patent 2401117.
[26] Burke S, Scheuer C, Mullard K (2009) Ophthalmic compositions comprising a terpene and a natural polymer. WO 2009018060 (A1). European Patent Office. http://worldwide.espacenet.com/publicationDetails/biblio?FT=D&date=20090205&DB=EPODOC&locale=ru_RU&CC=WO&NR=2009018060A1&KC=A1. Accessed on April 2005.
[27] Jung C, Enriquez P, Palmacci S, Josephson L, Lewis JM (1995) Arabinogalactan derivatives and uses thereof. US Patent Number 5478576.
[28] Cui L, Faraj A, el Alaoui A (1997) Arabinogalactan (9kDa)-9-β-D-arabinofuranosyladenine-5'-monophosphate, a novel liver-targeted conjugate that selectively inhibits hepatitis B virus replication in vitro. Antiviral Chemistry & Chemotherapy. 8: 529-536.
[29] Josephson L, Rutkowski JV, Paul K, Frigo T, Korba BE, Tennant B, Groman EV (1996) Antiviral activity of a conjugate of adenine-9-β-D-arabinofuranoside - 5'- monophosphate and a 9 kDa fragment of arabinogalactan. Antiviral Therapy. 1(3): 147-156.
[30] Groman EV, Menz ET, Enriquez PM, Jung C (1996) Delivery of Therapeutic Agents to Receptors using Polysaccharides. US Patent Number 5544386.
[31] Josephson L, Groman EV, Jung C, Lewis JM. (1994) Targeting of therapeutic agents using polysaccharides. US Patent Number 5336506.
[32] Nogawa M, Akaike T, Maruyama A (2001) Polyethylenimine/arabinogalactan conjugate as a hepatocyte specific gene carrier. S.T.P. Pharmaceutical Sciences. 11: 91-102.
[33] Ehrenfreund T, Golenser J, Frankenburg S, Domb AJ (2000) Water soluble amphotericin B - arabinogalactan conjugates for the treatment of experimental leishmaniasis. In: Proceedings of the 27th International Symposium on the Controlled Release of Bioactive Materials, 7-13 July 2000, Porte Maillot, Paris, France. pp578-579.
[34] Ehrenfreund-Kleinman T, Azzam T, Falk R, Polacheck I, Golenser J, Domb AJ (2002) Synthesis and characterization of novel water soluble amphotericin B - arabinogalactan conjugates. Biomaterials. 23: 1327-1335.
[35] Pozharitskaya ON, Urakova IN, Shikov AN (2008) Solid Nanocomposition for Delivering Biologically Active Substances. WIPO Patent Application WO/2008/123798.
[36] Reis RL (2008) Natural-based Polymers for Biomedical Application. Cambridge: Woodhead. 802 p.
[37] Enriquez P, Jung C, Groman E (1996) Directed delivery of radioprotectants using a receptor - specific carrier. US Patent Number 5490991.
[38] Domb AA (2001) Biodegradable Polycation Composition for Delivery of an Anionic Macromolecule in Gene Therapy. PCT Int. Application WO 2001007486 A1. See URL http://www.patentstorm.us/patents/7001891.html.

[39] Azzam T, Eliyahu H, Shapira L, Linial M, Barenholz Y, Domb AJ (2002) Polysaccharide-oligoamine based conjugates for gene delivery. Journal of Medicinal Chemistry. 45: 1817-1824.

[40] Dutta RC (2007) Drug carriers in pharmaceutical design: promises and progress. Current Pharmaceutical Design. 13: 761-769.

[41] Jain A, Gupta Y, Jain SK (2007) Perspectives of biodegradable natural polysaccharides for site-specific drug delivery to the colon. Journal of Pharmacy and Pharmaceutical Sciences. 10: 86-128.

[42] Kochetkov NK (1967) Carbohydrate Chemistry Methods. Moscow: Mir. 512 p.

[43] Domb A, Benita S, Polachek I, Linden G (2000) Drug conjugates with oxidized arabinogalactan or dextran. US Patent Number 6011008.

[44] Ponedelkina IY, Saitgalina EA, Tolstikov GA, Dzemilev UM, Tolstikova TG, Odinokov VN, Lukina ES (2009) Modified arabinogalactans and a method for the production thereof. Patent RU 2359975, WO 2008/123799.

[45] Mudarisova RKh, Shirokova EN, Badykova LA, Borisov IM, Tolstikova TG, Sorokina IV, Dolgikh MP, Monakova YB (2005) Synthesis and pharmacological activity of the oxidized fractions of arabinogalactan from siberian larch (*Larix sibirica L.*). Pharmaceutical Chemistry Journal. 39: 418-421.

[46] Badykova LA, Mudarisova RKh, Tolstikova TG, Borisov IM, Monakov YuB (2005) Arabinogalactan poly- and oligosaccharides modified with 5-aminosalicylic acid. Chemistry of Natural Compounds. 41(3): 272-275.

[47] Badykova LA, Mudarisova RKh , Khamidullina GS, Aminev KhK, Borisov IM, Monakova YB (2006) Modification of arabinogalactan and its oxidized forms with *p*-aminosalicylic acid. Russian Journal of Applied Chemistry. 79: 1626-1628.

[48] Badykova LA (2007) Reaction of arabinogalactan from Siberian larch with 5-aminosalicylic acid, 4-aminosalicylic acid and isonicotinic acid hydrazide. Abstract of candidate's PhD thesis. Ufa: Institution of the Russian Academy of Sciens Institute of Organic Chemistry of Ufa Scientific Centre of the Russian Academy of Sciences (IOChem USC RAS), Russia. 26 p.

[49] Medvedeva EN, Neverova NA, Fedorova TE, Babkin VA, Meteleva ES, Dushkin AV, Tolstikova TG, Khvostov MV, Dolgikh MP (2010) Structural transformations of arabinogalactan from the Siberian larch during mechanochemical processing and the biological properties of the products. Russian Journal of Bioorganic Chemistry. 36: 853-859.

[50] Tolstikova TG, Tolstikov GA, Dushkin AV, Meteleva ES (2008) Water Soluble Medical Composition and Synthetic Method Thereof. RU Patent 2 337 710.

[51] Dushkin AV, Meteleva ES, Tolstikova TG, Tolstikov GA, Polyakov NE, Neverova NA, Medvedeva EN, Babkin VA (2008) Mechanochemical preparation and pharmacological activity of water-soluble intermolecular complexes of arabinogalactan and pharmaceutical drugs. Russian Chemical Bulletin. 57: 1299-1307.

[52] Tolstikova TG, Khvostov MV, Bryzgalov AO, Dushkin AV, Tolstikov GA (2010) Plant polysaccharide arabinogalactan as new pharmacon clathrating agent. RAS Reports 433: 713-714.

[53] Babkin VA, Medvedeva EN, Neverova NA, Chipanina NN, Aksamentova TN, Meteleva ES, Dushkin AV (2009) Study of mechanocomposites on the basis of arabinogalactan and medical substances. In: Materials of the IV Russian Scientific Conference: New

Achievements in Chemistry and Chemical Technology of Plant Raw Material. Barnaul. 21-23 April 2009. pp176-178.
[54] Medvedeva EN, Neverova NA, Ostroukhova LA, Babkin VA, Gus'kov SA, Meteleva ES, Dushkin AV (2010) Mechanical composites based on biologically active compounds from larch wood. Chemistry of Natural Compounds. 42: 1-4.
[55] Trofimov BA, Sukhov BG, Aleksandova GP, Medvedeva SA, Grishchenko LA, Mal'kina AG, Feoktistova LP, Sapozhnikov AN, Dubrovina VI, Martynovich EF, Tirskii VV, Semenov AL (2003) Nanocomposites with magnetic, optical, catalytic and biologically active properties based on arabinogalactan. Doklady Chemistry. 393(4): 287-288.
[56] Sukhov BG, Aleksandrova GP, Grishchenko LA, Feoktistova LP, Sapozhnikov AN, Proidakova OA, T'kov AV, Medvedeva SA, Trofimov BA (2007) Nanobiocomposites of noble metals based on arabinogalactan: preparation and properties. Journal of Structural Chemistry. 48(5): 922-927.
[57] Aleksandova GP, Grishchenko LA, Fadeeva TV, Medvedeva SA, Sukhov BG, Trofimov BA (2006) Antimicrobial Preparation. RU Patent 2 278 969.
[58] Ohr L (2001) Arabinogalactan Adds More than Health Benefits. Prepared Foods. 170: 55.
[59] Babkin VA, Medvedeva EN, Sambueva ZG, Khobrakova VB, Neverova NA, Suyuncheva BO, Tatsyi AA (2012) Medical and biological properties of bakery and pastry products enriched with larch arabinogalactan. Khimia rastitel'nogo syr'ia. 2: 185-193.
[60] Ermakova MF, Chistyakova AK, Shchukina LV, Pshenichnikova TA, Medvedeva EN, Neverova NA, Belovezhets LA, Babkin VA (2010) Effect of arabinogalactan isolated from Siberian Larch on the baking value of soft wheat flour and bread quality. Russian Journal of Bioorganic Chemistry. 36: 951–956.
[61] Neretina OV, Medvedeva EN, Babkin VA (2011) Production Technology of Prophylactic Bakery and Compositions of Prophylactic Biscuit, Cheese and Sugar Cookies. RU Patent 2430614.
[62] Nesteruk VV, Vershchako VA, Likhova NV, Babkin VA, Medvedeva EN (2011) Production Technology of Oatmeal Cookies Containing Prophylactic Arabinogalactan Supplement and Composition Thereof. Patent application 2011113945.
[63] Utochkina EA (2011) Study and technology development of arabinogalactan-enriched cultured milk product. Abstract of PhD thesis. Blagoveshchensk: Far Eastern State Agrarian University; Russia.
[64] Galochkin VA, Galochkina VP, Maksimenko SV (2008) Non-specific resistance and meat productivity in calves at arabinogalactan-enriched feeding. Sel'skokhozyaistvennaya biologia. 4: 89-95.
[65] Maksimenko SV (2008) Influence of arabinogalactan upon productivity and non-specific resistance of calves. Problemy biologii produktivnykh zhivotnykh. 1: 82-86.
[66] Torshkov AA, Fomichev YuP (2010) Application of arabinogalactan in raising broiler chickens. Orenburg Agrarian University Reports. 1: 172-175.
[67] Torshkov AA, Gerasimenko VV (2010) Changes in chemical composition of broiler chicken meat at the use of arabinogalactan. Orenburg Agrarian University Reports. 2: 167-169.
[68] Moskovtsev NG, Belova TP, Valetov TA (1988) Isolation of water extracted substances from larch wood. Khimiya Drevesiny (Wood Chemistry). 1: 112-116.

[69] Kuznetsova SA, Kuznetsov BN, Aleksandrova NB, Danilov VG, Zhizhaev AM (2005) Obtaining arabinogalactan, dihydroquercetin and microcrystalline cellulose using mechanochemical activation. Chemistry for Sustainable Development. 13: 261-268.
[70] Kuznetsova SA, Mikhailov AG, Skvortsova GP, Aleksandrova NB, Lebedeva AB (2005) Intensifying the process of water extraction of arabinogalactan from larch wood. Khimia rastitel'nogo syr'ia. 1: 53-58.
[71] Antonova GF (1977) Study of fraction analysis of polysaccharide arabinogalactan from Siberian larch wood. Khimiya Drevesiny (Wood Chemistry). 4: 97-100.
[72] Medvedeva EN, Ostroukhova LA, Neverova NA, Onuchina NA, Babkin VA (2011) Phenolic admixtures in larch wood arabinogalactan. Khimiya rastitel'nogo syr'ia. 1: 45-48.
[73] Babkin VA, Kolzunova LG, Medvedeva EN, Malkov YuA, Ostroukhova LA (2005) Production method of arabinogalactan. RU Patent 2 256 668.
[74] Akselrud GA, Altshuler MA (1983) Introduction to Capillary-Chemical Technology. Moscow: Khimiya. 264 p.
[75] Malkov YuA, Ostroukhova LA, Babkin VA (2002) Application of mathematical modeling method for technology development of isolating extracted substances from larch wood. Khimiya rastitel'nogo syr'ia. 2: 133-138.
[76] Babkin VA, Malkov YuA, Medvedeva EN, Babkin DV, Ostroukhova LA, Kolzunova LG, Zolotar' RN, Nikolaev SM, Khobrakova VB (2005) Technology of arabinogalactan isolation from larch wood and research of its biological activity. In: Phytopharm 2005: proceedings of the Ninth International Congress, St. Peterburg. 22-25 June 2005. pp165-168.
[77] Kafarov VV, Vetokhin VN (1987) Fundamentals of Automatic Design of Chemical Production. Moscow: Nauka. 624 p.
[78] Mukhlenov IP (1986) Chemical-technological Systems. Leningrad: Khimiya. 425 p.
[79] Kafarov VV, Meshalkiin VP, Perov VL (1974) Principles of Mathematical Modeling of Chemical-technological Systems. Moscow: Khimiya. 624 p.
[80] Ovodov YuS, Golovchenko VV, Gunter EA, Kutchin AV (2009) Pectic Substances of Plants of the European North of Russia. Ekaterinburg: Uro RAN. 112 p.
[81] Tehnokonsalt Management. The organisation of production technology of pectin (in Russian). http://www.tc-m.ru/index.php/155.html. Accessed 2004 March.
[82] Minzanova ST, Mironov VF, Sosnina NA, Vyshtakalyuk AB, Lapin AA, Smolentsev AV, Mironova LG, Khirug SS, Konovalov AI (2007) Technological aspects of the manufacturing of amaranth fodder additives. Doklady Chemistry. 413(1): 75-78.
[83] Bubenchikov VN, Drozdova IL, Kochkarov VI (2002) Method of Isolation of the Polysaccharides Possessing Anti-inflammatory Activity. RU Patent 2177793.
[84] Ovodova RG, Golovchenko VV, Shazhkov AS, Popov SV, Ovodov YuS (2000) Structural investigation of pectin and physiological activity of lemnan and pectin from *Lemna minor L.* Bioorganicheskaya khimiya. 26:743-751.
[85] Ovodova RG, Bushneva OA, Golovchenko VV, Popov SV, Ovodov YuS (2000) Method of Isolation from Vegetative Raw Materials of the Polysaccharides Possessing Immunostimulating Action. RU Patent 2149642.
[86] Gelgay MK, Donchenko LV, Reshetnyak AI (2008) Innovative technology of pectin from secondary sources of raw material after processing coffee. New Technologies. 6:15-18.
[87] Graves F (1994) Process for Producing Pectin Through Ion Exchange. US Patent 5364851.

[88] Aleinikov IN, Sergeev VN, Rusakov AV, Merkulov YuG (2000) New approach to pectin production. Pishchevaya promyshlennost'. 1: 59-61.
[89] Aimukhomedova GB, Shelukhina NP (1964) Pectin Substances and Their Determination Methods. Frunze: Ilim. 119 p.
[90] Babkin VA, Ostroukhova LA, Ivanova NV, Malkov YuA, Ivanova SZ, Onuchina NA (2002) Phytocomplex of Antioxidant Activity and Production Method Thereof. RU Patent 2188031.
[91] Shelukhina NP, Abayeva RSh, Aimukhomedova GB (1987) Pectin and Parameters of its Production. Frunze: Ilim. 91 p.
[92] Kaisheva NSh, Shcherbak SN, Kompantsev VA, Krikova NI, Ivanova LI (1994) Analysis of protective action of pectins. Zhurnal analititcheskoi khimii. 49: 1158–1162.
[93] Babkin VA, Ivanova NV, Trofimova NN, Es'kova LA, Salyaev RK, Nurminski VN, Korzun AM, Feoktistova LP, Sapozhnikov AN, Lykhoshvai EV, Arsent'ev KYu (2010) Isolation of Pectin, Possessing Membrane Stabilizing Activity and Ability to Reduce Silver Ions, from Larch Wood. Silver Nanocomposites Stabilized by Pectin. RU Patent 2403263. http://www.nti.lin.irk.ru/bibl/rese.asp?n=2156&s=10 Accessed 29 June 2009.
[94] Nakanisi K (1965) Infrared Spectra and Structure of Organic Compounds. Moscow: Mir. p40.
[95] Odonomažig P, Balga D, Ebrigerova A (1992) Structures of pectin polysaccharides isolated from the Siberian apricot (*Armaniaca sibirica* Lam.). Carbohydrate Research. 226: 353-358.
[96] Grigoriuk GP (1998) Pectins from a bark of coniferous trees as biologically active components of food stuffs in the medical and practical purposes. In: International Conference on Natural Products and Physiologically Active Substances, 30 November-6 December 1998, Novosibirsk; Russia. p79.
[97] Kondratyuk TA (2009) Production and quality estimation of food products from Siberian larch bark. Abstract of PhD thesis; Krasnoyarsk: Krasnoyarsk State Trade and Economic institute, Russia; Kemerovo, 2009.
[98] Ministry of Agriculture of the Russian Federation. Manual on pectosorbitol application in veterinary. Registration PSP-2-2.0/00344. :Council on Veterinary Drugs of the Ministry of Agriculture of the Russian Federation.
[99] Lipson G, Still G (1972) Powder radiographinterpretation. Moscow: Mir. 384 p.
[100] Babkin VA, Ostroumova LA, D'yachkova SG, Svyatkin YuK, Babkin DV, Onuchina NA (1997) The complex wasteless Siberian and Dahurian larch biomass processing echnology. Khimiya v Interesah Ustoich. Razvit. (Chemistry for Sustainable Development) 5: 105–115.
[101] Levin ED, Denisov OB, Pen RZ (1978) Kompleksnaya pererabotka listvennitsy Complex Processing of Larch). Moscow: Lesnaya promyshlennost'. 224 p.
[102] Tsvetaeva IP, Yur'eva MK (1960) About water soluble larch wood substances. Izvestia Vuzov. Lesnoi Zhurnal. 1: 148–151.
[103] Fengel D, Vegener G (1988) Drevesina (Khimiya, ul'trastruktura, reaktsii) (Wood: Chemistry, Ultrastructure, Reactions)). Moscow: Lesnaya promyshlennost'. 512 p.
[104] Trofimova NN, Bichevina OB, Babkin VA (2004) The carbohydrates of the larch cellolignin. Khimiya rastitel'nogo syr'ya. 3: 11–14.

[105] Bichevina OB, Trofimova NN, Babkin VA (2001) Study of the aqueous prehydrolysis of cellolignin from Larch wood. In: Problems of Sustainable Regional Development, Abstracts of the Second School Workshop of Young Scientists of Russia, September 17-21, 2001, Ulan-Ude, Russia. pp24-25.
[106] Korol'kov II (1978) Percolation Hydrolysis of Row Plant Material. Moscow: Lesnaya promyshlennost'. 263 p.
[107] Fisher NP. Use of pentoses for manufacture of fodder yeast. In book: Problems of Using Pentose Containing Row Materials, Proceedings of the All Union Conference. Riga: Izdatel'stvo Akademii nauk Latviiskoi SSR, 1958. pp91-99.
[108] Trofimova NN, Gordienko II, Babkin VA (2005) Dependence of redusing substances yield on acid hydrolysis parameters of larch cellolignin. Khimiya rastitelnogo syr'ya. 4: 25-28.
[109] Glyukoza kristallicheskaya gidratnaya: Tekhnicheskie usloviya (Crystalline Hydrate Glucose: Technical Specification) GOST 97588.
[110] Sharkov VI, Sapotnitskii SA, Dmitrieva OA, Tumanov IF (1973) Technology of Hydrolytic Productions. Moscow: Lesnaya promyshlennost'. 480 p.
[111] Babkin VA, Ivanova SZ, Fedorova TE, Medvedeva EN, Malkov YuA, Ostroukhova LA, Trofimova NN, Ivanova NV (2007) Scientific bases of technology of complex processing of a biomass of a larch. Khimiya Rastiytel'nogo syr'ya 3: 9-21.
[112] Nikitin VM, Obolenskaya AV, Shchegolev VP (1978) Wood and Cellulose Chemistry. Moscow: Lesnaya promyshlennost'. 368 p.
[113] Rogovin ZA (1972) Khimiya tsellulosy. Moscow: Lesnaya promyshlennost'. 520 p.
[114] Trofimova NN, Babkin VA. (2003) The cellolignin residue of a larch wood as a drug for obtaining of a crystalline glucose. Khvoinye Boreal'noi Zony. 1: 116-122.
[115] Gert EV, Shishonok MV, Torgashov VI, Kaputskii FN (1988) Effect of dinitrogen etroxide on the structure and properties of cellulose. 8. Recrystallization of amorphous cellulose in the medium of nitrogen oxide (IV) and nitric acid resulting in the formation of the knecht compound. Khimiya Drevesiny (Wood Chemistry). 5: 52-56.
[116] Trofimova NN, Malkov YuA, Bichevina OB, Kurets IZ, Babkin VA (2003) The nvestigation of the sellulose hydrolysis pulp inversion process. Khimiya Rastitel'nogo syr'ya. 4: 11-15.
[117] Glazkova EL, Lebedev NV (1960) Sbornik trudov VNII, no. 8. 235 p;.
[118] Kurets IZ, Trofimova NN, Bichevina OB, Belonogova LN, Kashaev AA, Babkin VA 2003) The characteristics of glucose complex with sodium chloride crystals by X-Ray power diffraction and X-Ray stuctural analysis methods. Khimiya Rastitel'nogo syr'ya. 2: 15-18.
[119] Trofimova NN, Babkin VA, Chemeris MM (2002) Catalyzed steam hydrolysis of the cellolignin residue of a larch wood. Khimiya Rastitel'nogo syr'ya. 2: 53–56.
[120] Trofimova NN, Babkin VA (2010) Study of the acid hydrolysis of polysaccharides from arch wood for obtaining crystalline glucose. Russian Journal of Bioorganic Chemistry. 36(7): 847-852.
[121] Lowry OH, Roosebrough NJ, Farr AL, Randall RG (1951) Protein measurement with the Folin phenol reagent. Journal of Biological Chemistry. 193:265-275.

Chapter 7

Lichen Polysaccharides

Desiree Nedra Karunaratne, R.G.U. Jayalal and V. Karunaratne

Additional information is available at the end of the chapter

1. Introduction

Lichens are symbiotic organisms composed of a fungal partner, the mycobiont, and one or more photosynthetic partners, the photobiont, which may be either a green alga or a cyanobacterium (Nash III, 1996; Wolseley and Aguirre-Hudson, 1994 and Yoshimura, 2004). The mycobiont in this combination is an ascomycete or a basidiomycete and the photobiont a green alga or a cyanobacterium. Some lichen species can contain more than one green algal species as photobionts (Friedl and Gärtner, 1988; Friedl, 1989; Ihda *et al.*, 1993). At present, about 13,500 fungal species have been recognized to be involved in lichen symbiosis (Hawksworth *et al.* 1995; Kirk *et al.*, 2001). However, Sipman and Aptroot (2001) stated that this number could be as high as 20,000 after including "orphaned" species; according to Lumbsch *et al.*, (2011), about 10,000 of lichenized fungi have been estimated as undescribed species.

The hyphae of mycobionts are septate, branched, thin or thick walled, and the walls are colourless or variously coloured. It is necessary that the photobiont, which is essentially aquatic in nature, remains protected from desiccation in a lichenized terrestrial condition. The protection is provided by the mycobiont which forms the bulk of the thallus, by the development of specialized hyphal tissues in the form of a cover or cortex over the stratum of the photobiont. The development of the cortex is assumed to be stimulated by the photobiont (Ahmadjian, 1987). In some gelatinous lichens with cyanobacteria, the polysaccharidic sheath produced by the photobiont (cyanobacteria) contributes to water retention (Prieto *et al.*, 2008). Lichens have been used for ecological, medicinal and other economic purposes for over 100 years and these beneficial effects have been correlated to some extent with their polysaccharide content. Amongst the identified lichens so far, about 100 species have been studied for their polysaccharides and their composition (Cordeiro *et al.*, 2005).

2. Separation and characterization

The cell wall of fungi is composed mainly of polysaccharides such as lichenan, isolichenan, and galactomannan (fig.1) (Elix and Stocker-Wörgötter, 2008). Bernard and Latgé (2001)

described that the fungal wall is a complex structure composed typically of chitin, 1,3-β- and 1,6-β-glucan, mannan and proteins, although wall composition frequently varies markedly between the species of fungi.

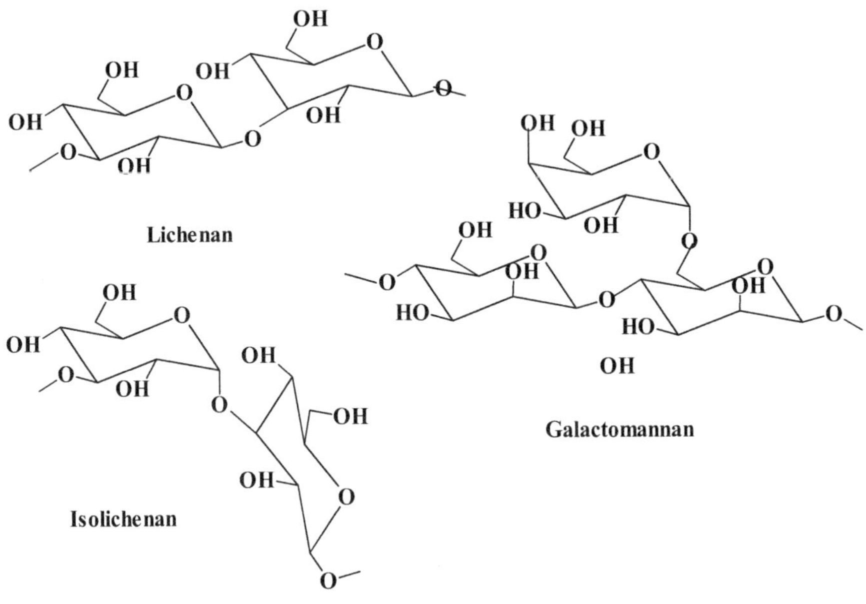

Figure 1. Structures of some polysaccharides contained in fungal cell wall

For the separation and isolation of lichen polysaccharides, traditional methods and modern techniques are used. Traditional methods basically involved freezing and thawing of the material originally extracted with hot water. Dialysis and ethanol precipitation has been employed for further purification. Also, alkali solutions have been used to extract these compounds. The polysaccharides present in the thalli of *Umbilicaria mammulata* were isolated by an exhaustive method starting with successive extraction with hot water at 100°C followed by extraction of the residue with hot 2% KOH at 100 °C. The alkaline extract was further treated with Fehlings solution to yield the 1,3-β- glucan from the supernantant and a 1,6-β- glucan was obtained from the precipitated Fehlings complex (Carbonero *et al.*, 2006). In our laboratory, we have used a similar alkaline extraction procedure to isolate a polysaccharide from *Usnea* cf. *cornuta* Körb. Chromatographic techniques and filtration devices were utilized to separate lichen polysaccharides (Paulsen *et al.*, 2002). However, different column chromatographic methods, including GP-HPLC, HPLC and ion exchange chromatography have been used recently to separate polysaccharides from lichens. Olafsdottir *et.al* (1999), fractionated the polysaccharides from *Cetraria islandica* on DEAE Sepharose CL-6B anion exchange columns with a 0 -1 M NaCl gradient. Further purification was performed by Sephacryl S-400 gel filtration. To determine homogeneity and M_r of these compounds, electrophoretic methods and gel permeation chromatography are used. For the determination of monosaccharides, TLC and GC methods are used. The monosaccharides could be derivatized into alditol acetates or into trimethylsilyl (TMS) ethers for analysis by GC. The linkage analysis is performed by using NMR spectroscopy (1D proton, 2D-COSY, NOESY, 2D-TOCSY, 1H 13C-HSQC, HMBC, H2BC and HSQC-NOESY) (Paulsen *et al.*, 2002 and Jensen *et al.*, 2010) and methylation analysis (Omarsdottir *et.al.*, 2005; Olafsdottir *et. al.*, 1999; Ruthes *et al.* 2010).

3. Biological source of the polysaccharide fraction

Whether these polysaccharides are produced by the mycobiont or photobiont separately or in symbiosis has been debated for a long time. Takahashi *et al.* (1979) showed that aqueous extracts from cultivated myco and photobiont had different monosaccharide composition and physical properties. It was also found that while the extracts of the mycobiont had a similar composition to that of the parent intact lichen, the photobiont fractions were different from those of the symbiotic thalli and its mycobiont. Complete structural analysis by Cordeiro *et al.* (2004) confirmed Takahashi's results, and showed that the nigeran, laminaran and the galactomannan, found previously in the symbiotic thalli of *Ramalina peruviana* (Cordeiro *et al.* 2003) were also generated by the aposymbiotically cultured fungal partner.

Three different polysaccharide structural types: β-glucans, α-glucans (linear or lightly substituted), and galactomannans (branched) (Carbonero *et al*, 2005a) are present in the fungal cell wall. According to Olafsdottir and Ingólfsdóttir (2001), all the polymers present in lichen thalli are categorized into glucan type [β-(1→3)(1→4)], lichenan type [α-(1→3)(1→4)] and pustulan [β-(1→6)]. Some of these types are depicted in Table 1. However, the recent discovery of a few additional complex heteroglycans, such as rhamnogalactofuranan (Olafsdottir, 1999), galactomannoglucans (Woranovicz-Barreira, 1999 and Carbonero *et al.*, 2002) and thamnolan (Carbonero *et al.*, 2005a) necessitated the closer examination of the origins of the polysaccharides. Studies carried out by Cordeiro *et al.*, (2005), revealed that the algal partner (*Trebouxia* sp.) also consists of some carbohydrates such as β-galactofuranan heteropolysaccharide. Later in 2007, new polysaccharides xylorhamnogalactofuranan and Xylan in the photobiont *Asterochloris* sp. were recorded by Cordeiro *et al.* Thereafter, Jensen *et al.* (2010), isolated another heteropolysaccharide called colleman from a cyanobacteria present in the lichen *Collema flaccidum*.

In the past, lichen polysaccharides have been extracted from the whole thallus without giving consideration to the origin of components such as the fungal partner or the photobiont (Gorin and Iacomini, 1984, 1985; Gorin *et al.*, 1993; Teixeira *et al.*, 1995). Later Cordeiro *et al.* (2005, 2007, 2008), found that, dissimilarities between the polysaccharides extracted from the cultivated photobionts *Trebouxia* and *Asterochloris* with those from their respective lichens could indicate that the photobiont or mycobiont was responsible for these differences. Galactoglucomannans were isolated from the mycobiont of *Parmotrema* species including *Parmotrema austrosinense, Parmotrema delicatulum, Parmotrema mantiqueirense,*

Polysaccharide	Main unit	Side chain	Lichen species	Mycobiont/ Photobiont	reference
Lichenan	Homoglucan with β(1→3) (1→4) linkage		*Cetraria ilandica, C. nivaris, C. richardsonii, Usnea barbata, U. lingissima, U. bayleyi, Parmelia tinctorum, P. conspersa, P. hypotrypella, P. nikkoensis, Alectoria sulcata, A. sarmentosa*	Whole thallus	Shibata, 1973
Isolichenan	Homoglucan with α(1→3) (1→4) linkage		-do-	-do-	Shibata, 1973
Pustulan	Glucan with β(1→6) linkage		*Lasallia pustulata, L. papulosa, Umbilicaria hirsuta, U. angulata, U. caroliniana, U. polyphylla,Gyrophora esculenta*	-do-	Shibata, 1973
Everniin	Glucan with α(1→3) (1→4) linkage		*Evernia prunastri*	-do-	Shibata, 1973
Acroscyphan	Homoglucan with α(1→3) (1→4) (1→6) linkage		*Acroscyphus sphaerophoroides*	-do-	Shibata, 1973
Galacto-glucomannans	a (1→6)-linked main chain of ◎-Manp	O-2 and O-4 by α-Galp and β-Galp nonreducing end-units	*Parmotrema austrosinense, P. delicatulum, P. mantiqueirense, P. schindlerii, P. tinctorum and Rimelia (R. cetrata* and *R. reticulata)*	mycobiont	Carbonero *et al.*, 2005b
Xylorhamno-galactofuranan	(1→3)-linked galactofuranosyl	galactofuranosyl units 5-*O* and 6-*O*-substituted, rhamnopyranosyl units 2-*O*, 3-*O* and 2,3-di-*O*-substituted in position 6,	*Cladina confusa*	Photobiont	Cordeiro *et al.*, 2007

Polysaccharide	Main unit	Side chain	Lichen species	Mycobiont/ Photobiont	reference
laminaran and pustulan and galacto-furanomannan	(1→3)-and (1→6)-linked β-glucans	(1→6)-linked α-mannopyranosyl	*Umbilicaria mammulata*	Mycobiont	Carbonero *et al.*, 2006
Hetero-polysaccharide	(1→5)-linked galactofuranosyl units	Complex	*Ramalina gracilis*	Photobiont	Cordeiro *et al.*, 2008
O-methylated mannogalactan	(1→6)-linked β-galactopyranose	at O-3 by β-Gal*p*, 3-OMe-α-Manp or α-Man*p* units	*Peltigera aphthosa*	Photobiont	Cordeiro *et al.*, 2010
Colleman like	Complex heteroglycan	2-OMe Man*p*, 2-OMe-Arap, Xyl*p* and Glc*pA*	*Collema flaccidum*	Whole thallus (cyanobacteria)	Jensen *et al.*, 2010
	(1→4)-linked β-D-xylan		*Peltigera canina*	Mycobiont	Ruthes *et al.*, 2010

Table 1. Summarized data of lichen polysaccharides and their origin

Parmotrema schindlerii, Parmotrema tinctorum and *Rimelia cetrata* and *Rimelia reticulata* by Carbonero *et al.*, in 2005b. These galactoglucomannans consisted of (1→6)-linked main chain of α-Manp units, which were substituted preferentially at O-2 and O-4 by α-Gal*p* and β-Gal*p* nonreducing end-units, respectively. Further, they also isolated two galactomannan fractions from the lichen, *Roccella decipiens*. One galactomannan fraction had a main chain with (1→4)-linked α-D-Man*p* units, substituted at O-2 with side chains containing a nonreducing end, 2-O- and 6-O-substituted α-Man*p* units. The other fraction had a similar α-D-Man*p* core structure, but with side chains containing nonreducing end, 5-O-, 6-O-, and 5,6-di-O-substituted β-D-Gal*f* units (Cordeiro *et al.*, 2005). Another heteropolysaccharide xylorhamnogalactofuranan was isolated from the lichen *Cladina confusa*. It consisted of (1→3)-linked galactofuranosyl units with side chains of galactofuranosyl units 5-*O* and 6-*O*-substituted, as well rhamnopyranosyl units 2-*O*, 3-*O* and 2,3-di-*O*-substituted at position 6. Nonreducing end units were composed of Xylose (Cordeiro *et al.*, 2007). In 2006, (1→3)-and (1→6)-linked β-glucans, namely laminaran and pustulan and galactofuranomannan which have a main chain of (1→6)-linked α-mannopyranosyl residues, partially substituted at O-2, O-4 were isolated from another lichen *Umbilicaria mammulata* (Carbonero *et al.*, 2006). Later in 2008, Cordeiro *et al.* found Galactofuranose-rich heteropolysaccharide from the lichen *Ramalina gracilis*. This polysaccharide has (1→5)-linked galactofuranosyl units at the main chain with very complex branched structures of side chains in position 6. They found that this polysaccharide arose from the algal symbiont *Trebouxia sp.* of the lichen *Ramalina gracilis*. An O-methylated mannogalactan was isolated from *Peltigera aphthosa* by Cordeiro *et al.* in 2010. This consisted of (1→6)-linked β-galactopyranose main chain partially substituted at O-3 by β-Gal*p*, 3-OMe-α-Manp or α-Man*p* units. The algal symbiont *Coccomyxa mucigena* of the lichen *Peltigera aphthosa* was thought to be the origin of this heteropolysaccharide since the lichen thallus yielded a polysaccharide of different structure. A colleman like heteropolysaccharide was isolated from the cyano lichen *Collema flaccidum*.

Colleman is a complex heteroglycan containing the unusual monosaccharides 2-OMe Man*p* and 2-OMe-Arap as well as Xyl*p* and Glc*pA* (Jensen *et al*., 2010). The presence of uronic acids has been reported previously from cyanobacterial polysaccharides. Since the structural features and sugar content of colleman is representative of polysaccharides of cyanobacterial origin, it is proposed that colleman originates from the cyanobacterial partner. Ruthes *et al.* (2010) were able to isolate (1→4)-linked β-D-xylan (an EPS) and heteropolysaccharide with a complex structure of β-L-Ara*p* and β-D-Xyl*p*-(1→4)-linked units from *Peltigera canina*. Again it was opined that the photobiont, *Nostoc muscorum* was the source of this heteropolysaccharide.

4. Polysaccharides as a taxonomic tool

The identification and classification is generally based on morphology of the organism. Taxonomy of lichen species have been corroborated by phylogenetic applications with the advances in DNA technology. Lichen polysaccharides have been used as a taxonomic tool and chemotaxonomic classification has resulted in clarification of conflicting taxonomic data.

The lichen-forming ascomycete order Lichinales comprises around 250 species and is distributed among 52 genera and four families (Eriksson, 2006). Earlier molecular studies (Wedin *et al*., 2005) did not confirm its phylogenetic relationships, although the order was treated as a separate class, *Lichinomycetes* (Hibbett *et al*., 2007). Since alkali and water-soluble polysaccharides from *Lichina pygmaea* and *L. confinis* reflect phylogeny in other ascomycetes (Prieto *et al.* 2008), an isolated polysaccharide was purified to investigate whether such polysaccharide compounds in the Lichinomycetes are distinctive. Results support molecular studies showing that lichen species are remote from Lecanoromycetes as the galactofuranose residues are in the α-configuration. That the Lichinomycetes were part of an ancestral lichenized group cannot be established from the present data because the extracted polysaccharide does not have the galactofuranose residue in the β configuration; however, the data suggests that an ancestor of the Lichinomycetes contained a mannan and was part of an early radiation in the ascomycetes. Polysaccharides present support the molecular data obtained recently that the Lichinales are distinct from other ascomycete groups and should be treated as a separate order in the separate class Lichenomycetes (Reeb *et al*., 2004). However, more representatives in the order must be subjected to molecular studies and more polysaccharides be investigated before confirming this hypothesis. Interestingly, the relative basal position of *Lichina* with respect to the water soluble polysaccharides agrees with the suggestion that ancestral lichens contained cyanobacteria as the photosynthetic partner (Hawksworth, 1982).

Investigation of mannose containing polysaccharides as a taxonomic tool centers around the structural diversity of the galactomannans isolated from several lichenized fungi. The taxonomic value of these galactomannans depends on the side-chain substituents on (1 →6)-linked α-D-mannopyranosyl main chains (Gorin and Lacomini, 1985).

Although classical taxonomy regarded *Cladina* as sub-species of *Cladonia,* lichenologists considered them to be distinct species. It was shown that galactomannans are important chemotypes in determining the taxonomy of *Clodonia* spp. and other related genera

(Woranovicz-Barreira *et al.*, 1999). However, based on molecular phylogentic results *Cladina* and *Cladonia* were confirmed as synonyms (Ahti and Depriest, 2001).

Lichen polysaccharides, can also be used as a taxonomic tool to differentiate some lichen species, since some of the heteropolysaccharides and their chemical characters are unique to certain groups of lichens (Carbonero *et al.*, 2006 and Cordeiro *et al.*, 2007). Further, the polysaccharides content of the lichen photobiont may be used as a marker in algal symbiont taxonomy (Cordeiro *et al.*, 2007).

5. Biological activities of lichen polysaccharides

Many lichens are known to have immunomodulating properties, potent antibiotic, antitumour, antiviral as well as antioxidant properties which are mostly attributed to the secondary metabolites (Malhotra *et al.*,2008, Behera, *et al.*, 2007). According to Yanaki *et al.*, (1986) and Bohn and BeMiller, (1995), functional activity of polysaccharides mainly depends on molecular weight, degree of branching, water solubility, structure and configuration. Hence they have different uses in different fields. The biological activities of lichen polysaccharides reported have been limited to anti-tumor, anti-inflamatory or immunomodulatory activity (Omarsdottir *et al.*, 2007).

Cordeiro *et al.*, (2008) reported that one of the β-galactofuranan polysaccharides isolated from *Trebouxia* sp., the algal symbiont of the lichen *Ramalina gracilis* expressed *in vitro* activity on peritoneal macrophages. Further studies carried out on 4 lichen polysaccharides by Omarsdottir, *et al.*, (2006) showed that, three heteroglycans namely Ths-4, Ths-5 and thamnolan and a β-glucan (Th-2) isolated from the lichen *Thamnolia vermicularis* var. subuliformis showed an effect on the human immune system. Thamnolan, a galactofuranorhamnan had less mitogenic effect than Ths-5 and Ths-2 indicating that its unusual galactofuranorhamnan structure may be responsible for its different immunomodulatory activity. In a study on the immunomodulatory activities of an aqueous lichen extract from *Cetraria islandica* , Freysdottir *et al.*, (2007) discovered that the extract was able to upregulate IL-10 secretion. Interestingly, when the individual components of this extract (lichenan and isolichenan and secondary metabolites protolichesterinic and fumarprotocetraric acids) were subjected to the same assay, only lichenan displayed antiinflammatory effects (Freysdottir, *et al.*, 2008).

The cytotoxic activity, phagocytic activity and antitumor activity of an α-D glucan from *Ramalina celastri* has been reported (Leão *et al.*, 1997 and Stuelp-Campelo *et al.*, 2002). Based on this, De Araújo *et al.*, (2011) found that a sulphated α-D glucan lichen polysaccharide extracted from *Ramalina celastri* exhibited antischistosomal activity. These α-D glucans with α(1–3) and α(1–4) linkages are linear and water soluble and are known for their ability to stimulate the mononuclear phagocyte system and improve host resistance to viral, bacterial and parasitic infections.

According to Nishikawa *et al.*, (1970), O-acetylated pustulan isolated from three species of *Umbilicaria* showed a significant antitumor effect against the implanted Sarcoma-180 in mice. The polysaccharide fractions isolated from six species of lichens were studied for antitumor activity (Nishikawa *et al.*, 1974). The active polysaccharide present in *Lasallia pensylvanica,* was the partially O-acetylated (1-6) β-glucan, pustulan while that of *Usnea rubescens,* was identified as a lichenan type polysaccharide. The other 4 polysaccharides were from the *Cladonia* sp.; *Cladonia crispata, Cladonia rangiferina* subsp. *grisea, Cladonia mitis* and *Cladonia squamosa* all of which showed significant antitumor activity. These *Cladonia* polysaccharides which were complex heteroglycans of the galactomannan type were found to exert moderate antitumor activity. However, previous studies have shown that the isolichenin type polysaccharides were not as effective in their antitumour effect. In another study, isolichenin isolated from the lichen *Usnea fasciata* showed moderate activity against Ehrlich tumor cells (Periera *et al.*, 1994). In 1989, Hirabayashi *et al.*, showed inhibitory effect of a lichen polysaccharide sulfate (GE-3-S), isolated from the lichen *Umbilicaria esculenta* on the replication of human immunodeficiency virus (HIV) in vitro.

Behera *et al.*, (2007) showed correlation between lichen protein/polysaccharide ratio and their antioxidant properties. Since cultured lichen extracts were used in the study, the effect of secondary metabolites as well as polyphenols present in the extract have not been evaluated. Thus the antioxidant activity cannot be solely attributed to the polysaccharide specially since polyphenols are known antioxidants. In addition, the mechanism of scavenging activity of polysaccharides on free radicals is not fully understood yet.

6. Conclusion

Polysaccharides from lichens unlike bacterial polysaccharides do not show a wide range of variation in sugar content. The predominant sugars are limited to glucose, galactose and mannose, with arabinose and xylose present in minor proportions in addition to others such as rhamnose. These polysaccharides however have been useful in chemotaxonomic studies due to this conservation of sugar structures and regular structural patterns. As far as biological activity is concerned, very few studies have been reported and it would be worthwhile further investigating the immunomodulating effects of the polysaccharides.

Author details

Desiree Nedra Karunaratne, R.G.U. Jayalal and Veranja Karunaratne
University of Peradeniya, Peradeniya, Sri Lanka

7. References

Ahmadjian, V. (1987). Coevolution in lichens. *Annals of the New york Academi of Sciences* 503, 307–315.

Ahti, T. and Depriest, P.T., (2001). New combination of *Cladina* epithets in *Caldonia* (Ascomycotina, Cladonaceae). *Mycotaxon*, 78, 499-502.

Behera, B.C., Verma, N., Sonone, A. and Makhija, U. (2007). Tissue culture of some lichens and screening of their antioxidant, antityrosinase and antibacterial properties. *Phytotherapy Research* 21, 1159–1170.

Bernard, M. and Latgé, J.P. (2001). *Aspergillus fumigatus* cell wall: composition and biosynthesis. *Med Mycol* 39 (1), 9–18.

Bohn, J.A. and BeMiller, J.N. (1995). (1–3)-β-Glucans as biological response modifiers: a review of structure-functional activity relationships. *Carbohydr Polym* 28, 3–14.

Carbonero, E.R., Cordeiro,L.M.C., Mellinger,C.G., Sassaki,G.L., Wörgötter,E.S., Gorin, P.A.J. and Iacomini, M. (2005a). Galactomannans with novel structures from the lichen *Roccella decipiens* Darb. *Carbohydrate Research* 340, 1699–1705.

Carbonero, E. R., Mellinger, C. G., Eliasaro, S., Gorin, P. A. J. and Lacomini, M. (2005b). Chemotypes significance of lichenized fungi by structural characterization of heteropolysaccharides from the genera *Parmotrema and Rimelia. FEMS Microbiology Letters* 246, 273–278.

Carbonero, E. R., Montai, A. V., Woranovicz-Barreira, S. M., Gorin, P. A. J. and Iacomini, M. (2002). Polysaccharides of lichenized fungi of three *Cladina* spp.: significance as chemotypes. *Phytochemistry* 61, 681–686.

Carbonero, E.R., Smiderle, F.R., Gracher, A.H.P., Mellinger, C.G., Torri, G., Ahti, T., Gorin, P.A.J. and Iacomini, M. (2006). Structure of two glucans and a galactofuranomannan from the lichen *Umbilicaria mammulata. Carbohydrate Polymers* 63, 13–18.

Cordeiro, L.M.C., Wörgötter, E.S., Gorin, P.A.J. and Iacomini, M. (2003). Comparative studies of the polysaccharides from species of the genus *Ramalina*-lichenized fungi of three distinct habitats. *Phytochemistry* 60, 967–975.

Cordeiro, L.M.C., Wörgötter, E.S., Gorin, P.A.J. and Iacomini, M. (2004). Elucidation of polysaccharide origin in *Ramalina peruviana* symbiosis. *FEMS Microbiol. Lett.* 238 (2004) 79–84.

Cordeiro, L.M.C., Carbonero, E.R., Sassaki, G.L., Reis, R.A., Wörgötter, E.S., Gorin, P.A.J. and Iacomini, M. (2005). A fungus-type β-galactofuranan in the cultivated *Trebouxia* photobiont of the lichen *Ramalina gracilis. FEMS Microbiol. Lett.* 244, 193–198.

Cordeiro, L.M.C., de Oliveira, S.M., Buchi, D.F. and Iacomini, M. (2008). Galactofuranose-rich heteropolysaccharide from *Trebouxia* sp., photobiont of the lichen *Ramalina gracilis* and its effect on macrophage activation. *International Journal of Biological Macromolecules* 42, 436–440.

Cordeiro, L.M.C., Sassaki, G.L., Gorin, P.A.J. and Iacomini, M. (2010). O-Methylated mannogalactan from the microalga *Coccomyxa mucigena*, symbiotic partner of the lichenized fungus *Peltigera aphthosa. Phytochemistry* 71, 1162–1167.

Cordeiro, L.M.C., Sassaki, G.L. and Iacomini, M. (2007). First report on polysaccharides of *Asterochloris* and their potential role in the lichen symbiosis. *International Journal of Biological Macromolecules* 41, 193–197.

De Araújo, I.W.F., Vanderlei, E.S.O., Rodrigues, A.R.G., Coura, C.O., Quinderé, A.L.G., Fontes, B.P., Queiroz, I.N.L., Jorge, R.J.B., Bezerra, M.M., Silva, A.A.R., Chaves, H.V., Monteiro, H.S.A., Paula, R.C.M. and Benevides, N.M.B. (2011). Effects of a sulfated polysaccharide isolated from the red seaweed *Solieria filiformis* on models of nociception and inflammation. *Carbohydr. Polym.*, 86: 1207-1215.

Elix, J. A. and Stocker-Wörgötter, E. (2008). Biochemistry and secondary metabolites. In Lichen Biology 2nd Edn. Vol. 1. ed. T.H. Nash III. Cambridge University Press, Chapter 7, 104-133.

Eriksson, O.E. (2006). Outline of Ascomycota - 2006. *Myconet* 12, 1–82.

Freysdottir,J., Omarsdottir, S., Ingólfsdóttir, K., Vikingsson, A. and Olafsdottir, E.S. (2007). Abstracts, Scandinavian Society for Immunology. 37th Annual Meeting Turku, Finland. Pp. 596.

Freysdottir,J., Omarsdottir, S., Ingólfsdóttir, K., Vikingsson, A. and Olafsdottir, E.S. (2008). In vitro and in vivo immunomodulating effects of traditionally prepared extract and purified compounds from *Cetraria islandica*. *International Immunopharmacology* 8: 423–430.

Friedl, T. (1989). Systematik und Biologie von *Trebouxia* (Microthamniales, Chlorophyta) als Phycobiont der Parmeliaceae (lichenisierte Ascomyceten). Ph.D. thesis, University Bayreuth, Bayreuth, 218.

Friedl, T. and Gärtner, G. (1988). *Trebouxia* (Pleurastrales, Chlorophyta) as a phycobiont in the lichen genus *Diploschistes*. *Arch. Protistenk.*143, 153–61.

Gorin, P.A.J., Baron, M., Silva, M.L.C., Teixeira, A.Z.A. and Iacomini, M. (1993). Lichen carbohydrates. *Ciência e Cultura* 45, 27–36.

Gorin, P.A.J. and Iacomini, M. (1984). Polysaccharides of the lichens *Cetraria islandica* and *Ramalina usnea*. *Carbohydr. Res.* 128, 119–132.

Gorin, P.A.J. and Iacomini, M. (1985). Structural diversity of D-galacto-D-mannan components isolated from lichens having ascomycetous mycosymbionts. *Carbohydr. Res.* 142, 253–267.

Hawksworth, D.L. (1982). Co-evolution and the detection of ancestry of lichens, *J. Hattori Botanical Laboratory* 52, 323-329.

Hawksworth, D.L., Kirk, P.M., Sutton, B.C. and Pegler, D.N. (1995). Ainsworth and Bisby's dictionary of the fungi. 8th edn. CAB International, Wallingford

Hibbett, D.S., Binder, M., Bischoff, J.F., Blackwell, M., Cannon, P.F., Eriksson, O.E., Huhndorf, S., James, T., Kirk, P.M., Lucking, R., Lumbsch, T., Lutzon, F., Matheny, P.B., McLaughlin, D.J., Powell, M.J., Redhead, S., Schoch, C.L., Spatafora, J.W., Stalpers, J.A., Vilgalys, R., Aime, M.C., Aptroot, A., Bauer R., Begerow, D., Benny, G.L., Castlebury, L.A., Crous, P.W., Dai, Y.C., Gams, W., Geiser, D.M., Griffith, G.W., Gueidan, C., Hawksworth, D.L., Hestmark, G., Hosaka, K., Humber, R.A., Hyde, K., Ironside, J.E., Koljalg, U., Kurtzman, C.P., Larsson, K.H., Lichtwardt, R., Longcore, J., Miadlikowska, J., Miller, A., Moncalvo, J.M., Mozley-Standridge, S., Oberwinkler, F., Parmasto, E., Reeb, V., Rogers, J.D., Roux, C., Ryvarden, L., Sampaio, J.P., Schussler, A., Sugiyama, J., Thorn, R.G., Tibell, L., Untereiner, W.A., Walker, C., Wang, Z., Weir, A., Weiss, M., White, M.M., Winka, K., Yao, Y.J. and Zhang, N. (2007). A higher-level phylogenetic classification of the Fungi. *Mycological Research* 111: 509–547.

Hirabayashi, K., Iwata, S., Ito, M., Shigeta, S., Narui, T., Mori, T. and Shibata, S. (1989). Inhibitory effect of a lichen polysaccharide sulfate, GE-3-S, on the replication of human immunodeficiency virus (HIV) in vitro. *Chem Pharm Bull (Tokyo)* 37(9), 2410-2412.

Ihda, T. A., Nakano, T., Yoshimura, I. and Iwatsuki, Z. (1993). Phycobionts isolated from Japanese species of *Anzia* (lichenes). *Archiv für Protistenkunde* 143, 163–172.

Jensen, J.S.R.E., Petersen, B.O., Veselinovic, T., Olafsdottir, E.S., Duus, J.Ø. and Omarsdottir, S. (2010). Structural characterisation of a new O-methylated heteroglycan, colleman, from the cyanolichen *Collema flaccidum*. *Carbohydrate Polymers* 80, 799 –807.

Kirk, P.M., Cannon, P.F., David, J.C. and Stalpers, J.A. (2001). *Ainsworth & Bisby's dictionary of the fungi*, 9th edition. CAB International, Wallingford.

Leão, A.M., Buchi, D.F., Iacomini, M., Gorin, P.A. and Oliveira, M.B. (1997). Cytotoxic effect against HeLa cells of polysaccharides from the lichen *Ramalina celastri*. *J Submicrosc Cytol Pathol.* 29(4), 503-9.

Lumbsch H. T., Ahti T., Altermann S., De Paz G.A., Aptroot A., Arup U., Peña A.B., Bawingan P.A., Benatti M.N., Betancourt L., Björk C.R., Boonpragob K., Brand M., Bungartz F., Cáceres M.E.S., Candan M., ChavesJ. L., Clerc P., Common R., Coppins B. J., Crespo A., Dal-Forno M., Divakar P.K., Duya M.V., Elix J.A., Elvebakk A., Fankhauser J.D., Farkas E., Ferraro L.I., Fischer E., Galloway D.J., Gaya E., Giralt M., Goward T., Grube M., Hafellner J., J.E. Hernández M., M. De L. A. H. Campos, Kalb K., Kärnefelt I., Kantvilas G., Killmann D., Kirika P., Knudsen K., Komposch H., Kondratyuk S., Lawrey J. D., Mangold A., Marcelli M. P., Mccune B., Messuti M. I., Michlig A., González R. M., Moncada B., Naikatini A., Nelsen M. P., Øvstedal D. O., Palice Z., Papong K., Parnmen S., Pérez-Ortega S., Printzen C., Rico V. J., Plata E. R., Robayo J., Rosabal D., Ruprecht U., Allen N. S., Sancho L., De Jesus L. S., Vieira T. S., Schultz M., Seaward M. R. D., Sérusiaux E., Schmitt I., Sipman H. J. M., Sohrabi M., Søchting U., Søgaard M. Z., Sparrius L. B., Spielmann A., Spribille T., Sutjaritturakan J., Thammathaworn A., Thell A., Thor G., Thüs H., Timdal E., Truong C., Türk R., Tenorio L. U., Upreti D. K., Van Den Boom P., Rebuelta M. V., Wedin M., Will-Wolf S., Wirth V., Wirtz N., Yahr R., Yeshitela K., Ziemmeck F., Wheeler T. and Lücking R.. (2011). One hundred new species of lichenized fungi: a signature of undiscovered global diversity. *Phytotaxa* 18, 1–127.

Malhotra, S., Subban, R. and Singh, A. (2008). Lichens- Role in Traditional Medicine and Drug Discovery. *The Internet Journal of Alternative Medicine* 5(2).

Nash. III, T. H. (1996). Photosynthesis, respiration, productivity and growth. *Lichen Biology* (ed. by T.H.Nash), Cambridge University Press, Cambridge, 88–120,

Nishikawa, Y., Tanaka, M., Shibata, S. and Fukuoka, F. (1970). Polysaccharides of lichens and fungi. IV. Antitumor active O-acetylated pustulan type glucans from the lichens of *Umbilicaria* species. *Chem. Pharm. Bull.* 18 (7), 1431-1434.

Nishikawa, Y., Ohki, K., Takahashi, K., Kurono, G., Fukuoka, F. and Emori, M. (1974). Studies on the water soluble constituents of lichens. II. Antitumor polysaccharides of *Lasallia, Usnea* and *Cladonia* species. *Chem. Pharm. Bull.* 22 (11), 2692-2702.

Olafsdottir, E.S. and Ingólfsdóttir, K. (2001). Polysaccharides from lichens: structural characteristics and biological activity. *Planta Medica* 67, 199-208.

Olafsdottir, E.S., Omarsdottir, S., Paulsen, B.S., Jurcic, K. and Wagner, H., (1999). Rhamnopyranosylgalactofuranan, a new immunologically active polysaccharide from *Thamnolia subuliformis*. *Phytomedicine* 6, 273-279.

Omarsdottir, S., Petersen, B.O., Barsett, H., Paulsen, B.S., Duus, J.Ø. and Olafsdottir, E.S. (2006). Structural characterisation of a highly branched galactomannan from the lichen *Peltigera canina* by methylation analysis and NMR-spectroscopy. *Carbohydrate Polymers* 63, 54–60

Omarsdottir, S., Freysdottir, J. and Olafsdottir, E.S. (2007). Immunomodulating polysaccharides from the lichen *Thamnolia vermicularis* var. *subuliformis*. *Phytomedicine* 14, 179–184.

Paulsen, B.S., Olafsdóttir, E.S. and Ingólfsdóttir, K. (2002). Chromatography and electrophoresis in separation and characterization of polysaccharides from lichens. *Journal of Chromatography* 967, 163–171

Periera, E. C, Nascimento, S. C. and Lima, R. C. (1994). Analysis of *Usnea fasciata* crude extracts with antineoplastic activity. *The Tokai journal of experimental and clinical medicine* 19 (1-2), 47-52.

Prieto, A., Leal, J.A., Bernabe, M. and Hawksworth, D.L. (2008). A polysaccharide from *Lichina pygmaea* and *L. confinis* supports the recognition of Lichinomycetes. *Mycological Research* 112, 381 – 388.

Reeb, V., Lutzoni, F. and Roux, C., (2004), Contribution of RPB2 to multilocus phylogenetic studies of the Euascomycetes (Pezizomycotina, Fungi) with special emphasis on the lichen forming acaromsporaceae and evolution of polyspory. *Molecular Phylogenetics and Evolution* 32, 1036-1060.

Ruthes, A.C., Komura, D.L., Carbonero, E.R., Sassaki, G.L., Gorin, P.A.J. and Iacomini, M. (2010). Structural characterization of the uncommon polysaccharides obtained from *Peltigera canina* photobiont *Nostoc muscorum*. *Carbohydrate Polymers* 81, 29–34.

Shibata, S. (1973). Polysaccharides of lichens. *Journal of the National Science Council of Sri Lanka* 1, 183-188.

Sipman, H.J.M. and Aptroot, A. (2001). Where are the missing lichens? *Mycological Research* 105, 1433–1439.

Stuelp-Campelo, P.A., Oliveira, M.B.M., Carneiro-Leão, A.M.A., Carbonero, E.R., Gorin, P.A.J. and Iacomini, M. (2002). Effect of a soluble α-D-glucan from the lichenized fungus *Ramalina celastri* on macrophage activity. *Int J Immunopharmacol* 2, 691-698.

Takahashi, K., Takeda, T. and Shibata, S. (1979). Polysaccharides of lichen symbionts. *Chem. Pharm. Bull.* 27, 238–241.

Teixeira, A.Z.A., Iacomini, M. and Gorin, P.A.J. (1995). Chemotypes of mannose containing polysaccharides of lichen mycobionts: a possible aid in classification and identification. *Carbohydr. Res.* 266, 309–314.

Wedin, M., Wiklund, E., Crewe, A., Döring, H., Ekman, S., Nyberg, Å., Schmitt, I. and Lumbsch, H.T. (2005). Phylogenetic relationships of Lecanoromycetes (Ascomycota) as revealed by analyses of mtSSU and nLSU rDNA sequence data. *Mycological Research* 109, 159–172.

Wolseley, P.A. and Aguirre-Hudson, B. (1994). Key to lichen genera in Thailand with special reference to epiphytic taxa, part I, macrolichens. *Natural History Bulletin*. Siam. Soc. 43, 303-335

Woranovicz-Barreira, S.M., Gorin, P.A.J., Sassaki, P.L., Marcelli, M.P. and Iacomini,M.(1999). Galactomannoglucans of lichenized fungi of *Cladonia* spp.: significance as chemotypes. *FEMS Microbiology Letters* 181(2), 313–317.

Yanaki, T., Itoh, W. and Tabata, K. (1986). Correlation between the antitumor activity of Schizophyllan and its triple helix. *Agric Biol Chem* 50, 2415–2416.

Yoshimura, I. (2004). Mysterious Lichens - An Introduction to their taxonomy, morphology, physiology, and chemistry. Proceedings of the 1st KoLRI International Symposium of Lichenology.

Chapter 8

Antiviral Levans from *Bacillus spp.* Isolated from Honey

Mona A. Esawy, Eman F. Ahmed, Wafaa A. Helmy,
Nahla M. Mansour, Waled M. El-Senousy and Mounir M. El-Safty

Additional information is available at the end of the chapter

1. Introduction

In recent years, significant progress has been made in discovering and developing new bacterial polysaccharides that possess novel and highly functional properties (Baird et al 1983). Although their ubiquitous role in biological processes and their versatility as biocompatible, environmentally friendly materials are beyond doubt, polysaccharides are still considered to be the "sleeping giant" of biotechnology.

Honey contained a great variety of dominant spores and in consequence their dominant spores are expected to be new expolysaccarides sources which could be isolated. This expectation comes from the honey constituents which is mainly fructose (about 38.5%) and glucose (about 31.0%) (Crosby and Alfred 2004). Aerobic spore forming Bacillus were the most frequently encountered microbes on the external surface, crop and intestine of the honey bees and consequently honey (Root, 1993, Esawy et al., 2011).

Most of the researches in the honey field focused on its antimicrobial, antioxidant and anticancer activities, also the identification of the dormant endospore inside it (Sabate, et al., 2009). None till now paid attention to the enzymatic products of these dormant endospores (Esawy et al., 2011). Osmophilic microorganisms survive environmental extremes of desiccation, pressure and acidity, it is expected that their biopolymers will also have some unique properties to adapt to such extreme conditions. This investigation concerned the question of whether honey collect bacteria that are good producers of levansucrase and levan yield. Recently, screening of 16 bacterial honey isolates for levansucrase production showed that all the tested isolates were levansucrase producers despite variations in the degree of activity (data not published yet). Levansucrase, one of the fructosyltransferases or glycansucrases, is produced by various microorganisms

(Iizuka et al.,1991; Hernandez, et al 1995; Kojima et al.,1993; Ben Ammar et al., 2002, Esawy et al., 2008). Bacterial levansucrases catalyze at least three different reactions: hydrolysis of sucrose, polymerization of fructose derived from sucrose and hydrolysis of levan. It is reported that levansucrase activity is involved in a variety of processes including survival of bacteria in soil (*B. subtilis*), phytopathogenesis (*Erwinia* and *Pseudomonas* species) and symbiosis (*Paenibacillus polymyxa*) of plant interactive bacteria (Hettwer et al.,1995). *Bacillus subtilis,* known as the hay bacillus or grass bacillus, is a Gram-positive, catalase positive bacterium commonly found in soil (Madigan & Martinko, 2005). Recently, Esawy et al 2011, Esawy [a] et al 2012 and Esawy [b] et al 2012 reported in novel *Bacillus subtilis* honey isolates as new sources of very important enzymes such as levansucrase, dextranase and.lipase.

Levan is one of two main types of fructans, which are natural homopolymers of fructose (Arvidson et al 2006). It is a naturally occurring polymer of β-D-fructofuranose with β (2→6) linkages between repeating five-member fructofuranosyl rings and branching at C-1 (Arvidson et al 2006, Barone and Medynets., 2007). Levans produced by different organisms differ in their molecular weight and degree of branching. Levans from plants generally have molecular weights about 2000 - 33.000Da (Rhee et al., 2002). The molecular weight of levan, and the fraction of residues incorporated in side chains, depends on both the source and the growth conditions, with plant levan and microbially-produced levan having very different characteristics (Arvidson et al 2006; Kasapis, and Morris 1994; Kasapis et al., 1994; Newbrun 1971; Stivala, and Bahary 1978; Huber et al., 1994). Recently it was reported in the *B. subtilis* NRC1aza levansucrase, the unique feature of this isolate its ability to produce two types of levan with different molecular weights (El Fattah et al, 2012). Bacterial levans are much larger than those produced by plants, with multiple branches and molecular weights (2-100 million Da) (Pontis and Del Campillo 1985 ; Keith et al., 1991). Levan is non-gelling, non-swelling in water, (Kasapis et al., 1994; Stivala, and Bahary 1978; Huber et al., 1994) and an unusual polysaccharide due to its relatively low intrinsic viscosity compared to other molecules of similarly high molecular weights. Levan can be used as food or a feed additive with prebiotic and hypocholesterolemic effects (Sanders, et al., 2003). Subsequently, there are a variety of potential industrial applications for levan such as a surfactant for household use due to its excellent surface-active properties, a glycol/levan aqueous two-phase system for the partitioning of proteins, etc. In addition, in vitro anti-tumor activity of levan produced from *Microbacterium laevaniformans, Rahnella aquatilis* and *Zymomonas mobilis,* has been shown against eight different tumor cell lines (Urdaci, et al 2004;Yoo,et al 2004; Yoon et al.,2004; Liu et al., 2012; El Fattah et al, 2012). Recently, Liu et al., (2012) and El Fattah et al, 2012 reported in the antioxidant activity of native levan and their derivatives. Dahech et al., (2011) reported that polysaccharide levan is efficient in inhibiting hyperglycemia and oxidative stress induced by diabetes and suggests that levan supplemented to diet may be helpful in preventing diabetic complications in adult rats. The market for levan will gradually increase in the various fields (Kang et al., 2009).

2. Experimental

2.1. Sources of honey and microorganisms

Three different honey samples were purchased; local honey bee collecting nectar from clover flower; Kashmiry honey, honey bee collecting nectar from desert flower (Saudi Arabian); and Gably honey, a honey bee collecting nectar from desert flower (Libya).

2.2. Isolation of bacterial strains from honey samples

One hundred micro liters of honey samples was spread on nutrient agar plates (g/L): beef ext., 1.0; yeast ext., 2.0; peptone, 5.0 and agar, 25.0. After drying for 20 min in a laminar flow hood, the plates were incubated at 50 °C to avoid the growth of any pathogenic spores for 24 h or until the colonies size was sufficient (approximately larger than 3–5 mm in diameter). The bacterial isolates were streaked onto agar plates and preserved at 4 °C. The purity of the isolates was assessed by colony morphology and microscopy.

2.3. Chromosomal DNA and plasmid extraction

Chromosomal DNA was prepared from overnight culture in LB,using AxyGEN Biosciences DNA extraction kit, according to manufacturer's instructions. Plasmid extraction was performed using Wizard mini prep. extraction kit (Promega) according to manufacturer's instructions with slight modification, where 50 L of lysozyme (200 mg/mL) were added to the resuspended buffer and incubated at 37 °C for 1 h then the protocol was carried on as described in the kit (O'Sullivan & Klaenhammer, 1993; Sambrook, et al., 1989).

2.4. PCR amplification for molecular identification

To amplify the 16S rRNA gene, a primer pair hybridizing to two conserved regions in 16S rRNA genes from *Bacillus spp.* was used: (bac-F and bac-R) (Ash et al., 1991; Kwon et al., 2009). For the amplification of the 16–23S intergenic region, a primer pair was used: L516SF and L523SR. While for recA gene, a primer pair corresponding to conserved regions in recA genes from *Bacillus sp.* was used: recA-F and recA-R. Speciesspecific primer set for *B. subtilis* corresponding for ytcP gene was used: ytcP-F and ytcP-R. All polymerase chain reaction amplifications were performed with the Taq DNA polymerase kit (Promega). Reaction mixtures consisted of 20 mm Tris–HCl (pH 8.4), 50 mm KCl, 3 mm $MgCl_2$, 50 mm of each of the four deoxynucleoside triphosphates (dNTP), 1 U Taq polymerase, 5 pmol of each primer and 1 L of template DNA in a final volume of 50 L. Samples were amplified in a GeneAmp polymerase chain reaction system 2700 (Applied Biosystems) programmed as follows: initial denaturation of DNA for 5 min at 94 °C, 30 cycles of 30 s at 94 °C, 30 s at 50 °C and 30 s at 72 °C. Polymerase chain reaction products were quantified by electrophoresis on a 1% (w/v) agarose gel containing ethidium bromide. Polymerase chain reaction products obtained from the selected isolates were purified, using QIA quick polymerase chain reaction purification KIT (Qiagen) and then sequenced commercially by Sigma-Egypt. The sequencings were performed and manually aligned, using DNAMAN software (version 4.0). Sequence homologies were examined by comparing the obtained sequence with those in the NCBI database and the NEB cutter V 2.0 database. NEB cutter V 2.0 is an on-line DNA sequence tool used to find large, non-overlapping, open-reading frames and works for all restriction enzymes. It provides a website, which allows users to check nucleotide sequences for restriction enzyme sites. The sequences were submitted and AluI enzyme was chosen for digestion. Finally gel photograph using 2% agarose was viewed (Ash et al., 1991).

2.5. Bacterial strains and growth conditions

Bacillus strains used in this work were cultivated in Luria-Bertani (LB) broth or agar at 37 °C. Defined medium was used for cellular production of levansucrase (Yanase et al., 1992). It had the following composition (g/L): yeast extract, 2.5; commercial sucrose, 80; $MgSO_4$, 0.2 and K_2HPO_4, 5.5. The medium was completed by the addition of 1 L distilled water and the pH was adjusted to 7.0 before autoclaving. The parameters included initial incubation temperature (25–45 °C); different concentrations of sucrose (80–160 g); incubation time (16–48 h); (50–150) rpm; pH (5-9) and NaCl (1–4% w/v) were studied. The sucrose was substituted with fructose, glucose, lactose (80 g/L) and beet molasses (equivalent to 80 g sucrose) to study their effects on enzyme production.

2.6. Cellular production

Cultivation was carried out in 250 mL Erlenmeyer flasks. Each flask contained 50 mL production medium and was autoclaved for 15 min. The flasks were then inoculated with 2.0 ml inoculum and incubated for 24 h at 30 °C. The culture broth was then centrifuged in a cooling centrifuge (K70; Janektzki, Germany) at 10,397×g to separate the bacterial cells from the supernatant.

2.7. Assay of levansucrase

Levansucrase assay was performed according to the method of Yanase et al. (1992) with some modification. 0.5 ml of culture filtrate was incubated with 1 ml 20% (w/v) sucrose and 0.5 mL 0.1 M acetate buffer at pH 5.2 and incubated at 37°C for min. The decreasing amounts of sugars produced were measured by glucose oxidase kits. One unit of enzyme activity was defined as the amount of enzyme that produced decreasing sugars equivalent to 1 _mol of glucose/min.

2.8. Separation of levan polymer

The levan producing organisms were cultivated on a defined medium as described above. After growth, the culture was centrifuged to remove bacterial cells; the levan was precipitated with two volumes of absolute ethanol. The precipitate was collected and dried under vacuum.

2.9. Chromatography

Paper chromatography was performed according to Block et al. (1995). Hydrolysate of products of levan were analyzed by either paper chromatography on whatman No. 1. The mixtures at the end of incubation time were boiled for 3 min to stop the reaction. Chromatographic development was carried out with a solvent system of n-butanol:acetone:water (4:5:1) and detected by spraying with aniline hydrogen phthalate. The acid hydrolysate of the polysaccharide produced by the six isolates was analyzed using high-pressure liquid chromatography (HPLC). A 7.8 mm ×300 mm PL-HI-PLEXPB column was linked to a differential refractometer. The column temperature was maintained at 80 °C. The aqueous mobile phase was delivered at a flow rate of 0.6 ml/min.

2.10. Determination of molecular weight

Different concentrations of levan and oligosaccharide were prepared and the flow time of equal volume for each concentration at 30°C was determined in a U-shaped Ostwald viscometer. Flow time of the same volume of distilled water was also determined as control. Thus, specific viscosity/C (gsp) was estimated. A plot of levan and oligosaccharide concentration (C) against intrinsic viscosity (C) (gsp/C) therefore yielded a straight line.

2.11. Antivirus detection

Two types of viruses were used, highly pathogenic avian influenza H5N1 virus Egyptian isolate, was used at titre of 106 EID50/mL (embryo infective dose per mL) and adenovirus type 40 with different doses $20^1 \times 10^4$, 1×10^5, and 1×10^6 infectious particles/mL obtained from the Holding Company for Biological Products & Vaccines (VACSERA).

2.12. Specific pathogen free (SPF) eggs

SPF embryonated chicken eggs were used at nine days old and inoculated via the allantoic sac route. SPF eggs (Brown et al., 2007) were obtained from Nile SPF Eggs, Koomoshiem, Fayoum, Egypt.

2.13. Cytotoxicity test

It was done according to Simoes et al., (1999) and Walum et al., (1990). Briefly, All samples (100 mg) were dissolved in 500 _L of water or ethanol. Samples A, E, M, and K were dissolved in ethanol while samples C and G were dissolved in water. Decontamination of samples was done by adding 12 _L of 100x of antibiotic–antimycotic mixture to 500 _L of each sample. Then, bi-fold dilutions were done to 100 _L of original dissolved samples and 100 _L of each dilutions were inoculated in Hep-2 cell line (obtained from the Holding Company for Biological Products & Vaccines VACSERA, Egypt) previously cultured in 96 multi well plates (Greiner-Bio one, Germany) to estimate the non toxic dose of the tested samples. Cytotoxicity assay was done using cell morphology evaluation by inverted light microscope and cell viability test applying trypan blue dye exclusion method.

2.14. Cell morphology evaluation by inverted light microscopy

Hep-2 cell cultures (2× 10^5 cells/mL) were prepared in 96-well tissue culture plates (Greiner-Bio one, Germany). After 24 h incubation at 37 °C in a humidified 5% (v/v) CO2 atmosphere cell monolayers were confluent, the medium was removed from each well and replenished with 100 _L of bi-fold dilutions of different samples tested prepared in DMEM (GIBCO BRL). For cell controls 100 _L of DMEM without samples was added. All cultures were incubated at 37 °C in a humidified 5% (v/v) CO2 atmosphere for 72 h. Cell morphology was observed daily for microscopically detectable morphological alterations, such as loss of confluence, cell rounding and shrinking, and cytoplasm granulation and vacuolization. Morphological changes were scored (Simoes et al., 1999).

2.15. Cell viability assay

It was done by trypan blue dye exclusion method (Walum et al., 1990). Hep-2 cell cultures (2×10^5 cells/mL) were grown in 12-well tissue culture plates (Greiner-Bio one, Germany). After 24 h incubation, the same assay described above for tested samples cytotoxicity was followed by applying 100 L of tested samples dilutions (bifold dilutions) per well. After 72 h the medium was removed, cells were trypsinized and an equal volume of 0.4% (w/v). Trypan blue dye aqueous solution was added to cell suspension. Viable cells were counted under the phase contrast microscope.

2.16. Haemagglutinating activity assay

This was applied for the allantoic fluids of the inoculated eggs and measured by micro technique of haemagglutination (HA) test (Takatsy, 1955).

2.17. Evaluation for antiviral activity

Three experiments were conducted.

2.17.1. Experiment 1

One hundred and five emberyonated chicken eggs (ECEs) were examined; equal volumes of HPAI H5N1 virus and original extracts were separately used at three levels:

Level 1: Equal volumes of HPAI H5N1 virus and the original undiluted samples were mixed and incubated at room temperature for 1 h then inoculated into the allantoic sac of five ECEs for each product sample at dose 0.2 mL/ECE.

Level 2: Equal volumes of HPAI H5N1 virus and the 1/5 dilution of each sample were mixed and preceded as level 1.

Level 3: Equal volumes of the virus and the 1/10 dilution of each sample were mixed and preceded as level 1. In addition, five ECEs were inoculated with the virus that mixed with equal volume of saline at a dose of 0.2 mL/ECE of saline alone (negative control). The ECEs are inoculated at 37 °C and candled every 2 h till all the positive control ECEs died.

2.17.2. *Experiment 2*

One hundred and five SPF ECEs were used in this experiment; equal volumes of HPAI H5N1 virus and the original samples were mixed with equal volume of the original samples and inoculated directly into the allontoic sac of five ECEs for each product sample at a dose of 0.20 mL/ECE for each product sample at a dose of 0.2 mL/ECE. Five ECEs were inoculated with equal volume of the HPAI H5N1 virus and saline at dose of 0.2 mL/ECE (positive control). Another five ECEs were inoculated with 0.20 mL/ECE of saline alone (negative control). All the ECEs were incubated at 37 °C and controlled every 2 h till the ECEs of the positive control died

2.17.3. *Experiments 3*

One hundred and five SPF ECEs of nine days old were used in this experiment. 0.10 mL of the HPAI H5N1 virus was inoculated via the allontoic sac of each ECE into 100 ECEs and then the inoculated ECEs were incubated for 1 h at 37 °C. The original samples were inoculated into five ECEs, which previously inoculated with the virus at a dose of 0.1 ml. Another five ECEs were inoculated with 0.2 ml/ECE of the mixed virus and saline. Other five ECEs were inoculated with 0.2 ml/ECE of saline alone. The ECEs were inoculated at 37 °C and candled every 2 h till ECEs of the positive control died.

2.18. Antiviral effect of tested samples on adenovirus type 40

Seventy five microliters of non toxic dilutions were mixed with 75 L of different doses 1×10^4, 1×10^5, and 1×10^6 infectious viral particles/mL of adenovirus type 40 provided by American Type Culture Collection (ATCC). Then the mixture was incubated overnight at 4 °C. Inoculation of 100 L of 10 fold dilutions of treated and untreated adenovirus was done into Hep-2 cell line in 12 multi well-plates. After 1 h incubation for adsorption at 37 °C, 1 mL medium (DMEM) was added to each well. The cell line was observed daily for one week then, three times freezing and thawing for tested plates were done. Nested PCR was done for confirmation of adenovirus (presence/absence) in each well (Puig et al., 1994).

3. Results

3.1. Molecular identification of the levansucrase producers strains

3.1.1. *16S rRNA sequences and their analogical electrophoresis*

Six levansucrase producers' bacteria were isolated from different honey sources. The isolates resembled each other in cell morphology where cells were rod-shaped, Gram-positive, motile and spore-forming. Colonies were circular, creamy, and no pigment was formed. They were identified as *Bacillus* spp. based on morphological examination. The identification was confirmed by molecular biological analysis, using 16S rRNA sequencing method. The results showed that the 6 isolates were identified as *Bacillus* spp. (99%), or *B. licheniformis* (99%), or *B. amyloliquefaciens* (99%) (Fig. 1).

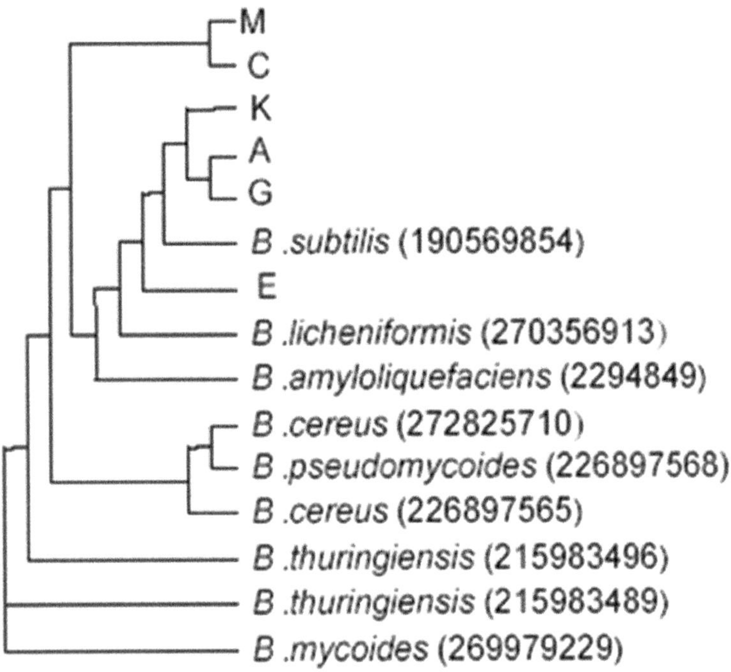

Figure 1. Phylogenetic neighbor-joining tree obtained by 16S rRNA sequence analysis of the tested isolates and other *Bacillus* spp. present in the gene bank database (accession numbers in parentheses).

The DNA of the isolates was extracted as described in Section 2 and the 1.5 kb 16S rRNA gene was amplified for each DNA by PCR using primers bac-F and bac-R. The PCR amplification, purification and sequencing were performed as described previously. The 1.5 kb obtained sequences were aligned and clustered with sequences from the NCBI database. 16S rRNA gene sequence analysis indicated that the six isolates (K, M, A, C, E, and G) were *Bacillus* spp. with 99% identity any of these three species *B. subtilis, or B. licheniformis, or B. amyloliquefaciens* and they clustered into a monophyletic line in a phylogenetic tree. To distinguish and clear identification of these strains on the species level the analogical electrophoresis, using NEB cutter was applied to identify the 16S rRNA results, which have been sequenced, as the strains of the same species expected to have almost the same sites when digested with AluI. Fig. 2 showed that the isolates M and G have the same size fragments as *B. subtilis* gi 269313996 while the other isolates A, C, E; K showed different AluIfragments which differ to the AluI fragments generated from 16S rRNA sequence of *B. subtilis, or B. licheniformis or B. amyloliquefaciens*. It was clear that 16S rRNA gene alone could not distinguish these three closely related Bacillus species

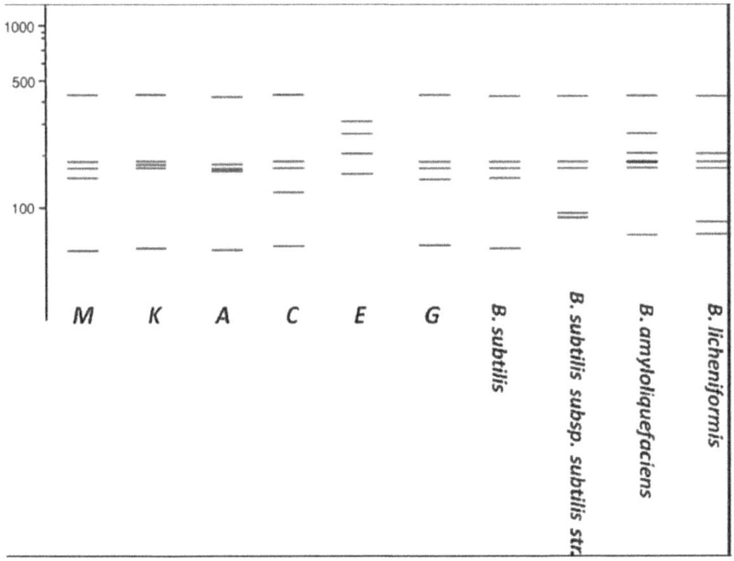

Figure 2. The analogical electrophoresis of Bacillus isolates compared to Bacillus strains (from the gene bank) by "AluI",using NEBcutter 2.0. The accession no. of the control Bacillus strains: *B. subtilis* gi|269313996|; *B. subtilis* subsp. subtilis str. AL009126.3; *B. amyloliquefaciens* gi|229484923|; and *B. licheniformis* gi|270356913|.

3.2. Identification by 16–23S intergenic region

To distinguish between these three closely related strains the 16–23S intergenic region was amplified by primers L516SF × L523SR, and then the sequence was determined for the six isolates. The homology results for the 16–23S intergenic region showed that: K and M strains showed 100% similarity to *B. subtilis* but strains A, C, E, G could be either *B. subtilis (99%) or B. amyloliquefaciens* (85%) **(Table. 1.).**

Target	Primer name	Oligonucleotide	Reference
16–23S intergenic region	L516SF L523SR	5′-TCGCTAGTAATCGCGGATCGGC-3′ 5′-GCATATCGGTGTTAGTCCCGTCC-3′	Yoon et al., 2001
recA gene	recA-F recA-R	5′-TGAGTGATCGTCAGGCAGCCTTAG-3′ 5′-CYTBRGATAAGARTACCAWGMACCGC-3′	Gun-Hee Kwon et al., (2009)
hypotheticl gene	ytcP-F ytcP-R′	5′-GCTTACGGGTTATCCCGC-3′ 5′CCGACCCCATTTCAGACATATC-3′	Gun-Hee Kwon et al., (2009)

Table 1. Primers used for gene amplification

3.3. Identification by recA sequence

Hence the recA gene has been used as a molecular chronometer in addition to rRNA genes. The 1.2 kb band was gel isolated and subjected to sequencing results and showed that the four strains A, C, E, and G were *B. subtilis* rather than *B. amyloliquefaciens*. Still these results need more confirmation.

3.4. Identification by specific-PCR for *B. subtilis*

To solve this problem, identification using specific-PCR for *B. subtilis* was described. Based on ytcP gene encoding a hypothetical protein, a PCR primer pair ytcF and ytcR were designed for *B. subtilis* species for specific amplification purpose. Using this primer pair, a 0.46 kb fragment was amplified only from *B. subtilis* strains, whereas no similar band was detected from *B. licheniformis or B. amyloliquefaciens* strains. These primers were subjected to the six isolates and PCR resulted in the 0.46 kb band as shown in Fig. 3 and this confirmed that all isolates A, C, E, G, K, M surely belong to *B. subtilis not B. amyloliquefaciens*.

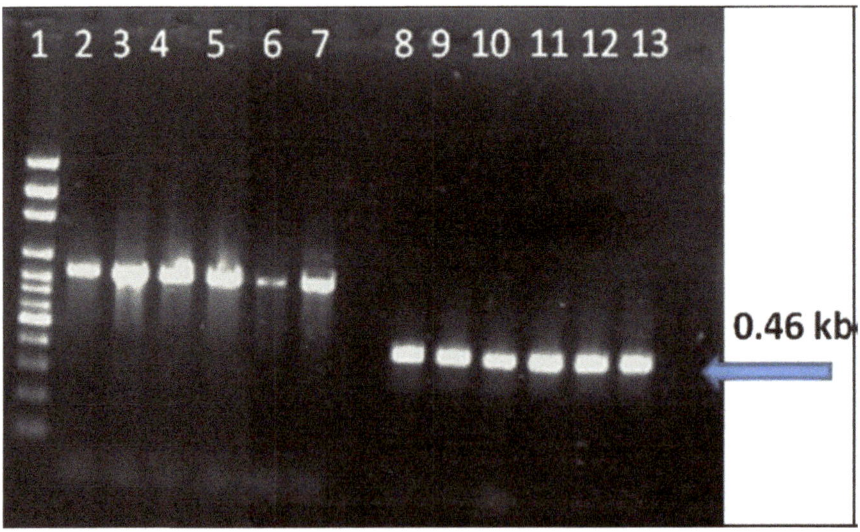

Figure 3. PCR using the chromosomal DNA of the 6 Bacillus isolates. Lane 1, 100 bp ladder; lane 2–7, using primers recR × recF; lane 8–13, using B. subtilis specific primers ytcF × ytcR.

3.5. Levan and levansucrase production

The six honey isolates were tested for production of levansucrase, the optimized conditions for the isolates ranged from 8 to 12% (w/v) commercial sucrose, 37–40 ºC, 24–28 h, 50–100 rpm and pH 6–7.0 (data not shown). Among all the tested isolates, M and K isolates showed the highest levansucrase activities (62 and 59 U/mL). The presence of

NaCl (1–4%, w/v) showed great influence in enzyme activity, the enzyme production increased from 2 to 3 folds according to the strain (Fig. 4). Paper chromatography of the product hydrolysate revealed the presence of only fructose and tiny traces of glucose, pointing to the levan nature of the product. Furthermore, the acid hydrolysate of the polysaccharide produced by the isolates was exclusively fructose, as revealed by HPLC. Levan was harvested by precipitation from the culture broth by addition of ethanol. The yield and consistency of the product varied according to the isolate. The final products were a brownish-white gummy material, which could be freeze-dried or vacuum-dried. The highest amount of levan was produced on the medium containing commercial sucrose, followed by beet molasses (Fig. 5). While a small amount of microbial polysaccharide (alcohol precipitate) was also produced when the organism was grown on lactose and glucose, it was worthy to record that no polysaccharides were produced on fructose. The amounts of levan decreased 40–50% in the presence of NaCl. Isolates M, K, A, C, E, and G produced 11, 16.25, 6.60, 1.81, 1.74, 6.6 g/L levan, respectively under the optimized conditions. The levan products consisted of one fraction characterized by high and low molecular masses (40.938, 71.887, 43.487, 154.638, 77.753 and 14.200 kDa for isolates K, M, A, C, E, and G, respectively) (Fig. 6).

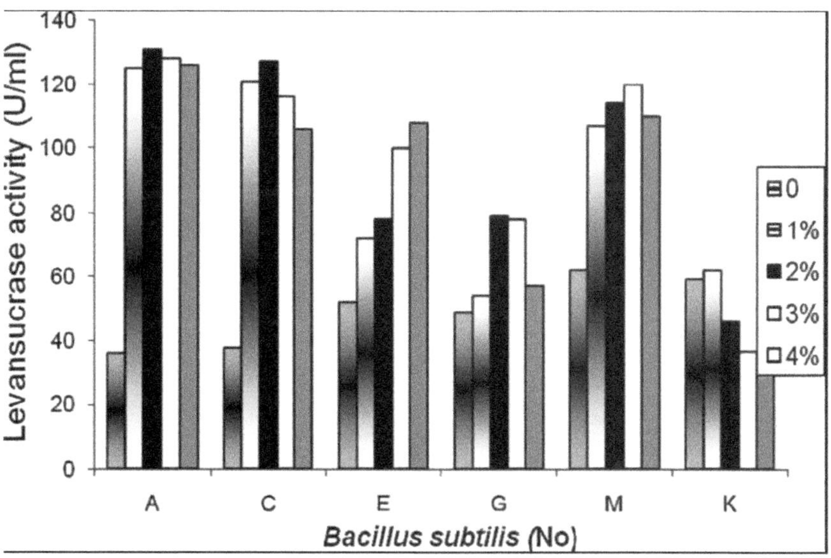

Figure 4. The effect of absence and presence of different concentrations of NaCl on levansucrase production from honey isolate. Note: NaCl conc (0–4%) represented from left to right.

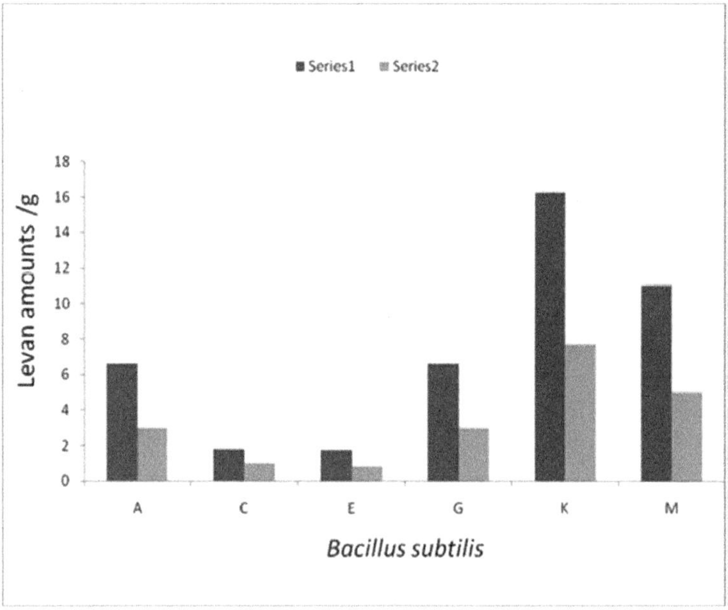

Figure 5. The difference between amounts of levan produced in sucrose medium (series 1) and in molasses medium (series 2)

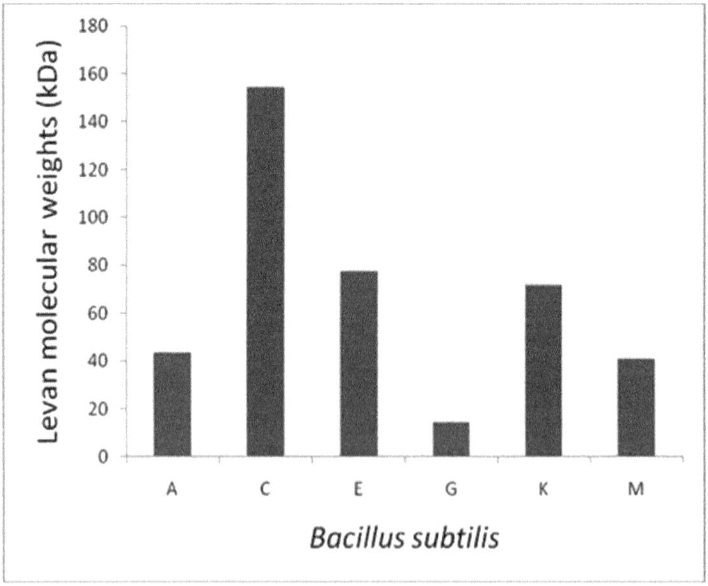

Figure 6. The levan molecular weights from *Bacillus subtilis* isolates

3.6. Cytotoxicity test

The non toxic doses for samples C and G were 0.5 mg/mL in water. On the other hand, the non toxic doses for samples A, E, and M were 0.2 mg/mL in ethanol while the nontoxic dose of sample K was 0.1 mg/ml in ethanol.

3.7. Anti-adenovirus type 40 assay

The samples (C, E, G and K) had weak effect on adenovirus 40 which did not exceed 10%. The two samples A and M revealed antiviral effect on adenovirus type 40 ranged from 50 to 60% as shown in Table 2.

Sample	Initial viral doses	Final viral doses	Percentage of reduction
A	$1x10^4$	$4x10^3$	60%
	$1x10^5$	$5x10^4$	50%
	$1x10^6$	$5x10^5$	50%
M	$1x10^4$	$4x10^3$	60%
	$1x10^5$	$4x10^4$	60%
	$1x10^6$	$5x10^5$	50%

Table 2. Effect of levan A and levan M on the infectivity of enteric adenovirus type 40 (DNA virus). viral infectivity.

3.8. Anti-H5N1 virus assay

All the embryos of the positive controls died and the allantoic fluid of each was positive for haemagglutination assay (HA), while all the embryo of negative control were not died and the allantoic fluid of each was negative for HA. Three levans of code K, M, E showed antiviral against HPAI H5N1. Each of these samples showed antiviral effect when inoculated with H5N1 virus 1 h before inoculation into nine days old ECEs, while they had no effect on the virus when inoculated simultaneously with the virus just after mixing or after the virus inoculation for 1 h (Table 3). These results also revealed that simultaneous inoculation of the levans product or even after infection was of no value.

Code of sample	Experiment 1						Experiment 2		Experiment 3	
	Level 1		Level 2		Level 3		NDE	+HA	NDE	+HA
	NDE	+HA	NDE	+HA	NDE	+HA				
con	5/5	5.0	5/5	5.0	5/5	5.0	5/5	5.0	5/5	5.0
M	0/5	0.0	5/5	5.0	5/5	5.0	5/5	5.0	5/5	5.0
C	5/5	5.0	5/5	5.0	5/5	5.0	5/5	5.0	5/5	5.0
E	0/5	0.0	5/5	5.0	5/5	5.0	5/5	5.0	5/5	5.0
K	0/5	0.0	5/5	5.0	5/5	5.0	5/5	5.0	5/5	5.0

Table 3. Showed that three levans of code K, M, E had antiviral against HPAI- H5N1 NDE: Number dead Haemagglutination assay

4. Discussion

The public health and the discovery of new drugs is a main objective of many research activities, however, sometime this type of research activity cost a lot of money. Although, in this proposed research we will be aiming towards the protection of the public health and introduce a new drugs contribute in solve the problem of serious diseases through products that will save a lot of money to our economy. Within this context, six mobile spore-forming, and Gram-positive facultative aerobic bacilli were isolated from different honey samples and identified as *Bacillus* spp. On the base of morphological, and molecular identification, using 16S rRNA sequence method. *B. subtilis* isolates are biologically and commercially important as producers of a great variety of secondary metabolites such as antibiotics, and enzymes (Desai & Banat 1997; Roberts et al., 1996). The16S rRNA sequence method could not identify the Bacillus isolates at the species level where revealed identical to any of three strains *B. subtilis or B. licheniformis or B. amyloliquefaciens.* In fact DNA-based identification methods such as 16S rRNA gene sequencing and 16S–23S intergenic region sequencing have been used widely for the purpose of identification and typing of microorganisms isolated from natural environments including fermented foods (Hansen et al., 2001; Levine, et al., 2005). But identification based on rRNA gene sequences fails to distinguish one species from the other if they share highly similar rRNA genes. This is true for some Bacillus species. It is difficult to distinguish *B. subtilis* from closely related *B. licheniformis* or *B. amyloliquefaciens* by rRNA gene sequences because of no significant differences in their rRNA sequences (Nakamura, 1989; Nakamura et al., 1999). Other genes such as recA (Rodriguez et al., 2007) and dnaJ (Shah et al., 2007) have been employed instead of rRNA genes. It is necessary to compare results from different identification methods as a whole before to reach a conclusion (Bourque et al., 1995).

To solve the ambiguity in differentiating them based solely on the 16S rRNA gene, it was turned to 16–23S intergenic region, recA gene, and ended with *B. subtilis* specific primers. It is specifically useful to distinguish organisms with highly similar rRNA genes recA gene was amplified from the chromosomal DNA of bacilli isolates by PCR, using primers (recF and recR) (Payne et al., 2005). When rRNA and recA gene sequences were considered together, it was possible to conclude that the G, M isolates are belonging to *B. subtilis* but the other isolates A, C, E needed further identification. For the accurate distinguish between the 6 isolates, the specific-PCR for *B. subtilis* based on ytcP gene was used and the results revealed that the six strains are belonging to *B. subtilis*. The present results proved high phenotypic and genotypic variability among *B. subtilis* isolates, where they showed different morphological and biological properties suggesting them as new different species of *B. subtilis* with valuable impact in the industry. Many authors reported in the production of levansucrase from *B. subtilis* (Euzenat et al., 1997; Le Gorrec et al., 2002). The various sugars, initial pH, fermentation temperature, and agitation speed affected the levansucrase production by *B. subtilis* (Abdel-Fattah et al., 2005; Shih et al., 2005). The result ensure the halophilc feature of *B. subtilis* levansucrases, this comes from its osmophilic character. Enhancements of levansucrase production in the presence of NaCl were reported (Euzenat et al., 2006; Poli et al., 2009). As far as we are aware no studies were reported on the effect of NaCl on the enzyme production. It seemed that levan production by the isolates was dependent mainly on commercial sucrose media, where the use of beet molasses, glucose, and lactose led to noticeable reduction in levan synthesis. Beet molasses was used as low-cost substitutes for sucrose in commercial levan yield (Han & Watson, 1992). The decrease in levan yield in the molasses medium (2.533 g/L) when compared to the commercial sucrose (21.685 g/L) was also reported (De Oliveira et al., 2007). Although higher Halomonas sp. AAD6 biomass concentrations were observed when glucose, maltose, fructose and galactose were used as carbon sources, levan levels were very low comparing with sucrose (Poli et al., 2009). The six isolates produced different levan weights, with wide range of molecular mass. On the other hand, it was reported that halophilic Halomonas sp. AAD6 cells grown in the presence of sucrose afforded the highest levan production levels (1.073 g/L) (Poli et al., 2009). Also, *B. polymyxa* produced about 40 g/L extracellular polysaccharide per liter in sucrose medium, which was about three times that produced by familiar levan producers (Han, 1989). Levan antitumor activity was reported by many authors (Yoon, Yoo, Cha, & Lee, 2004,) but as far as we are aware nothing was reported on antiviral activity of this fructose polymer. The present findings showed antiviral effects of K, M, and E levan products on H5N1 virus, While, A and M levan products showed antiviral effects on adenovirus type-40. It was obvious that the product M was entirely effective against both respiratory RNA virus (H5N1) and enteric adenovirus type 40 (DNA virus). It was apparent that each of effective levan showed antiviral effect when inoculated with H5N1 virus 1 h before inoculation into nine days old ECEs, while they had no effect on the virus when inoculated simultaneously with the virus just after mixing or after 1 h of inoculation.

The outcome of this study is the probable suitability of some types of levan as a safe and cheap natural product in antiviral treatments with applying the known roles concerning the use of these compounds. In addition, this article affords honey micro flora as a new and important sources of levansucrase enzymes, could be have biotechnological applications in pharmaceutical industries.

Author details

Mona A. Esawy*, Eman F. Ahmed and Wafaa A. Helmy
Department of Chemistry of Natural and Microbial Products,
National Research Centre, Dokki, Giza, Egypt

Nahla M. Mansour
Gut Microbiology and Immunology Lab,
Central of Excellence for Advanced Sciences (CEAS), National Research Centre, Dokki, Giza, Egypt

Waled M. El-Senousy
Water Pollution Research Department, National Research Centre, Dokki, Giza, Egypt

Mounir M. El-Safty
Central Laboratory for Evaluation of Veterinary Biologics (CLVB), Abbassia, Cairo, Egypt

---

* Corresponding Author

5. References

Abdel-Fattah, A. F., Mahmoud, D. A. R., & Esawy, M. A. (2005). Production of levansucrase from *Bacillus subtilis* NRC 33a and enzymic synthesis of levan and fructo-oligosaccharides. Current Microbiology, 51, 402–407.

Abdel-Fattah, A.M., Gamal-Eldeen, A.M., Helmy, W.A., & Esawy, M.A. Antitumor and antioxidant activities of levan and its derivative from the isolate *Bacillus subtilis NRC1aza. Carbohydrate Polymers xx (2012)89, 314-322*

Ash, C., Farrow, J. A. E., Dorsch, M. E., Stackebrandt, E., & Collins, M. D. (1991). Comparative analysis of Bacillus anthracis, Bacillus cereus and related species on the basis of reverse transcriptase sequencing of 16S rRNA. International Journal of Systemic Bacteriology, l41, 343–346.

Arvidson, S.A., Rinehart, B.T., & Gadala-Maria F. (2006). Concentration regimes of solutions of levan polysaccharide from *Bacillus* sp. Carbohydrate Polymers 65 144–149.

Baird[1], P. A. Sandford[1] & I. W. Cottrell. Industrial Applications of Some New Microbial Polysaccharides *Nature Biotechnology* 1, 778 - 783 (1983)

Barone, J.R., & Medynets, M., Thermally processed levan polymers, Carbohydrate Polymers 69 (2007) 554–561

Beine, R., Moraru, R., Nimtz, M., Na'amnieh, S., Pawlowski, A., & Buchholz, K. (2008). Synthesis of novel fructooligosaccharides by substrate and enzyme engineering. Journal of Biotechnology, 138, 33–41.

Ben Ammar, B. Y., Matsubara, T., Ito, K., Iizuka, M., Limpaseni, T., & Pongsawasdi, P., et al. (2002). Characterization of a thermostable levansucrase from Bacillus sp. TH4-2 capable of producing high molecular weight levan at high temperature. Journal of Biotechnology, 99, 111–119.

Block, R. J., Durrum, E. L., & Zweig, G. (1995). A manual of paper chromatography and paper electrophoresis. New York, NY: Academic Press Inc., pp. 127

Bourque, S. N., Valero, J. R., Lavoie, M. C., & Levesque, R. C. (1995). Comparative analysis of the 16 to 23S ribosomal intergenic spacer region sequences of *Bacillus thuringiensis* strains and subspecies and of closely related species. Applied Microbiology and Biotechnology, 61, 1623–1626.

Crosby, Alfred W. (2004). Ecological Imperialism: The Biological Expansion of Europe, 900–1900. Cambridge University Press, 188. ISBN 0-521-54618- 4. "There are many kinds of bees and other insects producing honey all round the world"

Dahech I, Belghith, K.S, Hamden, K, Feki, A., Belghith H, & Mejdoub, H. (2011). Antidiabetic activity of levan polysaccharide in alloxan-induced diabetic rats. International Journal of Biological Macromolecules. 2011 49:742-6.

De Oliveira, M. R., Da Silva, R. S. S. F., Buzato, J. B., & Celligoi, M. A. P. C. (2007, November). Study of levan production by *Zymomonas mobilis* using regional low-cost carbohydrate sources. Biochemical Engineering Journal, 37: 177–183.

Desai, J. D., & Banat, I. M. (1997). Microbial production of surfactant and their commercial potential. Microbiology and Molecular Biology Review, 61, 47–64.

EsawyI, M. A.; MahmoudI; D. A. R. & Abd- Fattah A. F. (2008). Immobilisation of *Bacillus subtilis* NRC33a levansucrase and some studies on its properties. Brazilian Journal of Chemical Engineering 25: 237-246

Esawy, M.A., Ahmed, E.F., Helmy, W. A., Mansour, N.M., El-Senousy,W.M., & El-Safty , M.M. (2011). Production of levansucrase from novel honey *Bacillus subtilis* isolates capable of producing antiviral levans. Carbohydrate Polymer, 36, 823-830

Esawy[a], M. A.; Mansour, S. H; Ahmed, E. F., Hassanein, N.M. & El Enshasy, H.A. (2012). Characterization of Extracellular Dextranase from a Novel Halophilic *Bacillus subtilis* NRC-B233b a Mutagenic Honey Isolate under Solid State Fermentation E-Journal of Chemistry 9: 1494-1510

Esawy[b], M.A.; Awad G.E.A., Mansour, S. H; Ahmed, E. F., & Danial, E.N. Evaluation of honey as new reservoir for probiotic bacteria. Advances in food science. V: 43 (2) in press.

Euzenat, O., Guibert, A., & Combes, D. (1997). Production of fructo-oligosaccharides by levansucrase from *Bacillus subtilis* C4. Process Biochemistry, 32, 237–243.

Euzenat, O., Guibert, A., & Combes, D. (2006). Production and purification of *Bacillus subtilis* C4 levansucrase kinetic characterization of the enzyme. Annals of the New York Academy of Sciences, 864, 288–294.

Han, Y. W. (1989). Levan production by *Bacillus polymyxa.* Journal of Industrial Microbiology, 4, 447–452.

Han, Y. W., & Watson, M. A. (1992). Production of microbial levan from sucrose, sugarcane juice and beet molasses. Journal of Industrial Microbiology, 9, 257–260.

Hansen, B. M, Leser, T. D., & Hendriksen, N. B. (2001). Polymerase chain reaction assay for the detection of Bacillus cereus group cells. FEMS Microbiology Letters, 202, 209–213.

Hernandez, L., Arrieta, J., Menendez, C., Vazquez, R., Coego, A., Suarez, V., et al. (1995). Isolation and enzymatic properties of levansucrase secreted by *Acetobacter diazotrophicus* SRT4, a bacterium associated with sugar cane. Biochemistry journal, 309, 113–118.

Hettwer, U., Gross, M., and Rudolph, K., (1995): Purification and characterization of an extracellular levansucrase from *Pseudomonas syringae* pv Phaseolicola. Journal of Bacteriology, pp. 2834–2839.

Huber, A. E., Stayton, P. S., Viney, C., & Kaplan, D. L. (1994). Liquid crystallinity of a biological polysaccharide: the levan/water phase diagram. Macromolecules, 27, 953–957.

Kang, (2009): "Levan: Applications and Perspectives". Microbial Production of Biopolymers and Polymer Precursors. Caister Academic Press. ISBN 978-1-904455-36-3.

Kasapis, S. & Morris, E., (1994) Conformation and physical properties of two unusual microbial polysaccharides: *Rhizobium trifolii* CPS and Levan. In Food Hydrocolloids: Structures, Properties and Functions, ed by K. Nishinari, Plenum Press, New York, pp 97-103.

Kasapis, S., Morris, E., Gross, M., & Rudolph, K., (1994): Solution properties of levan polysaccharide from *Pseudomonas syringae* pv. phaseolicola, and its possible primary role as a blocker of recognition during pathogenesis. Carbohydrate Polymers. 23: 55–64.

Keith, J., Wiley, B., Ball, D., Arcidiacono, S., Zorfass, D., Mayer, J., & Kaplan, D. (1991): Continuous culture system for production of biopolymer levan using *Erwinia herbicola.* Biotechnology and Bioengineering, 38: 557– 560.

Kojima, I., Saito, T., Iizuka, M., Minamiura, N., & Ono, S. (1993). Characterization of levan produced by Serratia sp. Journal of Fermentation and Bioengineering, 75, 9–12.

Kwon, H. K., Lee, H. A., Park, J. Y., Kim, J. S., Lim, J., & Park, C. S., et al. (2009). Development of a RAPD-PCR method for identification of Bacillus species isolated from Cheonggukjang. International Journal of Food Microbiology, 129, 282–287.

Le Gorrec, K., Connes, C., & Guibert, A., et al. (2002). Identification of three inducible and extracellular enzymatic activities working on sucrose in Bacillus subtilis NCIMB 11871 and 11872 supernatant. Enzyme and Microbial Technology, 31, 44–52.

Levine, S. M., Tang, Y. W., & Pei, Z. (2005). Recent advances in the rapid detection of *Bacillus anthracis*. Reviews in Medical Microbiology, 16, 125–133.

Lizuka, M., Tanaka, T., Yamamoto, S., et al. (1991). Synthesis of fructan and oligosaccharides by microbial and plant fructosyltransferases. Denpun Kagaku, 38, 217–222.

Liu, J., Luo, J., Ye, H., & Zeng, X. (2012). Preparation, antioxidant and antitumor activities in vitro of different derivatives of levan from endophytic bacterium *Paenibacillus polymyxa* EJS-3. Food and Chemical Toxicology, 50, 767–772.

Madigan, M. T., & Martinko, J. M. (2005). Brock Biology of Microorganisms (11th ed.). Pearson Prentice Hall.

Nakamura, L. K. (1989). Taxonomic relationship of black-pigmented Bacillus subtilis strains and a proposal for *Bacillus atrophaeus* sp. nov. International Journal of Systematic Bacteriology, 39, 295–300.

Nakamura, L. K., Roberts, M. S., & Cohan, F. M. (1999). Relationship of Bacillus subtilis clades associated with strains 168 and W23: A proposal for Bacillus subtilis subsp. subtilis subsp. nov. and *Bacillus subtilis* subsp. spizizenii subsp. nov. International Journal of Systematic Bacteriology, 49, 1211–1215.

Newbrun, E., and Baker, S., (1968): Physico-chemical characteristics of the levan produced by *Streptococcus salivarius*. Carbohydrate Research, 6: 165–170.

O'Sullivan, D. J., & Klaenhammer, T. R. (1993). Rapid mini-prep isolation of high quality plasmid DNA from Lactococcus and Lactobacillus spp. Applied and Environment Microbiology, 59, 2730–2733.

Payne, G. W., Vandamme, P., Morgan, S. H., et al. (2005). Development of a recA gene-based identification approach for the entire Burkholderia genus. Applied and Environmental Microbiology, 71, 3917–3927.

Poli, A., Kazak, H., & Gürleyenda˘g, B., et al. (2009). High level synthesis of levan by a novel Halomonas species growing on defined media. Carbohydrate Polymers, 78, 651–657.

Pontis, H. G., & Del Campillo, E. (1985). Fructans. In M. Dey, & R. A. Dixon (Eds.), Biochemistry of storage carbohydrates in green plants (pp. 205–277). Academic Press Inc.

Puig, M., Jofre, J., Lucena, F., Allard, A., Wadell, G., & Girones, R. (1994). Detection of adenovirus and enterovirus in polluted waters by nested PCR amplification. Applied and Environmental Microbiology, 60, 2963–2970.

Roberts, M.S., Nakamura, L.K., & Cohan, F.M. (1996). *Bacillus vallismortis* sp. nov., a close relative of *Bacillus subtilis*, isolated from soil in Death Valley, CA. Interational Journal of Systemic Bacteriology, 46, 470–475.

Rodriguez, H., de las Rivas, B., & Muňoz, R. (2007). Efficacy of recA gene sequence analysis in the identification and discrimination of Lactobacillus hilgardii strains isolated from stuck wine fermentations. International Journal of Food Microbiology, 115:70–78

Root, A.L., The ABC and XYZ of bee culture. London: Edward Arnold publication Ltd.; 1993. pp. 5–28.

Rhee, S., Song, K., Kim, C., Park, B., Jang, E., & Jang, K., (2002): Levan. In A.Steinbu¨ chel (Ed.), Biopolymers; polysaccharides , pages 351–377.

Sabaté, D. C., Carrillo, L., & Audisio, M. C. (2009). Inhibition of Paenibacillus honey samples. Research Microbiology, 160, 193–199.

Sambrook, J., Fritsch, E. F., & Maniattis, T. (1989). Molecular cloning: A laboratory manual. Cold Spring Harbor, NY: Cold Spring Harbor Laboratory.

Sanders, M. E., Morelli, L., & Tompkins, T. A. (2003). Sporeformers as human probiotics: Bacillus, Sporolactobacillus, and Brevibacillus. Comprehensive Reviews inFood Science and Food Safety (CRFSFS), 2, 101–110.

Shah, M. M., Lihara, H., Noda, M., Song, S. X., Nhung, P. H., & Ghkusu, K., et al. (2007). dnaJ gene sequence-based assay for species identification and phylogene grouping in the genus Staphylococcus. International Journal of Systematic and Evolutionary Microbiology, 57, 25–30.

Shih, I., Shieh, Y., Yu, C., & Hsieh, C. (2005). Selective production and characterization of levan by Bacillus subtilis (Natto) Takahashi. Journal of Agriculture Food Chemistry, 53, 8211–8215.

Simões, C., Amoros, M., & Girre, L. (1999). Mechanism of antiviral activity of triterpenoid saponins. Phytotherapy Research, 21, 317–325.

Stivala, S., and Bahary, W., (1978): Some dilute-solution parameters of the levan of *Streptococcus salivarius* in various solvents. Carbohydrate Research. 67: 17–21

Takatsy, G. (1955). The use of spiral loops in serological and virological micromethods. Acta Microbiologica Academic Science Hungarica, 3, 191.

Urdaci, M. C., Bressollier, P., & Pinchuk, I. (2004). *Bacillus clausii* antimicrobial and immunomodulatory activities. Journal of Clinical Gastroenterology, 38, 586–590.

Walum, E., Strenberg, K., & Jenssen, D. (1990). Understanding cell toxicology: Principlesand practice. NewYork: Ellis Howood., pp. 97–111.

Yanase, H., Iwata, M., Nakahigashi, R., Kita, K., Kato, N., & Tonomura, K. (1992). Purification, crystallization, and properties of extracellular levansucrase from *Zymomonas mobilis*. Bioscience, Biotechnology and Biochemistry, 56, 1335–1337.

Yoo, S. H., Yoon, E. J., Cha, J., & Lee, H. G. (2004). Antitumor activity of levan 536 polysaccharides from selected microorganisms. International Journal of Biological Macromolecules, 34, 37–41.

Yoon, E. J., Yoo, S. H., Cha, J., & Lee, H. G. (2004). Effect of levan's branching structure on antitumor activity. International Journal of Biological Macromolecules, 34, 191–194.

Section 2

Physical and Chemical Characteristics of Polysaccharides

The Molecular Structure and Conformational Dynamics of Chitosan Polymers: An Integrated Perspective from Experiments and Computational Simulations

Richard A. Cunha, Thereza A. Soares, Victor H. Rusu,
Frederico J.S. Pontes, Eduardo F. Franca and Roberto D. Lins

Additional information is available at the end of the chapter

1. Introduction

Chitosan is a semi-synthetic linear copolymer composed of a variable number of β-(1-4) linked units of 2-acetamide-2-deoxy-β-d-glucopyranose (GlcNAc) and 2-amino-2-deoxy-β-d-glycopyranose (GlcN) [1]. The two monomers differ with respect to the C2-substituent in the sugar ring, which is either an amino or acetamide group (Figure 1). Chitosan is obtained via the alkaline deacetylation of chitin. However, the deacylation reaction hardly proceeds completely in a normal heterogeneous reaction, leading to a random distribution of GlcNAc and GlcN residues in the chitosan polymer [2, 3]. The degree of acetylation of a polymer is a measure of the average number of GlcNAc per 100 chitosan monomers in percentile unit. The degree of acetylation governs important physical-chemical properties of the chitosan polymer such as solubility and conformation, being critical for the effectiveness of various technological applications [4-6]. The threshold for the conversion of chitin into chitosan depends on the solubility of the oligosaccharide in a slightly acid solution (0,1 mol/L of acetic acid). Conventionally, chitin polymers with a level of acetylation below 50% is considered as chitosan. The level of protonated amino groups in the glucosamine monomers dictates the solubility of chitosan, conferring the cationic nature to the polymer [7].

Chitin is the most abundant amino polysaccharide, being produced in the amount of one hundred billion tons per year in nature [8]. Its main source is the exoskeleton of crabs and shrimps, whose availability in nature makes chitin a renewable source of chitosan. In the last decades chitosan has emerged as a biomaterial with unique properties for advanced applications in green chemistry, biomedical, pharmaceutical, food [9] and agriculture [10].

The variety of applications of chitosan is determined by its chemical structure, which varies with respect to size (average molecular weight; MW), degree of acetylation (DA) and numerous chemical modifications [6, 11-15]. An increasing number of chitosan chemical modifications have been described in the literature [14, 15]. Chitosan is also a highly absorptive material used as heavy-metal chelators in water [16, 17]. The chelation of metals occurs via electrostatic interactions with chitosan reactive groups (hydroxyl, acetamide and primary amino groups). These functional groups are also responsible for properties such as high hydrophilicity, reactivity, and structural flexibility that make chitosan soluble in near-neutral acid solution [17]. The soluble adsorptive properties of chitosan enable its use for removal of pesticide and dyes from water, for adsorption of proteins, as flocculant agent, and even as a catalyst support for biodiesel production [18-24]. Besides being an abundant renewable resource, chitosan stands out due to some unique properties such as exceptional biocompatibility, biodegradability, non-toxicity and the easiness of production of chemically modified forms [14, 25, 26]. Furthermore, chitosan exhibits antimicrobial and antifungal activity [27-30]. These singular properties makes chitosan welll-suited for a wide range of biomedical applications such as drug delivery, platform for neural stem cell growth, tissue engineering (bone, cartilage, nerve, skin), immunoprophylaxis, gene therapy, wound healing and treatment of infections [31-44].

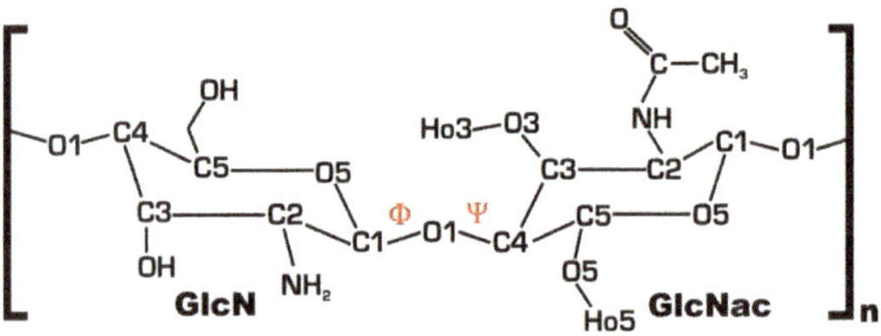

Figure 1. Schematic representation of a disaccharide formed by 2-amino-2-deoxy-β-d-glycopyranose (GlcN) bound to 2-acetamide-2-deoxy-β-d-glucopyranose (GlcNac). Φ and Ψ dihedral angles are shown in the scheme in red and here are represented by atoms: O5-C1-O1-C4′ and C1-O1-C4′-C5′, respectively. GlcN and GlcNac are the two units forming chitosan. Solubility is typically achieved when GlcN > 50%.

This short review will cover the structural dynamics of chitosan from a microscopic perspective, focusing on the interplay between its conformational variability and macroscopic properties such as solubility and aggregation. It is not the goal of the review to provide a detailed summary of the extensive literature on carbohydrate structure and its characterization. Excellent reviews on the subject can be found in the literature [45-47]. The text is organized in four main sections. First, we present an overview of chemical interactions between chitosan and biological materials, in particular lipid bilayers. In what follows, we describe the advantages and limitations of experimental and computational techniques used for the structural characterization of oligosaccharides, emphasizing the necessity of combining different approaches in order to obtain high-resolution structural data on chitosan. In this section we also introduce the theoretical principles underlying molecular dynamics (MD) simulations, which has been widely used to study the structural dynamics of carbohydrates in solution. The third section, we review the types of secondary structure observed for chitin and chitosan in the crystalline state. In the final section, we offer a detailed account of the structure and conformational dynamics of chitosan in solution as unveiled by computational simulations carried out in our group.

2. The molecular interactions underlying chitosan bioactivity

Chitosan is a very promising material with wide range of biomedical applications. This oligosaccharide incorporates highly sought properties for biomedical applications (biocompatibility, biodegradability and bioresorbability) with the easy processing into gels, membranes, nanofibers, beads, microparticles, nanoparticles, scaffolds and sponges forms [48-52]. Yet, chitosan has a flexible, hydrophilic helical structure with reactive amine groups, which offers a multitude of possible inter- and intra-molecular interactions. A detailed understanding of the effects of different materials and environmental conditions on such interactions can enable the design of novel chitosan-based technologies.

Chitosan amino groups are the major players in metal chelating processes. However, it has been previously shown that electrostatic interactions involving the protonated amino groups *per se* in low pH are not sufficient to explain the biological behavior of chitosan in presence of biological membranes [53]. A comparative study between chitosan and a fully cationic polymer has shown that hydrophobic interactions play an important role in the polysaccharide action. A study on the effect of the pH and the molecular weight of chitosan in multilamellar vesicles of dipalmitoylphosphatidylcholine (DPPC) has shown that increasing the biopolymer molecular weight (213 kDa) and decreasing of pH can lead to disruption of the membrane [54]. In contrast, another study has shown that the interactions between chitosan and DPPC lipids in liposomes led to an increase in thermodynamic stability of the composite. It has been proposed that this stabilization results from a shielding mechanism based on the electrostatic interactions between the chitosan chains and phospholipids polar heads [55]. In contrast, anti-fungal and bactericidal activity of chitosan has been attributed to the ability of chitosan to disrupt the inner and outer membranes of cells [56, 57]. The contribution of electrostatic, hydrophobic and hydrogen bond interactions between chitosan and three different lipids have been evaluated using Langmuir films to mimic the interaction of chitosan and bacterial membranes [29]. It has been shown that chitosan had a negligible effect on DPPC monolayers but it distinctly affected dipalmitoylphosphatidylglycerol (DPPG) and cholesterol monolayers [29]. The effect on DPPG was found to decrease with increasing pH, ascribed to the charge-mediating action of chitosan, whereas the pH did not affect the cholesterol monolayers where interactions occurs mainly via hydrogen bonding. A recent study has suggested that sensitivity of fungi to chitosan depends of the membrane fluidity and dynamics [58]. The same group has also suggested in a previous work that chitosan kills fungal cells by an unknown mechanism that does not involve endocytosis [59]. Although chitosan has been used in a variety of biologically relevant applications involving interactions with lipids, proteins, inorganic and organic compounds, a microscopic picture of these interactions remains lacking. The conformational flexibility of chitosan has hampered the acquisition of high-resolution structural data through of X-ray crystallography and NMR spectroscopy [46, 47]. However, current molecular modeling techniques can be used to bridge the gap of experimental resolution, thus providing complementary information to measurements.

3. Structural characterization of polysacchrides

Several analytical techniques have been applied to the characterization of a variety of oligosaccharide properties [60-63]. Among them, the molecular geometry is one of the most important properties that experimental data can provide on carbohydrates. Its characterization is critical for the understanding of the function and recognition mechanisms of carbohydrates in living organisms. However, sugars are inherently flexible, undergoing conformational changes in response to chemical modifications, complexation to biomolecules, changes in the pH, ionic strength and solvent type [2]. In solution, oligosaccharides tend to adopt a coiled conformation, which fluctuates between local and overall conformations, adopting an enormous variety of spatial arrangements around glycosidic linkages.

As a first approach to the complexity of the conformational flexibility of polysaccharides, let us assume that its monosaccharide units have a rigid ring structure. Thus, the determination of the conformation of oligosaccharide structures is reduced to the characterization of the glycosidic linkages between rigid monosaccharide monomers, *i.e.*, the description of two or three torsion angles for each glycosidic linkage would suffice to characterize the conformations of the oligosaccharide chain. However, the description of these torsion angles faces two major issues [46, 64]. First, the motions associated to each glycosidic linkage range across large-scale vibrations of a single well-defined conformation to transitions between several different conformations. Therefore, the accurate characterization of a given glycosidic linkage requires information on the number of conformations adopted, the time spent in each conformation and the flexibility of each conformation [46]. An additional difficulty is given by the fact that the conformational transitions in different linkages of an oligosaccharide chain are coupled. Second, the two experimental techniques most effective in providing atomic level structural information on biomolecules, namely X-ray diffraction and nuclear magnetic resonance (NMR) spectroscopy, have appreciable limitations when applied to oligosaccharides. In this section, we briefly overview the strengths and limitations of the most representative experimental techniques used for structural characterization of chitin and chitosan.

Mass spectrometry (MS) can be used to determine the total mass of a carbohydrate or differentiate distinct oligosaccharide as function of the respective weight [65]. Although MS cannot offer a detailed description of the oligosaccharide structure, it can identify the location of branch points [66-69]. Further fragmentation will not result in additional information, because the fragments can be alike. Despite the inadequacy of MS to generate information on the molecular geometry of oligosaccharides, MS can be coupled with separation techniques such as high performance liquid chromatography (HPLC) to differentiate between solutions containing different types of carbohydrates [65, 70-73].

The techniques of X-ray diffraction and NMR spectroscopy determine time and spatial averages of molecular properties in atomic coordinates measured from an ensemble of molecules corresponding the Avogadro's number [74-76]. Yet, the two techniques differ significantly with respect to the spatial distribution of molecules and time scale accessible to each one [74-76]. X-ray data represents an average over molecules arranged in a periodic crystal lattice over the second to hour timescale whereas NMR data represents an average over semi-randomly oriented molecules in solution over the nanosecond to second timescale. Despite the robustness of single X-ray crystallography in protein structure determination, the technique is not easily applicable to oligosaccharides. The difficulty to obtain highly crystalline samples for oligosaccharides imposes limits on the quality of the diffraction pattern. X-ray diffraction of carbohydrate structures larger than tetramers are rare and only seen when co-crystallized with proteins [18]. Due to the difficulty to obtain single crystals from oligosaccharides, oriented fibers have often been used for X-ray diffraction studies. Fibers exhibit helical symmetry rather than the three-dimensional symmetry seen in single crystals. Analysis of the diffraction pattern from orientated fibers allows deducing the helical symmetry of the molecule, in some cases also the structure. This is possible by constructing a model of the fiber and calculating the expected diffraction pattern. By comparing the calculated and observed diffraction patterns one eventually arrives at a better model. However, oligosaccharides in crystalline fibers can be affected by intra-molecular and crystal lattice packing, which may lock the structure in a conformation not representative of the conformational ensemble in solution.

On the other hand, NMR spectroscopy can provide information on the covalent structure and the complex conformational equilibria of oligosaccharides in solution [46, 47, 64, 77-79]. Moreover, it is the only available technique that can determine both the anomericities and linkages of a novel glycan. Another practical advantage of NMR spectroscopy is the possibility of measuring relative dilute solutions of oligosaccharides. Sample requirement amounts to as little as 1 mg, which is within the limits of enzyme-assisted synthesis [47]. NMR spectroscopy is probably the most used experimental tool to characterize the atomic structure of carbohydrates. For this reason, it has been the subject of numerous reviews [46, 47, 64, 78]. Biomolecular NMR spectroscopy has progressed appreciably in the last decades. Developments in the instrumentation, pulse sequences and spectral interpretation associated to molecular modeling techniques led to great advances in the determination of primary and three-dimensional structures of biomolecules in solution [17]. Such progress has been more noticeable in the structural characterization of proteins and nucleic acids. Notwithstanding, carbohydrates are not too far behind despite difficulties in proton assignment of each atom due to the structural similarity of the monosaccharide units [19-29].

The NMR spectroscopy data reflect primarily short-range through-bond interactions (J-coupling constants), short-range through-space interactions via the nuclear Overhauser effect (NOEs) or local perturbations to electronic shielding (chemical shifts). NOE is the main source of conformational information on carbohydrates. The strength of the NOE signal between two nuclei is proportional to the inverse sixth power of the distance between the atoms. However, a given distance between two protons is often consistent with a range of distinct conformations that will be represented by a set of NOE-derived distance constraints. The larger the number of available NOE constraints, the more consistent a single structure will be with this collection of spatial constraints. Nonetheless, the use of NOE constraints in structure determination of oligosaccharides is beset by a few issues (reviewed in [46]). For instance, the number of NOE constraints may not suffice to unambiguously define a conformation, particularly around the glycosidic linkage. In addition, NOE is sensitive only to short-distance nuclei (< 5-6 Å). For that reason, NOE constraints are obtained only between nuclei within a monosaccharide unit or across a glycosidic linkage. Due to the lack of long-range information, the accurate determination of the whole structure of oligosaccharides depends on combining the local conformations of the individual linkages. Such procedure leads to the cumulative addition of any uncertainties or errors in the local structure and its dissemination to the whole oligosaccharide structure (except if in the presence of sufficient sequential NOEs). A last issue concerns space-time ensemble averaging effects. Accordingly, different NMR parameters are averaged over time scales ranging from 50 ms to 1 s. In the case of oligosaccharides transitioning between several conformations, NOE constraints will represent average values that cannot be easily decomposed into each of the single conformation contributing to the average constraints. The conformational uncertainty ensuing from these issues can be minimized to some degree by the use of additional conformational constraints such as scalar coupling constants (J values), which are simple linear averages over the ensemble of individual conformers. ^{1}H-^{1}H J-coupling constants can be used to define ring conformations and dihedral angles. This information can also be obtained via NMR residual dipolar couplings measurements. Non-anomeric protons can be assigned through 2D homonuclear correlation COSY and/or TOCSY experiments. NOESY experiments can be used to provide monosaccharide sequence information due to the absence of coupling over the glycosidic linkage of the COSY and TOCSY spectra. ^{1}H-^{13}C HSQC and HMQC experiments provide important correlations in the determination of repeating units of polysaccharides [2]. Yet, the identification of distinct carbohydrate conformations requires combining complementary techniques. These techniques vary from other experimental methods to atomistic molecular dynamics simulations [46, 64, 80-82].

Classical molecular dynamics (MD) simulations can be used to complement incomplete experimental data and to provide detailed conformational distributions in time and space that experimental measurements can only obtain as averages [80, 82, 83]. It can also be used to predict properties under environmental conditions that may not be accessible to experimental measurements. As such, MD is an indispensable tool to interpret experimental data. However, the accuracy of MD simulations is intrinsically dependent on the quality of the empirical potential energy functions and the force-field parameters used. Robust force fields for MD simulations of carbohydrate-based systems are available. Some of the most used are CHARMM[84-87], GLYCAM/AMBER[88, 89], GROMOS[90] and OPLS-AA[91]. These force fields offer a realistic description of the structural dynamics of oligosaccharides within the limitations of the experimental data available, making MD simulations a reliable procedure for the prediction of molecular interactions [92-95]. Therefore, the importance of accurate measurements of the spatial arrangements of carbohydrates can never be overstated as they are the principal component in the development of physical chemical parameters (force fields) governing molecular simulations. The availability of high-quality experimental data is critical for biomolecular modeling [80, 81]. Classical force fields used for simulations of biomolecules are built from quantum chemistry calculations and/or experimental measurements. Without experimental measurements, the development of classical force fields would be extremely difficult as the expensive costs of quantum-chemical theoretical models limit their use in force-field construction [80, 90, 96-100]. In addition, quantum-mechanical data is not an ideal validation target as it only yields gas-phase quantities. Model validation and comparisons of biomolecular simulations are often best done against condensed-phase experimental data [82, 98, 101]. The availability of structural data on carbohydrates has made possible the creation of several databases like the SUGABASE, CarbBank, EUROCarbDB, Glycoconjugate DB, GLYCOSCIENCES.de, GlycoSuiteDB, JCGGDB, KEGG-Glycan, CFG-Glycan Database and GlycoBase. These databases represent a convenient tool for the building of molecular models as well as for comparison of atomistic simulations of carbohydrates against experimental data.

3.1. Theorical foundations of molecular dynamics simulations

The MD method has its foundations in the laws of classical mechanics [102, 103]. It allows the simulation of the time-dependent behavior of molecular systems according to Newton's laws of motion. The atom nuclei are treated classically as spheres connected to each other through a set of springs emulating chemical bonds. The forces acting on each atom, necessary to simulate their motion, are derived from a set of force field parameters, and the set of coordinates and velocities that mapped during the whole process comprise the phase space. In a simulation, the force *F* on each atom is expressed as a function of time, and is equal to the negative gradient of the potential energy *V* with respect to the position r_i of each atom, in a distinct expression of the more common form of the equation $F = ma$:

$$-\frac{dV}{dr_i} = m\frac{d^2r_i}{dt^2}$$

The MD method integrates iteratively and numerically the classical equations of motion for every atom in the system at time increments (Δt – *time step*) defined by the user. A number of algorithms exist for this purpose and are implemented in different computational codes [104]. There are several algorithms available for performing the numerical integration of the equations of motion. The Verlet-type algorithms (Verlet, velocity-Verlet and leap-frog) are widely used because it requires a minimum amount of computer memory and CPU time [105, 106]. The velocity Verlet, for instance, uses positions, velocities and accelerations at the current time step, which gives a more accurate integration than the original Verlet algorithm. Other algorithms, as the Beeman gives better energy conservation at the expense of computer memory and CPU time [107]. The Gear predictor-corrector algorithm predicts the next set of positions and accelerations, and then compares the accelerations to the predicted ones to compute a correction for the step [108]. Each step can thus be refined iteratively. Predictor-corrector algorithms give an accurate integration but are seldom used due to their large computational cost. In the classical Verlet algorithm, the positions in the next time step (*Δt*) are calculated from a given set of particles with coordinates r_i using a Taylor expansion:

$$r_{i+1} = r_i + \frac{\partial r}{\partial t}(\Delta t) + \frac{1}{2}\frac{\partial^2 r}{\partial t^2}(\Delta t)^2 + \frac{1}{6}\frac{\partial^3 r}{\partial t^3}(\Delta t)^3 + \ldots$$
$$r_{i+1} = r_i + v_i(\Delta t) + \frac{1}{2}a_i(\Delta t)^2 + \frac{1}{6}b_i(\Delta t)^3 + \ldots$$

where the last equation links the spatial coordinates with the velocities v_i (the first derivative of the positions in respect to time (dr_i/dt) at time t_i. The accelerations a_i (the second derivatives (d^2r/dt^2) at time t_i and so on. If the goal is to determine the positions for a small time step (Δt) earlier, the equation becomes:

$$r_{i-1} = r_i - v_i(\Delta t) + \frac{1}{2}a_i(\Delta t)^2 - \frac{1}{6}b_i(\Delta t)^3 + \ldots$$

Adding the last two equations makes it is possible to find a new equation that predicts the position at a chosen time step using the current and previous atom positions and current acceleration. The latter can be calculated from the force or potential.

$$r_{i+1} = (2r_i - r_{i-1}) + a_i(\Delta t)^2 + \ldots$$
$$a_i = \frac{F_i}{m_i} = -\frac{1}{m_i}\frac{dV}{dr_i}$$

At the beginning of the simulation, when the previous positions are not available, this quantity can be estimated from the following approximation:

$$r_{-1} = r_0 - v_0\Delta t$$

The time increment in MD simulation should be sufficiently small that errors in the integration equations keep small, preserving the conservation of the energy. Normally Δt is on the order of femtoseconds (10^{-15} s). This time order is one order of magnitude smaller than the fastest molecular process. Furthermore, because the forces F_i should be recalculated for every step, MD is a computation intensive task. Currently achieved timescale of MD simulations is on the order of multi-nanoseconds to a few microseconds. This time is shorter than many relevant phenomena, for this reason MD results should be analyzed under the point of view of the sample of the phase space close to the starting condition, in spite of this capacity of sampling different configurations. One strategy is to increase the time step value and to allow longer simulation times is to freeze the bond lengths related to hydrogen atoms. The fastest processes in molecules are stretching vibrations, especially involving hydrogen atoms. As these degrees of freedom have little influence on many properties, some algorithms were developed to keep frozen these chemical bonds as the SHAKE [109], RATTLE [110] and LINCS [111]. Alternative MD-based methodologies have also been recently developed aiming to partly overcome this limitation. The so-called enhanced sampling techniques artificially drive a given system according a set of pre-defined rules that result in a larger sampling of the configurational phase space within the same simulation time (e.g., simulated annealing, replica-exchange, parallel-tempering, local elevation search, metadynamics) [112, 113].

An accurate description of the aqueous medium that shapes the structure, dynamics and function of biological molecules is essential for the realistic reproduction of its kinetics and thermodynamics properties. It is known that simulations of a small arrangement of atoms do not reproduce satisfactorily the properties of bulk liquids due to surface effects suffered by a large fraction of the molecules. The obvious solution for this problem, which would be to increase the number of solvent molecules, can lead to issues in the evaluation of the force between the atoms. An alternative solution to treat explicit solvent molecules in MD simulations is the use of periodic boundary conditions [104]. In this approach the simulation box is replicated throughout the space to form an infinite lattice, where the number of molecules entering or leaving the simulation box is kept constant during the simulation and as a consequence, surface effects are canceled. There are currently numerous water models used in MD simulations. Some of the models currently implemented in major classical MD softwares are the SPC model[114] and the TIP3P, TIP4P, and TIP5P models [115, 116]. These models were parameterized assuming that a cut-off is applied to nonbonded interactions and treat water as a rigid molecule. Although bond stretching and bond-angle bending[117], or polarization effects and many-body interactions [118], have been introduced into water models, they involve a large increase of computational expense, which has limited their use as widely as the SPC or TIP models. The water models are usually parameterized at a single temperature (ca. 298 K) and therefore may not capture correctly the temperature dependence of properties such as the solvent density or diffusion coefficients [119].

The basic principle underlying the MD theory is that if one allows the system to evolve in time indefinitely, it will eventually pass through all possible states. Thus, MD simulations should cover time scales sufficiently long to generate enough representative conformations to satisfy this principle. In other words, the simulations must sample a sufficient amount of the phase space corresponding to the system in consideration. In that case, experimentally relevant information concerning structural, dynamic and thermodynamic properties can be calculated using a feasible amount of computational resources. The connection between theoretical results and experiments is made through the use of the Ergodic hypothesis. This fundamental axiom of statistical mechanics states that the average obtained by following a small number of particles over a long time is equivalent to averaging over a large number of particles for a short time. Exploring the limit of a sufficient large time scale, the Ergodic hypothesis implies that the time average over a single particle is equivalent to the average over a large number of particles at any given time. This theoretical justification in the scope of a MD simulation validates the calculation of thermodynamic averages for molecular systems if finite molecular dynamics trajectories are "long enough" in the ergodic sense.

$$\lim_{\tau\to\infty}\langle A(r,p)\rangle_{\tau}=\langle A(r,p)\rangle_{Z}$$

4. Chitosan molecular structure

In the solid state, chitosan is characterized by an ordered fibrillar structure with a high degree of crystallinity, and polymorphism [120, 121]. X-ray measurements of the chitosan polymer have shown an extended two-fold helix in a zigzag structure [122, 123]. The crystal packing is mainly formed by chitosan chains arranged in an antiparallel fashion (Figure 2A), and similar to the anhydrous form of the α-chitin structure. The structure of the α and β forms differ only in the arrangement of the piles of chains, which is alternately antiparallel in α-chitin and all parallel in β-chitin [92, 124]. The crystallographic structure of chitin and chitosan have also revealed that although both biopolymers exhibit a hydrated and anhydrous forms, chitin occurs exclusively in the conventional extended 2-fold helical conformation (Figure 2A) [123, 125-128]. The presence of free amino groups in the structure of chitosan gives rise to different types of helical conformations in acid (Figure 2) [128]. These structures can be classified in four main types: type I (anhydrous), type II (hydrated), type IIa (hydrated) and type III (anhydrous), which adopt a helical conformations in a two-fold helix, relaxed two-fold helix, a 4/1 helix and a five-fold helix, respectively (Figure 2) [128, 129]. The diversity of chitosan structural types depend on the experimental conditions (kind and concentration of acid, temperature and salt preparation) used for the conversion of chitin into chitosan [128]. The helical structure propensities can be determined according to the repeating unit and helical symmetry as observed in chitosan crystal structures [121, 123, 128, 130, 131]. A less common motif, classified as 3-fold, has also been identified (Figure 2B).

The type I salts are the anhydrous form of the unreacted chitosan crystal. The polysaccharide chains in these crystals have a 2/1 helical symmetry with a repeating pattern of 1.0 nm. This conformation is similar to that of chitin, and characterizes the two-fold helix (Figure 2A) [92, 132, 133]. Type II chitosan exhibits a hydrated crystal with a fiber repeat of about 4.08 nm long and an asymmetric repeating units consisting of tetrasaccharides. In this type, the helical conformation is composed of eight glucosamine residues with repeating units related by a 2/1 helical symmetry. This pattern suggests a two-fold helix even though the corresponding asymmetric unit is rather distinct from that of type I where the asymmetric unit has only one glucosamine residue. The main difference between the type I and type II conformations is that the latter is almost four times longer than chitosan, and originated the designation of relaxed two-fold helix (Figure 2E) [92, 134-136]. A type II salt variant, called Type IIa, has a similar fiber repeat (4.05 nm), but with an asymmetric unit consisting of a glucosamine dimer in a 4/1 helical symmetry. This right-handed helix comprised of four asymmetric subunits is classified as 4/1-helix conformation, being also called four-fold helix (Figure 2C) [121, 129]. The most recently discovered type III form has a chain repeat of 2.55 nm, a 5/3 helical symmetry, and an asymmetric unit of a single glucosamine residue. Type III helical conformation is classified as five-fold helix, and displays a less symmetric helicoidal conformation (Figure 2D) [129, 137].

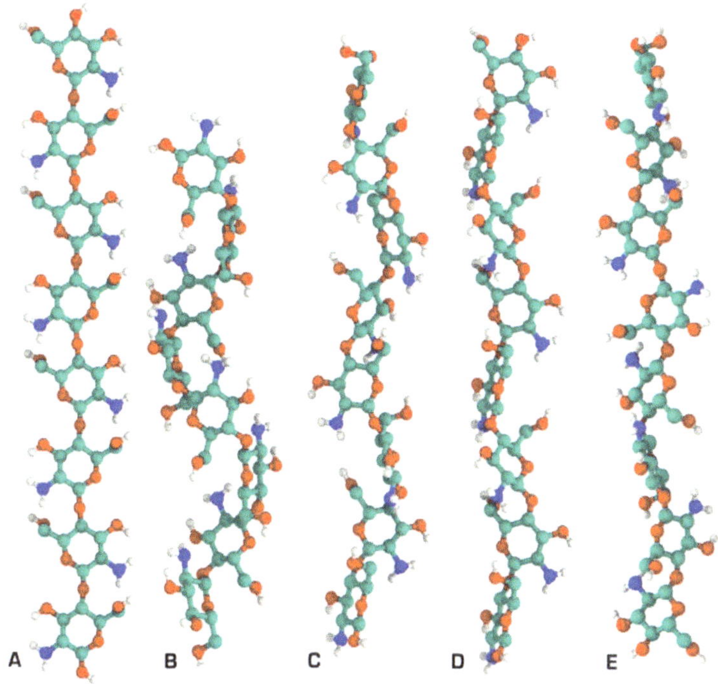

Figure 2. Chitosan secondary structures as determined by solid X-ray crystallography. A) two-fold; B) 3-fold; C) 4-fold; D) 5-fold; and, E) two-relaxed-fold.

In solid state, the two-fold helix pattern is stabilized by O3-HO3••• O5′ intra-chain hydrogen bonds across the glycosidic linkages [120]. In order to verify these helical properties in aqueous solution, MD simulations were carried out for chitin and chitosan [92, 93]. These simulations have shown that chitin chains assume exclusively a two-fold helix conformation which indeed is stabilized by the O3-HO3••• O5′ intra-chain hydrogen bonds [92]. However, chitosan chains can adopt several distinct conformations, including all the helical conformation observed in solid state. Helical preferences and conformational interchangeability were shown to be affected by the level of acetylation of the chitosan chains [92].

5. Structural dynamics of chitin and chitosan biopolymers

Structural characterization of chitin and chitosan conformations and their underlying interactions (intra- or inter-chain) has been largely determined by X-ray crystallography. The high flexibility of these oligosaccharides in solution has limited the acquisition of high-resolution structural data almost exclusively to X-ray diffraction of solid states (fiber, powder and tablet) (see section 3). Although NMR techniques are more suitable for structural characterization in solution, the flexibility of oligosaccharides makes NMR-derived geometrical constraints scant and limits the application of NMR spectroscopy to the determination of chitosan tridimensional structure [138]. Experimental data describing dynamic processes such as solvation, particle formation and aggregation remain limited to a macroscopic view, which is based on the measurement of chain stiffness and intrinsic viscosity [139]. Transmission electron microscopy has also been used as a complementary technique. Combining the latter with uranyl staining, electrostatic interactions involving chitosan protonated amino groups were attributed a major role on chitin and chitosan agglomeration in solution [140]. Therefore, the role of intra- and inter-chain hydrogen bonds, ionic strength and temperature on the structural dynamics of chitosan cannot be addressed exclusively by the means of experimental techniques [141]. Towards this end, MD simulations can be used to obtain information on the time-evolution of carbohydrate conformations at the atomic level and under varied environmental conditions that can be complementary to experimental measurements [92-95].

Chitosan conformational diversity influences its solubility/physical state (soluble, gel, aggregate), porosity, particle size and shape (fiber, nanoparticle, hollow fiber), ability to chelate metal ions and organic compounds, biodegradability and consequently its biological activity. The transition between these distinct conformational states is modulated by the percentage and distribution of acetyl groups. The level of chitosan acetylation and the distribution of N-acetyl groups along the chain have been shown to influence properties such as solubility [142, 143], biodegradability [144] and apparent pK_a values [145, 146]. Therefore, the percentage and distribution of acetyl groups are key parameters for determining if chitosan can effectively interact with biological systems [147]. The degree of acetylation can be experimentally determined by infra-red spectroscopy [148, 149], enzymatic reaction [150], ultra-violet spectroscopy [151], ^1H liquid-state NMR [152], and solid-state ^{13}C NMR [63, 153]. However, the interplay between chitosan acetylation and conformational transitions in solution cannot be characterized at high-resolution by experimental techniques. In these cases, atomistic MD simulation is a more suitable approach.

MD simulations in explicit solvent have been carried out for chitosan single chains and nanoparticle aggregates with varied percentage and distribution of acetyl groups [92, 93]. Four degrees of acetylation were considered: 0% (fully deacetylated chains), 40% (60% of the sites having a N-acyl group uniformily distributed), 40%-block (60% of the sites having a N-acyl group in two spatially located well-defined regions of the particle), 60% (40% of N-acyl uniformly distributed), 60%-block (40% of N-acyl groups spatially located in two well-defined regions of the particle), and 100% (fully N-acetylated nanoparticle), i.e., a chitin nanoparticle. Snapshots of molecular dynamics simulations after 40 ns for a chithin (100%) and fully deacetylated chitosan nanoparticles (0%) are shown in Figure 3. Both simulations started from aggregate crystal-like particles. It can be seen that chitin remain insoluble (in an aggregate form, Figure 3A), while chitosan chains separate apart one from another until each chain become fully hydrated (Figure 3B). Water molecules are not display in Figure 3 for clarity. These simulations have also shown a strong dependence of chitosan conformation and solubility with pH and degree of acetylation. An increase in the level of acetylation was shown to cause a progressive loss of flexibility and conformational interchangeability (Figure 4). Thus, acetylation promotes a shift from more flexible structural motifs such as 5-fold and relaxed 2-fold towards a 2-fold conformation (Figure 4). It was also shown that the spatial location of the N-acetyl groups influences significantly chitosan conformational preferences, and therefore its solubility (Figure 4). Analyses of the MD trajectories have also shown that a high degree of acetylation and/or an increase in pH leads to a 3-fold increase of the lifetime of O3-HO3•••O5′ intra-chain hydrogen bonds across the glycosidic linkages. The increase in the lifetime of this hydrogen bond was associated to a decrease in chitosan solubility. Chitosan with a high degree of acetylation favored the 2-fold conformation, but higher pH values did not affect significantly the secondary structure pattern of this oligosaccharide. In addition, we have also addressed the influence of spatial distribution of *N*-acetyl groups along the chitosan chain on swelling and the relative solubility of chitosan nanoparticles [93]. Simulations of chitosan with a uniform and block-wise distribution of *N*-acetyl groups along chains of a nanoparticle have shown that the latter displayed lower solubility [93]. The mechanism was attributed to the fact that 2-fold crystalline-like regions are created by the block distribution of acetyl groups, which is responsible to keep a more stable aggregate than its uniformly distributed counterpart.

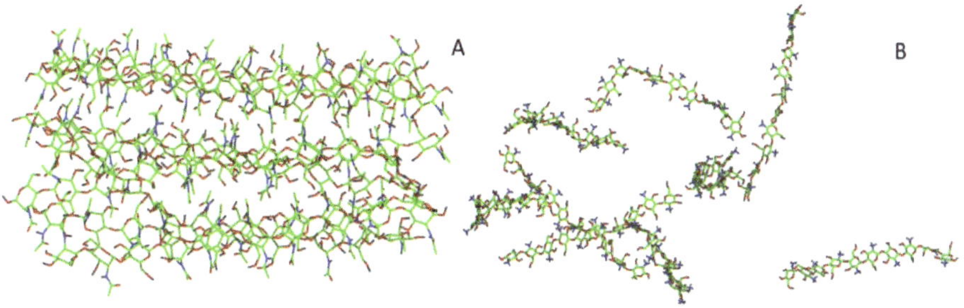

Figure 3. Molecular structure representation of nanoparticles of chitin (A) and chitosan in circumneutral pH (B) at the end of 40-ns molecular dynamics simulations in explicit water. Water molecules and hydrogen atoms have been removed for clarity. Structures are represented in stick model and atoms color coded as: green: carbon; red: oxygen, blue: nitrogen.

Analysis of the cumulative average water content around each chain in the nanoparticles illustrates the relative solubility of each system (Figure 5a). On average, there is 0.26 water molecule per monosaccharide a 0.5 nm radial distance from each chitin chain. That corresponds to one water molecule for roughly every four monosaccharides. The number of water molecules increases to one water molecule per monosaccharide within the same radial distance for fully *N*-deacetylated chitosan. As expected, nanoparticle swelling is directly proportional to its solubility. The relative swelling can be expressed as the average radius of gyration of each chain in a particle as a function of the degree of acetylation (Figure 5b). Chitosan particles with a degree of acetylation ≥ 60% did not display any significant swelling in water. At this level of acetylation, only a small increase in the relative solvation content of chitosan with a uniform distribution (ca. 0.13 water molecules per monosaccharide) than its counterpart with a block distribution was observed. This small difference in solvation did not affect the overall solubility of the particles, supporting the empirical observation of a solubility threshold around a level of 50% N-acetylation. Unexpectedly, water molecules within the insoluble chitosan particles were identified contributing for the maintenance of the regions in a 2-fold motif. The N-acetyl-glucosamine residues trapped water molecules between the chitosan chains, creating a hydrogen bond network between water molecules and the different chains without direct interaction between sheets. This finding substantiates a mechanism previously postulated by Ogawa and coworkers outlining the role of water molecules in chitin [123].

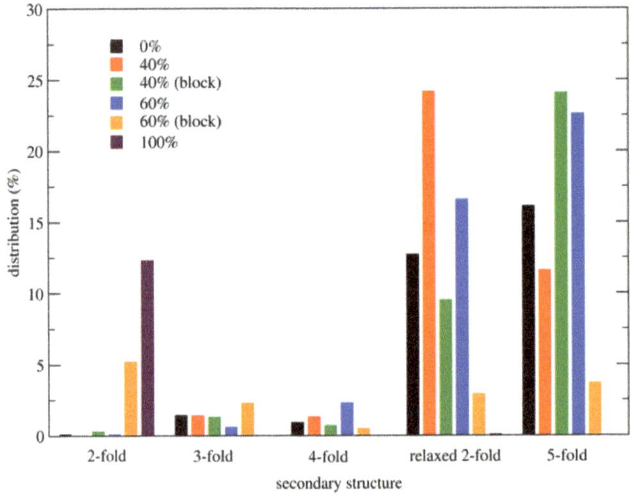

Figure 4. Secondary structure preferences for chitin and chitosan particles in water as a function of the degree of acetylation and spatial location of the N-acetyl groups. Results are the average of all chains in each nanoparticle, which are averaged over the last 5 ns of a 30-ns molecular dynamics simulation. The sum is each case does not reach 100% due to the increased flexibility and conformational diversity of these polymers in water. To be classified within a given secondary structure, as determined by solid X-ray crystallography, all dihedral angles in a chain should not present a deviation larger than 15% of the experimentally determined value. Data representation has been modified from [93].

Chitosan is a polyelectrolyte in acid medium. Its structure, physical state and conformational dynamics are greatly influenced by pH. The net charge of this cationic polyelectrolyte can be altered by its degree of acetylation [154]. Moreover, its apparent pK_a is directly related to the level of acetylation, varying from 6.1 to 7.32 units accordingly to proton concentration in the milieu [145, 155-158]. Based on these observations, it was proposed that aggregation occurs upon high levels of acetylation due to reduction of the biopolymer net charge [145], implying in a predictable behavior of chitosan chains depending on its electric charge distribution in aqueous solution [145, 146, 159]. It was also proposed that the low tendency of fully deacetylated chitosan to form aggregates is due to electrostatic repulsion among protonated amino groups. As result, chitosan electrostatic behavior was divided in three distinct patterns: i. DA < 20%, where it displays a polyelectrolyte behavior; ii. 20% < DA < 50%, where it is characterized by a counterbalance between hydrophilic and hydrophobic interactions;

and iii. DA > 50%, where associations of chitosan chains lead to the formation of stable aggregates. The results from atomistic molecular dynamics simulations in explicit water offer support to this hypothesis based on the accurate molecular description of the effect of the degree and distribution of *N*-acetyl groups on the swelling and aggregation stability of chitosan. Calculations of the hydrophobic and electrostatic contributions to the solvation free energy of the central chain in different particles as a function of acetylation are also consistent with the hypothesis (Figure 6) [129].

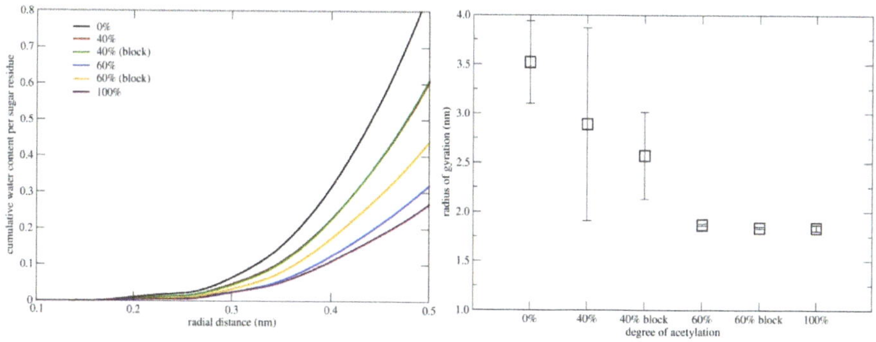

Figure 5. Relative solvation of chitin and chitosan nanoparticles as a function of the degree of acetylation and spatial location of the N-acetyl groups. A) Average coordination number of water molecules per residue as a function of distance (from 0.1 to 0.5 nm); B) Radius of gyration (and its variance, represented in bars) averaged per chain, over the last 5 ns of a 30-ns molecular dynamics simulation. Data representation has been modified from [93].

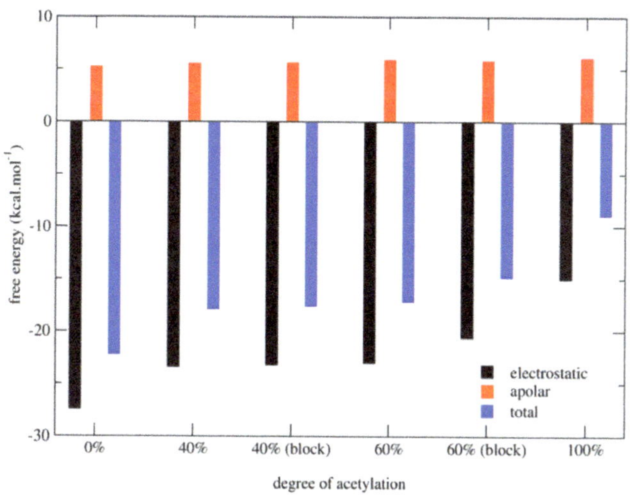

Figure 6. Average apolar and electrostatic contribution for the solvation free energy, per sugar residue, in chitin and chitosan nanoparticles as a function of the degree of acetylation and spatial location of the N-acetyl groups. Data representation has been modified from [92] and [93].

These contributions should be examined only as relative values as there are no experimental data for calibration or comparison of the calculated values. The apolar contribution remained nearly unaffected by the presence of water, while the electrostatic contribution is dominant even for insoluble chitin (100% acetylation). This finding suggests that hydrogen bond interactions, either intra-chains or between polymer chains and water molecules, play far a more important role in the solubility of chitin and chitosan than hydrophobic interactions. These results have further shown that fine tuning the electrostatic contributions in chitosan can be used to promote remodeling of its the physical state. Additional simulations have shown that the overall net charge and solubility of chitosan can be altered by changes in the pH. Comparison of the electrostatic response of a chitosan and chitin chains to pH changes shows a rather distinct surface charge profile for the two polymers. The electrostatic similarity between chitin and chitosan in basic pH aids to explain the loss of solubility of chitosan at high pH values (Figure 7). The positively charged character of chitosan chains in acid pH is shown by patches in blue (Figure 7D). On the other hand, chitin (Figure 7A) and chitosan chains in basic medium (Figure 7B) show a similar electrostatic potential at their molecular surfaces.

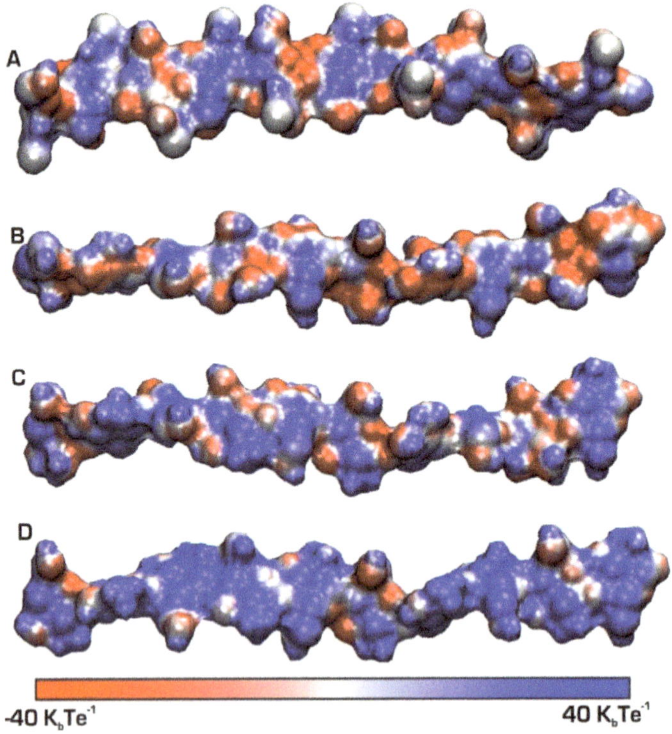

Figure 7. Electrostatic potential represented at the molecular surface of chitin (A) and chitosan (B-D) chains. Molecular structures were obtained from 20-ns molecular dynamics simulations. The different

chitosan chains were simulated at different pH values: B) basic (pH > 10); C circumneutral (pH = 6.5); and, D acid (pH < 3.5). Data representation has been modified from [92].

6. Final remarks and perspectives

Chitosan-based materials are involved in a plethora of medical, industrial and bioengineering applications such as bioremediation, radionuclide tissue decontamination and bone replacement to name a few. Due to the intrinsic flexibility and conformational variability of chitosan, the development of novel materials has been conducted mostly empirically. In this review, we have summarized the potential of using computer modeling to characterize in details the conformational behavior of chitin and chitosan. Understanding of the molecular properties of a given material allows for a more efficient/rational design. Therefore, this approach can be used to tailor these properties for specific needs. In this case, a systematic use of concerted experimental-theoretical information can provide a much clearer picture of the structural dynamics of polysaccharides and consequently can aid in such endeavor. This is still an emerging field that will benefit in the few years to come from the development of more accurate/extension of parameters for carbohydrate simulations, as well as novel models capable of better bridging the micro- and macroscopic scales.

Author details

Richard A. Cunha and Eduardo F. Franca*
Institute of Chemistry, Federal University of Uberlândia, Uberlândia, MG, Brazil

Thereza A. Soares, Victor H. Rusu, Frederico J.S. Pontes and Roberto D. Lins*
Department of Fundamental Chemistry, Federal University of Pernambuco, Recife, PE, Brazil

Acknowledgement

This work and the data herein contained was supported by the following sponsoring agencies: CNPq, FACEPE, FAPEMIG, INCT-INAMI (Brazilian National Science and Technology Institute for Integrated Markers), NanoBiotec-BR, CAPES, nBioNet and the Swedish Foundation for International Cooperation in Research and Higher Education (STINT). Part of the computational resources were provided by the Environmental Molecular Sciences Laboratory, an U.S. national scientific user facility sponsored by the U.S. Department of Energy located at the Pacific Northwest National Laboratory.

7. References

[1] Muzzarelli R A A (1973) Natural Chelating Polymers. Pergamon Press. 83.
[2] Toffey A, Samaranayake G, Frazier C E, Glasser W G (1996) Chitin derivatives. I. Kinetics of the heat-induced conversion of chitosan to chitin. Journal of Applied Polymer Science. 60: 75-85.

* Corresponding Authors

[3] Chang K L B, Tsai G, Lee J, Fu W-R (1997) Heterogeneous N-deacetylation of chitin in alkaline solution. Carbohydrate Research. 303: 327-332.
[4] Kasaai M R (2010) Determination of the degree of N-acetylation for chitin and chitosan by various NMR spectroscopy techniques: A review. Carbohydrate Polymers. 79: 801-810.
[5] Knaul J Z, Kasaai M R, Bui V T, Creber K A M (1998) Characterization of deacetylated chitosan and chitosan molecular weight review. Canadian Journal of Chemistry-Revue Canadienne De Chimie. 76: 1699-1706.
[6] Wang Q Z, Chen X G, Liu N, Wang S X, Liu C S, Meng X H, Liu C G (2006) Protonation constants of chitosan with different molecular weight and degree of deacetylation. Carbohydrate Polymers. 65: 194-201.
[7] Wang W, Du Y, Qiu Y, Wang X, Hu Y, Yang J, Cai J, Kennedy J F (2008) A new green technology for direct production of low molecular weight chitosan. Carbohydrate Polymers. 74: 127-132.
[8] Kurita K (2006) Chitin and Chitosan: Functional Biopolymers from Marine Crustaceans. Marine Biotechnology. 8: 203-226.
[9] Wei Z, Wang C, Zou S, Liu H, Tong Z (2012) Chitosan nanoparticles as particular emulsifier for preparation of novel pH-responsive Pickering emulsions and PLGA microcapsules. Polymer. 53: 1229-1235.
[10] Hernandez-Lauzardo A N, Velazquez-del Valle M G, Guerra-Sanchez M G (2011) Current status of action mode and effect of chitosan against phytopathogens fungi. African Journal of Microbiology Research. 5: 4243-4247.
[11] Abd-Elmohdy F A, El Sayed Z, Essam S, Hebeish A (2010) Controlling chitosan molecular weight via bio-chitosanolysis. Carbohydrate Polymers. 82: 539-542.
[12] Bagheri-Khoulenjani S, Taghizadeh S M, Mirzadeh H (2009) An investigation on the short-term biodegradability of chitosan with various molecular weights and degrees of deacetylation. Carbohydrate Polymers. 78: 773-778.
[13] Guo Z, Xing R, Liu S, Zhong Z, Ji X, Wang L, Li P (2008) The influence of molecular weight of quaternized chitosan on antifungal activity. Carbohydrate Polymers. 71: 694-697.
[14] Kean T, Thanou M (2010) Biodegradation, biodistribution and toxicity of chitosan. Advanced Drug Delivery Reviews. 62: 3-11.
[15] Vinsova J, Vavrikova E (2011) Chitosan Derivatives with Antimicrobial, Antitumour and Antioxidant Activities - a Review. Current Pharmaceutical Design. 17: 3596-3607.
[16] Muzzarelli R A A, Isolati A (1971) Methyl mercury acetate removal from waters by chromatography on chelating polymers. Water, Air, and Soil Pollution. 1: 65-71.
[17] Miretzky P, Cirelli A F (2009) Hg(II) removal from water by chitosan and chitosan derivatives: A review. Journal of Hazardous Materials. 167: 10-23.
[18] Chatterjee S, Lee M W, Woo S H (2010) Adsorption of congo red by chitosan hydrogel beads impregnated with carbon nanotubes. Bioresource Technology. 101: 1800-1806.

[19] Lu L C, Wang C I, Sye W F (2011) Applications of chitosan beads and porous crab shell powder for the removal of 17 organochlorine pesticides (OCPs) in water solution. Carbohydrate Polymers. 83: 1984-1989.

[20] Wan Ngah W S, Teong L C, Hanafiah M A K M (2011) Adsorption of dyes and heavy metal ions by chitosan composites: A review. Carbohydrate Polymers. 83: 1446-1456.

[21] Zeng D, Wu J, Kennedy J F (2008) Application of a chitosan flocculant to water treatment. Carbohydrate Polymers. 71: 135-139.

[22] Hoven V P, Tangpasuthadol V, Angkitpaiboon Y, Vallapa N, Kiatkamjornwong S (2007) Surface-charged chitosan: Preparation and protein adsorption. Carbohydrate Polymers. 68: 44-53.

[23] Krajewska B (2005) Membrane-based processes performed with use of chitin/chitosan materials. Separation and Purification Technology. 41: 305-312.

[24] da Silva R B, Lima Neto A F, Soares dos Santos L S, de Oliveira Lima J R, Chaves M H, dos Santos Jr J R, de Lima G M, de Moura E M, de Moura C V R (2008) Catalysts of Cu(II) and Co(II) ions adsorbed in chitosan used in transesterification of soy bean and babassu oils – A new route for biodiesel syntheses. Bioresource Technology. 99: 6793-6798.

[25] Muzzarelli R A A, Tanfani F (1982) N-(Ortho-Carboxybenzyl) Chitosan, N-Carboxymethyl Chitosan and Dithiocarbamate Chitosan - New Chelating Derivatives of Chitosan. Pure and Applied Chemistry. 54: 2141-2150.

[26] Pavinatto F J, Caseli L, Oliveira O N (2010) Chitosan in Nanostructured Thin Films. Biomacromolecules. 11: 1897-1908.

[27] Rabea E I, Badawy M E T, Stevens C V, Smagghe G, Steurbaut W (2003) Chitosan as antimicrobial agent: Applications and mode of action. Biomacromolecules. 4: 1457-1465.

[28] Ren J M, Liu J L, Li R C, Dong F, Guo Z Y (2012) Antifungal properties of chitosan salts in laboratory media. Journal of Applied Polymer Science. 124: 2501-2507.

[29] Krajewska B, Wydro P, Janczyk A (2011) Probing the Modes of Antibacterial Activity of Chitosan. Effects of pH and Molecular Weight on Chitosan Interactions with Membrane Lipids in Langmuir Films. Biomacromolecules. 12: 4144-4152.

[30] Kong M, Chen X G, Xing K, Park H J (2010) Antimicrobial properties of chitosan and mode of action: A state of the art review. International Journal of Food Microbiology. 144: 51-63.

[31] Shi W, Nie D, Jin G, Chen W, Xia L, Wu X, Su X, Xu X, Ni L, Zhang X, Chen J (2012) BDNF blended chitosan scaffolds for human umbilical cord MSC transplants in traumatic brain injury therapy. Biomaterials. 33: 3119-3126.

[32] Yannas I V, Burke J F, Orgill D P, Skrabut E M (1982) Wound Tissue Can Utilize a Polymeric Template to Synthesize a Functional Extension of Skin. Science. 215: 174-176.

[33] Muzzarelli R A A (2009) Chitins and chitosans for the repair of wounded skin, nerve, cartilage and bone. Carbohydrate Polymers. 76: 167-182.

[34] Leffler C C, Muller B W (2000) Chitosan-gelatin sponges for controlled drug delivery: the use of ionic and non-ionic plasticizers. Stp Pharma Sciences. 10: 105-111.

[35] Panos I, Acosta N, Heras A (2008) New drug delivery systems based on chitosan. Curr Drug Discov Technol. 5: 333-41.
[36] Nimesh S, Saxena A, Kumar A, Chandra R (2012) Improved transfection efficiency of chitosan-DNA complexes employing reverse transfection. Journal of Applied Polymer Science. 124: 1771-1777.
[37] Tikhonov V E, Stepnova E A, Babak V G, Yamskov I A, Palma-Guerrero J, Jansson H B, Lopez-Llorca L V, Salinas J, Gerasimenko D V, Avdienko I D, Varlamov V P (2006) Bactericidal and antifungal activities of a low molecular weight chitosan and its N-/2(3)-(dodec-2-enyl)succinoyl/-derivatives. Carbohydrate Polymers. 64: 66-72.
[38] Martinez-Camacho A P, Cortez-Rocha M O, Castillo-Ortega M M, Burgos-Hernandez A, Ezquerra-Brauer J M, Plascencia-Jatomea M (2011) Antimicrobial activity of chitosan nanofibers obtained by electrospinning. Polymer International. 60: 1663-1669.
[39] Liu H, Du Y M, Wang X H, Sun L P (2004) Chitosan kills bacteria through cell membrane damage. International Journal of Food Microbiology. 95: 147-155.
[40] Cho C H, Eliason J F, Matthew H W (2008) Application of porous glycosaminoglycan-based scaffolds for expansion of human cord blood stem cells in perfusion culture. Journal of Biomedical Materials Research Part A. 86: 98-107.
[41] Shi Z L, Neoh K G, Kang E T, Wang W (2006) Antibacterial and mechanical properties of bone cement impregnated with chitosan nanoparticles. Biomaterials. 27: 2440-2449.
[42] Suh J K F, Matthew H W T (2000) Application of chitosan-based polysaccharide biomaterials in cartilage tissue engineering: a review. Biomaterials. 21: 2589-2598.
[43] Cunha C, Panseri S, Antonini S (2011) Emerging nanotechnology approaches in tissue engineering for peripheral nerve regeneration. Nanomedicine-Nanotechnology Biology and Medicine. 7: 50-59.
[44] Tchemtchoua V T, Atanasova G, Aqil A, Filee P, Garbacki N, Vanhooteghem O, Deroanne C, Noel A, Jerome C, Nusgens B, Poumay Y, Colige A (2011) Development of a Chitosan Nanofibrillar Scaffold for Skin Repair and Regeneration. Biomacromolecules. 12: 3194-3204.
[45] Woods R J (1998) Computational carbohydrate chemistry: what theoretical methods can tell us. Glycoconjugate Journal. 15: 209-216.
[46] Wormald M R, Petrescu A J, Pao Y L, Glithero A, Elliott T, Dwek R A (2002) Conformational studies of oligosaccharides and glycopeptides: complementarity of NMR, X-ray crystallography, and molecular modelling. Chemical Reviews. 102: 371-86.
[47] Duus J, Gotfredsen C H, Bock K (2000) Carbohydrate structural determination by NMR spectroscopy: modern methods and limitations. Chemical Reviews. 100: 4589-614.
[48] Jayakumar R, Prabaharan M, Nair S V, Tamura H (2010) Novel chitin and chitosan nanofibers in biomedical applications. Biotechnol Advances. 28: 142-50.
[49] Jayakumar R, Menon D, Manzoor K, Nair S V, Tamura H (2010) Biomedical applications of chitin and chitosan based nanomaterials—A short review. Carbohydrate Polymers. 82: 227-232.
[50] Muzzarelli R A A, Muzzarelli C (2005) Chitosan chemistry: Relevance to the biomedical sciences. Polysaccharides 1: Structure, Characterization and Use. 186: 151-209.

[51] Cardenas G, Anaya P, von Plessing C, Rojas C, Sepulveda J (2008) Chitosan composite films. Biomedical applications. Journal of Materials Science-Materials in Medicine. 19: 2397-2405.

[52] Araujo J V, Lopes-Da-Silva J A, Almeida M M, Costa M E V (2006) Preparation and characterization of chitosan-based scaffolds for biomedical applications. Advanced Materials Forum Iii, Pts 1 and 2. 514-516: 1005-1009.

[53] Pavinatto A, Pavinatto F J, Barros-Timmons A, Oliveira O N (2010) Electrostatic Interactions Are Not Sufficient to Account for Chitosan Bioactivity. Acs Applied Materials & Interfaces. 2: 246-251.

[54] Fang N, Chan V, Mao H-Q, Leong K W (2001) Interactions of Phospholipid Bilayer with Chitosan: Effect of Molecular Weight and pH. Biomacromolecules. 2: 1161-1168.

[55] Mertins O, Lionzo M I Z, Micheletto Y M S, Pohlmann A R, da Silveira N P (2009) Chitosan effect on the mesophase behavior of phosphatidylcholine supramolecular systems. Materials Science & Engineering C-Biomimetic and Supramolecular Systems. 29: 463-469.

[56] Je J Y, Kim S K (2006) Chitosan derivatives killed bacteria by disrupting the outer and inner membrane. Journal of Agricultural and Food Chemistry. 54: 6629-6633.

[57] Helander I M, Nurmiaho-Lassila E L, Ahvenainen R, Rhoades J, Roller S (2001) Chitosan disrupts the barrier properties of the outer membrane of Gram-negative bacteria. International Journal of Food Microbiology. 71: 235-244.

[58] Palma-Guerrero J, Lopez-Jimenez J A, Perez-Berna A J, Huang I C, Jansson H B, Salinas J, Villalain J, Read N D, Lopez-Llorca L V (2010) Membrane fluidity determines sensitivity of filamentous fungi to chitosan. Molecular Microbiology. 75: 1021-1032.

[59] Palma-Guerrero J, Huang I C, Jansson H B, Salinas J, Lopez-Llorca L V, Read N D (2009) Chitosan permeabilizes the plasma membrane and kills cells of Neurospora crassa in an energy dependent manner. Fungal Genetics and Biology. 46: 585-594.

[60] Van de Velde K, Kiekens P (2004) Structure analysis and degree of substitution of chitin, chitosan and dibutyrylchitin by FT-IR spectroscopy and solid state C-13 NMR. Carbohydrate Polymers. 58: 409-416.

[61] Cardenas G, Cabrera G, Taboada E, Miranda S P (2004) Chitin characterization by SEM, FTIR, XRD, and (13)C cross polarization/mass angle spinning NMR. Journal of Applied Polymer Science. 93: 1876-1885.

[62] Jeffrey G A (1990) Crystallographic Studies of Carbohydrates. Acta Crystallographica Section B-Structural Science. 46: 89-103.

[63] Heux L, Brugnerotto J, Desbrieres J, Versali M F, Rinaudo M (2000) Solid state NMR for determination of degree of acetylation of chitin and chitosan. Biomacromolecules. 1: 746-751.

[64] Bubb W A (2003) NMR spectroscopy in the study of carbohydrates: Characterizing the structural complexity. Concepts in Magnetic Resonance Part A. 19A: 1-19.

[65] Zaia J (2004) Mass spectrometry of oligosaccharides. Mass Spectrom Rev. 23: 161-227.

[66] van der Hoeven R A M, Hofte A J P, Tjaden U R, van der Greef J, Torto N, Gorton L, Marko-Varga G, Bruggink C (1998) Sensitivity improvement in the analysis of

oligosaccharides by on-line high-performance anion-exchange chromatography/ion spray mass spectrometry. Rapid Communications in Mass Spectrometry. 12: 69–74.
[67] Ahn Y H, Yoo J S (1999) Effect of the labeling group in structural analyses of malononitrile-labeled oligosaccharides by matrix-assisted laser desorption/ionization time-of-flight mass spectrometry and electrospray ionization mass spectrometry. Rapid Commun Mass Spectrom. 13: 1985-90.
[68] Sato Y, Suzuki M, Nirasawa T, Suzuki A, Endo T (2000) Microsequencing of glycans using 2-aminobenzamide and MALDI-TOF mass spectrometry: occurrence of unique linkage-dependent fragmentation. Anal Chem. 72: 1207-16.
[69] Yamagaki T (1999) A new analytical method for oligosaccharides and glycoconjugates using post-source decay fragmentation by matrix-assisted laser desorption/ionization time-of-flight mass spectrometry. Trends Glycoscience Glycotechnology. 11: 227-232.
[70] Mechref Y, Novotny M V, Krishnan C (2003) Structural characterization of oligosaccharides using MALDI-TOF/TOF tandem mass spectrometry. Anal Chem. 75: 4895-903.
[71] Li M, Kinzer J A (2003) Structural analysis of oligosaccharides by a combination of electrospray mass spectrometry and bromine isotope tagging of reducing-end sugars with 2-amino-5-bromopyridine. Rapid Communications in Mass Spectrometry. 17: 1462-6.
[72] Cheng H L, Her G R (2002) Determination of linkages of linear and branched oligosaccharides using closed-ring chromophore labeling and negative ion trap mass spectrometry. J Am Soc Mass Spectrom. 13: 1322-30.
[73] Sturiale L, Naggi A, Torri G (2001) MALDI mass spectrometry as a tool for characterizing glycosaminoglycan oligosaccharides and their interaction with proteins. Seminars in Thrombosis and Hemostasis. 27: 465-72.
[74] Brunger A T (1997) X-ray crystallography and NMR reveal complementary views of structure and dynamics. Nature Structural Biology. 4 Suppl: 862-5.
[75] Wagner G, Hyberts S G, Havel T F (1992) NMR structure determination in solution: a critique and comparison with X-ray crystallography. Annu Rev Biophys Biomol Struct. 21: 167-98.
[76] Billeter M (1992) Comparison of protein structures determined by NMR in solution and by X-ray diffraction in single crystals. Quaterly Reviews in Biophysics. 25: 325-77.
[77] Dwek R A, Edge C J, Harvey D J, Wormald M R, Parekh R B (1993) Analysis of glycoprotein-associated oligosaccharides. Annu Rev Biochem. 62: 65-100.
[78] Bush C A, Martin-Pastor M, Imberty A (1999) Structure and conformation of complex carbohydrates of glycoproteins, glycolipids, and bacterial polysaccharides. Annual Reviews in Biophysics and Biomolecular Structure. 28: 269-93.
[79] Wang Y, Hollingsworth R I (1996) An NMR spectroscopy and molecular mechanics study of the molecular basis for the supramolecular structure of lipopolysaccharides. Biochemistry. 35: 5647-5654.
[80] van Gunsteren W F, Bakowies D, Baron R, Chandrasekhar I, Christen M, Daura X, Gee P, Geerke D P, Glattli A, Hunenberger P H, Kastenholz M A, Ostenbrink C, Schenk M,

Trzesniak D, van der Vegt N F A (2006) Biomolecular modeling: Goals, problems, perspectives. Angewandte Chemie - International Edition. 45: 4064-4092.
[81] van Gunsteren W F, Bakowies D, Burgi R, Chandrasekhar I, Christen M, Daura X, Gee P, Glattli A, Hansson T, Oostenbrink C, Peter C, Pitera J, Schuler L, Soares T A, Yu H B (2001) Molecular dynamics simulation of biomolecular systems. Chimia. 55: 856-860.
[82] Gulin V I, Samoilova R I, Dikanov S A, Tsvetkov Y D (1991) 2-mm Band EPR Spectroscopy of Semiquinone Radicals in Rigid Media. Applied Magnetic Resonance. 2: 425-434.
[83] Adcock S A, McCammon J A (2006) Molecular dynamics: Survey of methods for simulating the activity of proteins. Chemical Reviews. 106: 1589-1615.
[84] Guvench O, Hatcher E, Venable R M, Pastor R W, MacKerell A D (2009) CHARMM Additive All-Atom Force Field for Glycosidic Linkages between Hexopyranoses. Journal of Chemical Theory and Computation. 5: 2353-2370.
[85] Guvench O, Mallajosyula S S, Raman E P, Hatcher E, Vanommeslaeghe K, Foster T J, Jamison F W, MacKerell A D (2011) CHARMM Additive All-Atom Force Field for Carbohydrate Derivatives and Its Utility in Polysaccharide and Carbohydrate-Protein Modeling. Journal of Chemical Theory and Computation. 7: 3162-3180.
[86] Hatcher E, Guvench O, MacKerell A D (2009) CHARMM Additive All-Atom Force Field for Aldopentofuranoses, Methyl-aldopentofuranosides, and Fructofuranose. Journal of Physical Chemistry B. 113: 12466-12476.
[87] Hatcher E R, Guvench O, MacKerell A D (2009) CHARMM Additive All-Atom Force Field for Acyclic Polyalcohols, Acyclic Carbohydrates, and Inositol. Journal of Chemical Theory and Computation. 5: 1315-1327.
[88] Yongye A B, Tessier M B, Kirschner K, Tschampel S M, Gonzalez-Outeirino J, Daniels C R, Foley B L, Woods R J (2008) CARB 43-GLYCAM06, a generalizable biomolecular force field: Carbohydrates, lipids, lipid bilayers and glycolipids. Abstracts of Papers of the American Chemical Society. 235.
[89] Kirschner K N, Yongye A B, Tschampel S M, Gonzalez-Outeirino J, Daniels C R, Foley B L, Woods R J (2008) GLYCAM06: A generalizable Biomolecular force field. Carbohydrates. Journal of Computational Chemistry. 29: 622-655.
[90] Lins R D, Hunenberger P H (2005) A new GROMOS force field for hexopyranose-based carbohydrates. Journal of Computational Chemistry. 26: 1400-1412.
[91] Kony D, Damm W, Stoll S, van Gunsteren W F (2002) An improved OPLS-AA force field for carbohydrates. Journal of Computational Chemistry. 23: 1416-1429.
[92] Franca E F, Lins R D, Freitas L C G, Straatsma T P (2009) Characterization of Chitin and Chitosan Molecular Structure in Aqueous Solution. Journal of Chemical Theory and Computation. 4: 2141-2149.
[93] Franca E F, Lins R D, Freitas L C G (2011) Chitosan Molecular Structure as a Function of N-Acetylation. Biopolymers. 95: 448-460.
[94] Skovstrup S, Hansen S G, Skrydstrup T, Schiott B (2010) Conformational Flexibility of Chitosan: A Molecular Modeling Study. Biomacromolecules. 11: 3196-3207.

[95] Prathab B, Aminabhavi T M (2007) Molecular modeling study on surface, thermal, mechanical and gas diffusion properties of chitosan. Journal of Polymer Science Part B-Polymer Physics. 45: 1260-1270.
[96] Brooks B R, Brooks C L, Mackerell A D, Nilsson L, Petrella R J, Roux B, Won Y, Archontis G, Bartels C, Boresch S, Caflisch A, Caves L, Cui Q, Dinner A R, Feig M, Fischer S, Gao J, Hodoscek M, Im W, Kuczera K, Lazaridis T, Ma J, Ovchinnikov V, Paci E, Pastor R W, Post C B, Pu J Z, Schaefer M, Tidor B, Venable R M, Woodcock H L, Wu X, Yang W, York D M,Karplus M (2009) CHARMM: The Biomolecular Simulation Program. Journal of Computational Chemistry. 30: 1545-1614.
[97] Oostenbrink C, Soares T A, van der Vegt N F A, van Gunsteren W F (2005) Validation of the 53A6 GROMOS force field. European Biophysics Journal with Biophysics Letters. 34: 273-284.
[98] Soares T A, Daura X, Oostenbrink C, Smith L J, van Gunsteren W F (2004) Validation of the GROMOS force-field parameter set 45A3 against nuclear magnetic resonance data of hen egg lysozyme. Journal of Biomolecular Nmr. 30: 407-422.
[99] Soares T A, Hunenberger P H, Kastenholz M A, Krautler V, Lenz T, Lins R D, Oostenbrink C, Van Gunsteren W F (2005) An improved nucleic acid parameter set for the GROMOS force field. Journal of Computational Chemistry. 26: 725-737.
[100] Wang J M, Wolf R M, Caldwell J W, Kollman P A, Case D A (2004) Development and testing of a general amber force field. Journal of Computational Chemistry. 25: 1157-1174.
[101] van Gunsteren W F, Mark A E (1998) Validation of molecular dynamics simulation. Journal of Chemical Physics. 108: 6109-6116.
[102] Brooks C L, Karplus M, Pettitt B M (1988) Proteins: A Theoretical Perspective of Dynamics, Structure, and Thermodynamics. New York: John Wiley & Sons. p.
[103] McCammon J A, Harvey D J (1988) *Dynamics of Proteins and Nucleic Acids*. In. Cambridge: Cambridge University Press. pp. 248.
[104] Frenkel D, Smit B (1996) Understanding Molecular Simulation: From Algorithms to Applications. Orlando, FL, USA: Academic Press Inc. p.
[105] Verlet L (1967) Computer Experiments on Classical Fluids .I. Thermodynamical Properties of Lennard-Jones Molecules. Physical Review. 159: 98-&.
[106] Hockney R W (1970) The potential calculation and some applications. Methods Comput Phys. 9: 136-211.
[107] Beeman D (1976) Some Multistep Methods for Use in Molecular-Dynamics Calculations. Journal of Computational Physics. 20: 130-139.
[108] Gear C W (1971) Numerical Initial Value Problems in Ordinary Differencial Equations. New York: Prentice-Hall Englewoodcliffs. p.
[109] Ryckaert J P, Ciccotti G, Berendsen H J C (1977) Numerical integration of the cartesian equations of motion of a system with constraints: Molecular dynamics of *n*-alkanes. Journal of Computational Physics. 23: 327–341.
[110] Andersen A M, Norby P (2000) Ab initio structure determination and Rietveld refinement of a high-temperature phase of zirconium hydrogen phosphate and a new

polymorph of zirconium pyrophosphate from in situ temperature-resolved powder diffraction data. Acta Crystallogr B. 56: 618-25.

[111] Hess B, Bekker H, Berendsen H J C, Fraaije J (1997) LINCS: A linear constraint solver for molecular simulations. Journal of Computational Chemistry. 18: 1463-1472.

[112] Liwo A, Czaplewski C, Oldziej S, Scheraga H A (2008) Computational techniques for efficient conformational sampling of proteins. Current Opinion in Structural Biology. 18: 134-9.

[113] Trzesniak D, Lins R D, van Gunsteren W F (2006) Protein under pressure: Molecular dynamics simulation of the arc repressor. Proteins-Structure Function and Bioinformatics. 65: 136-144.

[114] Berendsen H J C, Grigera J R, Straatsma T P (1987) The Missing Term in Effective Pair Potentials. Journal of Physical Chemistry. 91: 6269-6271.

[115] Jorgensen W L, Chandrasekhar J, Madura J D, Impey R W, Klein M L (1983) Comparison of simple potential functions for simulating liquid water. Journal of Chemica Physics. 79: 926-935.

[116] Mahoney M W, Jorgensen W L (2000) A five-site model liquid water and the reproduction of the density anomaly by rigid, non-polarizable models. Journal Chemical Physics. 112: 8910-8922.

[117] Cornell W D, Cieplak P, Bayly C I, Gould I R, Merz K M, Ferguson D M, Spellmeyer D C, Fox T, Caldwell J W, Kollman P A (1995) A 2nd Generation Force-Field for the Simulation of Proteins, Nucleic-Acids, and Organic-Molecules. Journal of the American Chemical Society. 117: 5179-5197.

[118] Sprik M, Klein M L (1988) A polarizable model for water using distributed charge sites. Journal of Chemical Physics. 89: 7556-7561.

[119] Horn H W, Swope W C, Pitera J W, Madura J D, Dick T J, Hura G L, Head-Gordon T (2004) Development of an improved four-site water model for biomolecular simulations: TIP4P-Ew. Journal of Chemical Physics. 120: 9665-9678.

[120] Rinaudo M (2006) Chitin and chitosan: Properties and applications. Progress in Polymer Science. 31: 603-632.

[121] Mogilevskaya E L, Akopova T A, Zelenetskii A N, Ozerin A N (2006) The crystal structure of chitin and chitosan. Polymer Science Series A. 48: 116-123.

[122] Ogawa K (1991) Effect of Heating an Aqueous Suspension of Chitosan on the Crystallinity and Polymorphs. Agricultural and Biological Chemistry. 55: 2375-2379.

[123] Okuyama K, Noguchi K, Miyazawa T, Yui T, Ogawa K (1997) Molecular and crystal structure of hydrated chitosan. Macromolecules. 30: 5849-5855.

[124] Sugiyama J, Boisset C, Hashimoto M, Watanabe T (1999) Molecular directionality of beta-chitin biosynthesis. Journal of Molecular Biology. 286: 247-255.

[125] Blackwel J (1969) Structure of beta-chitin or parallel chain systems of poly-beta-(1-4)-N-acetyl-d-glucosamine. Biopolymers. 7: 281-298.

[126] Gardner K H, Blackwell J (1975) Refinement of Structure of Beta-Chitin. Biopolymers. 14: 1581-1595.

[127] Minke R, Blackwell J (1978) Structure of Alpha-Chitin. Journal of Molecular Biology. 120: 167-181.
[128] Ogawa K, Yui T, Okuyama K (2004) Three D structures of chitosan. International Journal of Biological Macromolecules. 34: 1-8.
[129] Franca E F, Leite F L, Cunha R A, Oliveira O N, Freitas L C G (2011) Designing an enzyme-based nanobiosensor using molecular modeling techniques. Physical Chemistry Chemical Physics. 13: 8894-8899.
[130] Ogawa K (1988) Chain Conformation of Chitin and Chitosan. Nippon Nogeikagaku Kaishi-Journal of the Japan Society for Bioscience Biotechnology and Agrochemistry. 62: 1225-1228.
[131] Ogawa K, Inukai S (1987) X-Ray-Diffraction Study of Sulfuric, Nitric, and Halogen Acid Salts of Chitosan. Carbohydrate Research. 160: 425-433.
[132] Yui T, Imada K, Okuyama K, Obata Y, Suzuki K, Ogawa K (1994) Molecular and Crystal-Structure of the Anhydrous Form of Chitosan. Macromolecules. 27: 7601-7605.
[133] Lertworasirikul A, Yokoyama S, Noguchi K, Ogawa K, Okuyama K (2004) Molecular and crystal structures of chitosan/HI type I salt determined by X-ray fiber diffraction. Carbohydrate Research. 339: 825-833.
[134] Okuyama K, Noguchi K, Kanenari M, Egawa T, Osawa K, Ogawa K (2000) Structural diversity of chitosan and its complexes. Carbohydrate Polymers. 41: 237-247.
[135] Okuyama K, Osawa K, Hanafusa Y, Noguchi K, Ogawa K (2000) Relaxed 2/1-helical conformation of type II chitosan has a tetrasaccharide motif. Journal of Carbohydrate Chemistry. 19: 789-794.
[136] Lertworasirikul A, Tsue S, Noguchi K, Okuyama K, Ogawa K (2003) Two different molecular conformations found in chitosan type II salts. Carbohydrate Research. 338: 1229-1233.
[137] Kawahara M, Yui T, Oka K, Zugenmaier P, Suzuki S, Kitamura S, Okuyama K, Ogawa K (2003) Fourth 3D structure of the chitosan molecule: Conformation of chitosan in its salts with medical organic acids having a phenyl group. Bioscience Biotechnology and Biochemistry. 67: 1545-1550.
[138] Brugnerotto J, Desbrieres J, Heux L, Mazeau K, Rinaudo M (2001) Overview on structural characterization of chitosan molecules in relation with their behavior in solution. Macromolecular Symposia. 168: 1-20.
[139] Weinhold M X, Thöming J (2011) On conformational analysis of chitosan. Carbohydrate Polymers. 84: 1237-1243.
[140] Pedroni V I, Schulz P C, Gschaider M E, Andreucetti N (2003) Chitosan structure in aqueous solution. Colloid and Polymer Science. 282: 100-102.
[141] DeMarco M L, Woods R J (2008) Structural glycobiology: A game of snakes and ladders. Glycobiology. 18: 426-440.
[142] Cho Y W, Jang J, Park C R, Ko S W (2000) Preparation and solubility in acid and water of partially deacetylated chitins. Biomacromolecules. 1: 609-614.
[143] Kubota N (1997) Permeation properties of chitosan transition metal complex membranes. Journal of Applied Polymer Science. 64: 819-822.

[144] Kristiansen A, Varum K M, Grasdalen H (1998) Quantitative studies of the non-productive binding of lysozyme to partially N-acetylated chitosans. Binding of large ligands to a one-dimensional binary lattice studied by a modified McGhee and von Hippel model. Biochimica Biophysica Acta. 1425: 137-50.

[145] Sorlier P, Denuziere A, Viton C, Domard A (2001) Relation between the degree of acetylation and the electrostatic properties of chitin and chitosan. Biomacromolecules. 2: 765-772.

[146] Sorlier P, Viton C, Domard A (2002) Relation between solution properties and degree of acetylation of chitosan: Role of aging. Biomacromolecules. 3: 1336-1342.

[147] Berth G, Dautzenberg H (2002) The degree of acetylation of chitosans and its effect on the chain conformation in aqueous solution. Carbohydrate Polymers. 47: 39-51.

[148] Brugnerotto J, Lizardi J, Goycoolea F M, Arguelles-Monal W, Desbrieres J, Rinaudo M (2001) An infrared investigation in relation with chitin and chitosan characterization. Polymer. 42: 3569-3580.

[149] Baxter A, Dillon M, Taylor K D A, Roberts G A F (1992) Improved Method for Ir Determination of the Degree of N-Acetylation of Chitosan. International Journal of Biological Macromolecules. 14: 166-169.

[150] Pelletier A, Lemire I, Sygusch J, Chornet E, Overend R P (1990) Chitin Chitosan Transformation by Thermomechanochemical Treatment Including Characterization by Enzymatic Depolymerization. Biotechnology and Bioengineering. 36: 310-315.

[151] Muzzarelli R A A, Rocchetti R (1985) Determination of the Degree of Acetylation of Chitosans by 1st Derivative Ultraviolet Spectrophotometry. Carbohydrate Polymers. 5: 461-472.

[152] Rinaudo M, Ledung P, Gey C, Milas M (1992) Substituent Distribution on O, N-Carboxymethylchitosans by H-1-Nmr and C-13 Nmr. International Journal of Biological Macromolecules. 14: 122-128.

[153] Raymond L, Morin F G, Marchessault R H (1993) Degree of Deacetylation of Chitosan Using Conductometric Titration and Solid-State Nmr. Carbohydrate Research. 246: 331-336.

[154] Lamarque G, Lucas J M, Viton C, Domard A (2005) Physicochemical behavior of homogeneous series of acetylated chitosans in aqueous solution: role of various structural parameters. Biomacromolecules. 6: 131-42.

[155] Wang Q Z, Chen X G, Liu N, Wang S X, Liu C S, Meng X H, Liu C H (2006) Protonation constants of chitosan with different molecular weight and degree of deacetylation. Carbohydrate Polymers. 65: 194-201.

[156] Delben F, Muzzarelli R A A, Terbojevich M (1989) Thermodynamic Study of the Protonation and Interaction with Metal-Cations of 3 Chitin Derivatives. Carbohydrate Polymers. 11: 205-219.

[157] Anthonsen M W, Smidsrod O (1995) Hydrogen-Ion Titration of Chitosans with Varying Degrees of N-Acetylation by Monitoring Induced H-1-NMR Chemical-Shifts. Carbohydrate Polymers. 26: 303-305.

[158] Domard A (1987) pH and CD measurements on a fully deacetylated chitosan - application to Cu-Li-polymer interactions. International Journal of Biological Macromolecules. 9: 98-104.
[159] Schatz C, Viton C, Delair T, Pichot C, Domard A (2003) Typical physicochemical behaviors of chitosan in aqueous solution. Biomacromolecules. 4: 641-648.

Chapter 10

Coupled Mass Spectrometric Strategies for the Determination of Carbohydrates at Very Low Concentrations: The Case of Polysaccharides Involved in the Molecular Dialogue Between Plants and Rhizobia

V. Poinsot, M.A. Carpéné and F. Couderc

Additional information is available at the end of the chapter

1. Introduction

Polysaccharides play important biological roles in all living beings. They act as membrane and cell wall constituents, afford physical and chemical protection from the environment, ensure nutrition storage or compose the antigens that increase or repress defenses during infection processes, protein folding, molecular recognition, cell adhesion, etc. [1,2]. With regard to their major biological roles it appears a priority to understand their structure/activity relationship. Unfortunately, even though their relevance has been clearly demonstrated, the structural analysis of such compounds remains difficult. Actually, even though they are often highly soluble in water, their polarity due to the presence for instance of hydroxyls, carboxyls and sulfates, makes them quite difficult to handle. These polar groups are involved in intra and inter-molecular hydrogen bonding, ionic interactions and for lipopolysaccharides (LPS) hydrophobic interactions. Such molecular interactions are sources of solubility problems and can lead to the formation of aggregates, making their analysis more difficult (enormous apparent size, broadening of NMR signals, ionization difficulties, and so on).

In the past, analysis of polysaccharides was mostly achieved through Gas Chromatography coupled to Mass Spectrometry (GC-MS) of the PS hydrolysates. This technique remains widely used, due to its simplicity, but the development of high field Nuclear Magnetic

Resonance (NMR) and cryoprobes have opened new avenues. Unfortunately, NMR presents major drawbacks such as sample purity, sample complexity and concentration requirements. However, in the last decade, the development of highly sensitive, high-throughput Mass Spectrometers (MS) and software able to make global approaches ("omics") on very complex mixtures, have induced a new interest for the MS analyses (especially for the study of protein glycosylation). This time, the instruments mostly use soft ionization techniques associated to powerful $MS^{2 \text{ or } n}$ capacity analyzers experiments on electrospray ionization coupled to tandem mass spectrometers (ESI-MS/MS) or matrix assisted laser desorption and ionization coupled to a time of flight analyser (MALDI-ToF) [3]. Finally, the recent appearance of ToF and FTICR (Fourier transform ion cyclotron resonance) analyzers on the market allowed access to high resolution measurements enabling the determination of the exact molecular masses (with ToF error of less than 5 ppm and FTICR less than 0.5 ppm). Exact masses give access to the elementary composition of the molecule that is extremely precious when the analyte is present in a complex biological matrix at low concentrations precluding study by MS/MS.

Beside the evolution of structural analysis, separations techniques have also been enhanced especially by using higher resolution techniques (narrow bore columns, high pressure tolerant or acido-basic tolerant chromatographic phases). This progress has made it possible to couple chromatographic and MS systems on-line.

In this chapter, we will describe the classical MS couplings like GC-MS or HPLC-MS, but also focus on unusual but useful polysaccharide analysis systems like capillary electrophoresis coupled with mass spectrometry (CE-MS) or high performance anion exchange chromatography coupled with mass spectrometry (HPAEC-MS) that we have developed. Finally, we will present the practical uses of these techniques in a fundamental application: the analysis of the exo-, lipo- and capsular polysaccharides of *Sinorhizobium meliloti.*

2. Structural analyses using mass spectrometry

2.1. Gas chromatography coupled to mass spectrometry

The analysis of polysaccharides is relatively easy to run using ESI-MS and ESI-MS/MS to identify the concatenation of polysaccharide in hexose, deoxy-hexose, pentose, acidic sugars (polyhydroxy carboxylic acids), but it does not provide the polysaccharide composition (glucose, mannose, fructose…) mainly because many monosaccharides have isobaric masses, and often lack an ion (X and A) to identify how the monosaccharides are branched. GC-MS using electron ionization (EI) and chemical ionization (CI) is well suited for such a purpose. Some recent reviews have been published which gather and summarize the different numerous strategies developed in the literature for sugar analysis [4, 5, 6].

Because polysaccharides are mainly composed of aldoses, ketoses, deoxymonosaccharide, amino-monosaccharide and acidic monosaccharide, GC-MS analysis is based on the hydrolysis of the polysaccharides followed by derivatization and GC/MS analysis to identify the different sugar units through their migration times and/or mass spectra. Among the hydrolyzed monosaccharides some cannot be observed like uronic acids, ulosonic acids (*e.g.* Kdo), 4-amino-monosaccharide, and other charged species. Acids cannot be observed because their sodium salts are not volatile. Their treatment must be adapted for study by GC-MS. We describe below the main treatments for the study of polysaccharides: hydrolysis, derivatization, and we evoke the case of the acidic sugars.

Scheme 1 presents the general strategy which is used in GC-MS to analyze polysaccharides by determining the nature of the sugars, the relative quantities and the mode of attachment.

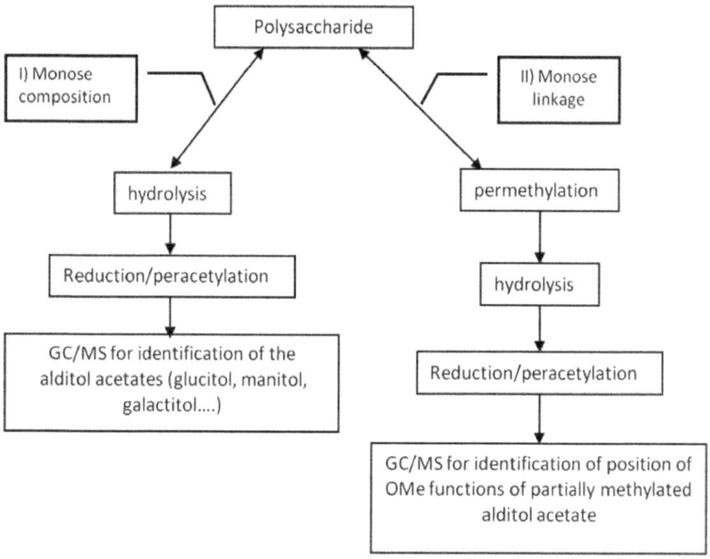

Scheme 1. General strategy used in GC-MS to analyze polysaccharides

2.1.1. *Quality and relative quantities of monosaccharides: alditol acetate*

To identify the nature of the monosaccharide, hydrolysis using H_2SO_4, HCl or trifluoroacetic acid (weaker than the others but easy to evaporate) is performed. After a reduction step using sodium borohydride [7], acetylation with acetic anhydride [8] results in alditol acetates which can be identified using GC retention time. GC-MS with chemical ionization using ammonia confirms the mass of the pseudo molecular ions $(M+H)^+$ and $(M+NH_4)^+$ of the alditol acetates. The chromatographic peaks help to identify the relative number of monomers composing the polysaccharide. Reduction with $NaBD_4$ allows easy identification of the anomeric carbon.

The difference in the hydrolysis rate between the sugars results in certain difficulties i.e. uronic acids being partially hydrolyzed, the monosaccharide linked to it is underrepresented. The same difficulties also exist for 2-acetamidohexoses which are partially *N*-deacetylated under excessive acid concentrations. The 2-aminohexoses obtained are not hydrolyzed and the carbohydrates linked to them are underrepresented [9].

For the acidic sugars, a modified strategy can be used, it consists first in methyl esterification of the uronic acids using DMSO and methanol or diazomethane. The methoxy group (or any other substitution) is a good leaving group for reduction of esters with sodium borohydride to yield aldehydes and, subsequently, their respective neutral sugars. This way, the reduction of the esters by $NaBD_4$ prior to hydrolysis yields a neutral polysaccharide allowing total hydrolysis and, after acetylation, identification by GC-EI-MS of the CD_2OCOCH_3 moiety bearing the monosaccharides [10,11,12] characteristic of the initial acidic sugars. Alternatively, methanolysis is also used to release monosaccharides as methylglycosides with esterified carboxyl groups. Conditions of methanolysis are presented in table 1.

Table 1 presents different hydrolysis conditions, reduction and acetylation on different kinds of polysaccharides.

Hydrolysis conditions	**Reduction and acetylation conditions**	**Polysaccharide**	**ref**
10 M HCl (80 °C, 30 min)	Excess of NaBH4 (2 h, 20 °C), acetylated with a 1:1 Ac2O–pyridine mixture (100 °C, 1 h)	LPS of *Providencia alcalifaciens* O12	[13]
100 μL sample, 50 μL of 0.7 M 4-methylmorpholine borane (MMB) and 100 μL of 6 M TFA , 80 °C, 30min.Then, 50 μL of 0.7M MMB. evaporation and addition of 200 μL 2 M TFA and 500 μL of acetonitrile. Dried again.	Dried sample +100 μL acetic anhydride and 100 μL TFA, reaction at 50 °C for 10 min. Then 1mL toluene to each dried sample . After dissolution in 1.5 mL CH_2Cl_2 and wash with 0.5M Na_2CO_3 and water. Sample are then dissolved in CH_2Cl_2 spiked with internal standards	Agar containing 3,6-anhydrogalactose	[14]
Sample was suspended in 2 N H_2SO_4 heated under vacuum at 100 °C for 3 h. 50 μL of water containing 500 ng monosaccharide internal standard and1 mL H_2O was added. Then,2 mL 50% *N,N*-dioctylmethylamine are added to neutralize the remaining acid. After centrifugation, the top layer was purified on a C18 column.	Add 50 μL of 100 mg/mL $NaBD_4$ and store at 4 °C overnight. Evaporate with methanol/acetic acid under vacuum. Then add 300 μL of acetic anhydride, react at 100 °C for 13–16 h. Quench with 750 μL H_2O for 1 h at room temperature. Extract with 1 mL of chloroform ($CHCl_3$)Remove aqueous phase, add 0.8mL of cold 80% NaOH, mix, elute with $CHCl_3$ on a Chem-Elut column.	Agar containing 3,6-anhydrogalactose	[14]

Hydrolysis conditions	**Reduction and acetylation conditions**	**Polysaccharide**	**ref**
2 mg sample treated with 0.5 mL TFA 2 M for 1 h at 121 °C	TFA removed by evaporation, then $NaBH_4$ reduction and then acetylation with Ac2O–pyridine (1:1, v/v; 2 mL) at room temperature for 12 h	Exopolysaccharide containing glucuronic acid	[15]
Methylesterification: 10mg sample dissolved in 5mL DMSO add diazomethane. Freeze- dry. For reduction, dissolve 2mL imidazol/HCl 1M pH 7 at 0°C, 2 drops of octanol add 200mg KBH_4 2h. Add acetic acic until pH 6, dialyze and freeze dry. Add 0.5mL TFA 4N at 100°C, 4h.	Evaporate sample, add NH_4OH 0.1N and 1mg KBH_4 2h 20°C. Then add acetic acid until pH 6, co-distillate with methanol. Acetylate dried residue with pyridine/acetic anhydride (Ac_2O) (50-200μL) 12h, 20°C. Dry the sample in presence of toluene.	Polysaccharide from Bacillus circulans	[12]
1 mg sample hydrolyzed with 1 M TFA at 100°C for 8 h, evaporate to dryness.	Reduce the dried sample with excess of $NaBH_4$ and acetylate with Ac_2O–pyridine (1:1, v/v; 2 mL) at room temperature for 12 h.	Exopolysaccharide from surface-liquid culture of Clonostachys rosea	[16]
1mg polysaccharide fraction in 2 M TFA at 100 °C for 8 h, followed by evaporation to dryness.	Dry residue reduced $NaBH_4$ or NaB_2H_4 and acetylated with Ac_2O-pyridine (1:1, v/v) at room temperature for 12 h.	Polysaccharide containing 3-O-methylated mannogalactan	[17]
1 mg mycelium 6 h at 100 °C with 3 mL 6 M HCl. Dry at 40°C	Add to dry sample, 20 μg inositol in 0.8 mL water, and 50 μl $NaBH_4$ (10 mg/ mL in 2 M ammonia), at 37 °C, 90 min. Add 2 mL acetic acid in methanol (1:200 v/v). After drying, add 300 μl Ac_2O, 100 °C for 15 h. In ice-bath add 750 μl of water. Extract with $CHCl_3$.	Chitin polysaccharide from mycelium.	[18]
Methanolysis	**Derivatization**		
4.5mg of freeze-dried fibers and pulps suspended in 2mL of 2M solution of HCl in anhydrous CH_3OH, 100 °C 4 h. Add 80μL pyridine. Add 1mL sorbitol 0.1mg/mL.	Dissolve dried sample in 70μL pyridine. Add 150μL hexamethyldisiloxane and 80μL trimethylchlorosilane for 12 h at room temperature.	O-acetyl-(4-O-methylglucurono)xylan from *Agave sisalana*	[19]

Hydrolysis conditions	Reduction and acetylation conditions	Polysaccharide	ref
Add MeOH 0.5M HCl (acetyl chloride 140mL + 4mL MeOH) at 80 ◦C for 16 h. dry under a stream of nitrogen gas.	300 µl of TriSilR reagent are added to dried material, at 80 ◦C for 20min and the reagents removed with a nitrogen stream.1mL hexane extraction of the sample, dry and add hexane.	Major grape polysaccharides	[20]
2mL of 0.01, 0.125, 0.25 or 0.5M HCl in MeOH added to 10mg sample, heated at85°C. Neutralize with Ag_2CO_3, dry the supernatant		Gelling carrageenans and agarose	[21]
Add 1.5mg sample to 0.5mL of MeOH/HCL (15mL MeOH and 0.4mL acetyl chloride). Heat at 80°C for 24h and dry.	Add to the dry sample 0.5mL of pyridine, hexamethyldisilazane , trimethylchlorosilane (9:3:1, v/v/v), 80°C 2h. Dry and add 0.5mL hexane.	Plant gums	[22]
Add 2mL 0.5M HCl/MeOH to 5mg sample at 80°C. Neutralize to pH 6.	30µL/30µLpyridine and *BSTFA* (N,O-bis(trimethylsilyl) trifluoroacetamide) added to dry sample, 2h 20°C.	Polysaccharides containing uronic acids.	[12]

Table 1. asd

Scheme 2 presents different ions obtained for alditol acetate in EI. Generally, no molecular ion and only very low abundance of $(M\text{-}CH_3COO\cdot)^+$ are observed in the spectrum. Stereoisomers give very near spectra and the base peak is the acetylium ion (CH_3CO^+) at m/z 43. The primary fragments are formed by cleavage of the carbon chain and then these odd mass number ions may lose an acetic acid (loss of 60 mass units) or a ketene (loss of 42 mass units) or an acetic anhydride (loss of 102 mass units) [23].

If deoxycarbons are present, alfa cleavage resulting in methyl radical loss is insignificant [24]. Some even mass number ions come from rearrangements resulting in the loss of acetaldehyde (loss of 44 mass units), acetic acid, acetic anhydride and ketene from the same mass number molecular ions [25].

When an acetamido group is present, the main ions are due to alfa cleavages induced by the N atom [26].

Consequently comparison of the mass spectrum (EI or CI) and the migration time to standards allows the monosaccharides to be identified.

Trimethylsilyl derivatives can also be used for alditol derivatives [27, 28]. No molecular peaks can be detected. As for most TMS derivatives in EI-MS, small M-15 ions are identified and m/z 103 ($(CH3)_3SiOCH_2^+$) and m/z (nx102)+1 (n being the number of carbon atoms in the chain of the alditol fragment) are recorded with their corresponding ion coming from a loss of 90 amu ((CH3)3SiOH).

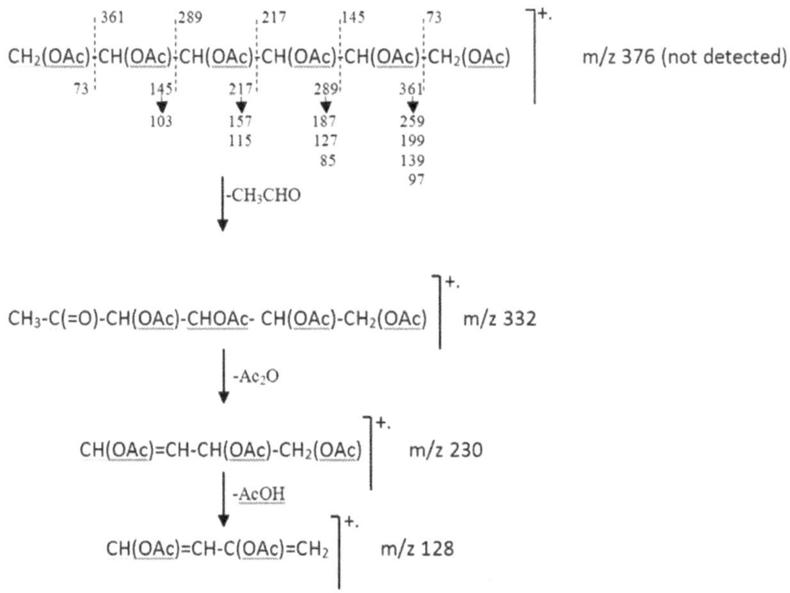

Scheme 2. Main fragmentation of hexitol peracetate.

Monosaccharides can also be analyzed prior to reduction by $NaBH_4$ or $NaBD_4$, under their cyclic forms, but the chromatograms are more complicated due to the hemiacetal group in sugars which leads to multiple structures [29].

2.1.2. *Quality and relative quantities of monosaccharides: derivatives of cyclic forms of monosaccharides*

Most of time the polysaccharide is submitted to a methanolysis and after the sample is silylated (see Table 1). With ammonia, CI can be used to identify the $(M+H)^+$ and $(M+NH_4)^+$ ions and can be identified by comparison of the migration times of standard the sugar. It is known that the relative intensity of the $(M+NH_4\text{-}CH_3OH)^+$ ion allows the assignment of pyranoside and furanoside structure [30]. Other fragmentation ions exist resulting from the loss by $(M+NH_4)^+$ or $(M+H)^+$ of CH_3OH or a single or two $(CH_3)_3SiOH$ molecules.

EI is frequently used and the resulting fragmentation has been extensively reported [30, 31]. If only very rare molecular ions are present in the spectrum, pentose and hexose are characterized by $(M\text{-}CH_3)^+$ ions and losses of CH_3OH and $(CH_3)_3SiOH$ from this last ion. Deoxyhexoses give $M^{+\cdot}$ and $(M\text{-}CH_3O\cdot)^+$. Glucuronic and galacturonic acids also exhibit $(M\text{-}CH_3)^+$ ions in the spectrum and losses of CH_3OH and $(CH_3)_3SiOH$ but also an *m/z* 234 ion is characteristic of acidic derivatives resulting from a complex McLafferty-type rearrangement of trimethylsilyl group to the carboxyl function. Finally, an *m/z* 204 ion is closely related to ring size, being favored by a six-atom cyclic structure [32].

For pyranose forms the relative intensity of this ion (compared to base peak) varies between 30 and 100% while for furanose it is lower (5%), while *m/z* 217 is very intense for furanose forms.

These fragmentations and migration times of the TMS derivatives will help to determine the species that are difficult to characterize owing to their migration times.

2.1.3. *The attachment mode*

2.1.3.1. Synthesis of partially methylated alditol acetate

The use of methylated alditol acetate is a convenient protocol to identify the mode of attachment, it has been well reviewed by Hellerqvist and Sweetman [33]. The first step of this study consists in a permethylation of the polysaccharide, using finely powdered sodium hydroxide and methyl-iodide. A recent article shows that presolubilization of 100-300μL of a solution of 1mg/mL polysaccharide with 5 μL of anhydrous glycerol dried prior to derivatization, offers better methylation yields [34]. Recently, solid-phase spin column permethylation and solid phase capillary permethylation were described [35, 36] and are presented in Table 2. These permethylation processes are derived from the classical method described by Costello [37, 38]. After the permethylation, the polysaccharide is hydrolyzed. For example: the polysaccharide is dried, then 1.8mL glacial acetic acid is added and the sample is briefly sonicated; 0.20mL of 2M sulfuric acid are added and heated at 100°C for 9h. Then acetic acid can be removed in a rotary evaporator several times by adding water and keeping the temperature below 40°C. An equimolar amount of $BaCO_3$ is added to remove the sulfuric acid, and the precipitate is filtered and washed in a minimum volume of water. Then washings and filtrates are combined and about 5mg of $NaBH_4$ (or $NaBD_4$) are added, the reaction is driven for at least 1h. Finally, per-acetylation is driven as reported in table 1.

GC/MS using ammonia CI and then EI are performed, to determine the molecular mass and the structure of the partially methylated acetate alditol respectively. Following this protocol, the positions of the *O*-methyl functions reflect the initial position of the free OH functions of the polysaccharides.

2.1.3.2. Fragmentation rules of partially methylated alditol acetate in EI

In EI, no molecular ions can be observed. The rules include amino sugars and predict the formation of ions formed by cleavage of the C-C chain. i) The major ion corresponds to the cleavage between acetyl-N-methylaminated carbon and *O*-methylated carbon, with charge retention predominantly on the aminylated carbon, ii) an intense ion comes from two *O*-methylated carbons with charge retention on either carbon (the intensities of both ions are inversely proportional to the size of the fragment), iii) weaker ions are due to the cleavage between two C atoms one acetylated the other methoxylated which keeps the charge. These ions may lose ketene, acetic acid or methanol molecules. Table 3 gives the different masses of the different possible ions. With the knowledge of the $(M+H)^+$ or/and $(M+NH_4)^+$ ion, obtained in GC:CI/MS and the fragment ions from the EI spectrum, the position of the methyl and the attachment of the different sugars can be identified. Combining these data with ESI-MS/MS gives the correct sequence of the polysaccharide.

Permethylation Conditions	**Polysaccharides**	**ref**
Dissolve in 200μL DMSO, powdered NaOH (5 mg) and 200μL CH_3I. Stir vigorously for 30 min and left overnight at room temperature. extract with $CHCl_3$.	matesaponin	[39]
Microspin column are placed in microcentrifuge vial, filled with acetonitrile. Pack with NaOH beads in acetonitrile. Centrifuge the spin column with acetonitrile then DMSO. Suspend dried polysaccharide in DMSO, CH_3I, H_20 (70.8; 26.4; 2.8%) spin the column at 800rpm for 30s. Recycle the 8x. Washed it with 100-μL aliquot of acetonitrile, at 10 000 rpm for 0.5 min. Extract column with $CHCl_3$.	Nglycans from serum proteins	[40,41]
Add 10 mg polysaccharide in 1mL DMSO to 1mL CH_3I and 20mg powdered NaOH for 30min. Stir vigorously and maintain at 25°C overnight. Stop with H_2O, neutralize with acetic acid, dialyse (8 kDaMr cut off) against distilled H_2O and freeze-dry. Extract with $CHCl_3$.	3-O-methylated mannogalactan from *Pleurotus pulmonarius*	[42]
Suspend 100-300μg of dry polysaccharide in 100μL DMSO. Mix 400μL CH_3OH and 100μL 50% NaOH, and 2mL DMSO and vortex. Centrifuge at 3500 rpm for 1 min to get the NaOH pellets. Wash 5x with 2 mL DMSO. Mix the polysaccharide and the 500μL NaOH slury solution with 100μL CH_3I. Evaporate, and clean with 2mL H_2O and 2mL CH_2Cl_2.	Pn6B, Pn14, Dextran polysaccharides	[43]

Table 2. Permethylation of polysaccharides.

$CH_2{=}O^+CH_3$ m/z 45	CH_2OCH_3 CHO^+CH_3 m/z89	CH_2OCOCH_3 CHO^+CH_3 m/z 117	CH_2OCOCH_3 $CH{=}NH^+(COCH_3)$ m/z 144	CH_2OCOCH_3 $CH{=}NH^+(COCH_3)CH_3$ m/z 158
CH_2OCH_3 $CHOCH_3$ $CH{=}O^+CH_3$ m/z 133	CH_2OCOCH_3 $CHOCH_3$ $CH{=}O^+CH_3$ m/z 161	CH_2OCH_3 $CHOCOCH_3$ $CH{=}O^+CH_3$ m/z 161	CH_2OCOCH_3 $CHOCOCH_3$ $CH{=}O^+CH_3$ m/z 189	CH_2CH_3 $CHOCOCH_3$ $CH{=}O^+CH_3$ m/z 131
CH_2OCH_3 $CHOCH_3$ $CHOCH_3$ $CH{=}O^+CH_3$ m/z 177	CH_2OCOCH_3 $CHOCH_3$ $CHOCH_3$ $CH{=}O^+CH_3$ m/z 205	CH_2OCH_3 $CHOCOCH_3$ $CHOCH_3$ $CH{=}O^+CH_3$ m/z 205	CH_2OCH_3 $CHOCOCH_3$ $CHOCOCH_3$ $CH{=}O^+CH_3$ m/z 233	CH_2OCOCH_3 $CHOCOCH_3$ $CHOCOCH_3$ $CH{=}O^+CH_3$ m/z 261

Table 3. Principal fragmentations in EI/MS of the C-C chain of permethylated alditol acetate
We do not present the ions which can lose CH_3OH, CH_2CO, CH_3COOH.

2.2. ElectroSpray ionisation mass spectrometry (ESI-MS) analyses

Using electrospray technology, the analyte is introduced into the mass spectrometer as a solution delivered by a syringe pump (direct input) or as a fraction eluted from a liquid chromatography (HPLC/UPLC). The analyte solution passes through a needle on which a high potential difference is applied (classically 3 kV). This produces a spray of droplets with a surface charge of the same polarity as the needle. The charged droplets shrink as the solvent evaporates. Charge promiscuity then produces continual explosions of the droplets into smaller ones until they reach the gas phase. During this process, the molecules present in the solution are concentrated and this often produces a suppression effect. This effect consists in the masking of one compound by another present in a mixture. The suppression effect is minimal when all the compounds together are chemically equivalent in terms of hydrophobicity or acido-basicity. As hydrophilic compounds stay longer in the dissolved form during the ESI process, the more hydrophobic compounds will generally suppress the more hydrophilic compounds. In the positive ion mode, basic compounds will easily suppress more acidic compounds. As carbohydrates are very hydrophilic and often acidic, only two possibilities exist to analyze them without suppression: a separation of the sugars based on their acidity (ion exchange chromatography) or derivatization of the hydroxyls to generate more hydrophobic compounds. The latter techniques include derivatization such as permethylation [44;45] or metal adduction [46;47] which has been extensively studied. Structural analysis of underivatized saccharides using negative mode ion electrospray ionization has also been investigated [48].

A systematic nomenclature for labeling fragment ions observed in MS/MS has been introduced by Domon and Costello [49] (Figure 1).

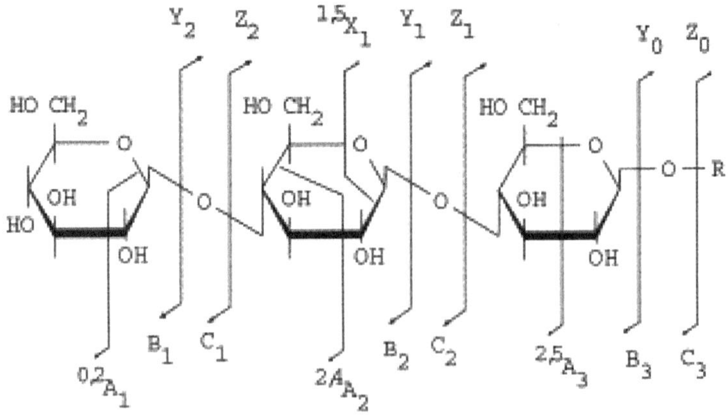

Figure 1. Systematic nomenclature for labeling fragment ions observed in MS/MS following Domon and Costello.

Acidic moieties strongly influence the fragmentation energetics and patterns of tandem mass spectrometric daughter ion formation. Non-sulfated compounds dissociate in the positive mode into abundant ions through B and Y type fragmentation resulting from the cationisation of the glycosidic oxygen [50], these fragmentations being facilitated on 4- and 6-linked reducing terminal residues [51]. In the negative mode deprotonated neutral carbohydrates produce C- and Z-type ion fragments [52], when the acidic sugars undergo B- and Y-type fragmentation [53]. Therefore, interpretation of the CID spectra implies knowledge of the fragmentation pathway, and the use of mass tables of the common monosaccharide building blocks is often essential to build the sequence of unknown structures. Of course, understanding the biosynthetic pathway of oligosaccharides helps to avoid making mistakes in the sequence [54].

2.2.1. *Direct Input (DI/MS)*

As explained just before, analyzing carbohydrate mixtures by direct introduction is a challenge due to the easy suppression of these compounds by numerous other natural compounds like proteins, aminoacids, lipids or other differently charged or decorated saccharides.

In comparison to MALDI, ESI produces less in-source fragmentation of acidic glucans and other fragile ions and is easily coupled on one side to tandem MS allowing structural investigations and on the other side on-line to liquid chromatography. As this ionization technique is very soft, true MS (without fragmentation) can be obtained. However, MALDI produces less complex charge state patterns and less multiple cation adducts and suppression effects. Actually, the analyte is not in solution with the salts and the deposit is quite heterogenic indicating places where suppression effects could be less important [55]. The polymeric complexity of carbohydrates often produces an overlap of the ESI charge state patterns making them extremely difficult to analyze [56] even with software such as MaxEnt (1 and 3). As a result, except for acidic glucans, ESI analysis of glycans requires chromatographic pre-separation of the sample.

2.2.2. *High-resolution liquid chromatography coupled to Mass Spectrometry (HPLC-MS)*

Normal phase HPLC on naked inorganic oxides was the technique first developed but it exhibited numerous drawbacks (peak tailing, retention time shifts, etc.), therefore, reversed phase HPLC was developed and proved to be suitable for most all bioanalytical solutes. However, a problem remained: how to create retention in RP-HPLC for polar compounds without dewetting the C_{18} phase [57]. Lack of retention of hydrophilic compounds is almost always due to solvophilicity. Actually, the polar functions enter more favorable dipolar interactions with the polar RP-HPLC eluent, than with the stationary phase [58]. When the lack of retention is due to charge, retention can easily be achieved on ion-exchange chromatography (see next paragraph) or by using ion pairing that does not require specific apparatus. Using this technique, glycosaminoglycans polymers up to a polymerization degree of 40 can be observed on reversed phase ion pairing HPLC [59]. Unfortunately, ion pairing has been shown to reduce ESI-MS sensitivity [60].

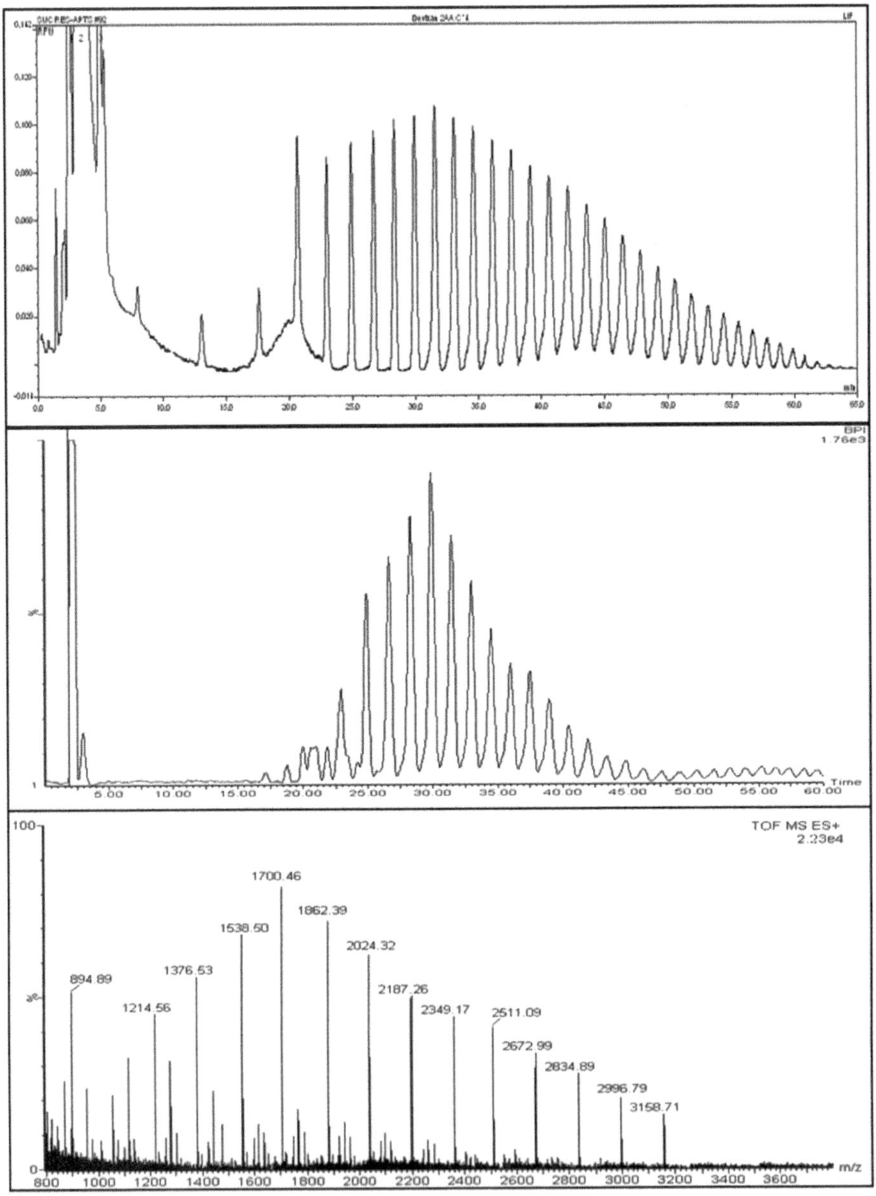

Figure 2. HILIC/LIF/ESI-MS analysis of 2-AB labeled glucose ladder. At the top: improved LIF detection of the neutral saccharides, in the middle: positive mode ESI-MS total ion current, at the bottom: spectrum obtained by summing all the spectra of the compounds eluting between 15 and 50 minutes.

The greatest problem concerns the highly hydrophilic but still neutral analytes. The classical way to analyze them is to convert them into more hydrophobic compounds by a chemical reaction [61]. In addition to the retention of the compounds, such organic derivatization improves their detection (UV absorption or fluorescence emission) (figure 2). The main drawbacks of this approach are that the separation of all the compounds is only based on the same hydrophobic tail, derivatization is time-consuming, and quantitative labeling cannot be systematically achieved. Therefore, a new strategy has been developed and is based on a new stationary phase. In practice, the stationary phase is polar and attracts the more polar part of the eluent that will act as the retentive phase. The complete eluent is relatively hydrophobic but is a sufficiently good solvent to allow distribution between the stationary and mobile phases. This technology is named hydrophilic interaction chromatography (HILIC). The quite neutral hydrophobic/hydrophilic balance of the eluent allows an easy interfacing with ESI-MS unlike the normal phase HPLC (NP-HPLC) working with totally organic solvents which are not compatible with the nebulization process [62].

2.2.3. *High Performance Anion Exchange Chromatography coupled to Mass Spectrometry (HPAEC-MS)*

High performance anion exchange chromatography (HPAEC) using sodium hydroxide or acetate based eluents is a well-established technique for determining underivatized carbohydrates [63;64] and high-performance separation of alditols, mono- and oligosaccharides ranging from 2 to 60 mers have been described using pulsed amperometric detection (PAD) [65]. The identification of individual carbohydrates is usually performed by comparison of their retention times with those of reference samples. In biological samples complex overlapping occurs: acidic sugars are eluted in the "oligosaccharide domain" making this approach very uncertain. Moreover, the nature of sugar moieties and the variability of glycosidic linkages makes the retention times of the oligosaccharides unpredictable. For these reasons, coupling of HPAEC-PAD with ESI-Q-ToF (ESI coupled to a hybrid tandem mass spectrometer constituted of a quadripole and a ToF) MS was required to collect structural information on the PS sequences. Two detections are necessary. PAD allows, from retention time, the nature of the saccharide (neutral, acidic, oligosaccharidic) to be determined while ESI-Q-ToF MS gives access to the composition. Coupling HPAEC with MS presents a technological challenge, due to the non-volatility and high conductance of the sodium hydroxide or acetate dissolved in water, used as eluent. To avoid this limitation, a commercial desalting device using a selective cation exchange membrane and a regenerant can be installed on-line between the column and the MS [71,66]. The use of this on-line desalter with oligosaccharide separations with MS has only been rarely described [67, 68, 69, 70, 71].

We have developed HPAEC and ESI-Q-ToF MS conditions to perform efficient on-line coupling of the two techniques [72]. PA1 ion exchange columns have been used for oligosaccharide analysis even though they are designed for monosaccharides, because the PA 200 column is not available in the narrowbore size. Two different gradients have been used at a flow rate of 0.3 mL/min. For neutral carbohydrates: during the first 15min NaOH concentration increased from 10 to 50mM, and during the following 5 min up to 80mM, finally, during the last 15 min the concentration of NaOAc increased from 0 to 90 mM while NaOH remains constant, return to the initial conditions was done over 10 min. A wait of 35 min between two injections is necessary to equilibrate the column. To separate acidic sugars, the gradient started with a NaOAc concentration ramp increasing from 0 to 90 mM in 15 min while the concentration of NaOH was kept constant at 80 mM, the concentration of NaOAc (90 mM) and NaOH (80 mM) were then held constant for 5 min then 5 min to return to initial conditions. A wait of 15 min between two injections was necessary to equilibrate the column.

Due to the low amounts of polysaccharide that can be dissolved, no post-column split could be used to couple ESI-Q-ToF with an analytical sized column. This methodology was never used for the analysis of heteropolysaccharides.

Performing HPAEC for carbohydrate analyses implies the use of a NaOH and NaOAc concentration gradient. At high pH (>13.5), carbohydrates are in the anionic form allowing their retention on the phase. Electrocatalytic oxidation mediated by NaOH at the surface of the gold electrode occurs by application of a positive potential. The current generated is then proportional to the carbohydrate concentration. Therefore obtaining a PAD signal, implies placing the suppressor after the PAD but before the MS (Scheme 3). Since a high ionic content is not compatible with electrospray ionization mode, the salt level was reduced using an on-line 4 mm ASRS (anion self-regenerating suppressor) desalter. The desalting efficiency and the pH of the mobile phase after desalting were checked. The peak broadening due to the additional void volume of the ASRS and the reference electrode cavity was then investigated. Desalting efficiency was followed through a conductometer as a second detector. The pH was followed each minute (using the combination pH reference electrode in the electrochemical cell of the PAD unit) to be sure that the residual conductivity was not due to acetic acid. When the regenerant used in the ASRS is pure water at a flow rate of 2 mL/min, it is possible to suppress NaOH up to 100 mM (at a flow rate 0.3 mL/min), but it does not provide the suppression of 100 mM NaOAc (the pH increased up to 13 at 25 min). In contrast, the use of 0.25% H_2SO_4 in water at a flow rate of 2 mL/min can maintain the conductivity below 300 μS for the entire analysis. Desalting of the NaOH eluent resulted in the measurement of a stable pH of 7; during the NaOAc gradient, the pH decreased regularly to pH 3. During conductimetric detection, a current is continuously applied to the membrane (50 mA). No effect of regeneration current on the membrane regeneration efficiency was observed.

The coupling of both PAD and MS detectors offers two advantages. First, the safety of the mass spectrometer, intrusion of salt in the source and in the mass analyzer can dramatically affect the performance of the mass spectrometer, because only a small quantity of salt produces signal suppression. Therefore, when signal is present on PAD but disappears on MS, it is an indication that desalting is no longer sufficient. The second advantage, as discussed in the next paragraphs, is that the MS and PAD sensitivity are complementary.

As we performed on line PAD and MS analyses, the first before and the second after desalting, we measured peak broadening, and separation efficiency. A 120 μL void volume was measured between the PAD and the ESI source with 30% peak broadening but without loss of efficiency. Since the flow rate of 0.3 mL/min is compatible with the ESI source, no split was necessary.

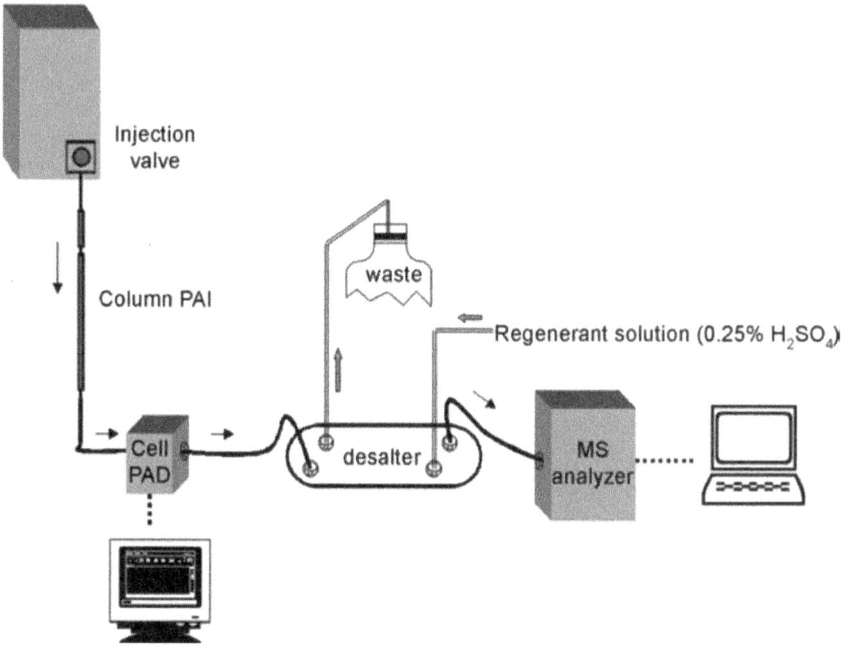

Scheme 3. Coupling scheme. The PAD is placed on-line with the MS, but before the desalting device to ensure the electrocatalytic process.

Analysis of acidic sugars opens up the possibility of using the positive or negative mode for MS detection. Theoretically, detection in the negative mode decreases the noise due to the increased specificity, and therefore increases the signal-to-noise ratio (S/N) and the sensitivity. The sugars we are looking for contain a carboxylic acid function and are easily ionisable in the negative mode. We studied the response factor of a commercial acidic oligosaccharide: 6′-sialyl-lactose; surprisingly, the TIC intensities were almost identical for ESI+ and ESI-. Only a sensitivity factor difference of two was measured at 5 μg/mL, and the positive ionization mode did not exhibit significantly more noise. In the positive mode, the presence of at most one sodium adduct was observed (as counter-ion of the acid). As both MS modes gave similar results, the behavior using MS/MS was investigated in both modes. However, collision energy is lower in the positive mode, which could be an advantage to obtain a better sensitivity, since high energy ions are difficult to refocus.

The detection sensitivity using HPAEC-PAD or HPAEC-ESI-Q-ToF MS is dependent on the nature of the sugar. The first fast oxidation step occurring on the gold electrode (PAD) involves the aldehyde of the carbohydrate, resulting in the formation of a carboxylic acid and the production of two electrons. The second fast oxidation reaction is the cleavage of the C1-C2 bond, followed by conversion of C2 and C6 to the corresponding carboxylates, resulting in the production of 6 electrons (most efficient response) [72]. Therefore the predictive response factors are: Hex>6-desoxyHex>HexA>2-desoxyHex>ulosonic acids (Figure 3). For ESI-Q-ToF MS, the ionization occurs on the glycosidic linkage and is often facilitated through the presence of acid functions close to the ionization site, implying the most sensitive response for the ulosonic acids.

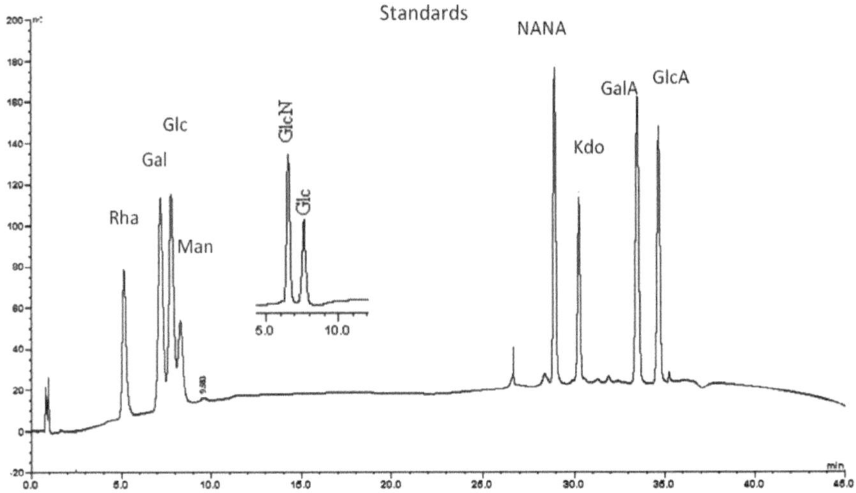

Figure 3. Chromatogram obtained by PAD detection of common sugars : Rhamnose (Rha), Galactose (Gal), Glucose (Glc), Mannose (Man), N-Acetyl Neuraminic acid (NANA), 2-Keto,3-deoxyoctuolosonic acid (Kdo), galacturonic acid (GalA) and glucuronic acid (GlcA).

The concentration response of ESI-MS is often not linear and is very variable from one sugar to the other. For this reason, a quantification curve and the limit of detection (LOD) for each type of saccharide standard were measured for the five standards in solutions ranging from 200 to 2μg/mL with an injection volume of 5 μl.

The HPAEC-ESI-Q-ToF MS response measurements surprisingly indicate that uronic acids respond weaker than expected, even less than hexoses. The LOD of GalA was not satisfying, unlike those obtained for all other saccharides.

2.3. FAB MS and FAB MS/MS of carbohydrates

The fast atom bombardment (FAB) ionization is less and less used due to the fact that the MS suppliers no more produce this type of sources. Nevertheless, this ionization process is of interest, because it is significantly more energetic than MALDI and ESI. In consequence, fragmentation can be observed in the ion source for natively charged glycans and be applied well to positive or ions. In the positive mode sodium cationized species are easily analyzed, while in negative mode it is the deprotonated ions that are observed with greatest abundances for those containing acid residues. Permethylated glycans are very easily ionized but do not result in "*in-source*" fragmentations, collision induced dissociation spectrum in MS/MS can be used to obtain such fragmentation, the nomenclature of the different fragmentations was proposed by Domon and Costello [49]. FAB ionization requires quantities of native glycans above 10nmol, and 10 times less for permethylated species [73].

Using FAB-MS and -MS/MS linkage position can be determined like with ESI, as well as the position of sulfate or other substitutions on the polysaccharide [74]. The internal fragmentation of the monosaccharide cycles (A and X) are useful for such topic.

In the eighties, periodate oxidation was reintroduced to address sequence problem. Oligosaccharides can be sequentially oxidized with periodate and reduced with sodium borodeuteride and permethylated or peracetylated [75]. The bindings of monosaccharide residues can be determined on the basis of the sequences of the primary and secondary ions.

An original attempt of determination of the anomeric configuration was done using methyl glycopyranosides and [M+Ag+]+ ions. It was observed that the unimolecular decompositions of the ions showed dramatic changes related to the relative positions of both the hydroxyl group on C2 and the anomeric methoxyl group. When these groups were in the *cis* position, methanol was exclusively expelled. When they were in the *trans* position, silver hydride was preferentially lost [76].

2.4. CE-MS

2.4.1. Capillary electrophoresis and laser induced fluorescence.

Capillary electrophoresis, due to its high resolving power and sensitivity, has been useful in the analysis of intact and derived oligosaccharides and polysaccharides affording concentration and structural characterization data essential for understanding their biological functions. As most glycans do not have any strong chromophores on their structures and have low extinction coefficient they are difficult to detect using UV absorbance. Indirect UV has been used, it is based on using a chromophore in the electrolyte resulting in negative peaks but because the chromophore must have a mobility very near that of the sugars it is often difficult to ensure an adequate detection limit [77, 78]. Labelling with a chromophore is very useful, because Laser induced fluorescence (LIF) [79,80] can be used on monosaccharides as well as on polysaccharides. CE experiments are comparable to GC experiments to identify monosaccharides. Acetylation or sylilalion of alditol is replaced by an amino reduction step using an amino fluorescent dye and $NaBH_3CN$. Most of the time 1-aminopyrene-3,6,8-trisulfonate (APTS) is used for monosaccharide, their migration time allow the identification of the different derivatized alditols, it is very convenient for small quantities of natural polysaccharides and specially polysaccharides extracted from DOC PAGE. An acidic compound is more easily prepared than with classical alditol strategies for GC analysis [81].

For polysaccharides the same labeling reaction can be used and the different compounds can be easily separated. Recently, optimization of the boric acid concentration (which increases the charge of the polysaccharide by complexation) and linear polyacrylamide concentration in the buffer yielded a separation of 12 polysaccharides of very similar structures, most with baseline separation in less than six minutes [85]. Using this technique, high throughput glycosylation can be performed [82]. Thirty-two IgG N-glycans were analyzed using a combination of weak anion exchange enrichment and exoglycosidase digestion. Aminobenzamide and Aminopyrentrisulfonate labeling of polysaccharide followed by a UPLC on 1.7 μm HILIC or CE-LIF respectively have been compared. Combination of these data demonstrated that complete structural annotation is possible within a total analysis time of 20 min due to the advantageous orthogonality of the separation mechanisms [83]. This work confirms the use of glycosidase in CE-LIF studies for the determination of the structure of polysaccharides. Significant labeling improvements were made by replacing the conventionally used acetic acid catalyst for $NaBH_3CN$ reduction with citric acid. Using a 1:10 glycan to fluorophore molar ratio resulted in a 95% derivatization yield at 55°C with only 50 min reaction time and negligible terminal sialic acid loss [84].

2.4.2. Capillary electrophoresis and mass spectrometry.

A large number of CE-MS and CE-MS/MS experiments were driven for polysaccharide structural determination and have been extensively reviewed recently by Y. Mechref [85]. Different interfaces between the CE-LIF instrument and MS are commercially available. Most works use APTS labeling and ESI ionization, although MALDI is also used but with lab-made instruments [86]. A very interesting study concerning polysaccharides from IgG was realized. Figure 4 presents a schema of the CE-MS system with on-line LIF detection adjusted 20 cm from the ESI tip. A sheath liquid consisting of a 50:50 isopropanol:water (0.2% ammonia) was added coaxially to the separation capillary at flow-rate 4 μL/min. CE was performed with PVA coated capillaries (60 cm total length x 50 cm effective length × 50 μm ID) and a running buffer of 40 mM *alpha*-aminocaproic acid, pH 4.5 (adjusted with acetic acid) + 0.02% hydroxypropylmethylcellulose (HPMC) with an applied voltage of -30 kV were used [87]. The inherent mass accuracy of the MS permitted the identification of major and minor glycans using their $(M\text{-}H)^-$ ions. This study demonstrated for the first time the ability to attain CE-MS separation efficiency somewhat comparable to that commonly observed in CE-LIF analysis. As seen in figure 4 the four early migrating species are clearly visible in both the online LIF and MS electropherograms. The MS electropherograms suffer from a shift in migration time and some loss in separation efficiency, which is due to dead volumes originating from the addition of the MS detector.

Neutral markers to label acidic polysaccharide can be used. Their main advantage will be to allow the separation of the acidic polysaccharide following the number of acid residues [88], 2-aminoacridone was proposed in place of APTS. Nakano *et al* [89] used 9-fluorenylmethyl chloroformate (Fmoc-Cl), during the digestion of a glycoprotein by PNGase which can label the released 1-amino function of the first GlcNac residue. It prevents reductive amination.

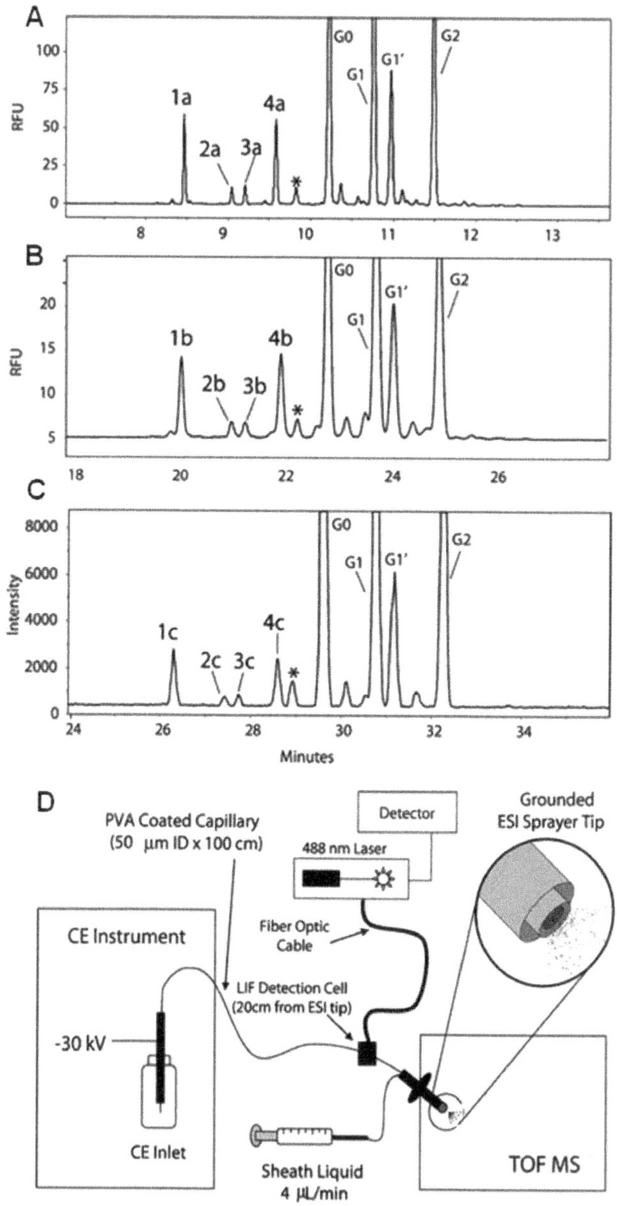

Figure 4. Expanded-scale electropherograms of rMAb 1: A) standard CE–LIF electropherogram using a 60 cm capillary, B) CE–LIF trace obtained on-line with MS detection and C) CE–MS base peak electropherogram. D) Scheme of the CE–MS system with on-line LIF detection.

Using a very simple 20 mM phosphate buffer (pH 8.5), the authors separated the different polysaccharides following the number of sialic acids, then MS and MS/MS spectra identified the composition of each polysaccharide thanks to $(M+H)^{2+}$ or $(M+H)^{3+}$ ions. An example of the mass electropherogram and the different mass spectra are presented in figure 4 which concerns the glycans from bovine fetuin. In this study the authors showed that this method can be used to identify polysaccharides from glycoprotein extracted from an SDS PAGE separation.

3. Polysaccharides of *Sinorhizobium meliloti*

Polysaccharides are commonly found at the surface of Gram negative bacteria (Figure 5). The aim of our work is to elucidate the structure of the polysaccharides from the surface of bacteria. Rhizobia are Gram negative bacteria living in soil and able to establish a symbiotic interaction with leguminous plants, known as nitrogen fixing symbiosis [90]. During this mutual interaction, bacteria bring combined nitrogen, in the form of ammonia directly transformed from atmospheric N_2, to the plant. In turn the plant provides hydrocarbons and develops new organs on its roots which host the bacteria: the nodules [91,92]. During the early stages of establishing symbiosis, a molecular dialogue takes place. First, the partners in the soil are recognized. The plant exudes flavonoid compounds and the neighboring rhizobia respond by secreting Nod Factors (lipochitooligomers) [93]. Nod factors play a major role during the physical contact between the bacteria and the root hairs, and also trigger the organogenesis of the nodule [94]. Then the bacterial threads can invade the root to colonize the nodule. At this stage, the rhizobial surface polysaccharides are essential [95,96,97,98]. To enlarge our knowledge about their role, it is necessary to determine their

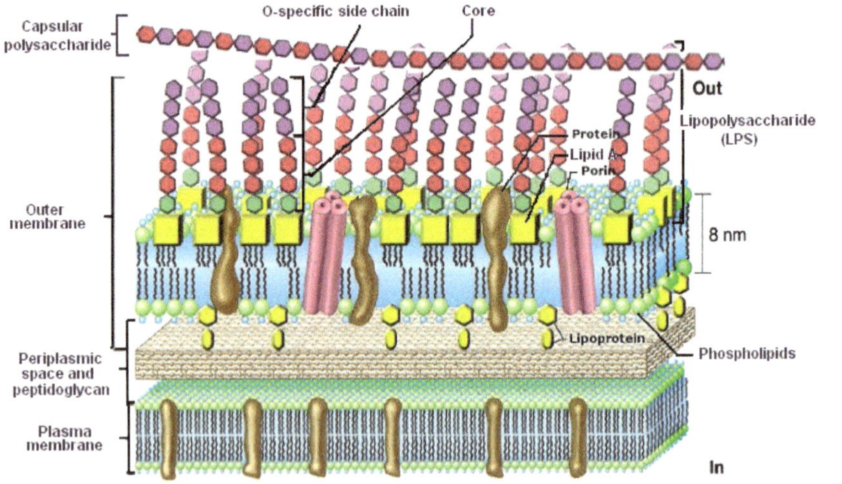

Figure 5. Scheme of the surface of Gram negative bacteria. The EPSs are not represented here; they are over the capsular polysaccharides.

structure. Their structural characterizations will be described below. They generally consist in analysis of composition by Gas Chromatography coupled with Mass Spectrometry (GC-MS), and sequence analysis made principally by Mass Spectrometry (MS). The choice of the MS coupling, mode or instrument depends on the nature of the polysaccharide (size and composition) as well as on the type of information we want to accede.

Sinorhizobium meliloti –the European model of rhizobia - has an external surface where 3 types of polysaccharides can be observed: exopolysaccharide (EPS), capsular polysaccharide (CPS) and lipopolysaccharide (LPS). Each class of PS have to be investigated alone, because their physico-chemical properties do not allow their simultaneous detection.

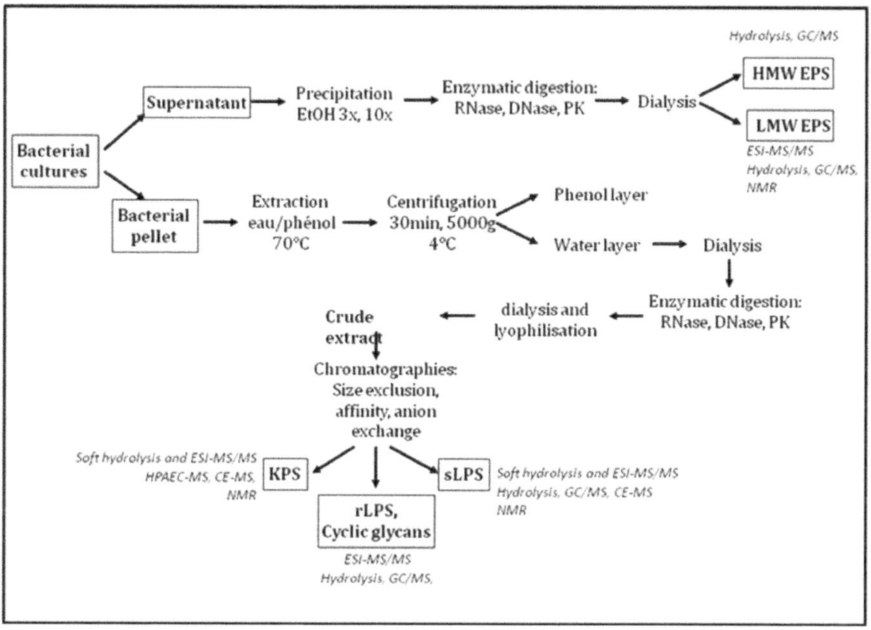

Figure 6. Isolation protocol for the different polysaccharides. In italic : caracterization methods.

3.1. Exopolysaccharides of *Sinorhizobium meliloti*

3.1.1. Isolation of the EPS

The EPS are generally composed of many repetitions of 8 to 12-mers of hexose-subunits in a linear and/or branched form. This can be more or less substituted by O-acetyl, succinyl and pyruvyl groups. Their composition and structure is species specific and depends on the growth conditions [99]. The EPS are produced by the bacteria during the stationary phase of growth [99].

When the EPS of *S. meliloti* were studied, the cultures were stopped 5 hours after the stationary phase, by centrifugation. The supernatant, containing the EPS, was isolated and lyophilized. When concentrated enough (increased viscosity but well dissolved), the EPS were precipitated, first with 3 volumes of ethanol. The pellet so formed was centrifuged and kept, the supernatant was concentrated. The pellet constitutes the high molecular weight (HMW) EPS fraction. The EPS left in the treated supernatant were then precipitated with 10 volumes of ethanol, the pellet was centrifuged and kept. This is the low molecular weight (LMW) EPS fraction. Each fraction was resuspended in water and subjected to proteinase K digestion (3.5 g/L at final concentration) for 4h at 36°C. The mixtures were then dialyzed against water. The pellets thus obtained were lyophilized, and once dried, used for structural analysis determination.

3.1.2. Characterization of the EPS

To learn about the monosaccharide composition, the polysaccharides must be first hydrolyzed. This chemical reaction is carried out in acidic conditions. A solution of EPS in water was acidified by 98% TFA to a final concentration of 10%, and the mixture was kept at 100°C for 2h. At the end of the treatment, the leaving acid has to be evaporated. To help this elimination in a N_2 stream, repetitive additions of isopropanol are necessary.

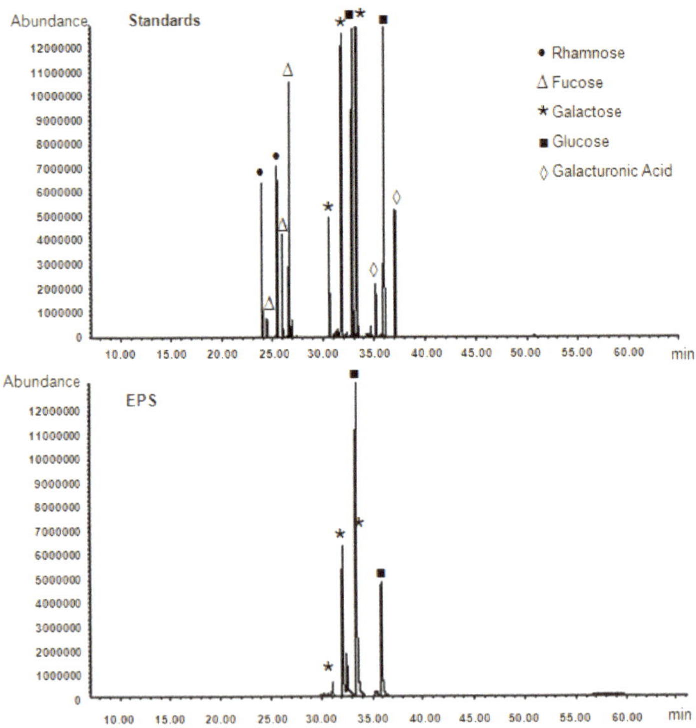

Figure 7. Chromatograms obtained by GC-MS for silylated standard saccharides and silylated hydrolyzed EPS LMW of *Sinorhizobium meliloti* 1021. EPS is made of Glucose and Galactose.

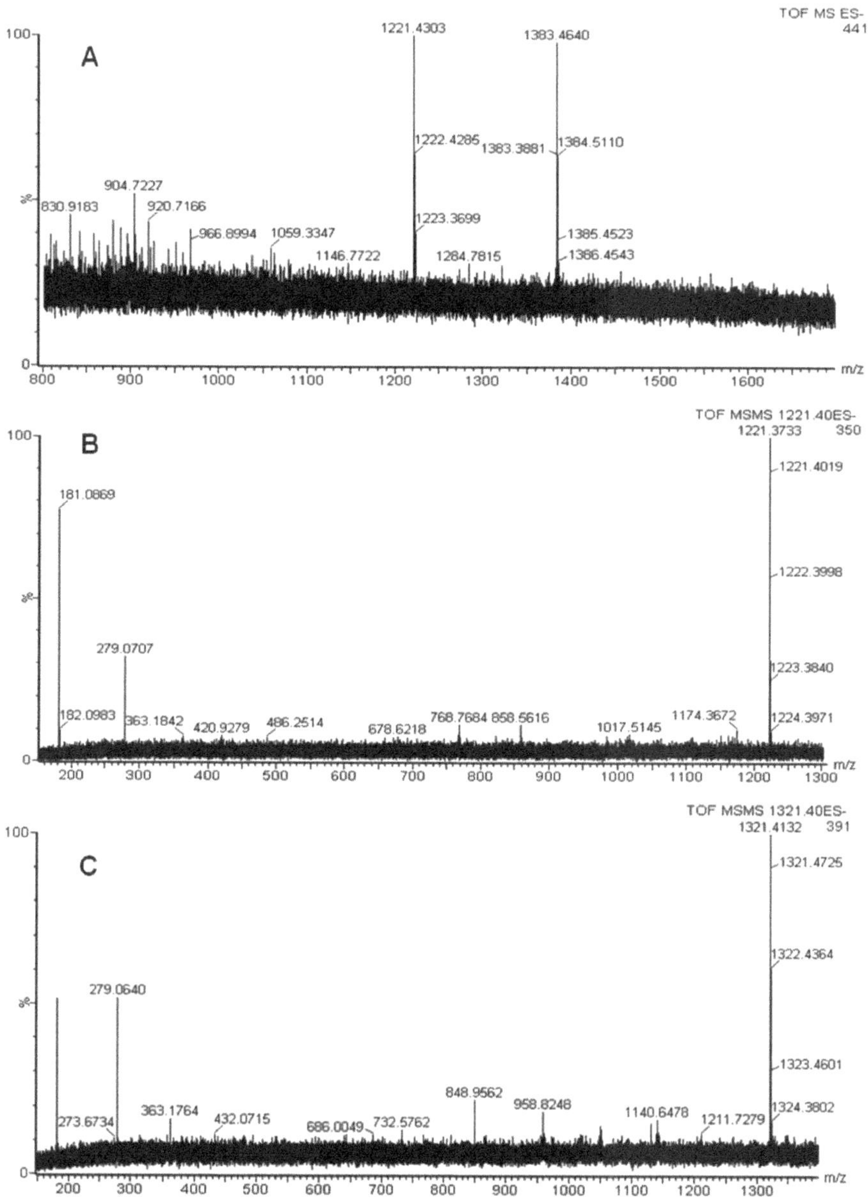

Figure 8. MS analyses of the LMW EPS of *Sinorhizobium meliloti* 1021A) ESI-MS analysis in negative mode B) ESI-MS/MS of ion *m/z* 1221.4 (the simplest form) amu. C) MS/MS of ion *m/z* 1321.4 amu.(the succinylated form), m/z 1383 corresponding to the succinylated and acetylated species.

After hydrolysis, the polysaccharides result in a mixture of monosaccharides and can be analyzed by GC-MS after derivatization. In the results presented here, we used the silylation of the sugar hydroxyl groups, which is a facile route. Actually, the protocol is simple, and a large databank is available for MS interpretations. In anhydrous pyridine, a mixture of HMDS and TMCS was added to the hydrolyzed EPS. This reaction was held for 30 min at 70°C [100]. After complete evaporation, the derivatized monosaccharides were injected directly as a hexane solution. The vector gas used was helium and the column was 95% dimethylsiloxane and 5% diphenylsiloxane. The temperature gradient was: 70°C hold for 3 min, increase at 5°C/min to 140°C and at 3°C/min to 240°C, then reach 300°C at 10°C/min, and hold at 300°C for 10 min.

This procedure applied to standard monosaccharides (glucose, galactose, mannose, glucuronic acid and galacturonic acid) allows interpretation of the chromatograms, determination of each sugar pattern (depending on alpha, beta, pyranoside, furanoside or lactone configuration) and the response factor for the different carbohydrates (Figure 7). The monosaccharide composition of the repeated units of LMW and HMW EPSs of *S. meliloti* was thus obtained.

The preparation of a simple solution of intact EPS in acidic methanol allows its ESI-MS analysis. This kind of analysis shows the purity of the EPS and the disparity of the sugars (Figure. 8A). Actually, the polysaccharide diversity is due to a variable degree of polymerization (DP) and to non-carbohydrate substituents. Here, the substituent is probably a succinyl group, adding 100 amu to ion *m/z* 1221.4 amu to yield the ion *m/z* 1321.4 amu. The ESI-MS/MS analysis of the ions found in the MS spectrum allows the sequence to be assessed by studying the fragmentation ions (Figure 8 B and C).

3.2. Lipopolysaccharides of *Sinorhizobium meliloti*

LPS have three structural domains: the lipid A, the oligosaccharide core and the O-antigen polysaccharide. The LMW LPS, named rough LPS are composed of lipid A and core polysaccharides, whereas the HMW LPS (smooth LPS) are made of the 3 associated parts [101].

3.2.1. Isolation of the LPS

As LPS are anchored in the outer membrane by lipid A, after growth of *S. meliloti* to the stationary phase, it was necessary to keep the pellet of centrifuged cultures (1L, grown until OD (600nm) 1.5). The pellet resuspended in about 40 mL of water was extracted by phenol at 60°C for 1h [102]. The extraction mixture was centrifuged and the upper phase (aqueous) contained the LPS, and all the other hydrosoluble molecules (DNAs, RNAs, proteins, carbohydrates). Enzymatic digestions were performed to eliminate DNAs, RNAs and proteins, remaining the carbohydrates. Affinity chromatography allowed the isolation of LPS. Then, gel filtration separated the LMW and HMW LPS (respectively rLPS and sLPS).

3.2.2. Characterization of the LPS

In this work, it was the HMW LPS that were studied, so both types of structure had to be determined, the saccharide parts and the lipid A. Soft hydrolysis, in 1% acetic acid at 100°C for 1h, isolated lipid A, separating out of the aqueous solution. A liquid-liquid extraction, with $CHCl_3$:CH_3OH (3:1), separated the lipidic part from the sLPS.

Structural analysis of lipid A was performed by ESI-MS. The deconvoluted spectrum of doubly-charged ions revealed two series of monocharged ions, with a mass resolution that allowed the identification of the monoisotopic molecular masses (and not only the average molecular masses), i.e.: *m/z* 2050.41, 2038.39/2036.3, 2022.41, 2010.39 for one series and 1950.41, 1938.39/1936.3, 1922.37, 1910.37 for the second series with sufficient accuracy to allow elemental composition analysis (Figure 9, note that the spectrum presented is not deconvoluted). Accurate mass analysis combined with GC-MS data led to the proposal of a first general structure for lipid A : a di-phosphorylated penta-acylated diglucosamine comprising two 3OH-C14 fatty acids, one 3OH-C18, one 3OH-C19:1 and C27OH-C28 fatty acids for a compound giving a singly charged molecular ion of *m/z* 1950.4 amu (Figure 10)

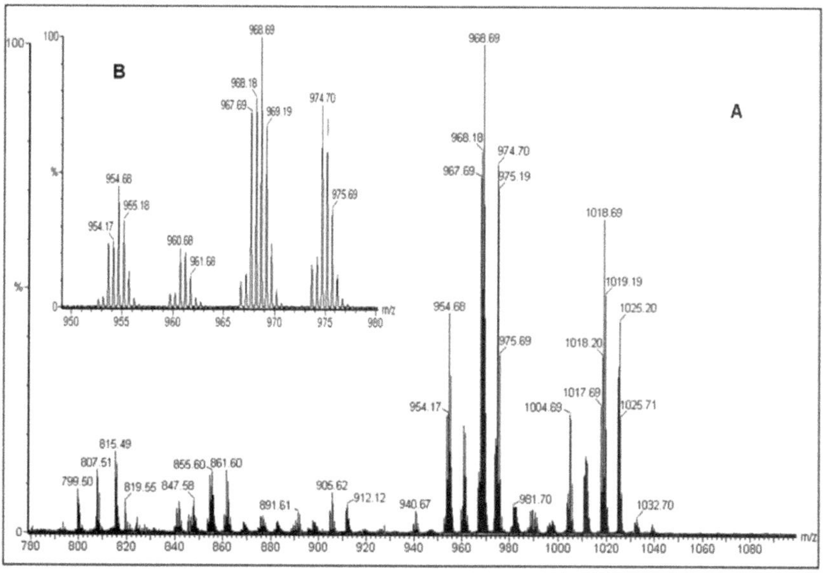

Figure 9. A) Direct ESI-MS spectrum of LPS lipid anchor. Ions ranging from 954 to 976 are assumed to be molecular species and ions observed between *m/z* 1004 and 1026 are native methoxybutyrate derivatives. Insert B) high-resolution spectrum indicating the double charged state of lipidA.

The difference of 100 amu (50 amu in spectrum shown figure 9) can be attributed to the presence of a 3-methoxybutyrate in the first series. Actually, as the second lipid A distribution was hypothesized to be 3-O-methoxy-butyrate substitution of the 27OH-C28, exact mass measurements were performed. In the mean mass difference found between the respective members of the two distributions was 100.03 amu, corresponding with high significance to a $C_5O_2H_8$. To confirm this hypothesis, MS/MS analyses of the double charged ion at *m/z* 1018.2 (Figure 11) have been performed. At low collision energy (15V) *m/z* 1018.2 generated three major fragments, respectively *m/z* 2004.4, 1992.4 and 1936.3 amu. *M/z* 2004 corresponds to the usual neutral loss of methanol (-32 amu), 1992 could be interpreted as a neutral loss of CO_2 (44 amu) after rearrangement. Finally *m/z* 1936 corresponds to the ketene loss of the 3-methoxy butyrate. Increased collision energy (up to 40V) resulted in the production of the same fragments as observed directly from the parent ion that is not substituted by a 3-O-methoxy-butyrate (*m/z* 1935.4). The MS analyses of the complete LPS provided, by comparison to lipid A, the mass of the saccharide part.

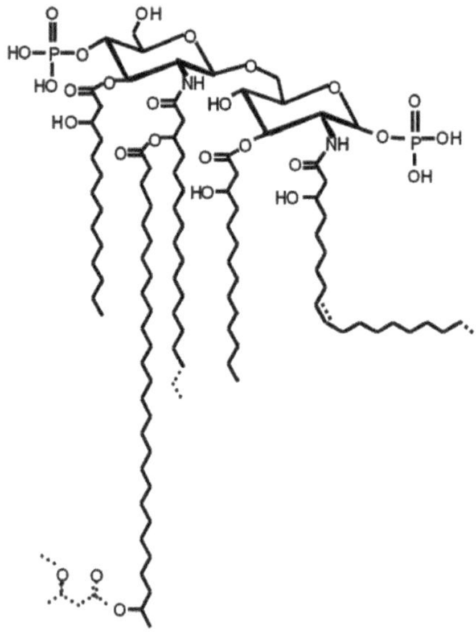

Figure 10. General Lipid A structure of *Sinorhizobium meliloti* 1021's LPS. The dotted lines correspond to the substituent variation found in the sample.

Composition analysis of the saccharide part of the sLPS was performed as for the structural characterization of EPSs. After hydrolysis of the polysaccharide, the monosaccharides obtained were derivatized into alditol acetates, because the O-antigen usually exhibits great sugar diversity.

The process consisted of a reduction by sodium borohydride in ammonia for 1h at room temperature, followed by washing with acetic acid, and an acetic acid in methanol solution [103]. A supplementary wash was made in methanol only, before acetylation. After adding acetic anhydride and pyridine, the reaction stood for 1h at 70°C [104]. Liquid-liquid extraction in water against dichloromethane allowed the isolation of the so-formed alditol acetates in the organic phase. The dried mixture of derivatized monosaccharide was directly analyzed by GC-MS. The temperature gradient started at 110°C and increased at 3°C/min to reach 300°C. Of course, the same process was performed on standard monosaccharides to establish a short database, helpful for interpreting the chromatogram (Figure 12).

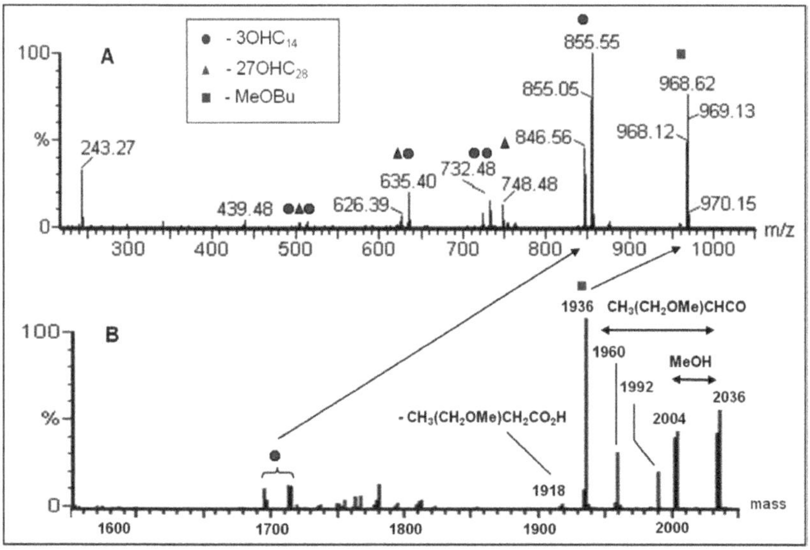

Figure 11. A) CID MS/MS high energy (Ecoll 40V) fragmentation of ion at *m/z* 1018 results principally in *m/z* 968 (loss of methoxybutyrate : -MeOBu) double charged species. B) deconvoluted spectrum indicating a characteristic methoxybutyrate fragmentation at lower collision energy (18V).

3.3. Capsular polysaccharides of *Sinorhizobium meliloti*

In rhizobia, capsular polysaccharides are generally composed of a dimer repeating unit, composed of a hexose and a 1-carboxy-2-keto-3-deoxysugar, like Kdo (2-keto-3-deoxy octulosonic acid). Such a structure looks like the K-antigen found in *E. coli*, and was therefore named Kdo-rich capsular polysaccharide (KPS) for *Rhizobia* [105,106].

3.3.1. Isolation of the K type CPS (KPS)

To isolate KPS from bacterial surface molecules, the same protocol as for the isolation of LPS was executed. Surprisingly, the *S.meliloti* 1021 KPSs were also retained by affinity chromatography. Gel filtration allowed enrichment in KPS, but their size is too close to the size of LPS, to be isolated from each other.

3.3.2. Characterization of the EPS

The analysis of carboxy-sugars such as Kdo is extremely difficult with GC-MS, due to their molecular weight and their instability. So, the composition of the KPS of *S. meliloti* was determined by GC-MS and mainly by HPAEC-PAD-MS. The latter technique requires an

HPLC system, an anion exchange column and a desalter, which is required to couple the ion chromatograph with the MS (see part II.2.c). The device is detailed above, in the methodological part. Analysis of sugars with HPAEC needs no derivatization, better for the stability of Kdo. PAD detection is not sufficient to precisely determine the composition of the mixture analyzed, so the system is coupled to a mass spectrometer. The combination of HPAEC-PAD and MS is a challenge, because of the quantity of salt used in HPAEC-PAD, hence the presence of the desalter is essential. The challenge is rewarded by the information provided by this coupling for determining the saccharidic composition, with simple and non-destructive sample preparation.

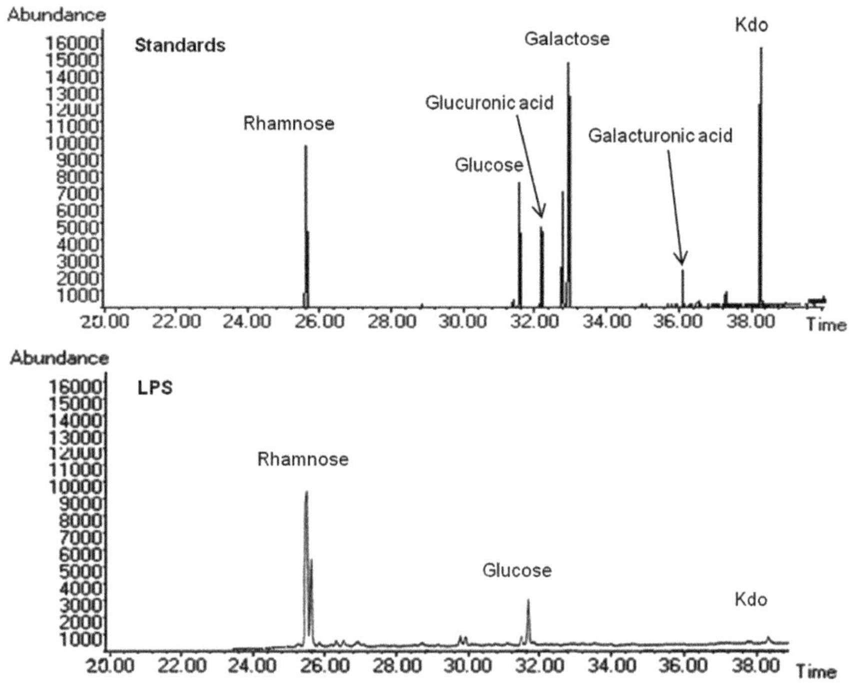

Figure 12. Chromatograms obtained by GC-MS for alditol acetates of standard saccharides and alditol acetates of hydrolyzed LPS of *Sinorhizobium meliloti* 1021. LPS is made of Rhamnose, Glucose and Kdo.

For GC-MS analysis, as the acidic sugar content was high, derivatization with heptafluorobutyric anhydride (HFBA) was necessary [107]. The methanolyzed polysaccharides were dried and resuspended in anhydrous acetonitrile, a solution of HFBA was added. The reaction was heated to 60°C for 30 min. Evaporation of the leaving HFBA involved several co-evaporations with anhydrous acetonitrile. The final solution was made up in anhydrous acetonitrile and injected directly into the GC-MS. The temperature gradient was 70°C for 3 min, 5°C/min to 100°C, 3°C/min to 240°C and 5°C/min to reach 300°C and hold for 10 min. Figure 13 indicates that the KPS enriched fraction is exclusively composed of Kdo.

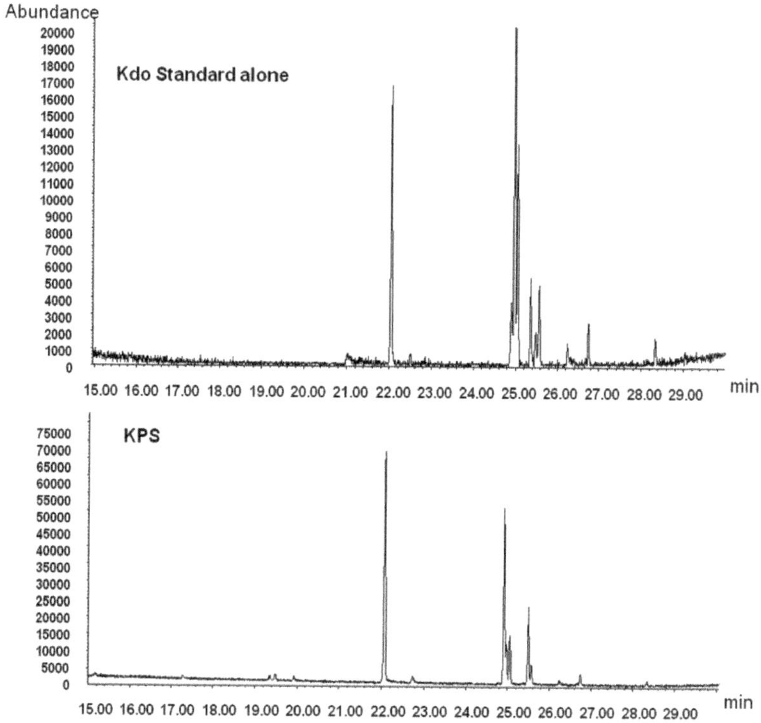

Figure 13. Chromatograms obtained by GC-MS for Kdo standard and methanolyzed KPS of *S. meliloti* 1021, both derivatized by HFBA. KPS is only composed of Kdo.

Unlike GC-MS analyses, in HPAEC-PAD-MS analyses, only soft hydrolysis of the polysaccharide mixture from the size exclusion chromatography (SEC) fractions containing KPS was implemented (1% acetic acid, 100°C, 1h). During the soft hydrolysis, a pellet appeared, indicating the presence of fatty acid on the KPS. MS analysis described bellow will detail this. The HPAEC-PAD-MS analysis revealed free Kdo and oligo-Kdo, characterized in MS by a mass difference of *m/z* 220 amu, when ion *m/z* 221 amu is extracted (Figure 14). The other compounds present in this sample are mostly composed of hexose dimers (*m/z* 324 amu), revealed when ion *m/z* 325 amu is extracted. So, the HPAEC-PAD chromatogram allows simultaneous analysis of other polysaccharides in this mixture: glycans, and substituted glycans [72].

A MS analysis confirmed that KPS is a homopolymer of Kdo. Actually, direct ESI-MS in negative mode (Figure 15 A) indicated the presence of many charged compounds. Zooming between *m/z* 1020 and *m/z* 1060, multi-charged ions (*e.g.*:1035.30 to 1036.05 are quadri-charged ions and 1028.54 to 1029.04 are discharged ions) clearly appear (Figure 15 B). A deconvolution process recovers the native mass of the compounds (Figure 16 A). This reveals a mass difference of 220 amu between the members of the series, typically corresponding to Kdo. The MS/MS analysis of one of these polymers confirms that it is exclusively composed of Kdo, but also revealed the presence of an anchor explaining the discontinuity between Kdo_{15} and Kdo_{17} (Figure 16 B).

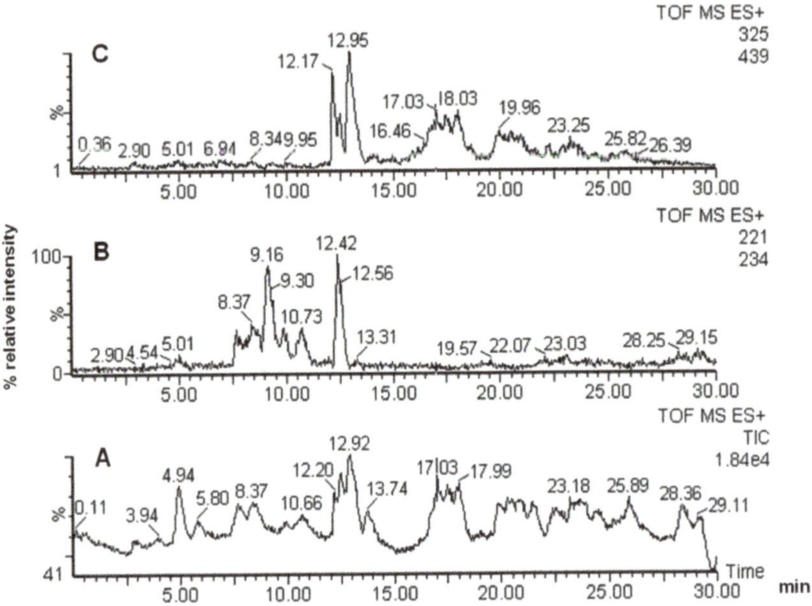

Figure 14. Chromatograms obtained by HPAEC-PAD-MS. A) Total ion current measured by MS. B) Chromatogram obtained when ion *m/z* 221 is extracted. C) Chromatogram obtained when ion *m/z* 325 is extracted. These chromatograms revealed that the KPS fraction is composed of polymers of Kdo and of glycans.

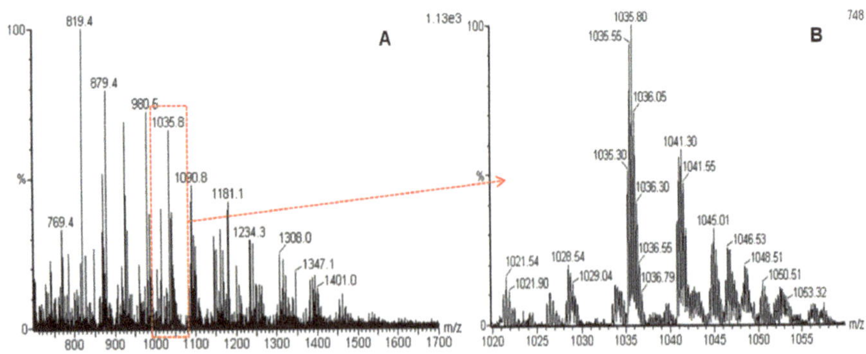

Figure 15. A) ESI-MS spectrum of KPS of *S. meliloti* 1021. B) Zoom on a group of ions revealing the multi-charged ions.

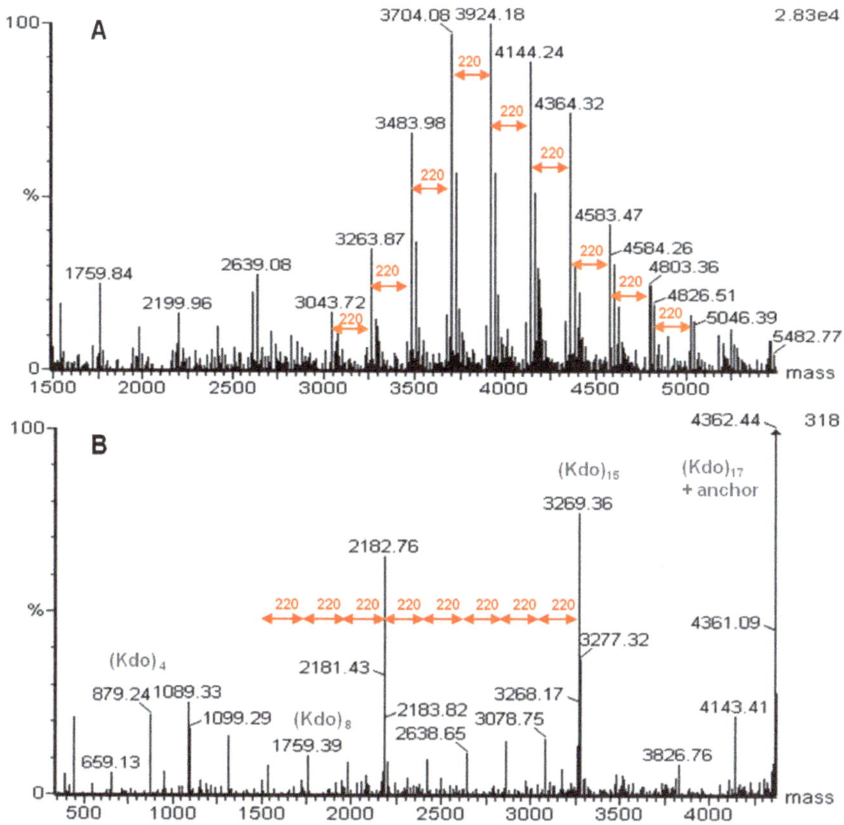

Figure 16. A) deconvoluted spectrum of KPS of S.mmeiloti 1021. B) MS/MS analysis of one Kdo-polymer, revealing an anchor on the KPS.

A MS/MS analysis of the smallest Kdo-oligomer revealed the mass of the anchor at *m/z* 622.4 amu. This ion has been studied separately by MS/MS, after mild hydrolysis on the KPS fraction (Figure 17). Ions m/z 530 corresponds to the loss of the glycerol and 548 to the loss of glycerol minus one water molecule. The interpretation of the spectra, in negative and in positive mode, determined that the anchor was lipidic and composed of a glycerol and phosphoglycerol unit, leading to the structure detailed in figure 18.

For structural characterization of polysaccharides, classical techniques are mostly used. Here, we reported results obtained with chromatography coupled to MS, MS alone and MS/MS. Depending on the characteristics, the type of polysaccharide analyzed, and which data are sought from those samples, it is necessary to adapt the coupled techniques. Advanced techniques, like NMR, are increasingly used for precise characterization. But NMR spectra of polysaccharides are very complex, therefore not easy to interpret. Actually, the heterogeneity in substitution and size is principally the origin of this complexity, especially in 2D NMR.

Moreover, because of the interpretation difficulties, detailed mapping of NMR spectra of polysaccharides in the literature are seldom found. Lastly, as detailed in part I, note that polysaccharide solubility problems can lead to poorly resolved NMR spectra. While NMR can provide useful information about the polysaccharide, MS analysis is not obsolete, because the information provided with MS is unambiguous and confirms NMR spectrum interpretation. In many examples, MS analysis is necessary for an easier interpretation of the NMR spectra. Bacterial polysaccharides are highly complex molecules and many variations occur in one family. The role of polysaccharides from the rhizobia family during nitrogen fixing symbiosis has been demonstrated, as well as the activity of other bacterial polysaccharides during pathogen infection. However, little is known about their structure/activity relationships [108], which implies a long life for polysaccharide structural characterization.

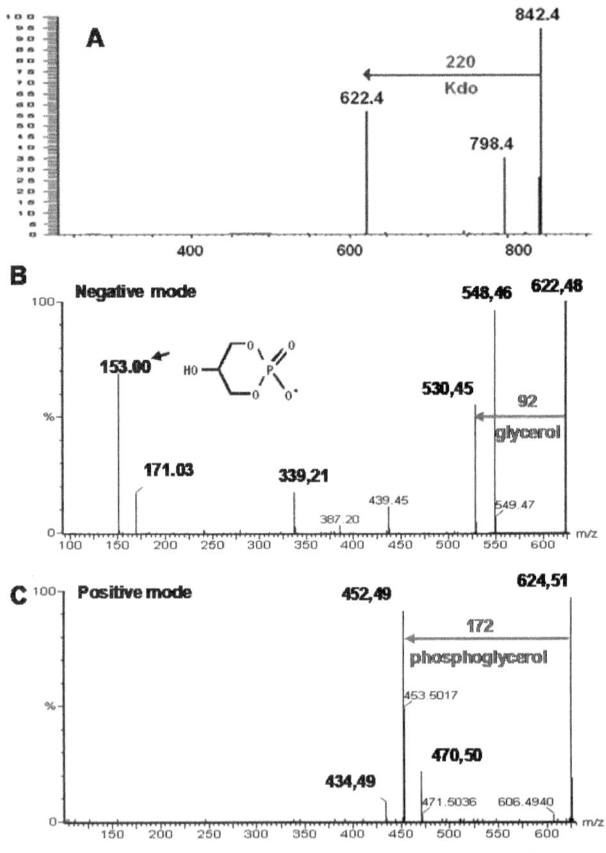

Figure 17. Top, MS/MS analysis of the smallest compound containing Kdo in the KPS, revealing the anchor at *m/z* 622 amu (A). MS/MS Fragmentation of the lipidic anchor (ion *m/z* 622 amu) obtained after soft hydrolysis in negative (B) and positive (C) mode. The presence of glycerol and phosphoglycerol in this lipid anchor was thus demonstrated.

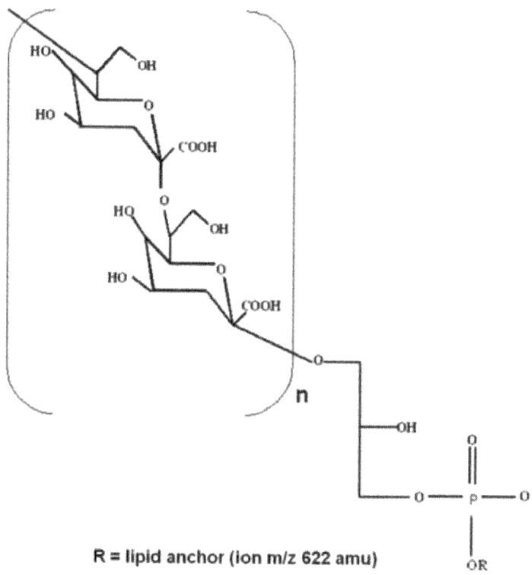

Figure 18. General structure of KPS of *S. meliloti* 1021

Author details

V. Poinsot, M.A. Carpéné and F. Couderc
Laboratoire IMRCP, UMR 5623, Université Paul Sabatier, Toulouse, France

4. References

[1] Varki A (1993) Biological roles of oligosaccharides: all of the theories are correct. Glycobiology, 1993, 3: 97-130
[2] Dweck RA (1996) Glycobiology: toward understanding the function of sugars. Chem. Rev. 96: 683-720
[3] Matamoros Fernandez LE (2007) Introduction to ion trap mass spectrometry: Application to the structural characterization of plant oligosaccharides. Carbohydr. Polym. 68: 797-807
[4] Molyneux RJ, Gardner DR, James LF, Colegate SM (2002) Polyhydroxy alkaloids: chromatographic analysis. J. Chromatogr. A. 967-957.
[5] Sanz ML, Martínez-Castro I (2007) Recent developments in sample preparation for chromatographic analysis of carbohydrates. J. Chromatogr. A. 1153: 74-89.
[6] Ruiz Matute AI, Hernández Hernández O, RodríguezSánchez S, Sanz ML, Martínez-Castro I (2011) Derivatization of carbohydrates for GC and GC–MS analyses. J Chromatogr. B, 879: 1226–1240.
[7] Wolfrom ML, Thompson A (1963) Reduction with sodium borohydride. Meth. Carbohydr. Chem. 2: 65–68.
[8] Wolfrom ML, Thompson A (1963b) Acetylation. Meth. Carbohydr. Chem. 2: 211–215.

[9] Bromund WH, Herbst RM (1945) The synthesis of oxazoline derivatives of monosaccharides and their relationship to the amino sugars. J. Org. Chem. 10: 267–276.
[10] Kim JB, Carpita NC (1992) Changes in Esterification of the Uronic Acid Groups of Cell Wall Polysaccharides during Elongation of Maize Coleoptiles, Plant Physiol. 98: 646-653.
[11] Maness NO, Ryan JD, Mort AJ (1990) Determination of the degree of methyl esterification of pectins in small samples by selective reduction of esterified galacturonic acid to galactose. Anal. Biochem. 185 : 346-352.
[12] Fontaine T, Fournet B, Karamanos Y (1994) A new procedure for the reduction of uronic acid containing polysaccharides. J. Microbiol. Meth. 20: 149-157.
[13] Parkhomchuk AA, Kocharova NA, Białczak-Kokot M, Shashkov AS, Chizhov AO, Knirel YA, Rozalski A. (2010) Structure of the O-polysaccharide from the lipopolysaccharide of Providencia alcalifaciens. O12.Carbohydr Res. 345: 1235-1239.
[14] Wunschel DS, Colburn HA, Fox A, Fox KF, Harley WM, Wahl JH, Wahl KL (2008) Detection of agar, by analysis of sugar markers, associated with Bacillus anthracis spores, after culture. J. Microbiol. Methods. 74: 57-63.
[15] Lima LFO, Habu S, Gern JC, Nascimento BM, Parada JC, Noseda MD, Gonçalves A, Nisha VR , Pandey A , Soccol VT, Soccol CR (2008) Production and Characterization of the Exopolysaccharides Produced by Agaricus brasiliensis in Submerged Fermentation. Appl. Biochem. Biotechnol. 151:283–294
[16] Viccini G, Martinelli TR , Cognialli RCR, de Faria RO, Carbonero ER , Sassaki LK (2009) Mitchell DA Exopolysaccharide from surface-liquid culture of Clonostachys rosea originates from autolysis of the biomass. Arch. Microbiol. 191:369–378
[17] Smiderle FR, Olsen LM, Carbonero ER, Marcon R, Baggio CH, Freitas CS, Santos ARS, Torri G, Gorin PAJ, Iacomini M (2008) A 3-O-methylated mannogalactan from Pleurotus pulmonarius: Structure and antinociceptive effect, Phytochemistry 69: 2731–2736
[18] Vesentini D, Steward D, Singh AP, Ball R, Daniel G, Franich R (2007) Chitosan-mediated changes in cell wall composition, morphology and ultrastructure in two wood-inhabiting fungi. Mycol. Res. 111 :875-90.
[19] Marquesa G, Gutiérreza A, del Ríoa JC, Evtuguin DV (2010) Acetylated heteroxylan from Agave sisalana and its behavior in alkaline pulping and TCF/ECF bleaching. Carb. Polym. 81: 517–523.
[20] Ayestarán B, Guadalupe Z, León D (2004) Quantification of major grape polysaccharides (Tempranillo v.) released by maceration enzymes during the fermentation process, Anal. Chim. Acta 513: 29–39.
[21] Quemener, Lahaye M, Metro F (1995) Assesment of methanolysis for the determination of composite sugars of gelling carrageenans and agarose by HPLC. Carbohydr. Res. 266: 53-64.
[22] Bleton J, Mejanelle P, Sansoulet J, Goursaud S, Tchapla (1996) Characterisation of neutral sugars and uronic acids after methanolysis and trimethylsilylation for recognition of plant gums, J Chromatogr. A 720: 27-49.

[23] Golovkina LS, Chizhov OS, Wul'fson NS Acetate of polyols. In"Mass spectrometric investigation of carbohydrate".vol 9. Bull. Acad Sci USSR, 1966.

[24] Mc Lafferty FW and Turecek F, In Interpretation of Mass Spectra, University Scince boof, Mill Valley, CA 1993

[25] Jansson PE., Lindberg B (1980) Mass spectrometry of alditol acetates: origin of the fragments having even mass numbers. Carbohydr. Res. 86: 287–292.

[26] Patoureau-Promé D, Promé JC, Carbohydrate, in GC/MS applications in Microbiolog. pp106-156, G Odham, L Larsson, PA Mardh, Plenum Press, New York 1984.

[27] Dizdaroglu M, Henneberg D, Von Sonntag C, Schuchman MN (1977) Mass spectra of trimethylsilyl-di-O-methyloximes of aldosuloses and dialdoses Org. Mass spectrum., 8: 335.

[28] Petersson G (1969) Mass spectrometry of aldonic and deoxyaldonic acids as trimethylsilyl derivatives, Tetrahedron, 25: 4437-4443.

[29] Bleton J, Mejanelle P, Sansoulet J, Goursaud S, Tchapla A (1996) Characterization of neutral sugars and uronic acids after methanolysis and trimethylsilylation for recognition of plant gums, J. Chromatogr. A, 720: 27-49.

[30] De Jongh DC, Radford T, Hribar JD, Hanessian S, Bieber M, Dawson G, Sweeley CC (1969) Analysis of trimethylsilyl derivatives of carbohydrates by gas chromatography and mass spectrometry J. Am. Chem. Soc. 91: 1728-1740.

[31] Kennedy JF, Robertson SM (1978) The structure of Cryptococcus neoformans galactoxylomannan contains β-D-glucuronic acid Carbohydr. Res. 67: 1-15.

[32] Kochetkov NK, Chizhov OS (1966) Mass Spectrometry of Carbohydrate Derivatives, Adv. Carbohydr. Chem. 21, 39-93.

[33] Hellerqvist CG and Sweetman BJ Mass spectrometry of Carbohydrates, pp 90-143. In Biomedical applications of Mass spectrometry Suelter CH, Throck Watson Ed Interscience Publication John Wiley & Sons New York 1990.

[34] Kim JS, Reuhs BL, Michon F, Kaiser RE, Arumugham RG. (2006) Addition of glycerol for improved methylation linkage analysis of polysaccharides. Carbohydr Res., 341: 1061-1064.

[35] Kang P, Mechref Y, Novotny MV. (2008)High-throughput solid-phase permethylation of glycans prior to mass spectrometry. Rapid Commun Mass Spectrom. 22: 721-734.

[36] Kang P, Mechref Y, Klouckova I, Novotny MV. (2005) Solid-phase permethylation of glycans for mass spectrometric analysis. Rapid Commun Mass Spectrom. 19: 3421-3428.

[37] Ciucanu I, Kerek F. (1984) A simple and rapid method for the permethylation of carbohydrates. Carbohydr. Res. 131: 209–217

[38] Ciucanu I, Costello CE (2003) Elimination of oxidative degradation during per-Omethylation of carbohydrates.. J. Am. Chem. Soc. 125: 16213-16219.

[39] de Souza LM, Cipriani TR, Serrato RV, da Costa DE, Iacomini M, Gorin PA, Sassaki GL. (2008) Analysis of flavonol glycoside isomers from leaves of Maytenus ilicifolia by offline and online high performance liquid chromatography-electrospray mass spectrometry. J Chromatogr A., 1207: 101-109.

[40] Kang P, Mechref Y, Novotny MV. (2008) High-throughput solid-phase permethylation of glycans prior to mass spectrometry. Rapid Commun Mass Spectrom.;22: 721-734.

[41] Mechref Y, Kang P, Novotny MV. (2009) Solid-phase permethylation for glycomic analysis. Methods Mol. Biol. 534:53-64.
[42] Smiderle FR, Olsen LM, Carbonero ER, Marcon R, Baggio CH, Freitas CS, Santos AR, Torri G, Gorin PA, Iacomini M. A (2008) 3-O-methylated mannogalactan from Pleurotus pulmonarius: structure and antinociceptive effect. Phytochemistry, 69: 2731-2736.
[43] Kim JS, Reuhs BL, Michon F, Kaiser RE, Arumugham RG. (2006) Addition of glycerol for improved methylation linkage analysis of polysaccharides. Carbohydr Res. 34:, 1061-1064
[44] Weiskopf A.S., Vouros P., Harvey P. (1997) Characterization of oligosaccharide composition and structure by quadrupole ion trap mass spectrometry. Rapid Commun Mass Spectrom, 11: 1493-1504
[45] Visieux N, De Hoffmann E., Domon B. (1998) Structural assignment of permethylated oligosaccharide subunits using sequential tandem mass spectrometry. Anal. Chem., 70: 4951-4959
[46] Hofmeister G.E., Zhou Z., Leary J.A. (1991) Linkage position determination in lithium-cationized disaccharides: tandem mass spectrometry and semiempirical calculations J. Am. Chem. Soc., 113: 5964-5970
[47] König S., Leary J.A (1998) Evidence for linkage position determination in cobalt coordinated pentasaccharides using ion trap mass spectrometry J. Am. Soc. Mass Spectrom., 9: 1125-1134
[48] Chai W., Lawson A.M., Piskarev V.J. (2002) The structural motif in chondroitin sulfate for adhesion of Plasmodium falciparum-infected erythrocytes comprises disaccharide units of 4-O-sulfated and non-sulfated N-acetylgalactosamine linked to glucuronic acid J. Am. Soc. Mass Spectrom, 13: 670-679
[49] Domon B. and Costello C.E. (1988) A systematic nomenclature of carbohydrate fragmentation in FAB-MS/MS spectra of glycoconjugates. Glycoconj. J., 5: 397-409
[50] Ngoka L.C., Gal J.F., Lebrilla C.B. (1994) Effects of cations and charge types on the metastable decay rates of oligosaccharides. Anal. Chem., 66: 692-698
[51] Caroll J., Willard D., Lebrilla C. Energetics of cross-ring cleavages and their relevance to the linkage determination of oligosaccharides. Anal. Chim. Acta, 1995, 307: 431-447.
[52] Dell A., Morris H.R., Egge H., von Nikolai H.Strecker G. (1983) Fast-atom-bombardment, negative-ion mass spectrometry of the mycobacterial O-methyl-D-glucose polysaccharide and lipopolysaccharides. Carbohydr. Res., 115: 41-52
[53] McClellan J.M., Costello C.E., O'Connor P.B., Zaia J. (2002) Influence of charge state on product ion mass spectra and the determination of 4S/6S sulfation sequence of chondroitin sulfate oligosaccharides. Anal Chem, 74: 3760-37771
[54] Tseng K, Hedrick J.L., Lebrilla C.B. (1999) Catalog-library approach for the rapid and sensitive structural elucidation of oligosaccharides, Anal Chem., 71: 3747-3754
[55] Zaia J. Mass spectrometric ionization of carbohydrates. In Encyclopedia of Mass Spectrometry, vol.6. M.G. Gross and r.M. Caprioli, eds. (Elsevier, New-York)
[56] Deery M.J., Stimson E., Chappell C.G. (2001) Mass spectrometry and glycomics. Rapid Commun mass Spectrom., 15: 2273-2283.

[57] Walter T.H., Iraneta P., Caparella P. (2005) Mechanism of retention loss when C8 and C18 HPLC columns are used with highly aqueous mobile phases J. Chromatogr. A, 1075: 177-183.

[58] Hemström P, Irgum K. (2006) Hydophilic interaction chromatography J. Sep. Sci, , 29: 1784-1821.

[59] Kuberan B. Lech M., Zhang L., Wu Z.L., Beeler D.L., Rosenberg R.D. (2002) Analysis of heparan sulfate oligosaccharides with ion pair reversed phase capillary HPLC-microelectrospray ionization time-of-flight spectrometry J. Am Chem Soc., 124:8707-8718

[60] Gustavsson, S.A., Samskog J., Markides K., Langström B.(2001) Studies of signal suppression in liquid chromatography-electrospray ionization mass spectrometry using volatile ion-pairing reagents.J. Chromatogr. A, 937: 41-47

[61] Lunn G., Hellwig L.C. Handbook of derivatization reaction for HPLC, Wiley-Interscience, New-York, 1998

[62] Wang X.D., Li W.Y., Rasmussen H.T. (2005)Orthogonal method development using hydrophilic interaction chromatography and reversed-phase high-performance liquid chromatography for the determination of pharmaceuticals and impurities. J. Chromatogr. A, ,1083: 58-62

[63] Lee Y.C. (1990) High-performance anion-exchange chromatography for carbohydrate analysis. Anal. Biochem, 189: 151-162.

[64] Rocklin R.D., T.R. Tullsen, M.G. Marucco, (1994) High-performance anion-exchange chromatography for carbohydrate analysis, J. Chromatogr. A , 671: 109-114.

[65] Thayer J.R., J.S. Rohrer, N. Avdalovic, R.P. Gearing, (1998) Improvements to in-line desalting of oligosaccharides separated by high-pH anion exchange chromatography with pulsed amperometric detection. Anal. Biochem, 256: 207-216.

[66] Torto N., Hofte A., Tjaden U.R., Gorton L., Marko-Varga G., Bruggink C., Van der Greef J. (1998) Microdialysis introduction high-performance anion-exchange chromatography/ ionspray mass spectrometry for monitoring of on-line desalted carbohydrate hydrolysates. J. Mass Spectrom., 33: 334-341.

[67] Van der Hoeven R.A.M., U.R. Tjaden, J. Van der Greef, W.H.M. Van Casteren, H.A. Schols, A.G.J. Voragen, C. Bruggink (1998), Recent progress in high-performance anion-exchange chromatography/ionspray mass spectrometry for molecular mass determination and characterization of carbohydrates using static and scanning array detection J. Mass Spectrom., 33: 377-386.

[68] Rumbold K., Okatch H., Torto N., Siika-Aho M., Gubitz G., Robra K.H., Prior B. (2002) Monitoring on-line desalted lignocellulosic hydrolysates by microdialysis sampling micro-high performance anion exchange chromatography with integrated pulsed electrochemical detection/mass spectrometry. Biotechnol. Bioeng, 78, 822-828.

[69] Bruggink C., R. Maurer, H. Herrmann, S. Cavalli, F. Hoefler (2005) Analysis of carbohydrates by anion exchange chromatography and mass spectrometry. J. Chromatogr. A, 1085: 104-109.

[70] Bruggink C., Wuhrer M., Koeleman C.A., Barreto V., Liu Y., Pohl C., Ingendoh A., Hokke C.H., Deelder A.M. (2005) Oligosaccharide analysis by capillary-scale high-pH

anion-exchange chromatography with on-line ion-trap mass spectrometry. J. Chromatogr. B., 829: 136-143.
[71] Johnson D.C., Lacourse W.R. (1990) LC with pulsed ECD at gold and platinum Anal. Chem., 62: 589A-597A.
[72] Chataigné G, Couderc F, Poinsot V. (2008) Polysaccharides analysis of sinorhizobial capside by on-line anion exchange chromatography with pulsed amperometric detection and mass spectrometry coupling. J Chromatogr A. 1185:241-250.
[73] Peter Katalinic J. (1994), Analysis of glycoconjugates by fast atom bombardment mass spectrometry and related MS techniques, Mass Spectrom. Rev., 13: 77-98.
[74] Treilhou M., Ferro M., Monteiro C., Poinsot V., Jabbouri S., Kanony C., Promé D., Promé JC. (2000) Differentiation of O-acetyl and O-carbamoyl esters of N-acetyl-glucosamine by decomposition of their oxonium ions: Application to the structure of the non-reducing end of Nod factors. *JAS mass*, 11: 301-311.
[75] Angel AS., Lindh F., Nilsson B. (2007) Determination of binding positions in oligosaccharides and glycosphingolipids by fast-atom-bombardment mass spectrometry. Carbohydr Res. 16:, 15-31.
[76] Berjeaud JM., Couderc F., Promé JC. (1993), Stereochemically controlled decomposition of silver-cationized methyl glycosides, Org. Mass Spectrom, 28, 455-458.
[77] Grimshaw J (1997) Analysis of glycosaminoglycans and their oligosaccharide fragments by capillary electrophoresis Electrophoresis, 18, 2408-2414.
[78] Lagane B, Treilhou M, Couderc F (2000) Capillary Electrophoresis: Theory, Teaching Approach and Separation of oligosaccharides Using Indirect UV Detection BMBE, 28: 251-255.
[79] El-Rassi Z, Postlewait J, Mechref Y, Ostrander GK. (1997) Capillary electrophoresis of carboxylated carbohydrates. III. Selective precolumn derivatization of glycosaminoglycan disaccharides with 7-aminonaphthalene-1,3-disulfonic acid fluorescing tag for ultrasensitive laser-induced fluorescence detection. Anal Biochem. 244: 283-290.
[80] Chen FT, Evangelista RA. (1995) Analysis of mono- and oligosaccharide isomers derivatized with 9-aminopyrene-1,4,6-trisulfonate by capillary electrophoresis with laser-induced fluorescence. Anal Biochem. 230: 273-280.
[81] Fraysse N, Verollet C, Couderc F, Poinsot V. (2003) Capillary electrophoresis as a simple and sensitive method to study polysaccharides of Sinorhizobium sp. NGR234. Electrophoresis. 24: 3364-3370.
[82] Szabo Z, Guttman A, Bones J, Karger BL. (2011) Rapid high-resolution characterization of functionally important monoclonal antibody N-glycans by capillary electrophoresis. Anal Chem. 83: 5329-5336.
[83] Mittermayr S, Bones J, Doherty M, Guttman A, Rudd PM. (2011) Multiplexed analytical glycomics: rapid and confident IgG N-glycan structural elucidation. J Proteome Res. 10: 3820-3829.
[84] Szabo Z, Guttman A, Rejtar T, Karger BL. (2010) Improved sample preparation method for glycan analysis of glycoproteins by CE-LIF and CE-MS. Electrophoresis., 31: 1389-95.

[85] Mechref Y (2011) Analysis of glycans derived from glycoconjugates by capillary electrophoresis-mass spectrometry Electrophoresis, 32: 3467–3481
[86] Tegeler TJ, Mechref Y, Boraas K, Reilly JP, Novotny MV. (2004) Microdeposition device interfacing capillary electrochromatography and microcolumn liquid chromatography with matrix-assisted laser desorption/ionization mass spectrometry. Anal. Chem., 76: 6698-6706.
[87] Gennaro, L. A., Salas-Solano, O., On-line CE-LIF-MS technology for the direct characterization of N-linked glycans from therapeutic antibodies. (2008) Anal. Chem., 80: 3838–3845.
[88] Militsopoulou M, Lecomte C, Bayle C, Couderc F, Karamanos NK. (2003) Laser-induced fluorescence as a powerful detection tool for capillary electrophoretic analysis of heparin/heparan sulfate disaccharides. Biomed Chromatogr., 17: 39-41.
[89] Nakano M, Higo D, Arai E, Nakagawa T, Kakehi K, Taniguchi N, Kondo A. (2009) Capillary electrophoresis-electrospray ionization mass spectrometry for rapid and sensitive N-glycan analysis of glycoproteins as 9-fluorenylmethyl derivatives. Glycobiology., 19: 135-143.
[90] Spaink HP (2000) Root nodulation and infection factors produced by rhizobial bacteria. Ann. Rev. Microbiol. 54:257-288.
[91] Robertson JG, Wells B, Brewin NJ, Wood E, Knight CD, Downie JA (1985) The legume-Rhizobium symbiosis: a cell surface interaction. J. Cell Sci. Suppl. 2: 317-331.
[92] Gage DJ (2004) Infection and invasion of roots by symbiotic, nitrogen-fixing rhizobia during nodulation of temperate legumes. Microbiol. Mol. Biol. Rev. 68: 280-300.
[93] Dénarié J, Debellé F, Rosenberg C (1992) Signaling and host range variation in nodulation. Annu. Rev. Microbiol. 46: 497-531.
[94] Murray JD (2011) Invasion by invitation: rhizobial infection in legumes. Mol. Plant Microbe Interact. 24:631-639.
[95] Carlson RW (1982) Surface chemistry, 199-234, In W. J.Broughton (ed.), Nitrogen fixation: Rhizobium, vol. 2. ClarendonPress, Oxford.
[96] Pellock BJ, Cheng HP, Walker GC (2000) Alfalfa root nodule invasion efficiency is dependent on Sinorhizobium meliloti polysaccharides. J. Bacteriol. 182: 4310-318.
[97] Fraysse N, Couderc F, Poinsot V (2003) Surface polysaccharide involment in establishing the Rhizobium-legume symbiosis. Eur. J. Biol. Chem. 270:1365-1380.
[98] Skorupska A, Janczarek M, Marczak M, Mazur A, Krol J (2006) Rhizobial exopolysaccharides: genetic control and symbiotic functions. Microbial Cell Factories. 5:7, 1-19.
[99] Freitas F, Alves VD, Reis MA (2011) Advances in bacterial exopolysaccharides: from production to biotechnological applications. Trends Biotechnol. 29:388-398.
[100] Rojas-Escudero E, Alarcon-Jimenez AL, Elizalde-Galvan P, Rojo-Callejas F (2004) Optimization of carbohydrate silylation for gas chromatography. J. Chromatogr. A. 1027: 117-120.
[101] Carlson RW, Forsberg LS, Kannenberg EL (2010) Lipopolysaccharides in Rhizobium-legume symbioses. Subcell. Biochem. 53:339-386.

[102] Carlson RW, Garci F, Noel D, Hollingsworth R (1989) The structures of the lipopolysaccharide core components from Rhizobium leguminosarum biovar phaseoli CE3 and two of its symbiotic mutants, CE109 and CE309. Carbohydr. Res. 195:101-110.

[103] Abdel-Akher M, Hamilton JK, Smith F (1951) The reduction of sugars with sodium borohydride. J. Am. Chem. Soc. 73: 4691-4692.

[104] York WS, Darvill AG, MCneil M, Stevenson TT, Albersheim P (1985) Isolation and characterization of plant cell walls and cell wall components. Methods in Enzymol. 118: 3-40.

[105] Reuhs BL, Carlson RW, Kim JS (1993) Rhizobium fredii and Rhizobium meliloti produce 3-deoxy-D-manno-2-octulosonic acid-containing polysaccharides that are structurally analogous to group II K antigens (capsular polysaccharides) found in Escherichia coli. J. Bacteriol. 175 :3570-3580.

[106] Fraysse N, Lindner B, Kaczynski Z, Sharypova L, Holst O, Niehaus K, Poinsot V (2005) Sinorhizobium meliloti strain 1021 produces a low-molecular-mass capsular polysaccharide that is a homopolymer of 3-deoxy-D-manno-oct-2-ulosonic acid harboring a phospholipid anchor. Glycobiology. 15 :101-108.

[107] Zanetta JP, Timmerman P, Leroy Y (1999) Determination of constituents of sulphated proteoglycans using a methanolysis procedure and gas chromatography/mass spectrometry of heptafluorobutyrate derivatives. Glycoconj. J. 16: 617-627.

[108] Becker A, Fraysse N, Sharypova L (2005) Recent advances in studies on structure and symbiosis-related function of rhizobial K-antigens and lipopolysaccharides. Mol. Plant Microbe Interact. 18: 899-905.

Chapter 11

Concept of Template Synthesis of Proteoglycans

Aleksandr N. Zimnitskii

Additional information is available at the end of the chapter

1. Introduction

It is accepted in modern genomics that only a minor portion (1-5%) of the genome contains information about the primary structure of protein molecules. The higher is the organism on the evolutionary ladder, the lower the density of genetic information, in the modern sense of the term, per kilobase (kb). For instance (Table 1), this parameter is 1-1.7 kb per gene in bacteria and more than 30 kb per gene in human.

Organism	Genome size, Mb	Total genes, 10^3	Density, kb/gene
Bacteria	0.5-5	0.47-4.29	1-1.7
Yeasts	12	6	2
Nematode	97	19	5
Human	3000	80-100	>30

Table 1. Comparison of the genome size and gene number (Kiselev, L.L., 2000)

The genome can be conventionally divided into three groups of nucleotide sequences: unique sequences, moderate repeats (DNA segments repeated frequently), and fast repeats (satellite DNA). This division is exclusively conventional; the assignment to a particular group is determined by the capability of fragmented and denatured DNA to find complementary regions upon reassociation in solution. Yet such a division makes it possible to study the functional role of individual genome regions. It was shown, for instance, that unique DNA sequences are responsible mostly for encoding the primary structure of protein molecules. Such sequences occur in a few copies in the genome. Part of information of repetitive DNA systems (histone, rRNA, and tRNA genes) is encoded by so-called slow DNA repeats

The role of fast repeats and satellite DNA is still enigmatic to a great extent. Such DNA is often called selfish or waste, suggesting a lack of information valuable for the organism.

Surprisingly, it is this DNA that increases in amount in organisms with the highest level of organization (for instance, repetitive sequences account for more than 95% of the human genome). Data accumulating in recent decades for inverted repeats, regulatory gene regions, centromeric and telomeric satellite DNAs, introns, and mobile genetic systems still fail to improve the understanding of the functional role of tandem repeats, which quantitatively constitute the basis of the genome. The hypothesis that repeats are just dilutors of genetic information is below the level of current knowledge, testifying to a poor understanding of the phenomenon rather than helping to elucidate their role (Singer M., Berg P. 1998).

Some authors believe that tandem repeats play an important, though still unknown, part in evolutionary improvement of organisms. Thus repeats show species specificity (Mednikov B.M., et al., 2001) and occur in hundred of thousands or even millions of copies in the genome (Table 2). Simple tandem repeats, such as 5-CA, 5-GA, 5-AT, and 5-GC, are present in virtually all eukaryotes and have huge copy numbers.

Repeat	Organisms	Purine:pyrimidine ratio
5-GGAAG	birds	2:0
5-CA	many eukaryotes	1:1
5-GA	many eukaryotes	2:0
5-GT	many eukaryotes	1:1
5-AT	many eukaryotes	1:1
5-CAAA GTTTT GTTTGA	*Xenopus laevis*	1:0.9
5-TCTCC	birds	0:2

Table 2. Some tandem repeats dispersed through the genomes of various eukaryotes (Singer, M., Berg, P., 1998)

In terms of base groups, the sequences shown in Table 2 can be presented as multiple repeats of (Pyr-Pyr), (Pur-Pur), and (Pur-Pyr), where Pyr is a pyrimidine (T or C) and Pur is a purine (A or G). Thus, the genome contains extended monopurine and/or monopyrimidine arrays as well as tandem repeats whose averaged unit contains Pur and Pyr bases in nearly equal proportions. It is also beyond doubt that excess information is read from DNA during transcription yielding gnRNA, part of which is excised during RNA maturation. Information is transferred from DNA to mRNA almost directly, without processing, in prokaryotes, whereas the situation observed in higher organisms is similar to that with DNA repeats. The higher the position of an organism on the evolutionary ladder, the greater the amount of information that is transmitted to RNA and is senseless in terms of the polypeptide sequence (Singer M., Berg P., 1998).

While DNA determines the ontogenetic and phylogenetic development of organisms and is structurally described as a polymer of dimeric units, a natural question is whether the cell possesses another structurally similar biopolymer (apart from RNA). As is well known, proteins are unfit for this role, because the triplet code of polypeptides requires at least three different monomers for their implementation. There is only one group of tandem biopolymers playing a key role in the functioning of higher organisms as multicellular entities. This group is polysaccharides.

Polysaccharides are biopolymers that consist of monosaccharides linked to each other. It was believed until the 1960s (Colman Y., Rem C.-G. 2000), that carbohydrates serve only as an energy source (monosaccharides and storage polysaccharides such as starch and glycogen) and structural material (e.g., cellulose in plants and chitin in insects). Interest in carbohydrates was moderated by their extreme structural complexity. While there is only one way of linking together monomers of nucleic acids (nucleotides) or proteins (amino acids), monosaccharide units of oligosaccharides and polysaccharides can be linked in several ways at several different positions. Thus only 256 tetranucleotides can be obtained with four different nucleotides, whereas four different monosaccharides can form 35,560 unique tetrasaccharides (Colman Y., Rem C.-G. 2000). Consequently, biological polymorphism of polysaccharides is immeasurably higher than that of proteins or nucleic acids (NA).

By composition, polysaccharides are usually divided into homopolysaccharides (homoglycans) and heteropolysaccharides (heteroglycans). In almost all cases, the chemical structure can be reduced to a dimeric unit repeated many times (Fig. 1, Table 3). Glycosaminoglycans (GAG) are the most common natural heteroglycans.

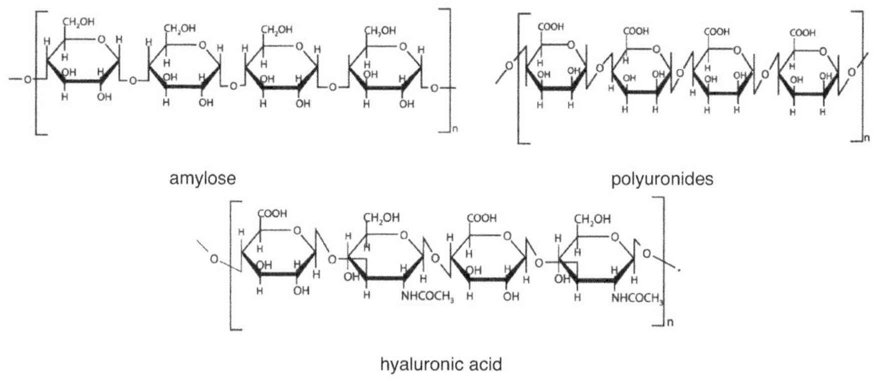

Figure 1. Structures of common polysaccharides.

GAG is a polymer that consists of hexuronic acid and hexosamine residues linked by O-glycoside bonds. The three major GAG classes are nonsulfated hyaluronic acid (HA), moderately sulfated chondroitin sulfates (CS), and highly sulfated heparan sulfates (HS). GAG of the last two classes belong to proteoglycans. GAG contained in proteoglycans play various vital parts in the intercellular matrix and within the cell (Zimnitskii A.N., Bashkatov S.A. 2004).

PS, source	PS formula	Uronic acid:hexose ratio
Polysaccharides of microorganisms	GlcUA - Glc	1:1
Polyuronides of plants	GlcUA - GlcUA	2:0
Amylose and cellulose of plants, glycogen of animals	Glc-Glc	0:2
Chitin of insects and crustaceans	GlcNAc – GlcNAc	0:2
Heparan of animals	GlcUA - GlcN	1:1
Hyaluronic acid of animals	GlcUA - GlcNAc	1:1
Dermatan sulfate of animals	IdoUA - GalNAc	1:1
Chondroitin sulfate (CS) of animals	GlcUA - GalNAc	1:1
CS A of animals	GlcUA – GalNAc4SO_4	1:1
CS C of animals	GlcUA – GalNAc6SO_4	1:1
CS of animals	GlcUA3S – GalNAc4SO_4	1:1
CS of animals	GlcUA2S – GalNAc6S	1:1
CS of animals	GlcUA - GalNAc(4,6-diSO_4)	1:1
CS of animals	IdoUA – GalNAc4SO_4	1:1
CS of animals	IdoUA2(3)SO_4 - GalNAc	1:1
CS of animals	IdoUA2(3)SO_4 - GalNAc4SO_4	1:1
CS of animals	IdoUA2SO_4 – GalNAc6SO_4	1:1
CS of animals	IdoUA – GalNAc(4,6-diSO_4)	1:1
CS of animals	IdoUA2(3)SO_4 – GalNAc(4,6-diSO_4)	1:1
Heparan sulfate (HS) of animals	IdoUA - GlcNSO_4	1:1
HS of animals	IdoUA – GlcN$SO_4$6SO_4	1:1
HS of animals	IdoUA2(3)SO_4 - GlcNSO_4	1:1
HS of animals	IdoUA2(3) – bGlcN$SO_4$6SO_4	1:1
HS of animals	GlcUA - GlcNSO_4	1:1
HS of animals	GlcUA – GlcNAc6SO_4	1:1
HS of animals	GlcUA – GlcNS6SO_4	1:1
HS of animals	GlcUA2SO_4 – GlcN$SO_4$6SO_4	1:1
HS of animals	GlcUA2SO_4 - GlcNSO_4	1:1
HS of animals	GlcUA2SO_4 -GlcN$SO_4$$SO_4$	1:1

Table 3. Uronic acid/hexose ratio in storage and structural polysaccharides (PS)

HA biosynthesis was studied in group A hemolytic streptococci (Dorfman A 1958, 1964,1965). Similar synthetic processes occur in higher animals (Glaser L., Brown D.H. 1955). Roseman (Roseman S., Ludowieg J., et. al. 1953, Roseman S., Moses F.E., et. al. 1953, Roseman S., Moses F.E, et. al. 1954) have shown that D-glucose is a precursor of glycosamine and glucuronic acid, which are HA monomers. The immediate precursors of HA are uridine diphosphate (UDP) derivatives of glucuronic acid and N-acetylglucosamine, which act as donors of carbohydrate residues during HA synthesis (Cifonelli J.A., et al. 1957, Markovic O., et al. 1964, Markovitz A., et al. 1958, Lorenzel I. 1959). A cell-free system allowing HA synthesis in the presence of UDP-glucuronic acid, UDP-glucosamine, and Mg^{2+} was obtained by sonicating a microbial culture producing HA (Markovic O., Huttl S. 1964). The enzyme responsible for HA synthesis sedimented at 100.000 g. A similar enzyme was found in rat embryo skin by Schiler (Schiller S. 1964, Schiler S. 1965) and in various chicken, rodent, and human tissues by Altschuler (Altschuler C.H., et al. 1963). As in microorganisms, nucleotide derivatives result from the reactions catalyzed by pyrophosphorylases in higher animal tissues producing HA (Strominger J.L. 1964). Acetylation of UDP-glucose involves acetyl coenzyme A. Glucose is transformed into glucosamine before being linked to UDP.

As Slutskii (Slutskii L.I. 1969). showed in the late 1970s, when biosynthesis of heteroglycans of the intracellular matrix were studied most intensely in higher animals, and modern data demonstrate, the enzyme involved in synthesis of polysaccharide fragments of GAG is associated with membrane structures in the intact cell. One is forced to accept the fact that little is still understood about the mechanism of final steps of GAG biosynthesis: how it is possible that a strict alternation of acetylglucosamine and glucuronic acid residues in the macromolecular glycan chain is achieved in the absence of a template.

Studies showed that synthesis of sulfated GAG is essentially similar to synthesis of nonsulfated HA and that D-glucose is a precursor of monosaccharide residues as in the case of HA. However, a necessary step is epimerization of glucosamine into galactosamine in the case of CS and glucuronic acid into iduronic acid in the case of dermatan sulfate. Several works demonstrated epimerization of UDP-monosaccharides in the presence of $NADPH^{+}$ and specific epimerases (Maley F., et al. 1959, Roden L., et al. 1958, Rondle C.J.M., et al. 1955). White (White B.N., et al. 1965) observed the reversible character of epimerization with the example of transformation of galactosamine into glucosamine during GAG synthesis. During synthesis of HS and, in particular, heparin, nucleotide derivatives of hexuronic acids and hexosamines form not only β-glycosidic bonds, characteristic of HA and CS, but also α-glycosidic bonds, characteristic of HS (Silbert J.J. 1963).

According to current views, the first step of proteoglycan synthesis is translation of the protein core mRNA on ribosomes of the rough endoplasmic reticulum. Then glycosylation of the protein core is initiated: a site is generated at which the prospective GAG is to be linked to the polypeptide chain. This process has not been localized so far; it is known only that the protein core already has a polysaccharide fragment when delivered into the endoplasmic reticulum (Zimina N.P., et al. 1992, Zimina N.P., et al. 1987, White J, et al. 1978, Silbert J.E. et al, 1995). It is thought that synthesis of GAG chains is template-independent:

glycosyltransferases just consecutively add monosaccharides of donor UDP-saccharides to the growing carbohydrate chain. The substrate specificity of glycosyltransferases is determined by the monosaccharide sequence. The growth of GAG chains is initiated by xylosyltransferase. However, only some serine residues of the protein core are subject to xylosylation. The site of attachment of a GAG chain is presumably selected depending on the amino acid sequence of the region to be glycosylated (Zimina N.P., et al. 1992, Zimina N.P., et al. 1987, Jeffrey D Esko, et al. 1996). After a xylose residue is linked to the protein, a carbohydrate chain is formed via adding consecutively two galactose residues and one glucuronic acid residue by galactosyltransferases I and II and glucuronyltransferase I. The newly synthesized tetrasaccharide serves as an acceptor in the first reaction of GAG synthesis. Polymeric GAG chains result from multiple reactions performed alternately by two enzymes associated with the membrane Golgi complex. These enzymes are N-acetylgalactosaminyltransferase and glucuronyltransferase II in the case of CS and dermatan sulfates. In the case of HS and heparin, polymerization is catalyzed by N-acetylglucosaminyltransferase and glucuronyltransferase (Zimina N.P., et al. 1992, Zimina N.P., et al. 1987, Sugahara K., et al. 2000, Silbert J.E. 1982). The structure of the protein core is critical for the structure of GAG chains (Jeffrey D Esko, et al. 1996).

The immediate substrates of GAG synthesis are nucleotide (UDP) derivatives of monosaccharides; these derivatives are generated in the reactions catalyzed by pyrophosphorylases. Glucuronic acid is produced from UDP-glucose via two-step oxidation of C_6, with transformation of the hydroxyl group into a carboxyl group. In dermatan sulfate synthesis, UDP-glucuronic acid is epimerized into UDP-iduronic acid by specific epimerase.

Thus, during GAG biosynthesis, one enzyme complex elongates the heteroglycan chain by adding consecutively monosaccharide units. Treatment with ribonucleases or deoxyribonucleases did not prevent chain elongation in a model HA-producing system. This finding was interpreted as demonstrating the template-independent character of GAG synthesis. We think, however, that template independence of GAG synthesis is not evident from this finding, because the glycans produced in the model system were not compared with natural glycans. It is beyond doubt that, in the absence of a template, the enzyme complex is capable of producing a heteroglycan chain at random, as is the case with NA synthesis in *in vitro* systems.

To eliminate the contradiction between the clearly ordered GAG structure (Zimina N.P., et al. 1992, Zimina N.P., et al. 1987) and the concept of non-template GAG synthesis, attempts were made over the past decade to associate the ordered character of the GAG structure with template synthesis of the protein core of proteoglycans; i.e., the amino acid sequence of the protein core was considered as a kind of a template for GAG synthesis. Although some achievements were made in the field, they mostly revealed statistical, rather than cause-and-effect, relationships. It seems that this situation has been accepted as satisfactory, and the concept of non-template GAG synthesis associated with the protein core is now described in textbooks. Yet the concept is shattered completely by the fact that that HA, reaching a molecular weight of 10^7 Da, is an individual GAG lacking any protein core, in contrast to CS and HS.

Current views of proteoglycan biosynthesis are related to studies of the role of intracellular membranes. It is commonly known that cell proteoglycans are almost always associated with membrane structures. The processes of proteoglycan biosynthesis, intracellular transport to organelles, and transfer onto the cell surface in the intercellular matrix are coupled with the functioning of membranes. GAG accumulation in lysosomes and subsequent cleavage to monosaccharides are also connected with the functioning of membrane structures. Free proteoglycans and GAG that are not associated with membranes are detectable only at the last degradation stages in lysosomes and some other cell structures such as the nucleus and mast cell granules.

Scarce, if any, information is available concerning the early steps of GAG synthesis and GAG transport to the sites where polysaccharide chains are generated (Silbert J.E. et al. 1995). As the views of the structure of the protein core are generally discrepant, it is still unclear whether the protein core determines the formation of proteoglycans. The primary structure of a core protein with a potential serine xylosylation site suggests that the protein molecule plays no crucial role in recognition of the xylosylation site. Hence it is probable that the conformation of membrane structures is a factor determining the recognition of xylosylation sites in protein molecules of proteoglycans. This assumption can be advanced because the main determinant of xylosylation is probably associated with membrane structures of the endoplasmic reticulum and Golgi complex, where glycosyltransferase activity is high (Silbert J.E. et al. 1995).

Synthesis of the linker tetrasaccharide (an uronic acid residue is added to the trisaccharide) initiates addition of a particular N-acetylhexosamine and thereby determines which GAG chain is to be generated. As a result, the nascent proteoglycan is transferred into the corresponding site, containing either galactose N-acetyltransferases or glucose N-acetyltransferases; this determines the chain to be synthesized. It is the amino acid sequence of the peptide moiety that determines the position of the nascent proteoglycan on membrane structures and, consequently, the direction of synthesis and modification of the polysaccharide chain by the corresponding enzyme systems.

The primary structure of GAG in proteoglycan molecules is an important factor determining their function. For instance, the GAG chain size, the number and location of sulfo groups, and epimerization affect the properties and biological role of proteoglycans. The extent of modification of the polysaccharide chain and its size probably depend on the structure of membrane-associated complexes responsible for proteoglycan synthesis. Membrane enzyme complexes play a key role in epimerization and/or sulfation of GAG polysaccharide chains (Silbert J.E. et al. 1995).

In 1987, Lindahl (Lindahl U., et al. 1987) showed convincingly that the results of GAG biosynthesis *in vitro* cannot be extrapolated to the *in vivo* situation. Studying heparin biosynthesis, these researchers observed that polysaccharides obtained with intact microsomes contain extended sequences of both N-sulfated and N-acetylated disaccharides, suggesting a nonrandom character of their synthesis. In contrast, products obtained with a solubilized microsomal system displayed a random distribution of such groups with a greater portion of N-acetylated and N-sulfated disaccharides (Lindahl U., et al. 1987). In fact, this finding demonstrates that synthesis of the glycoside moiety of proteoglycans is genetically determined *in vivo*, because their primary structure is tissue- and species-specific.

A basic unit monomers of polysaccharides is glucose, which occurs in solution both in the α and in the β form owing to the mutarotation reaction. In addition, glucose can be epimerized to other hexoses and be oxidized to yield glucuronic and other hexuronic acids. Owing to such lability of glucose, biosynthetic systems always contain sufficient amounts of the α and β forms of hexoses and hexuronic acids. Modified with UDP at C_1, these monosaccharides provide the main components for synthesis of polysaccharide chains.

As Fig. 1 and Table 3 demonstrate, the majority of known structural and storage polysaccharides each consist of two monomers, a hexose and hexuronic acid, which occur in the α or β form, are linked through (1-3), (1-4) O-glycoside bonds, and are modified to a various extent at various carbon atoms as a result of acetylation, amination, sulfation, etc. All these biopolymers can be combined in one group with a universal structure of multiply repeated (A-B), (A-A), or (B-B) units, where A is a hexose and B is a hexuronic acid.

We think that the periodicity of the primary structure is similar between polysaccharides and DNA tandem repeats. In view of this structural similarity, it was justified and important to study a possible complementarity between monosaccharides of glycans and bases of NA.

To check the hypothesis of complementarity of NA bases to hexoses and hexuronic acids, quantum chemical methods were used in our lab for a particular case of glucose and glucuronic acid contained in the heteropolysaccharide HA. It is clear that such an approach is computational and that the relevant conclusions need experimental verification. In view of this, it was necessary to obtain empirical information supporting or contradicting the results of quantum chemical computations. For this purpose, we employed UV spectrophotometry and dot hybridization, which allow detection of specific complexes of biopolymers.

Since the mechanism initiating GAG biosynthesis is unclear, it was expedient to study glycan synthesis in the rat liver upon administration of elevated doses of glucose. The use of $^{35}SO_4^{2-}$ as a radioactive label is inadequate for studying GAG synthesis, and we decided to label a precursor of the glycan polysaccharide chain. We used glucose as such a precursor: glucose is transformed into UDP-glucose and then into UDP-glucuronic acid, which is utilized in GAG synthesis. It should be noted that ribose 5-phosphate, which is formed from glucose 6-phosphate, is incorporated in NA. Glucose is converted into ribose in the pentose phosphate cycle by eliminating C1, which is released as carbon dioxide. Hence, we used [^{14}C]glucose labeled at C_1 to prevent generation of radiolabeled NA in our experiments. We studied the composition of rat liver polysaccharides in cell nuclei, microsomes, and in a total liver homogenate. In addition, the nuclear and microsomal fractions were used to monitor the accumulation of radiolabeled saccharides, which are polysaccharide precursors.

2. Experimental

Quantum chemical modeling of biological structures, in particular, the geometric and electron structures of NA and polysaccharides, requires that the methods used report adequately the effects of weak intermolecular interactions, such as hydrogen bonds. The MP2 *(ab initio)* procedure, which utilizes bases with diffuse and polarization functions, meets this requirement quite well (Cybulski S.M., et al. 1989, Latajka Z. et al. 1990) but is hardly suitable for our objects because of their size: their analysis would be extremely time-consuming and requiring excessive computational resources. A possible alternative in this situation is provided by the corresponding semiempirical methods. Early semiempirical methods (MINDO/3 and MNDO) considerably underestimated the energy of hydrogen bonds and, consequently, were unsuitable for studying biopolymers (Williams I.H. 1987). To eliminate such drawbacks, the AM1 (Dewar M.J.S., et al. 1985) and PM3 (Stewart J.J.P. 1989) methods were developed on the basis of neglect of diatomic differential overlap (NDDO). These methods were expected to adequately describe systems with hydrogen bonds. This was not the case with AM1: while computed energies of hydrogen bonds agreed well with experimental estimates, geometric parameters failed to represent the facts (Jurema J.M.W., et al. 1993). Parametrization was performed with a far greater body of experimental data in PM3 than in other semiempirical methods, which allowed PM3 to describe well the geometric structure of molecules and the heat of their generation. PM3 is indeed the first method yielding semiempirical estimates that agree with the results of experiments and *ab initio* calculations for hydrogen-bonded systems (Kallies B., et al.1995).

2.1. Method of calculation *Ab initio*

Ab initio calculations are the main computational procedure in quantum chemistry and consists in solving Hartree-Fock one-electron equations. As initial data, the method utilizes the charges of nuclei, their positions in the molecule, and Slater- or Gaussian-type basis function sets. The method involves none of the observed physico-chemical properties of substances and, accordingly, is known as unempirical calculations. *Ab initio* calculations most commonly employ the MO LCAO (molecular orbitals as a linear combination of atomic orbitals) approximation, which takes account of all electrons of the system in question. This method is most accurate in quantum chemical computations, especially with intermediate (6-31G*) and large (6-31**) bases, and allows correct estimation of the electron and geometric structures of hydrogen bonds, which play an important role in biological objects.

Both unempirical *ab initio* calculations with basis 6-31G* and the PM3 semiempirical method were employed in theoretical computations in this work. Geometric parameters and energy characteristics were computed by the unempirical and semiempirical methods for interacting nucleotides and saccharides and by the semiempirical methods for structures containing several pairs of nucleotides and saccharides. All computations were performed with complete optimization of geometrical parameters, using the GAMESS program (Schmidt M.W., et al. 1993). A global minimum of the total electron energy was sought by the Newton-Raphson method with an energy gradient of 0.010 kcal/mol·Å, starting from various initial approximations of the complex structure.

2.2. Isolation of oat polyuronides

Fresh oat seedlings grown for 10–12 days were thoroughly ground and hydrolyzed with 2 M NaOH (1:10 w/v) at room temperature with occasional stirring for 1 day. The hydrolysate was neutralized with concentrated HCl to pH 7.0 and filtered through filter paper. Gel filtration on Sephadex G-25 was used to remove salts and low-molecular-weight components and to isolate a high-molecular-weight fraction containing polysaccharides. The preparation collected to obtain polyuronides was fractionated by anion-exchange chromatography on DEAE cellulose. The preparation was applied onto a column with fibrous DEAE cellulose (a chloride form) equilibrated with distilled water. The column was washed consecutively with distilled water and with 0.15 and 0.5 M NaCl to the minimal A_{220} of the eluate . Then, polyuronides were eluted with 1.5 M NaCl. To remove NaCl, the resulting fraction was subjected to gel filtration on Sephadex G-25; the presence of chlorides was checked using silver nitrate. The preparation was dried under vacuum at 45^{0}C and proved to contain glucuronoxylans with a purity of no less than 95% and a proportion of hexuronic acids of no less than 95%. The quality of the product was checked by the Dishe test for uronic acids. The product was dissolved to 0.1 mg/ml. A 0.5-ml aliquot of the solution was combined with 3 ml of concentrated sulfuric acid containing 0.025 M sodium tetraborate and thoroughly mixed. The mixture was heated in boiling water for 10 min, chilled to room temperature, combined with 0.1 ml of 0.1% carbazole (ethanol solution), heated in boiling water for 15 min, and chilled. A_{530} was measured against a control sample. The polyuronide concentration (mg/ml) was estimated using a calibration plot, which was constructed using aqueous solutions of glucuronic acid.

2.3. Isolation of garlic glucuronoxylans

Unpeeled garlic bulbs were thoroughly ground and hydrolyzed with 2 M NaOH (1:10 w/v) at room temperature with occasional stirring for 1 day. The hydrolysate was neutralized with concentrated HCl to pH 7.0 and filtered through filter paper. Gel filtration on Sephadex G-25 was used to remove salts and low-molecular-weight components and to isolate a high-molecular-weight fraction containing polysaccharides. The preparation collected to obtain glucuronoxylans was fractionated by anion-exchange chromatography on DEAE cellulose. The preparation was applied onto a column with fibrous DEAE cellulose (a chloride form) equilibrated with distilled water. The column was washed consecutively with distilled water and with 0.15 and 0.3 M NaCl to the minimal A_{220} of the eluate. Then, glucuronoxylans were eluted with 0.5 M NaCl. To remove NaCl, the resulting fraction was subjected to gel filtration on Sephadex G-25; the presence of chlorides was checked using silver nitrate. The preparation was dried under vacuum at 45^{0}C, yielding glucuronoxylans in a dry form. The quality of the product was checked by the Dishe test for uronic acids as above. The purity of the product was no less than 95%; the proportion of hexuronic acids was 20%.

2.4. Isolation of GAG (HA, CS, and HS) from human placenta

Placenta was thoroughly ground and hydrolyzed with 2 M NaOH (1:10 w/v) at room temperature with occasional stirring for 3 days. The hydrolysate was neutralized with concentrated HCl to pH 7.0. To precipitate nonhydrolyzed proteins, concentrated trichloroacetic acid (TCA) was added slowly to a final concentration of 5%. The preparation was incubated at 4^0C for 1 h, and the precipitate was removed by centrifugation at 1500 g for 10 min. Concentrated NaOH was added to the supernatant to pH 10-11, because some proteins withstanding TCA treatment precipitated in an alkaline milieu. The precipitate was allowed to form and was removed by centrifugation (1500 g, 10 min). The supernatant was filtered through filter paper. Gel filtration on Sephadex G-25 was used to remove salts and low-molecular-weight components and to isolate a high-molecular-weight fraction containing GAG.

The preparations used to isolate HA and CS were fractionated by anion-exchange chromatography on DEAE cellulose. Each preparation was applied individually onto a column with fibrous DEAE cellulose (a chloride form) equilibrated with distilled water. The column was washed consecutively with distilled water and 0.15 M NaCl to the minimal A_{220} of the eluate. Then HA was eluted with 0.25 M NaCl; CS, with 0.7 M NaCl; and HS, with 1.5 M NaCl. To remove NaCl, each fraction was subjected to gel filtration on Sephadex G-25; the presence of chlorides was checked with silver nitrate. The resulting fractions were dried under vacuum at 45^0C to obtain HA, CS, and HS in a dry form. The purity was no less than 95%. The quality of the products was checked by the Dishe test for uronic acids as above.

2.5. NA and amylose

The single-stranded probe $d(T)_{16}$ was obtained from Sintol (Russia). Garlic DNA was isolated from root meristem tissue of seedlings. Human DNA was isolated from the placenta by phenol-detergent extraction (Belozersky A.N. 1970). The purity and concentration of DNA were checked by optical density and by Hoechst 33258 fluorescence, using a DNA minifluorimeter (Hoefer Scientific Instruments, United States). Fragmented calf thymus double-stranded DNA (MW ~1 x 10^6 Da) and single-stranded polynucleotides $d(A)_{350}$, $d(C)_{350}$, and $d(GC)_{350}$ were purchased from Amersham Biosciences. Amylose was purchased from Serva Feinbiochemica GMBH.

2.6. Spectrophotometric assays

Working solutions of polynucleotides were prepared in a quartz cuvet with a light path of 2 mm. Small (5-25 µl) aliquots of a stock polynucleotide solution were added to 0.3M NaCl, 0.002M Na-phosphate (pH 6.8) to the final volume 400 µl. After recording the absorption or circular dichroism (CD) spectra of a polynucleotide solution, polynucleotide-polysaccharide complexes were obtained. For this purpose, 400 µl of the polynucleotide solution were combined in the cuvet with an aliquot of a stock polysaccharide solution with continuous stirring. Then the absorbance or CD spectra were recorded again.

The annealing of double-stranded DNA with the corresponding polysaccharide was carried out in a closed quartz cuvet with a light path of 2 mm. The cuvet containing DNA and the polysaccharide in certain proportions was heated in boiling water for 2 min and immediately chilled in ice-cold water (0°C) for 30 s. Then the absorbance or CD spectra were recorded.

The absorbance spectra were obtained using a Specord M-40 spectrophotometer. The CD spectra of complexes were recorded with an SKD-2 portable dichrometer (Institute of Spectroscopy, Russian Academy of Sciences, Troitsk). The CD spectra were presented as a wavelength dependence of the CD value $\Delta A = A_L - A_R$ (Dunin V.V., et al. 1979).

2.7. NA radiolabeling

NA were 5'-endlabeled *in vitro* with γ-[^{32}P]ATP and T4 polynucleotide kinase (Maxam A., et al. 1977). Dephosphorylation was carried out in 50 mM Tris-HCL (pH 9.0), 10 mM $MgCl_2$, 10 mM spermidine, 1 mM $ZnCl_2$, 0.1 units of alkaline phosphatase at 37°C. The enzyme was removed by treating the samples twice with phenol. NA was precipitated with ethanol and collected by centrifugation in a minicentrifuge at 12,000 rpm for 5 min. The pellet was washed with 70% ethanol to remove phenol, dried, and dissolved in a necessary volume of bidistilled water. The resulting solution was combined with a kinase buffer (50 mM Tris-HCl, pH 7.6), 10 mM $MgCl_2$, 50 mM dithiothreitol, 1 mM spermidine), 1-2 MBq of γ-[^{32}P]ATP, and 2 units of T4 polynucleotide kinase. The mixture was incubated at 37°C for 30 min. The reaction was terminated by adding EDTA to 20 mM, and the enzyme was removed by phenol treatment. NA was precipitated with ethanol, and the nonbound label was removed by repeated precipitation with ethanol or by gel filtration on Sephadex G-25.

2.8. Dot hybridization

BA83 and BA85 filters (Schleicher & Shuell, Germany) with immobilized DNA and polysaccharides (20 μg per dot) were hybridized with radiolabeled NA probes at 37°C for 18-24 h. The hybridization buffer contained 5 x Denhardt's solution (0.02% bovine serum albumin, 0.02% polyvinylpyrrolidone, 0.02% Ficoll), 0.1% SDS (Denhardt D.T. 1966). Before being applied onto a filter, DNA was denatured in boiling water for 5 min. Filters were dried and exposed with a PM-B X-ray film for 1-7 days.

2.9. Neutral sugar assays

Neutral sugars (monosaccharides contained in protein-polysaccharide complexes) were assayed by the anthrone method according to (Jermyn M.A. 1975, Chaplin M.F., et al. 1986). A sample (0.5 ml) was combined with 4 ml of an anthrone reagent (20 mg of anthrone, 100 ml of 80% sulfuric acid). The mixture was heated in boiling water for 5 min, chilled in water to room temperature, and tested for A_{630}. As a control, we used an aqueous solution water treated with the anthrone reagent as above. Neutral sugars were quantitated using a calibration plot, which was constructed with galactose solutions.

To study the GAG composition, a homogenate and the nuclear fraction of the liver were obtained from intact adult rats after fasting over 24 h. To estimate the contents of glycans (HA, CS, and HS), alkaline lysis, deoxyribonuclease treatment, gel filtration on Sephadex G-25, and ion-exchange chromatography on DEAE cellulose were performed as above. Polysaccharide elution profiles were obtained with a gradient of NaCl concentration (0-1.5 M). DNA was assayed in nuclei and liver homogenates as in (Slutskii L.I. 1969).

In the next series of experiments, non-inbred white rats weighing 190-210 g were injected intraperitoneally with [^{14}C]glucose, labeled at C1, at 1 g per kg body weight (4 x 10^8 Bq per rat). Rats were sacrificed and the liver frozen in liquid nitrogen 15, 30, 60, 120, and 360 min after the injection. Differential centrifugation was used to obtain the nuclear and microsomal fractions (Orekhovich V.N. 1968), which were tested for radioactivity. Tissue homogenates were obtained similarly; their GAG fractions were isolated and tested for radioactivity.

The results were statistically processed using the software package STATISTICA 6.0 for WINDOWS. The processing included computation of the mean and standard deviation, tests for significance of differences; and correlation, regression, and factor analyses as well as analysis of variance.

3. Results

3.1. Quantum chemical analysis

3.1.1. Analysis of the electron and geometric structures of hydrogen bonds in complementary pairs of NA by the PM3 method

As a control, computations were performed for the interactions of bases in classical complementary pairs AT and CG, which are DNA components. Selection of computation models showed that substitution of methyl groups for the sugar-phosphate backbone has no effect on the parameters of hydrogen bonds, but considerably reduces the computation time. It was found that the PM3 method adequately describes the geometric and electron structures of hydrogen bonds in AT and CG pairs. The two interacting molecules are lying approximately in the same plane.

In the AT pair, the bases form two hydrogen bonds, one between the amino and keto groups and the other between nitrogen atoms of the purine and the pyrimidine. The mean length of the bonds is about 1.8 Å, and their total energy is –5.55 kcal/mol (Fig. 2). In the CG pair, three hydrogen bonds are formed, including two between the amino and keto groups of the bases and one between nitrogen atoms. The mean length of the hydrogen bonds is also about 1.8 Å (Fig. 2), similar to that reported in (Singer M., Berg P., 1998). The total energy of the three hydrogen bonds is –11.73 kcal/mol (Fig. 2). Thus, the PM3 method can be used to study other types of hydrogen bonds formed by nucleotide bases.

The results of computations performed by the PM3 methods for pairs of noncomplementary nucleotides are shown in Table 4.

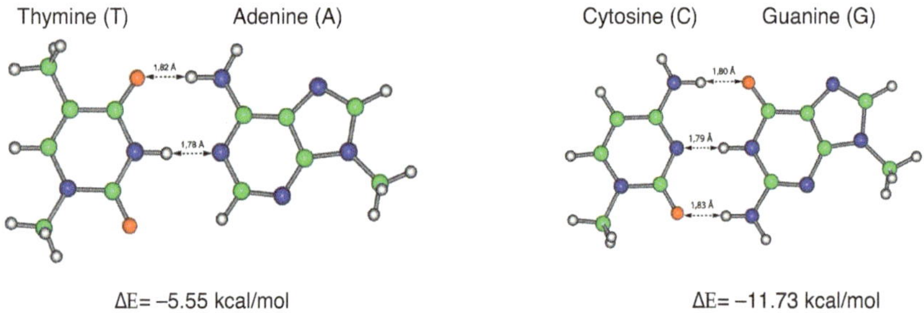

Figure 2. Hydrogen bonding in the AT and CG pairs of complementary nucleotides. Quantum chemical analysis by the PM3 method. Here and below, carbon atoms are shown green or black; oxygen atoms, red; nitrogen atoms, blue; and hydrogen atoms, gray.

	T	A	C	G
T	-2.79			
A	-5.55	-2.71		
C	-4.46	-6.07	-9.24	
G	-6.12	-7.08	-11.73	-10.57

Note: Energy (kcal/mol) was computed by the PM3 method.

Table 4. Energy of hydrogen bonds in all possible pairs of thymine (T), adenine (A), cytosine (C), and guanine

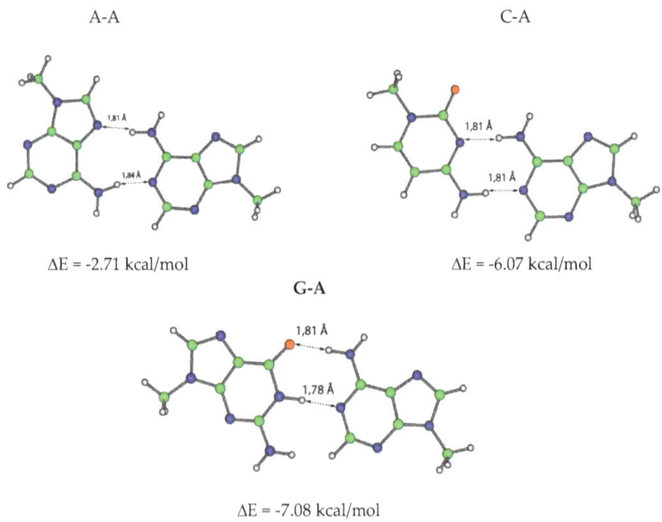

Figure 3. Hydrogen bonding in the noncomplementary pairs AA, GA, and CA. Quantum chemical computations by the PM3 method.

As follows from Table 4, A is theoretically capable of forming pairs with a total energy of -2.71 (A), -5.55 (T), -6.07 (C), and -7.08 (G) kcal/mol. Based on the energy, the preferential partner of A is G or at least C, but it is actually T. Let us consider in more detail the results of computations performed for the interactions of A with A and G with C (Fig. 3).

As Fig. 3 demonstrates, hydrogen bonding between noncomplementary nucleotides is possible, but the pairs are formed so that the angle between the interacting nucleotides is distorted in the case of AA and GA (Fig. 4) or the necessary distance between two strands in the NA double helix is not maintained in the case of CA (Fig. 4). The distance is higher in CA (Singer M., Berg P., 1998).

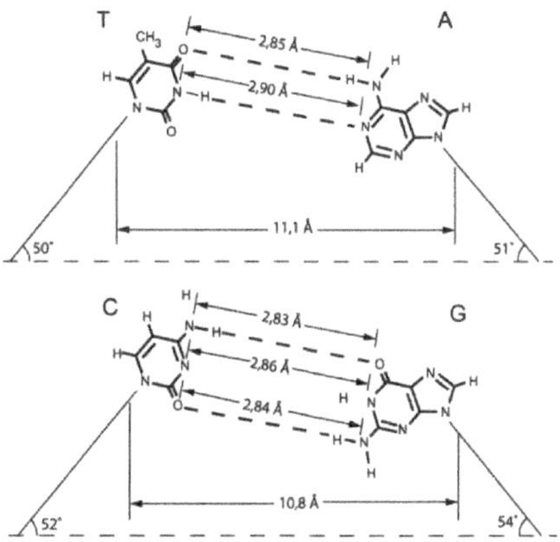

Figure 4. Geometric structure of hydrogen bonds in the complementary pairs AT and CG.

The above data allow the following conclusions.

1. As found by the PM3 method, complementarity in the AT and GC pairs is determined not only by the energy of hydrogen bonding.
2. Complementarity is mostly due to the orientation of hydrogen bonds formed between the interacting nucleotides and the distance between deoxyribose residues in the NA double helix.

3.1.2. Analysis of the electron and geometric structures of hydrogen bonds in complementary NA pairs by non-empirical ab initio calculations in the 6-31G basis set*

Calculations were performed again by the ab initio method with the 6-31G* basis set with complete optimization of the geometric structure (Fig. 5).

Ab initio calculations in the 6-31G* basis also demonstrated the formation of proper hydrogen bonds in complementary pairs. The energy estimates of the bonds were about twice higher than with the PM3 method, the lengths of the bonds were about 0.2 Å greater, but their bond orders remained the same. Electron density analysis according to Mulliken

(Mulliken R.S.,1955) showed that the charge of the hydrogen atoms is positive, about +0.5e, while the interacting O and N atoms have a negative charge ranging from –0.68 to –0.83e. As above, these results suggest quite stable hydrogen bonds.

Then, we performed *ab initio* calculations for noncomplementary pairs of adenine with other nucleotides (Table 5).

T-A C-G

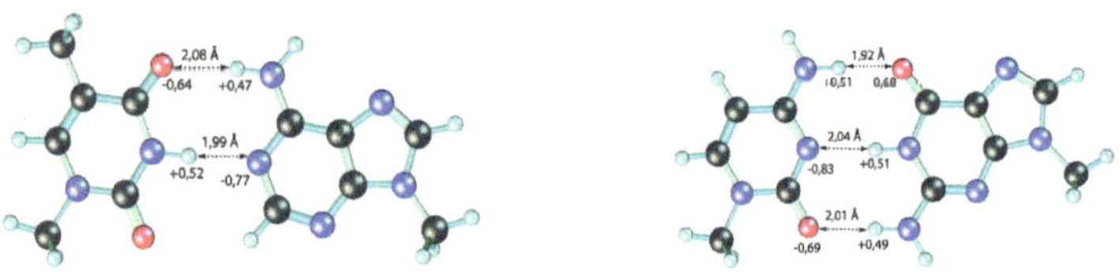

ΔE = -11.78 kcal/mol ΔE = -25.39 kcal/mol

Figure 5. Hydrogen bonding in the complementary pairs AT and CG. Quantum chemical computation by the *ab initio* method in the 6-31G* basis. The charges computed according to Mulliken are shown for the atoms involved in hydrogen bonding.

Nucleotides	T	A	C	G
A	**-11.78**	**-8.93**	**-12.51**	**-13.84**

Note: Energy (kcal/mol) was estimated by *ab initio* calculations in the 6-31G* basis.

Table 5. Energy of hydrogen bonds in pairs of A with T, A, C, and G

As Table 5 demonstrates, selection of complementary T for A is not most advantageous in terms of energy. The interactions AC and AG are more advantageous. The selectivity of nucleotide matching is due to the specific geometric structure of the NA double helix (Fig. 4) (Singer M., Berg P. 1998). Thus, *ab initio* calculations in the 6-31G* basis confirmed that complementarity in the AT and CG pairs is determined mostly by the orientation of hydrogen bonds formed between the interacting nucleotides and the distance between deoxyribose residues in the NA double helix, rather than by the energy of hydrogen bonding.

3.1.3. Analysis of the electron and geometric structures of hydrogen bonds in complementary pairs within the NA double helix by the PM3 method

Let us construct a model NA double helix of six nucleotide pairs (C-G, A-T, T-A, G-C, C-G, and A-T) and fully optimize their electron structure by the PM3 method.

The results showed that a right-turn helix is produced (Fig. 6), which agrees with experimental data. The lengths of the hydrogen bonds in the DNA double helix coincide within 0.02 Å with those computed for isolated nucleotides. The energy of the hydrogen bonds is ΔE = -8.82 kcal/mol per nucleotide pair, which is approximately the average of the energies in two different nucleotide pairs (-5.55 and -11.73 kcal/mol). Thus, the results of computations make it possible to assume that only the above interactions in nucleotide pairs occur in DNA and that there are no other significant interactions.

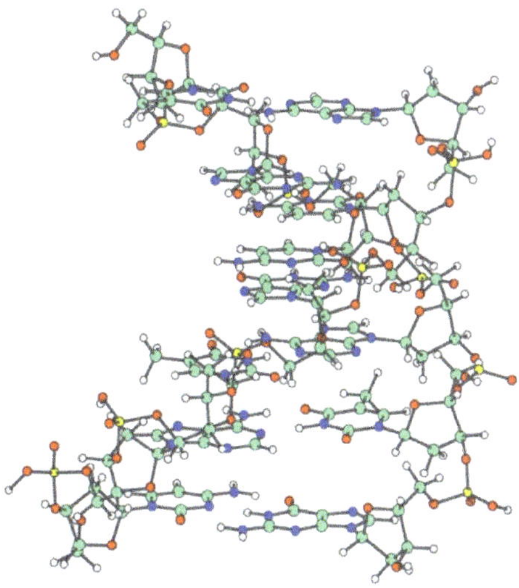

ΔE = -8.82 kcal/mol per nucleotide pair

Figure 6. Hydrogen bonding in complementary pairs AT and CG contained in the DNA double helix (exemplified by a 6-bp duplex). Quantum chemical computation by the PM3 method.

Computations performed for mismatching nucleotide pairs contained in the NA double helix yielded the following picture. In the case of two pyrimidines (T and C), the energy of interaction is equal to zero, because the internucleotide distance is too high. In the case of two purines (A and G), there is not sufficient room for the nucleotides: their rings are bent and the sugar-phosphate backbones forced apart. This is disadvantageous in terms of energy; the loss in electron energy is about 120 kcal/mol. The interactions of T with G and A with C are also disadvantageous, because the parallel arrangement of nucleotides is distorted.

Thus, the quantum chemical computations allow the following conclusions.

1. Hydrogen bonds are formed in the complementary pairs AT and GC, but nucleotide selectivity during NA synthesis is not due to the difference in energy of hydrogen bonds.
2. Selectivity is due to the geometric structure of the NA double helix, which allows the interaction for only those nucleotides that match each other by geometric parameters (AT and GC pairs). All other interactions are disadvantageous in terms of energy.

At the next step, we studied the possibility of complexation of NA bases with carboxyl or hydroxymethyl groups of polysaccharides. The questions were whether hydrogen bonds can be formed between NA bases and sugars and how the four NA bases interact with polysaccharide structural units (UDP-glucose and UDP-glucuronic acid). In addition, it was important to analyze how selectivity may be achieved in template synthesis of a polysaccharide (e.g., HA) on NA.

3.1.4. Quantum chemical analysis of the electron and geometric structures of hydrogen bonds in complementary pairs of NA and polysaccharides by the PM3 method

First, a model was selected for calculations. We found that substitution of glucuronic acid for UDP-glucuronic acid does not affect the hydrogen bonding parameters, but substantially reduces the computation time. Hence the UDP moiety attached to a sugar was omitted in further computations.

Let us assume that NA interacts with the carboxyl or hydroxymethyl group linked to C5 of a monosaccharide unit. Complete optimization of the geometric structure was performed for complexes formed with various initial arrangements of NA bases and glucuronic acid, and several local minima of potential energy were revealed. Among the resulting structures, those with the highest energy of interactions between the components were selected for each NA base–glucuronic acid pair (Fig. 7).

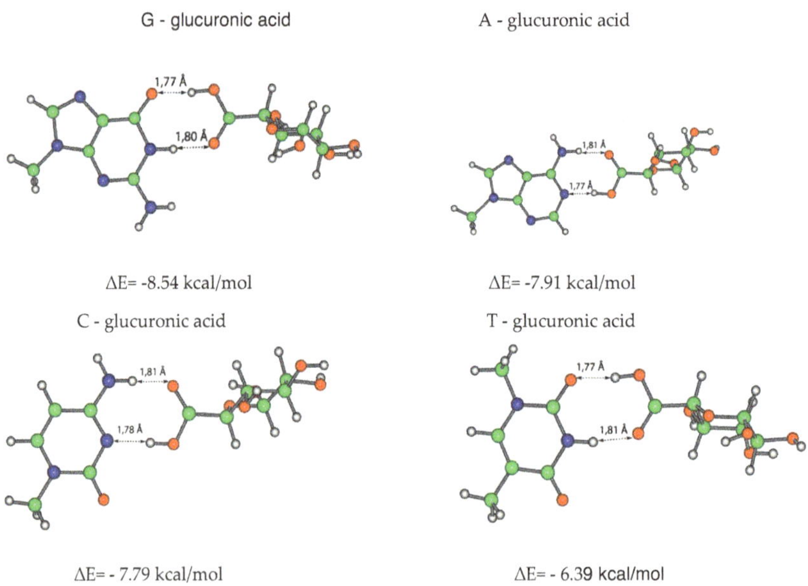

Figure 7. Formation of hydrogen bonds between the carboxyl group of glucuronic acid and DNA bases. Quantum chemical computations by the PM3 method.

The complex structure shows clearly that two hydrogen bonds are formed and are nearly in the same plane, as in the case of the T–A interaction. The most important in this situation is that the bonds have approximately the same lengths and binding energies as in the complementary pairs AT and GC. In addition, Fig. 7 shows that the interactions of the carboxyl group with the nucleotides are nonequivalent. The interactions with G and A are more efficient than with T and C, which is determined by the geometry of the hydrogen bonds. A hydrogen bond should be of a certain length and orientation relative to other atoms, and the conditions for hydrogen bonding are better in some cases and poorer in some others.

Then, the PM3 method was used to study the possibility of interactions between the bases and the hydroxymethyl group of N-acetylglucosamine. Several initial arrangements of a base and N-acetylglucosamine were analyzed for each of the four bases. Optimization of the initial arrangement allowed several modes of binding. Among the resulting structures, those with the highest energy of interactions were selected. The results are shown in Fig. 8.

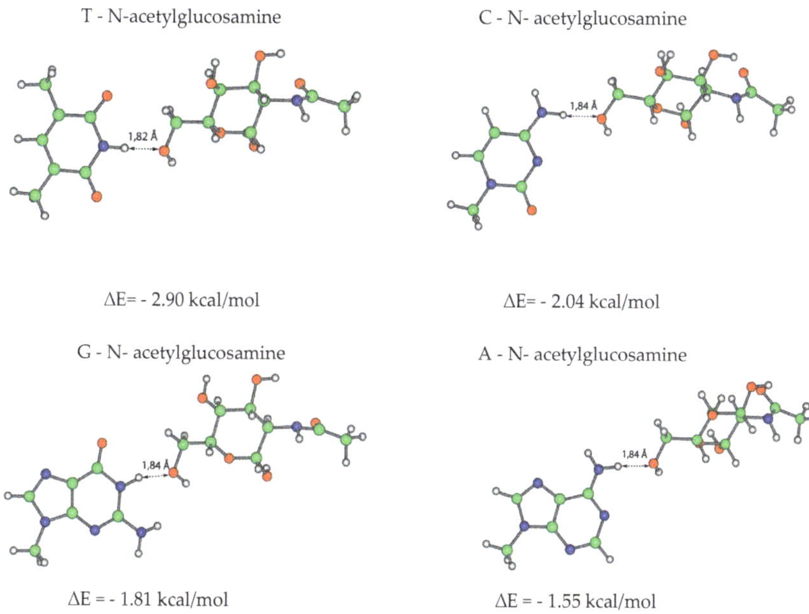

Figure 8. Formation of hydrogen bonds between the hydroxymethyl group of N- acetylglucosamine and the DNA bases. Quantum chemical computations by the PM3 method.

Let us consider the specifics of hydrogen bonding between the nucleotides and the hydoxymethyl group of N-acetylglucosamine. It is known that hydrogen bonds formed by the hydroxymethyl group are less stable than those formed by the carboxyl group (Roberts J., et al. 1978). As complete optimization of the geometric structures showed, the energy of hydrogen bonds was indeed higher in the complexes of the NA bases with the hydroxymethyl group of N-acetylglucosamine (Fig. 8) than in the complexes with the carboxymethyl group (Fig. 7). Optimization of the geometric structure for complexes of the bases with N-acetylglucosamine with various initial arrangements of the interacting groups yielded complexes with one or two hydrogen bonds. Complexes containing one hydrogen bond were most advantageous in terms of energy according to PM3 computations. The nucleotides proved to vary in energy of interaction with the hydroxymethyl group: the interaction was more efficient with T and C than with G and A. The conclusions are as follows.

1. Quantum chemical computations showed that NA bases and the carboxyl or hydroxymethyl group of sugars are capable of forming hydrogen bonds, which are comparable in energy with those occurring in the complementary AT and GC pairs.
2. The bonds are nonequivalent. The carboxyl group forms tighter bonds with purines and less tight bonds with pyrimidines. In contrast, the hydroxymethyl group forms more stable hydrogen bonds with pyrimidines and less stable bonds with purines.
3. The differences in energy computed for hydrogen bonds are insufficient for selection of monosaccharides with necessary structures.

3.1.5. Quantum chemical analysis of the electron and geometric structures of hydrogen bonds in complementary pairs of NA bases and polysaccharides by Ab initio calculations in the 6-31G basis*

Let us optimize the geometry of the above complexes by *ab initio* calculations in the 6-31G* basis (Table 6).

	T	A	C	G
Carboxyl group of glucuronic acid	-12.95	-13.77	-15.98	-17.89

Note: Energy (kcal/mol) was obtained by *ab initio* calculations in the 6-31G* basis.

Table 6. Energy of hydrogen bonds between the nucleotides T, A, C, and G and the carboxyl group of glucuronic acid

In general, *ab initio* calculations confirmed the results obtained by the PM3 method and described in the previous section. The interaction of the carboxyl group of glucuronic acid and the nucleotides is possible, and its energy is sufficient for hydrogen bonding similar to that in the classical complementary DNA pairs. However, the energy of interaction changes with nucleotide in another order: the interaction with the carboxyl group of glucuronic acid is more advantageous with C and G and less advantageous with T and A (Table 6). As for the interaction of the hydroxymethyl group of N-acetylglucosamine with NA bases, the results differed from those obtained by the PM3 method. If the two fragments were initially bonded by one hydrogen bond, like in the case of the interaction between the carboxymethyl group and hydrogen of a base (Fig. 8), then, during optimization, the groups were rotated relative to each other so that the interaction involved the atoms of the monosaccharide ring that by no means can bind to the nucleotide. This is probably explained by the fact that *ab initio* calculations yield higher estimates of hydrogen bond energy as compared with the PM3 method. As a result, greater energy estimates are obtained for the interactions of H with distant O and N and the atoms are brought close together during optimization to form new hydrogen bonds, which are impossible in the double helix.

Thus, *ab initio* calculations in the 6-31G* basis confirmed that the carboxyl group of glucuronic acid is capable of hydrogen bonding to the bases T, A, G, and C; the two bonds formed in each case are similar to those occurring in the complementary pairs AT and GC.

3.1.6. PM3 Analysis of the electron and geometric structures of hydrogen bonds in complementary pairs of a NA–glucuronic acid double helix. Selectivity in template synthesis of polysaccharides

Consider the possibility of polysaccharide synthesis on single-stranded NA. It is clear that, even if some hydrogen bonds are formed between NA bases and monosaccharides, such bonds are not necessarily formed in a polysaccharide–NA double helix because of geometric limitations.

Let us construct two oligomers each consisting of five monomeric units: NA (GCGCA) and oligosaccharide (a HA fragment). As NA, we use the chain analyzed above (Fig. 6). Constructing a chain of five HA monosaccharides, let us rotate the units around the bonds between them so that the carboxyl and hydroxymethyl groups of the resulting chain can be hydrogen bonded to nucleotides. Let us bring the two chains close together so that the distance between groups was 1.5-2.5 Å as in Figs. 9 and 10, with G and A interacting with the carboxyl group (Fig. 9) and C, with the hydroxymethyl group (Fig. 10). Let us fully optimize the resulting double helix by the PM3 method. The energy of hydrogen bonding in the resulting construct (Figs. 9, 10) is -7.04 kcal/mol, rather high and almost reaching the energy of interaction in the classical DNA double helix (-8.82 kcal/mol, Fig. 6).

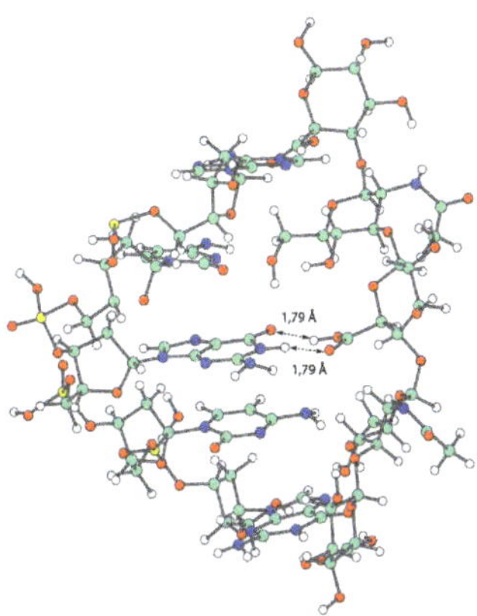

ΔE = -7.04 kcal/mol per complementary pair

Figure 9. Hydrogen bonding in the complementary pairs G–glucuronic acid, A–glucuronic acid, and C–N-acetylglucosamine in a DNA–polysaccharide double helix exemplified by a five-unit chain. Quantum chemical computation by the PM3 method.

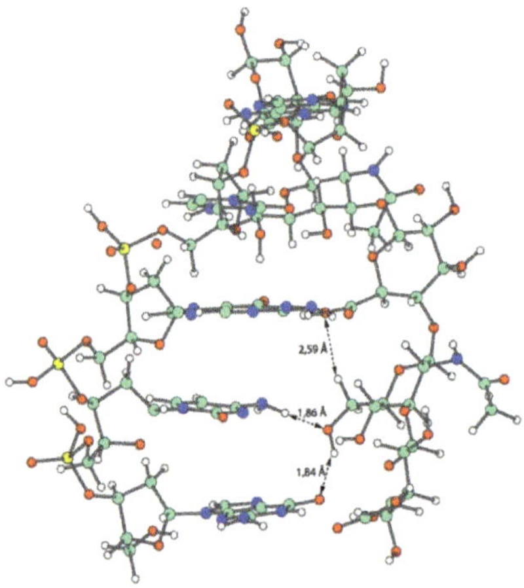

ΔE = -7.04 kcal/mol per complementary pair

Figure 10. Hydrogen bonding in the complementary pairs G–glucuronic acid, A–glucuronic acid, and C-N-acetylglucosamine in a DNA–polysaccharide double helix exemplified by a five-unit chain. Quantum chemical computation by the PM3 method. (Another view point).

The interaction purine–glucuronic acid in the NA–polysaccharide double helix was much the same as in isolation: the lengths of the two hydrogen bonds were 1.79 Å. However, interesting differences were observed for the pyrimidine–N-acetylglucosamine interaction in the double helix and in isolation. In addition to the only hydrogen bond (1.86 Å) formed in isolation (Fig. 10), another hydrogen bond (1.84 Å) was formed with O of the neighboring nucleotide in the double helix. Moreover, there was one more, weaker hydrogen bond (2.59 Å) directed oppositely. Thus, N-acetylglucosamine also forms two hydrogen bonds with nucleotides of an NA strand, which ensures its sufficiently tight and highly selective interaction with nucleotides of a native NA molecule. In addition, computations were performed for incorrect sequences of monosaccharides. As in the case of incorrect sequences of bases in the DNA double helix, low-energy hydrogen bonds were obtained in this variant or the NA–polysaccharide double helix was not formed at all because the bonds were disadvantageous in terms of energy. The conclusions are as follows.

1. Quantum chemical calculations showed that hydrogen bonds are formed in the complementary pairs purine–glucuronic acid and pyrimidine–N-acetylglucosamine, allowing the formation of a NA–polysaccharide double helix.
2. Geometric parameters of the interacting NA and polysaccharide strands play the major role in selection of particular monosaccharides during putative template synthesis of polysaccharides. The difference observed for the energy of hydrogen bonds is insufficient for selection of monosaccharides with necessary structures.
3. Quantum chemical modeling confirmed that purines are complementary to the carboxyl group and pyrimidines, to the hydroxymethyl group of UDP-monosaccharides.

3.2. Spectral analysis and dot hybridization

Table 7 shows the polysaccharide composition determined biochemically. It should be noted that all polysaccharides were end products of modification, that is, mature glycans. Polyuronides, which consist mostly of uronic acid residues, and amylose, which consists only of glucose residues, can be considered as virtually nonmodified glycans. The hydroxymethyl group at C5 of monosaccharide is substituted with hydrogen in glucuronoxylans. HA and CS have the acetoamide group at C2 of hexose residues. In CS, hydrogen of hydroxyl at C6 is substituted with the sulfo group, which provides an additional partial negative charge to the molecule (Table 7.)

Polysaccharide	mkg on mg				in %			
	UA	NS	AAH	OM	UA	NS	AAH	OM
Glucuronoxylan	220	771	0	9	22	77	0	1
Polyuronide	747	33	0	220	75	3	0	22
Hyaluronic acid	535	0	465	0	54	0	46	0
Chondroitin sulfate	413	0	587	0	41	0	59	0
Amylose	0	1021	0	0	0	100	0	0

Note: UA, uronic acid (on Dishe); NS, neutral sugar (on anthrone method); AAH, aminoacetylhexoses(AAH=1000-UA); OM, other monosaccharides (OM=1000-UA-NS).

Table 7. Characterization of the composition of polysaccharides used in experiments

Electrophoretic analysis of polysaccharides used in experiments is illustrated in Fig. 11. It is seen that polysaccharides were sufficiently pure, which was important for our study.

The results shown in Figs. 12A, 12B, and 13A demonstrate that the absorption and CD spectra remained much the same when poly(dC) was combined with oat polyuronides or poly(dA), with amylose. This finding suggests that these polynucleotides and polysaccharides do not form sufficiently tight complexes.

When poly(dA) was combined with oat polyuronides or poly(dC), with amylose, a decrease in amplitude was observed on both the absorption and CD spectra (Figs. 12C, 13B, 13C). The decrease in the peak intensity of the CD spectrum and the simultaneous decrease in the amplitude of peak absorption (a hypochromic effect) can result from an increase in the degree of spiralization and the twist of the polynucleotide (Blagoi Yu.P., et al. 1999), suggesting the formation of a single-stranded polynucleotide–polysaccharide complex. It should be noted that the amplitude of the absorption spectra changed by 9-11% in this case. Since the hyperchromic effect of double-stranded DNA (poly(dA)–poly(dT)) in 0.3 M NaCl was 27-30% (Figs. 14A, 14B), which agreed with published data (Yang J.T., et al. 1969), the change observed in our experiment accounted for 30-35% of the hyperchromic effect characterizing complete base pairing.

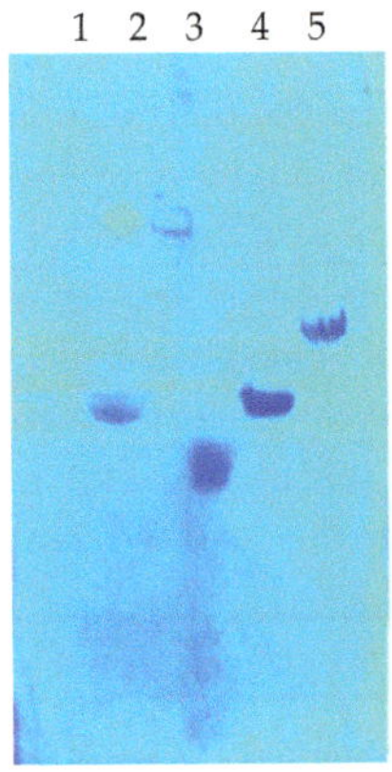

1, oat polyuronides
2, garlic glucuronoxylans
3, HS
4, CS
5, HA

Figure 11. Electrophoresis of polysaccharides in cellulose acetate bands in 0.1 N HCl.

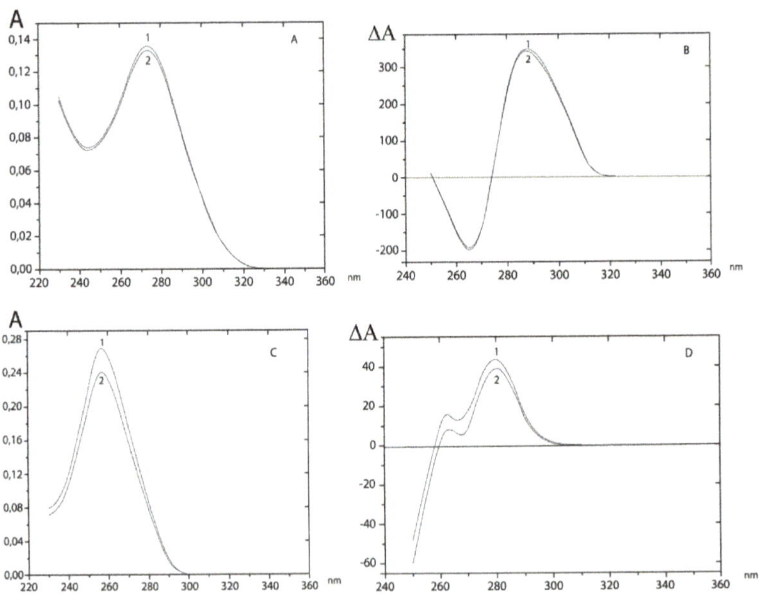

Figure 12. A. Absorption spectrum of poly(dC) in the absence (curve 1) or presence (curve 2) of polyuronide (PU), r = 2. (r - is hereafter the ratio of polysaccharide units to the molar concentration of nucleotides.) B. CD spectrum of poly(dC) in the absence (curve 1) or presence (curve 2) of PU, r = 2. C. Absorption spectrum of poly(dA) in the absence (curve 1) or presence (curve 2) of PU, r = 1. D. CD spectrum of poly(dA) in the presence (curve 1) or absence (curve 2) of PU, r = 1.

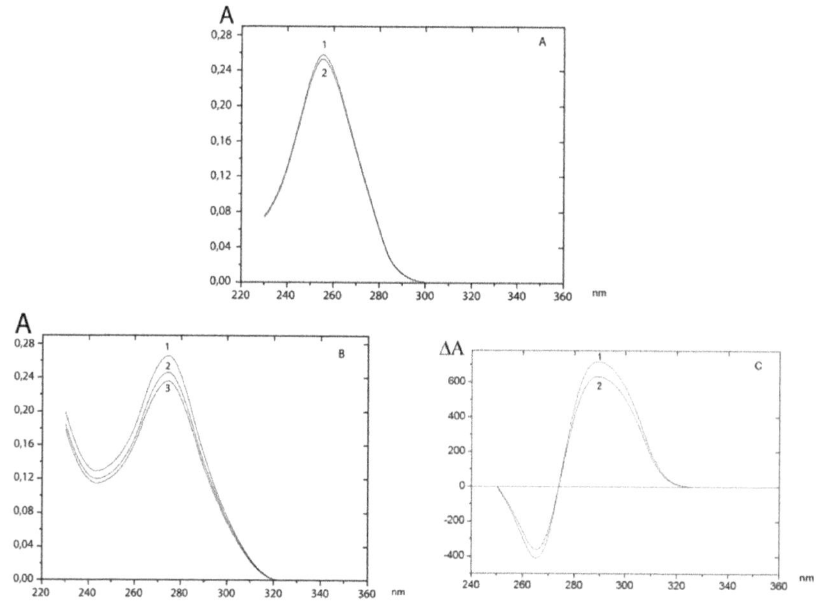

Figure 13. A. Absorption spectrum of poly(dA) in the absence (curve 1) or presence (curve 2) of amylose (AM), r = 2.B. Absorption spectrum of poly(dC) in the absence (curve 1) or presence (curves 2, 3) of AM; r = 0.5 (1), r = 1 (2), r = 2 (3). C. CD spectrum of poly(dC) in the absence (curve 1) or presence (curve 2) of AM, r = 2.

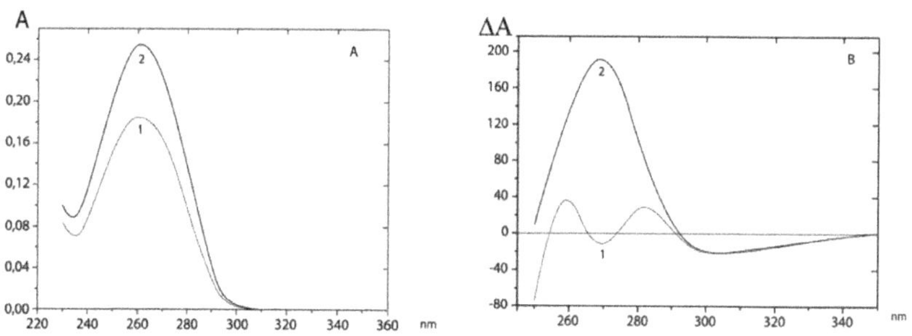

Figure 14. A. Absorption spectra of poly(dA)–$(dT)_{16}$ (curve 1) and an equimolar mixture of poly(dA) and $(dT)_{16}$ (curve 2), r =1. B. CD spectra of of poly(dA)–$(dT)_{16}$ (curve 1) and an equimolar mixture of poly(dA) and $(dT)_{16}$ (curve 2), r = 1.

The results shown in Figs. 15A and 15B demonstrate that CS does not form a tight complex with DNA, because the DNA absorption spectrum was almost completely restored after annealing in the presence of CS. Another situation was observed with HA (Fig. 15A): a hyperchromic effect of about 10-11% was observed in this case. This finding suggests complexation of HA with DNA regions, because such complexation should prevent the DNA duplex from being restored after annealing in the presence of HA.

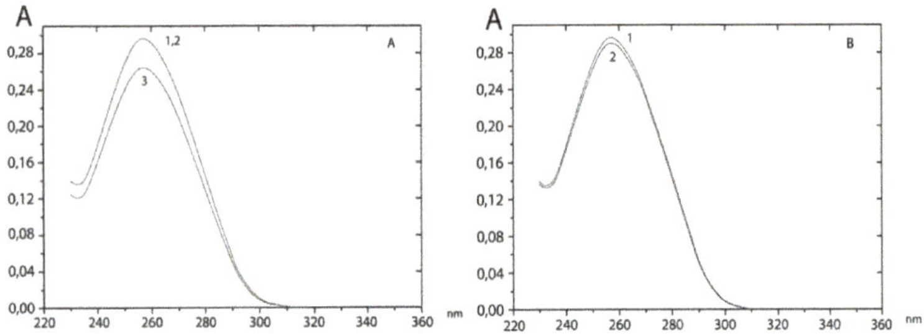

Figure 15. A. Absorption spectra of calf thymus DNA in the absence (curve 1, before annealing; curve 2, after annealing) or presence (curve 3, after annealing) HA; r = 1. B. Absorption spectra of calf thymus DNA in the presence of CS before (curve 1) and after (curve 2) annealing, r = 1.

Dot hybridization showed that DNAs of all organisms specifically binds with oligodeoxyribonucleotide probes (Fig. 16). Although the method is semiquantitative, it is possible to state that the GC tandem repeat has more complementary sequences in DNA as compared with monopurine (A) or monopyrimidine (C) homopolymers. For instance, poly(dT) and poly(dGC) arrays are rather abundant in human DNA, while poly(dG) occur in a smaller amount. Garlic DNA showed a greater amount of poly(dG) as compared with human DNA. This result agrees with the modern views of the plant genome (Zelenin A.V. 2003).

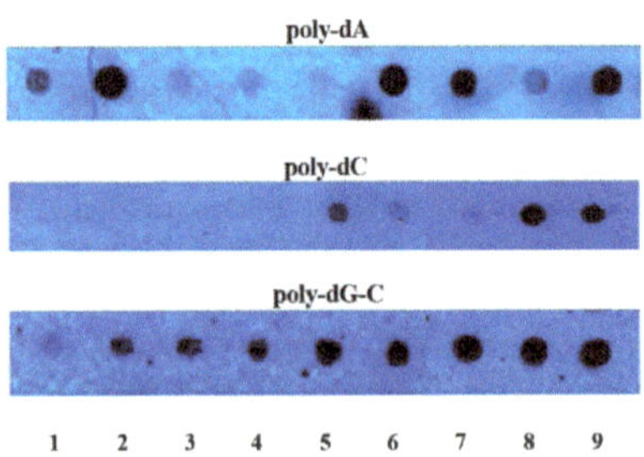

1. glucuronoxylans
2. polyuronides
3. CS
4. HA
5. amylase
6. garlic DNA
7. human DNA
8. fish DNA
9. calf thymus DNA

Figure 16. Dot hybridization of polysaccharides and DNA with oligodeoxyribonucleotide probes. The membrane was autoradiographed.

Since we aimed at studying the possibility of specific bonding between NA and polysaccharides, the results obtained with DNA were used exclusively as a control.

Biopolymer in a dot	Dot radioactivity, cpm			% of maximal count		
	Poly(dA) probe	Poly(dC) probe	Poly(dGC) probe	Poly(dA) probe	Poly(dC) probe	Poly(dGC) probe
Glururonoxylans	165	47	40	33	5	26
Polyuronides	496	18	107	**100**	2	69
HA	102	8	155	20	0	**100**
CS	2	65	97	0	7	63
Amylose	61	868	40	12	**100**	69
Garlic DNA	860	466	390	34	15	38
Human DNA	925	182	394	36	6	38
Fish DNA	2556	2990	1034	100	100	100
Calf thymus DNA	2161	1570	979	85	52	95

Table 8. Radioactivity of dots of polysaccharides and DNA after hybridization with purine-pyrimidine DNA probes

The results of autoradiography (Fig. 16, Table 8) demonstrated that the poly(dA) probe specifically hybridized with oat polyuronides to an extent comparable with that of DNA (garlic DNA). The poly(dC) probe specifically hybridized to a high extent with potato amylose. Since oat polyuronides consist almost exclusively of hexuronic acid residues and potato amylose, of hexose residues, we observed significant specific hybridization of pyrimidines (exemplified by C) with polyhexoses and purines (exemplified by A) with polyuronic acids.

The data obtained with glucoronoxylans were interpreted according to the same principle. It is known that glucuronoxylans contains about one carboxyl group per four neutral saccharide residues; i.e., the content of carboxyl groups is three or four times lower than in polyuronides. Hybridization of glucuronoxylans with poly(dA) was indeed about threefold lower than that of polyuronides.

The regular tandem poly(dGC) probe efficiently hybridized with all polysaccharides and especially with GAG. The highest hybridization efficiency was observed with HA. This finding agrees with the tandem structure of HA consists of glucose–glucuronic acid dimers, which interacted with the purine–pyrimidine dimers of the synthetic probe NA.

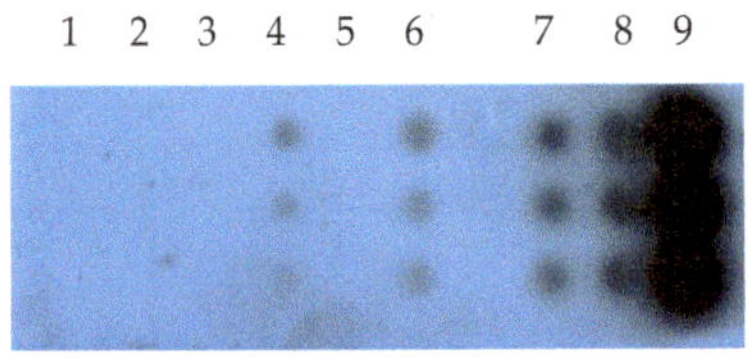

1, garlic glucuronoxylans
2, oat polyuronides
3, animal CS
4, animal HA
5, potato amylose
6, garlic DNA
7, fish DNA
8, phage DNA
9, calf DNA

Figure 17. Dot hybridization of polysaccharides and DNAs with calf thymus DNA. The membrane was autoradiographed.

Figure 17 shows the results of dot hybridization of polysaccharides and DNAs of various organisms with fragmented calf thymus DNA. The highest radioactivity was observed in dot 9, i.e., for self-hybridization of calf genomic DNA. In addition, repetitive elements of calf thymus DNA detected homologous sequences in virtually all genomic DNAs examined. Of all polysaccharides, only HA (dot 4) had matching structures in calf thymus DNA and formed specific complexes similar in radioactivity to DNA–DNA complexes.

3.3. Glycan biosynthesis in rat liver

The profile of elution of rat liver and marker polysaccharides from DEAE cellulose with a NaCl gradient is shown in Fig. 18. As markers, we used homopolymers of uronic acid (oat polyuronides) and glucose (potato amylose).

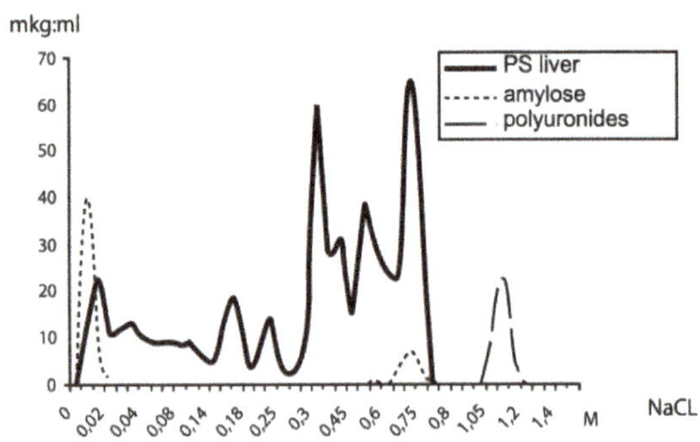

Figure 18. Profiles of elution of rat liver and marker polysaccharides from DEAE cellulose with a NaCl gradient.

As evident from the elution profile, the rat liver contains a complex spectrum of polysaccharides. Analysis with the marker homoglycan amylose, consisting of glucose residues, showed that polymers with a similar structure of neutral sugars are absent from rat liver glycans isolated by our method (such polymers were eluted from the column during washing). Glucose homoglycans were eluted with 0.01 M NaCl. The first peak of liver glycans was detected at 0.02-0.03 M NaCl. With NaCl increasing to 0.15 M, we eluted polysaccharides with a negative charge greater than in polymers of neutral sugars and lower than in the 0.25 M NaCl fraction, which mostly consisted of HA, as commonly known (Zimnitskii A.N., et al. 2004). The fractions 0.7 and 1.5 M NaCl exemplified a classical separation of CS and HS. As expected, highly charged oat polyuronides were eluted at the highest ionic strength, at about 1.2 M NaCl.

The results of quantitating uronic acids, hexosamines, and NS in the chromatographic fractions are summarized in Table 9.

Fraction	Uronic acids	Hexosamines	Neutral sugars
0.15 M	34.1±0.3	2.0±0.1	48.5±0.3
0.25 M	8.3±0.1	7.1±0.9	-
0.7 M	61.5±0.7	63.7±0.6	2.1±0.2
1.5 M	64.2±0.5	57.9±6.4	2.2±0.2

Table 9. Concentrations (μg/ml) of uronic acids, hexosamines, and NS in fractions of liver polysaccharides of intact rats (rats were not subjected to external influence)

In the 0.25, 0.7, and 1.5 M fractions, the ratio uronic acids : hexosamines was about 1:1, which is characteristic of HA, CS, and HS and agrees with published data. A minor amount of NS detected in the 0.7 and 1.5 M fractions could be explained by the presence of a trisaccharide fragment linking GAG with the core protein in proteoglycans.

Subfraction	Nuclei	Microsomes	Homogenate
0.02 M	265±18 / 67±7	108±12 / 52±7	128±15 / 67±8
0.06 M	36±5 / 7±2	24±3 / 40±6	34±6 / 22±3
0.15 M	20±3 / 6±2	18±2 / 50±6	33±5 / 81±11

Table 10. Contents (μg per g tissue) of NS/uronic acids in the 0.15 M NaCl saccharide fractions of cell nuclei, microsomes, and a homogenate of the liver of intact rats

Subfraction	Nuclei	Microsomes	Homogenate
0.02 M	0.76	1.40	1.60
0.06 M	1.44	5.0	1.94
0.15 M	1.57	8.30	7.4

Table 11. Uronic acids per neutral trisaccharide in the 0.15 M NaCl saccharide fractions of cell nuclei, microsomes, and a homogenate of the liver of intact rats

The 0.15 M fraction contained glycans enriched in neutral sugars (Table 9). Three subfractions could be isolated in this fraction: the first peak eluted up to 0.02 M NaCl, the second peak eluted up to 0.06 M NaCl, and a plateau eluted up to 0.15 M NaCl. The primary structure of polysaccharide fragments of this fraction is characterized in Tables 10 and 11. As evident from Table 11, nuclear oligosaccharides had the NS : uronic acid ratio similar to 3:1, suggesting the presence of neutral trisaccharides containing no more than one uronic acid residue. Both microsomal and homogenate fractions harbored neutral oligosaccharides containing more than one uronic acid residue per trisaccharide, the proportion of uronic acids increasing sixfold from the 0.02 to the 0.15 M subfraction. This finding testifies that uronic acid is linked to trisaccharides mostly in cell microsomes. Only one uronic acid residue per trisaccharide is linked in cell nuclei.

Substance	Content
DNA_n	9.26±0.30
$0.15\ M_n$ fraction of PS (Fig. 18)	1.42±0.12
HA_n	1.46±0.14
CS_n	2.56±0.16
HS_n	0.68±0.07
DNA_h	167.61±6.94
$0.15\ M_h$ fraction of PS (Fig. 18)	42.02±4.13
HA_h	23.74±3.35
CS_h	23.60±3.31
HS_h	13.45±1.22

Note: Subscripts n and h correspond to the nuclear fraction and the homogenate, respectively.

Table 12. Contents (μg per g tissue) of DNA and GAG in the nuclear fraction and the homogenate of the liver of intact rats

The DNA content and the GAG composition in the nuclear fraction and the homogenate of the rat liver are shown in Table 12. The total GAG content was 6.12 μg per g tissue in the nuclear fraction and 102.81 μg per g tissue in the homogenate. Nuclear GAG contained 0.15 M fraction PS to 23.20%, HA to 23.86%, CS to 41.83%, and HS to 11.11%. GAG of the homogenate contained 0.15 M fraction PS to 40.87%, HA to 23.09%, CS to 22.95%, and HS to 13.08%.

Our isolation procedure yielded about 5% of nuclei in the intact form. The remaining nuclei, along with their contents, were in the liver homogenate.

Correlation analysis was performed to study the association between the biochemical parameters and the time after a glucose load. Since the empirical sample was small ($n<30$), we computed Pearson parametric empirical correlation coefficients and Spearman nonparametric rank correlation coefficients. The results are summarized in Tables 13-15.

As Table 13 shows, a strong positive correlation was observed between time and HS content in the homogenate ($r = 0.81$) and between the contents of HS in nuclei and the 0.15 M fraction in the homogenate ($r = 0.95$) and a strong negative correlation between the contents of the 0.15 M fraction and DNA in nuclei ($r = -0.71$).

	t (min)	DNAn	0.15 Mn	HAn	CSn	HSn	DNAh	0.15 Mh	HAh	CSh	HSh
t (min)	1.00	-0.01	-0.28	-0.31	-0.35	-0.54	-0.53	-0.34	-0.24	-0.58	**0.81**
DNAn	-0.01	1.00	**-0.71**	-0.41	0.62	0.54	-0.09	0.53	-0.19	0.09	0.35
0.15 Mn	-0.28	**-0.71**	1.00	0.45	-0.08	-0.12	0.29	-0.05	-0.13	0.37	-0.47
HAn	-0.31	-0.41	0.45	1.00	0.24	0.07	-0.18	-0.09	0.30	0.37	-0.55
CSn	-0.35	0.62	-0.08	0.24	1.00	0.66	0.06	0.54	0.03	0.50	-0.30
HSn	-0.54	0.54	-0.12	0.07	0.66	1.00	0.61	**0.95**	0.24	0.69	-0.24
DNAh	-0.53	-0.09	0.29	-0.18	0.06	0.61	1.00	0.64	0.37	0.43	-0.42
0.15 Mh	-0.34	0.53	-0.05	-0.09	0.54	**0.95**	0.64	1.00	0.12	0.62	-0.03
HAh	-0.24	-0.19	-0.13	0.30	0.03	0.24	0.37	0.12	1.00	-0.18	-0.50
CSh	-0.58	0.09	0.37	0.37	0.50	0.69	0.43	0.62	-0.18	1.00	-0.34
HSh	**0.81**	0.35	-0.47	-0.55	-0.30	-0.24	-0.42	-0.03	-0.50	-0.34	1.00

Note: Significant correlation coefficients are in bold.

Table 13. Pearson empirical correlation coefficients (r) between the time after a glucose load and the DNA and GAG contents in the rat liver

	t (min)	DNAn	0.15 Mn	HAn	CSn	HSn	DNAh	0.15 Mh	HAh	CSh	HSh
t (min)	1.00	-0.15	-0.15	-0.15	-0.34	-0.47	-0.68	-0.15	-0.32	-0.49	0.63
DNAn	-0.15	1.00	**-0.74**	-0.36	0.54	0.48	-0.20	0.35	0.01	-0.01	0.34
0.15 Mn	-0.15	**-0.74**	1.00	0.53	0.07	-0.10	0.31	0.10	-0.04	0.41	-0.40
HAn	-0.15	-0.36	0.53	1.00	0.45	0.23	0.19	0.12	0.40	0.40	-0.50
CSn	-0.34	0.54	0.07	0.45	1.00	0.67	0.14	0.62	0.26	0.50	-0.07
HSn	-0.47	0.48	-0.10	0.23	0.67	1.00	0.30	**0.84**	-0.04	**0.84**	0.16
DNAh	-0.68	-0.20	0.31	0.19	0.14	0.30	1.00	0.19	0.61	0.38	-0.45
0.15 Mh	-0.15	0.35	0.10	0.12	0.62	**0.84**	0.19	1.00	-0.24	**0.74**	0.36
HAh	-0.32	0.01	-0.04	0.40	0.26	-0.04	0.61	-0.24	1.00	-0.17	-0.49
CSh	-0.49	-0.01	0.41	0.40	0.50	**0.84**	0.38	**0.74**	-0.17	1.00	0.00
HSh	0.63	0.34	-0.40	-0.50	-0.07	0.16	-0.45	0.36	-0.49	0.00	1.00

Note: Significant correlation coefficients are in bold.

Table 14. Spearman rank correlation coefficients (R) between the time after a glucose load and the DNA and GAG contents in the rat liver

Table 14 shows three strong correlations: a negative one between the contents of the 0.15 M fraction and DNA in nuclei and positive correlations between the HS content in nuclei and the CS content in the homogenate and between the contents of the 0.15 M fraction and CS in the homogenate ($R = 0.74$).

To identify latent variables uniting the parameters under study, factor analysis was performed using the Varimax procedure (Table 15).

Parameter	Factor 1	Factor 2	Factor 3
t (min)	-0.490	0.494	-0.473
DNAn	0.656	0.656	-0.189
0.15 Mn	-0.064	**-0.851**	-0.009
HAn	0.054	**-0.768**	0.019
CSn	**0.804**	-0.064	-0.063
HSn	**0.913**	0.061	0.361
DNAh	0.386	-0.106	**0.728**
0.15 Mh	**0.863**	0.159	0.279
HAh	-0.112	0.001	**0.866**
CSh	**0.776**	-0.500	-0.041
HSh	-0.146	0.692	-0.573
Proper factor value	3.687	2.762	2.083
Variance portion accounted for by the factor	0.335	0.251	0.189

Note: Significant correlations between the parameter and the factor are in bold.

Table 15. Factor analysis (Varimax) of the time after a glucose load and the DNA and GAG contents in the rat liver

As evident from Table 15, Factor 1 included CSn, HSn, the 0.15 Mh fraction, and CSh. Factor 2 included the 0.15 Mn fraction and HAn. Factor 3 included DNAh and HAh.

The most important result of correlation analysis is that strong linear correlations were observed between the time after a glucose load and the content of HS in the liver homogenate and between the contents of the 0.15 M fraction and DNA in the nucleus. In the first case, we record an increase in the time of synthesis of GAG modified to the greatest extent (at least two sulfo groups per disaccharide fragment). This result was conformed by factor analysis, since the time and the HS content in the homogenate are components of one factor. In the second case, we possibly establish a functional relationship between synthesis of nuclear oligosaccharides contained in the 0.15 M fraction and the genetic apparatus of the cell. The two biopolymers involved belong to different classes, but their biosynthesis utilizes the same initial substrate, glucose. The strong negative correlation suggests competition for the initial substrate between the two biosynthetic pathways. Since the nuclear DNA content was constant under the conditions of our experiments, the negative correlation suggests that synthesis of saccharides belonging to the 0.15 M fraction increases when free glucose enters cells in considerable amounts. This assumption is supported by the results shown in Fig. 19.

The composition of the 0.15 M fraction and GAG in nuclei, microsomes, and the homogenate is shown in Fig. 20. It is seen that nuclear GAG are enriched in CS (about 42%), which agrees with the accepted views (Silbert J.E. Sugumaran G ., 1995). Homogenate GAG consist mostly of HA and CS (about 23%). In microsomes, all GAG types occur in nearly equal proportions (about 12%). The results obtained for the 0.15 M fraction, which consists of three main subfractions (Table 11), are shown in Fig. 21. The fraction occurred at a high content (about 60%) in microsomes, while its content in the nucleus was no more than 23%. The homogenate was intermediate between nuclei and microsomes in the content of this fraction.

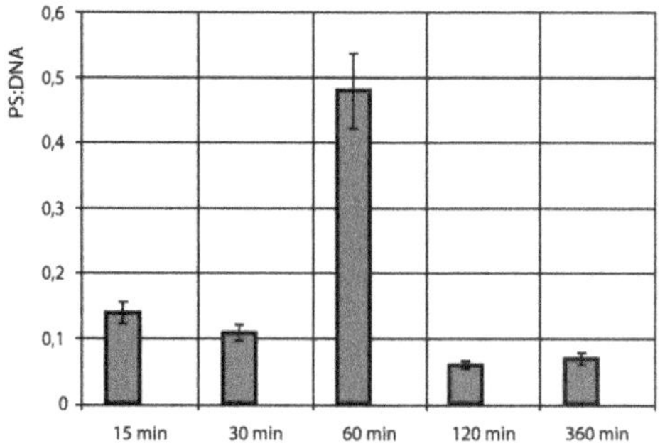

Figure 19. Accumulation of polysaccharides (PS) of the 0.15 M fraction in the nucleus. The content of the fraction was normalized with respect to the DNA content.

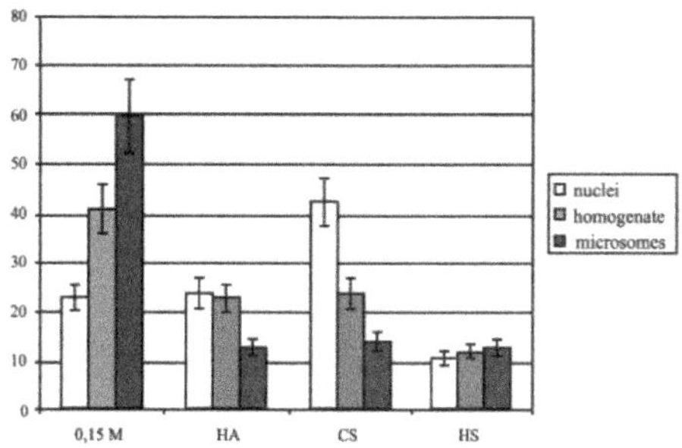

Figure 20. Relative contents (%) of polysaccharides in nuclei, microsomes, and the homogenate of the rat liver.

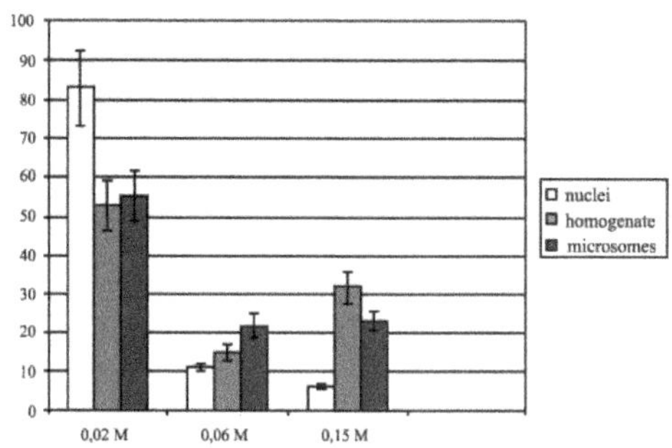

Figure 21. Relative contents (%) of polysaccharides differing in the portion of neutral sugars in the 0.15 M fractions of nuclei, microsomes, and the homogenate of the rat liver.

To study biosynthesis of polysaccharide fragments, we analyzed the time course of incorporating radiolabeled glucose in polysaccharide chains (Fig. 22).

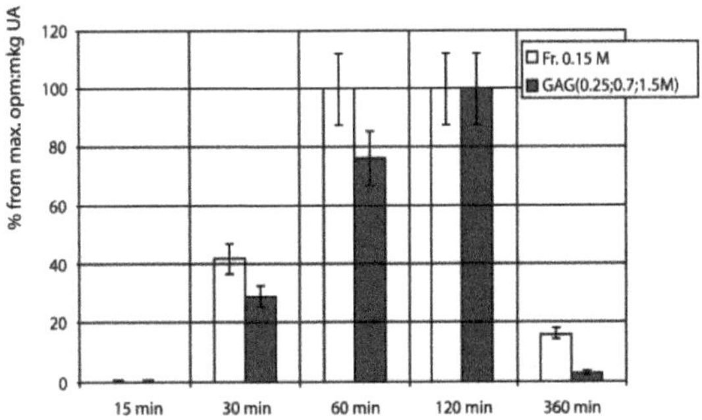

Figure 22. Growth to the maximal specific radioactivity of polysaccharides as dependent on the time of label presence in rats. The maximal specific radioactivity was taken as 100% for each fraction.

The highest radioactivity of the 0.15 M fraction was detected 60-75 min after a glucose load, while mature GAG (fraction 0.25-1.5 M) showed the highest radioactivity after 120 min. Thus, initiation of glycan synthesis is probably associated with the 0.15 M fraction, enriched in neutral oligosaccharides, which is consistent with common views of proteoglycan biosynthesis.

To identify the subcellular structures where generation of a polysaccharide chain is initiated, we studied the dynamics of label incorporation in the nuclear and microsomal fractions of the rat liver. The results of this experiment are shown in Fig. 23.

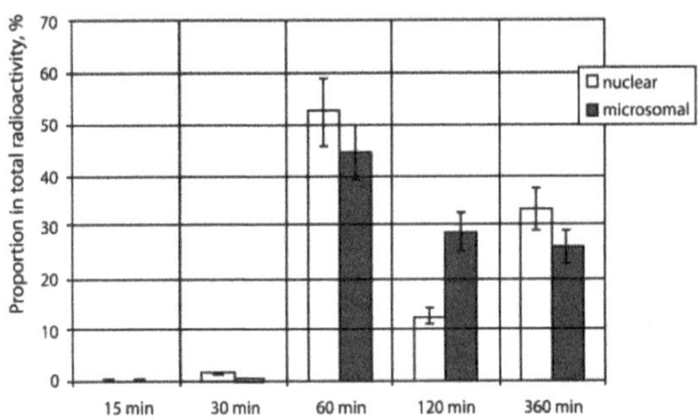

Figure 23. Dynamics of accumulation of radiolableled glucose in the nuclear and microsomal fractions of the rat liver. As 100%, we used the radioactivity averaged over all fractions and all time points.

As Fig. 23 demonstrates, the label became detectable in cell structures 30 min after a glucose load. The label accumulated to the highest content in both fractions within 1 h of the experiment. It should be noted that, in the first hour, the accumulation rate was higher in the nuclear fraction (radioactivity 53% compared with 44% in the microsomal fraction).

After 120 min of the experiment, the microsomal fraction accumulated a greater amount of the label as compared with nuclei (29 and 12%, respectively). After 3 h, the label was again accumulated more intensely in nuclear structures (33% compared with 26% in microsomes).

4. Discussion

Our results demonstrate that polysaccharides are similar to NA in many aspects. To a great extent, this is a consequence of the fact that NA is a polymer of ribose or deoxyribose having a base as a side-chain substituent. Some polysaccharides (e.g., HA, amylose, cellulose) are themselves capable of forming helical structures similar to NA.

Quantum chemical computations revealed the possibility of selective bonding between UDP-uronic acids and purines and between UDP-hexoses and pyrimidines contained in NA. The bonding strength per monomer unit in the polysaccharide–NA complex is similar to that computed for the NA double helix. This selectivity allows us to assume a binary genetic code for polysaccharides on the basis of the above principles in addition to the commonly accepted genetic code for amino acids. Quantum chemical computations were supported by the results of dot hybridization and spectral analysis of NA–polysaccharide complexes.

The spectrophotometric studies showed that amylose selectively binds with polypyrimidines (poly(dC)) while polyuronides bind with polypurines (poly(dA)). As well known, amylose is a homopolysaccharide of hexose and polyuronide, of hexuronic acid. The difference between these monosaccharides is that glucose contains the hydroxymethyl group at C5, while hexuronic acid has the carboxyl group at C5. Apart from this, the two monosaccharides are identical. Hence the observed difference in physicochemical properties of the above polysaccharides can be interpreted in terms of the structural difference at C5 of their units. Consequently, our results suggest that purines (exemplified by A) selectively bind to the carboxyl group of uronic acid residues, while pyrimidines interact with the hydroxymethyl group of hexose residues contained in polysaccharides. These experimental findings agree with the results of quantum chemical computations performed for the NA–polysaccharide complexes, testifying again to their adequacy.

The bonding of nucleotides and glycans was detected in experiments aimed at studying the interaction between mature GAG (HA and CS) with fragmented calf thymus DNA. In primary structure, HA is a multiple tandem repeat of a unit consisting of a hexose and an uronic acid residue. In turn, tandem repeats of purine–pyrimidine account for a considerable proportion of DNA in higher organisms, including calf thymus DNA (Kiselev L.L. 2000, Singer M., et al. 1998). As our results demonstrate, HA did find complementary regions in calf thymus DNA. CS showed no complementary interactions with calf thymus DNA. It is known that C6 of hexose residues of CS is modified with the sulfo group, which provides an additional partial negative charge. This modification of hexoses dramatically affects the physicochemical properties of the polysaccharide and probably prevents hydrogen bonding.

Dot hybridization confirmed possible complementarity of hexoses to pyrimidines and uronic acid to purines. The specific bonds between NA and polysaccharides are comparable in strength with the bonds between complementary nucleotides in DNA.

Thus, analysis of the interactions between polysaccharides and NA showed that purines of NA are complementary to uronic acids of polysaccharides and that pyrimidines are complementary to hexoses. The relationships between the genetic apparatus of the cell and polysaccharides deserve further studying in terms of the above complementarity principle. Such studies will probably yield a fundamentally novel view as to whether microheterogeneity of polysaccharide moieties of proteoglycans is genetically determined and related to purine–pyrimidine DNA repeats. Yet it is clear that the nature of these relationships need additional comprehensive studies.

The idea that polysaccharide moieties of proteoglycans contain information is not new. For instance, Zimina (Zimina N.P., et al. 1992, Zimina N.P., et al. 1987, Zimina N.P., et al. 1986, Zimina N.P., Rykova V.I., Dmitriev I.P. 1987) demonstrated that carbohydrate chains of proteoglycans are chemically heterogeneous and structurally irregular and concluded, quite justifiably, that the information content of proteoglycans provides a chemical basis for their intricate and highly specific functions in the cell. There is still no method for GAG sequencing, and data on the structure of their chains are circumstantial. Since chemical heterogeneity is widespread, these data make it possible to assume that proteoglycan chains are of irregular structure with a cluster arrangement of disaccharides. The clustering of bonds sensitive to testicular hyaluronidase and, consequently, of glucuronic acid residues was demonstrated for pig skin dermatan sulfate. The clusters are arranged along the chain without any distinct regularity (Fransson L.A., et al. 1982). Likewise, glucuronic acid-containing disaccharides are clustered at random along the chain of dermatan sulfates from the pig intestinal mucosa and the umbilical cord (Fransson L.A., et al. 1982). The clustering of various uronic acid residues is characteristic of dermatan suflate from the human uterine neck. The irregularity of dermatan sulfate chains was suggested from chromatographic elution profiles of dermatan sulfate isolated from the bovine aortic intima and digested with chondroitinase AC (Oegema T.R., et al. 1979). Structural studies revealed a cluster organization of HS and heparin chains. Their molecules include extended regions consisting of low-sulfated disaccharides that contain glucuronic acid residues as a main component and only a minor amount of glucose residues (Bjork I., et al. 1982). Clusters sulfated to a high extent were also found: they consist of highly charged disaccharides harboring glucose residues as a main component. These clusters are possibly separated by less ordered regions where alternating sugars occur in similar proportions. Molecules vary in size of highly and low-charged regions.

It should be noted that the structure of oligosaccharide sites is an informative element of GAG as opposed to NA, which contain information in the form of a strict sequence of monomeric units (nucleotides). For instance, the capability of self-association of proteoHS and proteodermatan sulfates is due to so-called contact zones, specific regions with alternating disaccharides containing glucuronic acid and glucose residues with a certain arrangement of sulfo groups (Franson L.-A. 1982). It is noteworthy that polysaccharides are characterized by a determined structural–functional interdependence similar to that of NA (Zimina N.P., et al. 1992).

The concept of nontemplate synthesis of polysaccharide components of proteoglycans does not allow a genetically grounded explanation of microheterogeneity (polymorphism) of proteoglycans. Yet data are continuously accumulating that polysaccharide components of proteoglycans are highly polymorphic and that their polymorphism shows distinct tissue, organ, and species specificities. It should be noted that glycans are still poorly understood and, consequently, the current views of glycans are similar to the views of NA in early molecular biology. Compounds belonging to one class strikingly differ in primary structure. We know today that, for instance, every mRNA is a unique element in realization of genetic information, but such functions are still to be elucidated in the case of glycans.

We characterized the fraction composition of polysaccharides contained in the rat liver homogenate. As Fig. 18 shows, rat liver polysaccharides include polymers of neutral sugars, hexoses, and hexuronic acids and vary in proportions of monomeric units and the degree of their modification (amination, sulfation, epimerization, etc.).

Correlation analysis revealed an association between time and synthesis of HS, which are GAG modified to the greatest extent. This result is beyond doubt because their modification is known to result in a high degree of sulfation and, consequently, their maturation takes more time as compared with maturation of other GAG groups.

In addition, the analysis results (Pearson empirical correlation coefficients and Spearman rank correlation coefficients) showed a strong linear correlation between cell nuclear structures (DNA) and the fraction of nuclear saccharides eluted from DEAE cellulose until 0.15 M NaCl. Such a correlation was not observed for total DNA and the 0.15 M fraction of the homogenate (oligosaccharides of microsomes, lysosomes, nuclei, other cell structures, and the intercellular matrix). The correlation testifies that nuclear synthesis of saccharides of this group increases after a glucose load, especially 1 h after loading. This finding was confirmed by quantitating polysaccharides with normalization with respect to the DNA content (Fig. 19).

The 0.15 M fraction of the homogenate distinctly correlated with the contents of mature proteoglycans in the homogenate and the nucleus, implicating saccharides of this fraction in GAG maturation. In other words, an increase in synthesis of this saccharide fraction of the homogenate increased synthesis of proteoglycans (CS and HS). The fact that proteoglycan synthesis starts with this fraction was evident from the data on incorporation of radiolabeled glucose in GAG. It should be noted that the 0.15 M fraction varies in composition among cell structures. The nuclear fraction harbors neutral sugar components and uronic acids at 3:0.75, suggesting incomplete addition of uronic acids to neutral trisaccharide units, which takes place in the nucleus. The 0.15 M fraction of nuclear saccharides consists mostly (to 83%) of components eluted at 0.02 M NaCl. The 0.06 and 0.15 M subfractions occur at 11 and 6%, respectively. In microsomes, the proportion of the 0.02 M subfraction is decreased to 55%, while the proportions of the 0.06 and 0.15 M subfractions are increased to 22 and 23%, respectively. The changes observed in the homogenate are similar to those in microsomes, but the proportion of the 0.15 M subfraction is greater (up to 32%).

These findings associate the genetic apparatus of the cell with synthesis of oligosaccharides similar in composition to the universal tetrasaccharide of proteoglycans. The association is also evident from the results obtained with radiolabeled glucose: the label accumulation rate in nuclear structures was higher than in other structures 1 h after a glucose load.

Oligosaccharides with a neutral sugar : uronic acid ratio of 3:1 are most probably synthesized in the nucleus. Heteroglycan chains with a tandem arrangement of uronic acid–hexose units (1:1) are formed in structures of the microsomal fraction.

Analysis of incorporation of radiolabeled glucose in the nuclear and microsomal fractions of the cell (Fig. 23) showed that the incorporation rate is more intense in the nucleus early (at the end of the first hour) after a glucose load. After 2 h, microsomes accumulate radiolabeled glucose to a high extent, which is probably due to a transfer of radiolabeled oligosaccharides from the nucleus and the formation of GAG chains. After 3 h, the content of radiolabeled saccharides in the cell nucleus increased again owing most probably to a transfer of mature GAG from the EPR. This scenario agrees with modern views of proteoglycan synthesis.

It is of interest that the label appeared first in the 0.15 M fraction and then in mature GAG (the 0.25-1.5 M fraction) of the homogenate (Fig. 22). The initially high radioactivity of this fraction can be explained by the presence of glycans involved in initiating synthesis of the polysaccharide chain of GAG. GAG synthesis starts with xylosylation of core proteins and generation of a linker tetrasaccharide; then, the polysaccharide chain is formed by consecutive addition of hexoses and uronic acids in almost equal proportions.

Comparison of the data on radiolabeled glucose incorporation in GAG and fractions of cell structures showed that the high rate of label accumulation in the nucleus early (within the first hour) after a load was associated with the dynamics of label incorporation in polysaccharides eluted until 0.15 M NaCl. This was not the case with mature GAG, which showed the highest radioactivity only 2 h after a load. At this time, the label content is minimal in the nucleus and higher in microsomes. Hence, we can state that generation of the GAG chain in the cell is associated with microsomes, while synthesis of the linker tetrasaccharide, which belongs to the 0.15 M fraction, is associated with nuclear structures.

Saccharides of the 0.15 M fraction differ from classical GAG in having another hexosamine content and probably harbor fragments with linker tetrasaccharides. Note that, until uronic acid–hexosamine ratio reaches 1:1 (as characteristic of HA), such fragments, having a considerable portion of neutral sugars, are eluted from the column at an ionic strength lower than necessary for HA elution (0.15 M NaCl). Since proteoglycan synthesis starts with generation of the linker tetrasaccharide, the radioactivity of this fraction should be higher early after a radiolabeled glucose load. Our results fully agreed with this expectation.

Thus, the glycoside moiety of proteoglycans is probably synthesized in a stepwise manner in the cell. The linker tetrasaccharides are synthesized in structures associated with the cell nucleus, which is also evident from the results of correlation analysis. Then the fragments are transferred into the EPR, where the main glycan chain is synthesized, biochemically modified, and used to form proteoglycans. The GAG chain is synthesized in the EPR. Mature GAG are delivered into the nucleus in a small amount and are mostly exported into the intercellular space to produce the intercellular matrix. This scenario agrees with data of many studies that core proteins entering the EPR and Golgi system already have the linker tetrasaccharide (xylose-galactose-galactose-uronic acid) (Colman Y., et al. 2000, Zimina N.P., et al. 1992,

Zimina N.P., et al. 1987, Silbert J.E. et al. 1995). These data indirectly support our assumption that DNA plays a role in synthesis of the universal tetrasaccharide of proteoglycans.

Based on our data, we suggest a conceptual mechanism underlying the functional association between the nuclear apparatus and the formation of the tetrasaccharide linking the protein core with the GAG moiety in proteoglycans.

Taken together, the results obtained with different methods allow us to propose a concept of template synthesis of proteoglycans with the involvement of tandem DNA repeats.

It is clear that proteoglycan synthesis starts with DNA transcription to yield pre-mRNA. The informative region of the mRNA for the core protein is formed according to the commonly accepted mechanism up to the serine codon. When the serine codon is followed by a triplet complementary to the trisaccharide uronic acid-hexose-hexose, the mode of transcription changes. It should be noted that the complementarity requirements are met only by eight DNA triplets (ACC, ACT, ATC, ATT, GCC, GCT, GTC, and GTT). The energy of bonding during NA synthesis is distributed as follows among these triplets: it is about 11.7 kcal/mol in one triplet (GCC), 9.6 kcal/mol in three triplets (GCT, GTC, and ACC), 7.6 kcal/mol in three triplets (GTT, ACT, and ATC), and 5.5 kcal/mol in one triplet (ATT). Thus, the ATT triplet has the bonding energy of only 5.5 kcal/mol per monomer, lower than in the triplet duplex DNA-polysaccharide (more than 7 kcal/mol).

Under certain conditions, this situation with the ATT codon may lead to appreciable competition with UDP-sugars in the formation of complementary pairs during RNA synthesis on DNA. As a result, synthesis of a trisaccharide may be more advantageous in terms of energy than addition of three structural units to RNA. Hence the following process is possible. Glycosyltransferase utilizes the first uronic acid residue bound to DNA adenine via complementary interactions and generates a bond between the saccaride and the last ribonucleotide of the serine codon of RNA, simultaneously converting uronic acid to xylose via decarboxylation at C_5. Decarboxylation initiates separation of the saccharide unit from DNA. Then, Gal transferases generate glycoside bonds between xylose and two hexose (galactose) residues, which are hydrogen-bonded to the thymine tandem of DNA according to the above scheme.

It is clear that some pre-mRNAs synthesized *in vivo* under such conditions have the trisaccharide xylose-galactose-galactose in place of a ribonucleotide triplet at the site of RNA branching on the ATT codon of DNA. It is well known that the ATT triplet determines a stop codon (UAA) terminating synthesis of polypeptide chains, which indirectly supports our assumption. If this sequence is followed by tandem repeats (purine-pyrimidine, such as CA) with a strength of bonding in NA synthesis about 8.8 kcal/mol, synthesis of oligosaccharide is disadvantageous and the RNA is extended according to the DNA-RNA complementarity rule, which is more advantageous in terms of energy. Although the energy of bonds plays an important role in this situation, we assume that the process is still far more intricate.

Thus, it is possible that hnRNA contains pre-mRNAs that have a trisaccharide followed by a tandem ribonucleotide repeat consisting of more than 300 monomers (which is sufficient for generating the glycoside moiety of proteoglycans) and starting from 5′-P-terminal guanosine at the nuclear RNA branching site. This can explain why the corresponding RNA regions block reverse transcription and are resistant to some RNases. Guanine acts as a hnRNA branching site and, as a purine, is capable of ensuring further addition of the first uronic acid residue, which always follows the trisaccharide. With the above system of complementary interactions, the trisaccharide is analogous to UAA and, when immediately followed by GU, allows the processing and splicing of pre-mRNA according to the commonly accepted scheme. Like adenines, the two galactose residues may interact with two uridines of the spliceosome U-RNA through their hydroxymethyl groups. Thus, the sequence Gal-Gal-G-U provides a binding site for spliceosome structures, as characteristic of the splicing tetranucleotide consensus sequence (AAGU).

In such an RNA strand, saccharides are linked by the 3′-5′ bond up to the end of the serine codon. C_2 of the saccharide of the last nucleotide of the serine codon interacts with C_5 of xylose, which is thereby capable of binding through C_1 with serine when xylosylation is initiated. Then xylose C_4 binds with C_1 of a galactose dimer. C_2 of the last galactose interacts with C_5 of the saccharide moiety of guanosine. Thus, C_3 of the last nucleotide of the serine codon and C_3 of the last galactose are free for bonding, which allows a 3-1 bond with the first glucuronic acid residue. Uridine following guanosine in an RNA intron determines the addition of a hexose to the glucuronic acid residue through the 4-1 (or 3-1) bond during template synthesis of a polysaccharide (Fig. 24).

Invariant GG at the 5′ end of the next exon, continuing the protein-coding sequence, represents the first two nucleotides of a glycine codon (GGA, GGG, GGC, or GGU). This GG is in the trisaccharide site (the RNA branching site) and, as the flanking intron is excised in the processingosome during RNA maturation, continues the mRNA coding region after the serine codon in the 5′-3′ direction. The obligatory presence of invariant splicing-site GG in the triplet following the serine codon can explain the conservation of the xylosylation site (serine-glycine tandem) among core proteins, because only glycine codons start with two guanines.

The resulting transcript is capable of directing synthesis of a polypeptide chain wherein serine is covalently bound to the trisaccharide and then with glycine. Tandem RNA sequences, representing the intron side chain following the tetrasaccharide (xylose-galactose-galactose-uronic acid), allow a polysaccharide fragment to be synthesized according to the NA base–monosaccharide complementarity by glycosyltransferases that form 4-1 or 3-1 glycoside bonds, starting from the last saccharide. After splicing occurs and the two guanines find themselves in the exon part of the molecule, the tandem RNA fragment is linked to C_2 of uronic acid through the terminal uridine. The fragment determines an ordered arrangement of monosaccharides for synthesis of a polysaccharide chain. This process is advantageous (bonding energy is about 7 kcal/mol) in systems having an excess of UDP-saccharides and glycosyltransferases and lacking mononucleotides and NA polymerases, as characteristic of membrane structures of EPR and the Golgi complex. We propose that the process is termed glycotranscription, because information contained in NA is directly transferred to the polysaccharide chain.

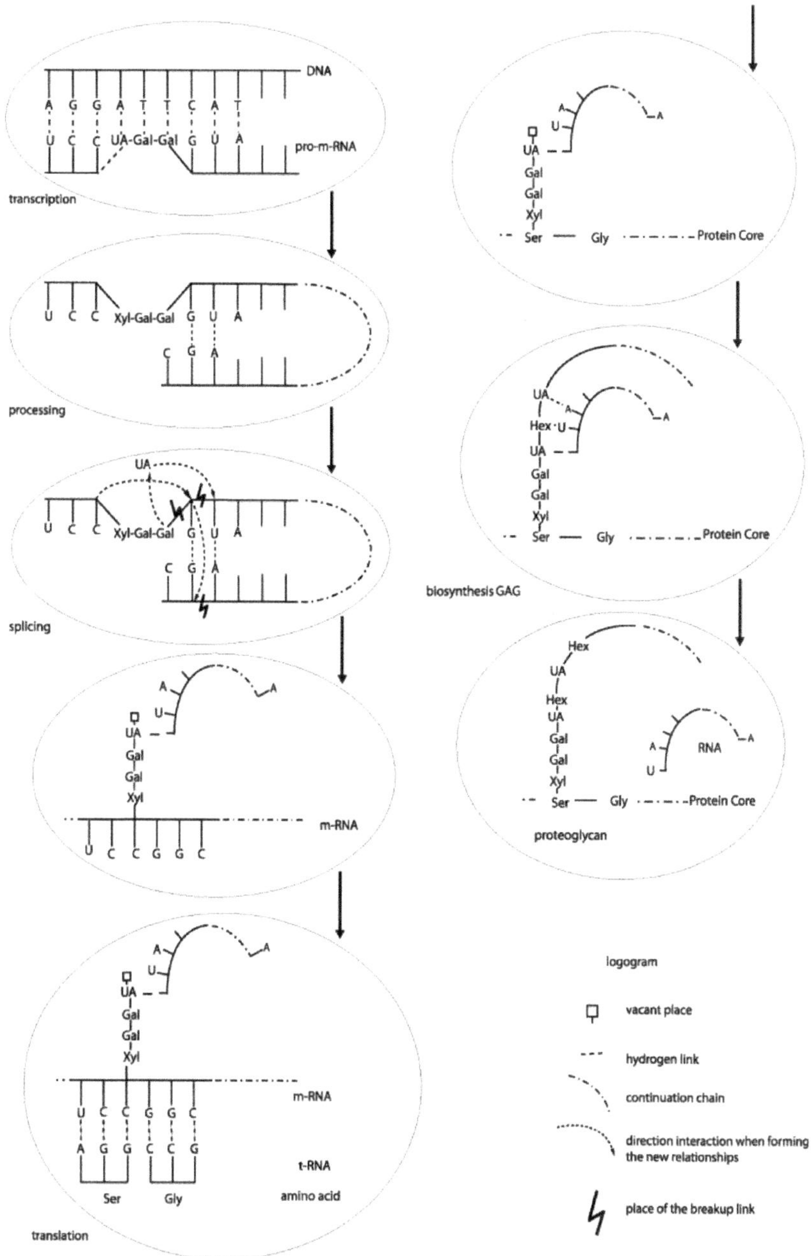

Figure 24. Hypothetical scheme of genetically determined template synthesis of proteoglycans.

It is clear that synthesis of the core protein chain on ribosomes is quite possible, because the tetrasaccharide does not occupy the 3′-5′ bond of the last nucleotide of the serine codon and acts as a spacer linking the RNA intron. As a result, the newly synthesized core protein contains the linker tetrasaccharide and the RNA intron attached to the serine. The protein is delivered after synthesis into the site where the corresponding polysaccharide fragment is generated in membrane structures of the EPR and the Golgi complex. Serine is xylosylated in the EPR during biosynthesis of the core protein on ribosomes. This scenario does not contradict the modern views of GAG synthesis (Silbert J.E. et al. 1995).

It should be noted that pre-mRNA splicing is tissue-specific. Moreover, selection of the splicing site depends on the developmental stage in some cases (Singer M., et al. 1998). In other words, the primary structure of mRNA introns varies with tissue and ontogenetic stage. The same is true for the primary structure of the glycoside moiety of proteoglycans. It is known that proteoglycans are completely absent from unicellular and prokaryotic organisms, as well as pre-mRNA processing is. This fact suggests a relationship between glycan biosynthesis and mRNA maturation in terms of biological significance for the cell.

This assumption is supported by data obtained for mRNAs of CS core proteins. Translated in a cell-free wheat germ system, cartilage mRNA directed synthesis of a 340-kDa core protein. Immediately after translation, the protein already contained glycosylation signals for subsequent glycan synthesis (Hook M., et al. 1984), i.e., the linker tetrasaccharide was already present.

Based on quantum chemical analysis of the advantage of bonding, the hypothesis of proteoglycan synthesis of the NA template (Fig. 24) agrees with the views of the processes involved in realizing genetic information (genome structure, transcription, hnRNA processing and splicing, translation) (Singer M., Berg P. 1998) and eliminates the main contradictions of the existing concept of proteoglycan metabolism. The hypothesis explains why the site where the linker tetrasaccharide is attached to the protein core has not been identified in more than fifty years of studies on proteoglycans: the tetrasaccharide is already contained in RNA before translation. There is convincing evidence that the tetrasaccharide finds its way in the EPR as covalently bound to the core protein (Colman Y., et al. 2000, Silbert J.E. et al. 1995). A role of protein core structures in determining the serine xylosylation site was rejected in recent studies. A "vitalistic" hypothesis has been formulated that ascribes this role to intracellular membranes. The hypothesis is based on the fact that glycosyltransferases are mostly in membrane structures of the EPR and Golgi complex (Silbert J.E. et al. 1995). Yet this fact alone does not prove synthesis of the linker tetrasaccharide in these structures, because glycosyltransferases are also detectable in nuclei (nuclear membranes). Our biochemical studies implicate structures of the cell nucleus in initiation of synthesis of the linker tetrasaccharide.

According to our hypothesis, a serine is subject to xylosylation only when its codon is followed in DNA consequently by the ATT stop codon and CA (GU in RNA) of an intron, responsible for hnRNA branching and mRNA processing.

The above scheme of proteoglycan synthesis allows generation of a linear heteroglycan of a particular size (about 300 monomers). The question is how template synthesis following this scheme proceeds in the case of HA, consisting of thousands of monomers. It is clear that RNA templates of a corresponding size are absent from EPR structures. It is possible that GAG synthesis in this case utilizes lasso-like intron RNA, which result from cis-splicing and are potentially capable of directing cyclic synthesis of HA with any number of monomeric units. It is known that mRNA trans-splicing yields linear introns. Such RNA structures may direct template synthesis of branched homoglycans whose branching period is a multiple of the full NA helix turn (10-12 units). Monosaccharide 10 (12) overlies monosaccharide 1 of the nascent strand under these conditions, which allows an additional bond between them. Then synthesis proceeds through several such rounds to yield a branched polysaccharide structurally similar to glycogen. This hypothetical mechanism generating branched polysaccharides is supported by the fact that their branching period is a multiple of the full NA helix turn.

RNAs with the above structures may be components of small cytoplasmic RNA, the role of which is still poorly understood. It is known that scRNA account for less than 1% of total cell RNA. Some scRNAs contain tandem repeats, in particular, *Alu* sequences. These scRNAs are associated with membrane structures of the EPR. Some scRNAs (7SL RNA) are involved in transmembrane transport of polypeptides across the EPR lipid bilayer. The scRNA size varies from 90 to 330 nt, which falls well into our concept. It should be noted that synthesis of storage homoglycans probably requires no template and proceeds through a simpler mechanism, because their chains carry no information.

Many glycoproteins act as antigens on the plasma membrane. Immunohistochemical comparison of glycoproteins isolated from hepatocyte membranes of intact, embryonic, and regenerating liver and from hepatoma showed that the protein component of antigens is detectable on all hepatocytes regardless of the state of the liver. Moreover, the protein component is tissue-nonspecific and universal for most cell groups. The carbohydrate component of antigens proved to be specific. It is the structure of the polysaccharide component that determines the immunogenicity of a proteoglycan. Zimina (Zimina N.P., et al. 1987) studied the specifics of GAG synthesis in the liver of adult rats and embryos and in hepatoma and showed that GAG of tissues with active cell proliferation are sulfated to a lower extent as compared with the corresponding GAG of quiescent tissues. That is, polysaccharide fragments of the same proteoglycans differ depending on the state of hepatocytes.

According to the concept of nontemplate synthesis of GAG, the protein core is synthesized on polysomes of the rough EPR. The same protein components of proteoglycans were synthesized in all above cases. The identity of the protein core suggests that the glycoside components of these proteoglycans are synthesized in one site of the smooth EPR and, consequently, are also identical. Experimental results contradict this assumption. The variation of glycoside components is possible only when glycosyltransferase complexes of EPR membranes vary depending on the physiological state of hepatocytes. A glycosyltransferase complex of the smooth EPR should contain about 300 molecules of the enzyme generating the glycoside bond in a certain sequence. For instance, HS synthesis requires a complex of about 300 N-acetylglucosaminylatransferases and glucuronyltransferases occurring in equal proportions. Each of the 150 enzyme molecules, e.g., N-acetylglucosaminyltransferases, should be encoded by a separate gene with a nonconserved sequence coding for the site of enzyme attachment to a strongly specific membrane site of the smooth EPR. If GAG synthesis is template-independent, at least three types of such complexes are necessary for HS generation. About 450 N-acetylglucosaminyltransferase genes are required for this process. Since the enzyme is also involved in synthesis of heparin and other polysaccharides, the number of N-acetylglucosaminyltransferase genes should occur at thousands of copies per genome. Yet it is known that DNA regions complementary to mRNAs have unique sequences and occur at a few copies per genome. Glycosyltransferase genes are not exceptions to this rule. Thus, the commonly accepted hypothesis of nontemplate synthesis of GAG cannot explain the genetically determined heterogeneity of polysaccharide components of proteoglycans. This hypothesis does not stand up, since recent works have demonstrated the informative value of the glycoside moiety of proteoglycans (Zimina N.P., et al. 1992, Zimina N.P., et al. 1987).

Our scheme of template synthesis of GAG solves this problem, because only a few glycosyltransferase genes are sufficient for generating any diversity of polysaccharide chains in this case.

Synthesis of some glycans with a high information content according to the glycotranscription principle makes it possible to assume, by analogy with RNA, the existence of reverse glycotranscription, whereby information is transferred from polysaccharide to RNA and then, by reverse transcriptase, to DNA fragments, which can be inserted into the genome to preserve the acquired information about new glycans in the genome structure. Such information is contained in specific tandem DNA repeats, which are unique for each individual. The relevant genetic systems are probably capable of being transmitted to the progeny and being fixed as a hereditary or acquired character by selection. It seems that both nontemplate and genetically determined template synthesis of polysaccharides exist in nature and are closely associated with each other through information flows.

It should be noted in this connection that Ronichevskaya and Rykov (Ronichevskaya G.M., Rykova V.I. 1977) observed that GAG suppress DNA replication in proliferating cells in a keilon-like manner. This observation can be explained by homology of polysaccharides to NA. We think that polysaccharide fragments are capable of blocking DNA polymerases via specific binding to complementary DNA repeats. As reported earlier, the substances examined in (Ronichevskaya G.M., et al. 1977) were isolated from RNA preparations, which provides indirect evidence for a role of RNA in glycan synthesis. According to our results, HA fragments form stable bonds with NA.

Further studies of proteoglycan polymorphism (species and tissues specificity, age- and pathological changes) and development of sequencing technique will probably demonstrate the specificity of proteoglycans at the level of individual organisms. By analogy, polymorphism of highly repetitive genome sequences is now beyond doubt and is widely used in genome fingerprinting. We assume that it is genomic repeats that are responsible for the information structure of glycans, whose polymorphism (microheterogeneity) has come to be commonly recognized.

The results of many studies implicate proteoglycans in intricate processes regulating cell proliferation (Ronichevskaya G.M., et al. 1977) and differentiation (Kinoshita S., et al. 1979, Kinoshita S., Yoshii K. 1979), cell recognition, and the organization of intercellular interactions (Henkart P., et al. 1973). In evolution, polysaccharides played a key role in the transition from unicellular organisms to multicellular entities. It is owing to polysaccharides that the individual cell functions as a structure with a particular function in the cell ensemble of the total organism. In the ontogeny, proteoglycans are associated with the aging of cells, tissues, and organs (Zimnitskii A.N., et al. 2004). A genetic defect in synthesis of GAG polysaccharide chains leads to severe hereditary disorders (Zimnitskii A.N., Bashkatov S.A. 2004). It is beyond doubt that the relevant processes are controlled by DNA, which carries genetic information.

To conclude, life is a mode of existence of not only protein and nucleic, but also of polysaccharide bodies, because all three biopolymers determine life that we observe. Analysis of the distinct association of NA and polysaccharides will clarify the role of both polysaccharides and DNA repeats. The latter cannot be described now as waste or selfish, because some of them are potentially capable of directing polysaccharide synthesis.

We are grateful to prof. N.K. Yankovskii (Institute of General Genetics, Russian Academy of Sciences, Moscow) and Dr. V.I. Salyanov (Institute of Molecular Biology, Russian Academy of Sciences, Moscow) for help in collecting and interpreting empirical data.

Author details

Aleksandr N. Zimnitskii
«NanoDerm-profi» Ltd., Moscow, Russia

5. References

Altschuler C.H., Kinsman G., Bareta J.(1963). Connective tissue metabolism: I. Hyaluronic acid synthesis by cell-free extracts of human tissues. *Arch Pathol.* Feb;75:206-211.

Belozersky A.N. (1970). Methods for Studying Nucleic Acids. Moscow: Mir.

Blagoi Yu.P., Galkin V.L., Gladchenko G.O., Kornilova S.V. (1999). Metallocomplexes of Nucleic Acids in Solutions. Kiev: Naukova Dumka, 199p.

Bjork I., Lindahl. U. (1982). Mechanism of the anticoagulant action of heparin. *Mol Cell Biochem.*18:161-82..

Chaplin M.F., Kennedy J.F.(1986). Carbohydrate Analysis: A Practical Approach. Oxford, Washington.

Cifonelli J.A., Dorfman A.J.(1957). The biosynthesis of hyaluronic acid by group A streptococcus: V. The uridine nucleotides of group A streptococcus. *J Biol Chem.* Oct; 228(2):547-57.

Colman Y., Rem C.-G. (2000). Illustrative Biochemistry. Moscow: Mir.

Cybulski S.M., Scheiner S.J.(1989). Hydrogen bonding and proton transfers involving the carboxylate group. *J Am Chem Soc.*, 111, p.23-31.

Denhardt D.T.(1966). A membrane-filter technique for detection of complementary DNA. *Biochem Biophys Res Commun.*, 23:641-52.

Dewar M.J.S., Zoebisch E.G., Healy E.F., Stewart J.J.P. (1985). Development and use of quantum mechanical molecular models. 76. AM1: A new general purpose quantum mechanical molecular model. *J Am Chem Soc.*,107, p.3902-3909.

Dorfman A. (1958). Studies on the biochemistry of connective tissue. *Pediatrics.* Sep;22(3):576-89.

Dorfman A. (1964). Metabolism of acid mucopolysaccharides. *Biophys J.* Jan; 71:SUPPL 155-65.

Dorfman A.(1965). In: Structure and Function of Connective and Skeletal Tissue. London,178, 297.

Dunin V.V., Rukhadze E.G., Potapov V.M.(1979). Generation and Analysis of Optically Active Compounds. Moscow: Mosk. Gos. Univ.

Fransson L.A., Havsmark B. (1982). Structural requirements for heparan sulphate self-association. *Carbohydr Res.*, 16:135-44.

Franson L.-A.(1982). Structural features of the contact zones for heparan sulphate self-association. *Carbohydr Res.*,110:127-33.

Glaser L., Brown D.H.(1955). Purification and properties of d-glucose-6-phosphate dehydrogenase. *Proc. Natl. Acad. Sci. USA*, 41:253.

Henkart P., Humphreys S., Humphreys T.(1973). Characterization of sponge aggregation factor: A unique proteoglycan complex. *Biochemistry.*, 12:3045-50.

Hook M., Kjellen L., Jahansson S., Robinson J.(1984). Cell-surface glycosaminoglycans. *Annu Rev Biochem.*, 53:847-69.

Jeffrey D Esko, Zhang L.(1996). Influence of core protein sequence on glycosaminoglycan assembly *Curr Opin Struct Biol.* 5:663-70.

Jermyn M.A.(1975). Increasing the sensitivity of the anthrone method for carbohydrate. *Anal Biochem.*, 68:332-5

Jurema J.M.W., Shields G.C.(1993). Ability of the PM3 quantum-mechanical method to model intermolecular hydrogen bonding between neutral molecules. *J Comput Chem.* 14, p.89-104.

Kallies B., Mitzner R. (1995). The ability of the semiempirical PM3 method to model proton transfer reactions in symmetric hydrogen bonded systems. *J Mol. Model.*, 1, p.68-78.

Kinoshita S., Yoshii K.(1979). The role of proteoglycan synthesis in the development of sea urchins: II. The effect of administration of exogenous proteoglycan. *Exp Cell Res.*,123:361-9.

Kinoshita S., Saiga H. (1979). The role of proteoglycan in the development of sea urchins: I. Abnormal development of sea urchin embryos caused by the disturbance of proteoglycan synthesis. *Exp Cell Res.*, 123:229-36.

Kiselev L.L. (2000). Human genome and biology of the 21st century. *Vestn Ross Akad Nauk.*, 70: 414-24.

Latajka Z. Scheiner S.J. (1990). Structure, energetics, and vibrational spectrum of ammonia...water. *J Phys Chem.* 94, p.217-221.

Lindahl U., Kjellen L.(1987). Biology of Proteoglican. Wight T.N., Mecham R.P., Eds., Orlando: Acad. Press. p. 59.

Lorenzel I.(1959). Epinephrine-induced alterations in connective tissue of aortic wall in rabbits. *Proc Soc Exp Biol Med.* Nov;102:440-2.

Maley F., Maley G.F. (1959).The enzymic conversion of glucosamine to galactosamine. *Biochim Biophys Acta.* Feb;31(2):577-8.

Markovic O., Huttl S.(1964). Contribution to a methodical approach to determination of protein and glycoprotein components in synovial fluid. *Cas Lek Cesk.* May 8;103:522-5. Czech.

Markovitz A., Cifonelli J.A., Dorfman A. (1958). Biosynthesis of hyaluronic acid by cell-free extracts of group-A streptococci. *Biochim Biophys Acta.* May;28(2):453-5.

Maxam A., Gilbert W.(1977). A new method for sequencing DNA. *Proc Natl Acad Sci USA.* 74:560-4.

Mednikov B.M., Shubina E.A., Mel'nikova M.N. (2001). Molecular mechanisms of genetic isolation. *Priroda,* 5: 19-28.

Mulliken R.S.(1955). Electron population analysis on LCAO-MO molecular wave functions: II. Overlap populations, bond orders, and covalent bond energies. *J Chem Phys.* 23:1841-6.

Oegema T.R., Jr, Hascall V.C., Eisenstein R.(1979). Characterization of bovine aorta proteoglycan extracted with guanidine hydrochloride in the presence of protease inhibitors. *J Biol Chem.* 254:1312-8.

Orekhovich V.N., (1968). Ed Modern Methods in Biochemistry., Moscow: Meditsina.

Roberts J., Kaserio M.(1978). Basics of Organic Chemistry. Moscow: Mir.

Roden L., Dorfman A. (1958). The metabolism of mucopolysaccharides in mammalian tissues: V. The origin of L-iduronic acid. *J Biol Chem.* Nov;233(5):1030-3.

Rondle C.J.M., Morgan W.T. (1955). The determination of glucosamine and galactosamine. *Biochem J.* Dec;61(4):586-9.

Ronichevskaya G.M., Rykova V.I. (1977). Organic specificity of the antimitotic effect of glycopeptides isolated from RNA preparations. Dokl. Akad. Nauk SSSR., 237:481-3.

Roseman S., Ludowieg J., Moses F., Dorfman A.(1953). The biosynthesis of the glucuronic acid moiety of hyaluronic acid. *Arch Biochem Biophys.* Feb; 42(2): 472-3.

Roseman S., Moses F.E., Ludowieg J., Dorfman A. (1953).The biosynthesis of hyaluronic acid by group A streptococcus: I. Utilization of 1-C14-glucose. *J Biol Chem.* Jul; 203(1):213-25.

Roseman S., Moses F.E, Ludowieg J, Dorfman A.(1954). The biosynthesis of hyaluronic acid by group A streptococcus: II. Origin of the glucuronic acid. *J Biol Chem.*Feb; 206(2):665-9.

Schiller S.(1964). Synthesis of hyaluronic acid by a soluble enzyme system from mammalian tissue. *Biochem Biophys Res Commun.* Mar 26;15(3):250-255.

Schiler S.(1965). Structure and Function of Connective and Skeletal Tissue. London, p.302.

Strominger J.L.(1964). Nucleotide intermediates in the biosynthesis of heteropolimeric polysaccharides. *Biophys J.* Jan;71:SUPPL139-53.

Schmidt M.W., Baldridge K.K., Boatz J.A., Elbert S.T., Gordon M.S., Jensen J.H., Koseki S., Matsunaga N., Nguyen K.A., Su S.J., Windus T.L., .Dupuis M., Montgomery J.A.(1993). General atomic and molecular electronic structure system. *J Comput Chem.* 14, p. 1347-1363.

Silbert J.J. (1963). Incorporation of 14C and 3H from nucleotide sugars into a polysaccharide in the presence of a cell-free preparation from mouse mast cell tumors. *J Biol Chem.* Nov;238:3542-6.

Silbert J.E. Sugumaran G .(1995). Intracellular membranes in synthesis, transport, and metabolism of proteoglycans. *Biochim Biophys Acta.* 1241(3):371-84.

35. Silbert J.E. (1982). Structure and metabolism of proteoglycans and glycosaminoglycans. *J Invest Dermatol.* 79(1):31-7.

Singer M., Berg P. (1998). Genes & Genomes. A changing perspective. University science Books, Mill Valley, California.

Slutskii L.I.(1969). Biochemistry of Normal and Pathologically Changed Connective Tissue. Leningrad: Meditsina.

Strominger J.L.(1964). Nucleotide intermediates in the biosynthesis of heteropolimeric polysaccharides. *Biophys J.* Jan;71:SUPPL139-53.

Stewart J.J.P.(1989). Optimization of parameters for semiempirical methods: II. Applications. *J Comput Chem.* 10, p.221-261.

Sugahara K., Kitagawa H.(2000). Recent advances in the study of the biosynthesis and functions of sulfated glycosaminoglycans. *Curr Opin Struct Biol.* 10(5):518-27.

White J, Handler P., Smith E.L., Hill R.L., Lehman I.R., (1978). Principles of Biochemistry, 6th ed. , New York: McGraw-Hill, p.215.

Williams I.H.(1987). Theoretical modeling of specific solvation effects upon carbonyl addition. *J Am Chem Soc.* 109, p.6299-6307.

White B.N., Shetlar M.R., Shurley H.M., Shilling J.A. (1965). Incorporation of (1-14C)Glucosamine into mucopolysaccarides of rat connective tissue. *Biochim Biophys Acta.* Mar 1;101:97-105.

Yang J.T., Samejima T.(1969). Optical rotatory dipersion and circular dichroism of nucleic acids. In Progress in Nucleic Acids Research and Molecular Biology, Lavidson J.N., Cohn W.E., Eds., New York: Academic, pp.223-38.

Zelenin A.V.(2003). Plant genome. *Vestn Ross Akad Nauk.* 73:797-806.

Zimina N.P., Rykova V.I.(1986). Proteoglycans of animal tissues and their effect on DNA synthesis. Biokhimiya. 51:1555-61.

Zimina N.P., Dmitriev I.P., Rykova V.I.(1987). Composition and sulfation degree of glycosaminoglycans from tissues of various animals: Heterogeneity and tissue specificity of heparan sulfates. *Biokhimiya.* 52(6):984-90.

Zimina N.P., Rykova V.I., Dmitriev I.P.(1987). Comparison of composition and sulfation extent for glycosaminoglycans of quiescent and actively proliferating tissues. *Biokhimiya*, 52:856-61.

Zimina N.P., Rykova V.I., Arkhipov I.A. (1992). Animal tissue proteoglycans as irregular biopolymers: Informativeness of structure and control of biosynthesis. *Usp Sovrem Biol.* 112(4):571-88.

Zimnitskii A.N., Bashkatov S.A., Petrenko E.G., Khusnutdinova S.B.(2004). Age-related changes in skin glycosaminoglycan content in women, Biomed Khim., 50:309-13.

Zimnitskii A.N., Bashkatov S.A.(2004). Glycosaminoglycans in biochemical mechanisms of adaptation to some physiological and pathological states. Moscow, Globus: Farmatsevticheskii Byulleten.

Chapter 12

Diatom Polysaccharides: Extracellular Production, Isolation and Molecular Characterization

Ranieri Urbani, Paola Sist, Galja Pletikapić,
Tea Mišić Radić, Vesna Svetličić and Vera Žutić

Additional information is available at the end of the chapter

1. Introduction

Diatoms (phylum Heterokontophyta, class Bacillariophyceae) are photoautotrophic protists derived from a secondary endosymbiosis involving a heterotrophic eukaryote host and a red alga endosymbiont. The annual primary production of diatoms is estimated to be 20 Pg (10^{15} g) of carbon, equivalent to 40% of marine and one-fifth of total global primary production and diatoms are responsible for more than half of the organic carbon flux to the deep ocean. DOM (dissolved organic matter) represents the largest pool of organic matter in the sea, with substantial part still uncharacterized [1].

The extracellular polysaccharide production by marine diatoms is a significant route by which photosynthetically produced organic carbon enters the trophic web and may influence the physical environment in the sea as observed for example when massive aggregation events on basin scale occur. Macroscopic dimensions and the massive appearance of gelatinous macroaggregates known as 'mucilage events' occurring episodically in Northern Adriatic are illustrated in Figure 1.

Despite many advances in organic carbon concentration in marine diatoms, the biophysical and biochemical mechanisms of extracellular polysaccharide production remains a significant challenge to resolve at molecular scale in order to proceed towards the understanding of its function at the cellular level and interactions and fate in the ocean. Chemical characterization of diatom extracellular polymer substance (EPS) isolated from laboratory cultures revealed that EPS are predominantly heteropolysaccharides that contain substantial amounts of uronic acid and sulphate residues [4-7]. There are of course other substances derived by disruption or decay of algal cells. Such substances are both of low and high molecular weight, present mainly as dissolved or particulate matter. Microscopic (transmission electron and atomic force microscopy) and NMR studies reveal that fibrillar polysaccharides formed the bulk of oceanic DOM [8].

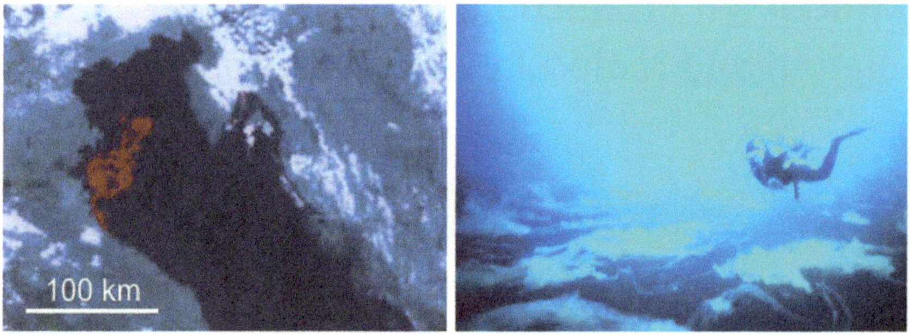

Figure 1. Macroscopic phenomenon of extracellular polysaccharide gellation in the Northern Adriatic Sea: (**a**) remote sensing by satellite showing gel phase in red color (adopted from [2]); and (**b**) at 10 m depth captured by a scuba-diver [3].

Many papers showed that the aldose signatures of marine DOM obtained from different seawater samples around the world is similar to that determined on cultured phytoplankton DOM [9,10] and that the carbohydrate production could be very different among the species selected, growth and environmental conditions [5,7,11-17]. These results are very important in order to understand the role of algal exudation in the aggregation processes observed in all of the seas and in general in carbon cycling in the euphotic zone. Many authors showed that cultured diatoms growth in P-limiting condition determines an increase of polysaccharides exudated by different diatoms species [4,6,7,11,12,15,17,18] both pelagic and benthic.

2. Techniques for primary structure characterization

Among the three major classes of biopolymers, the polysaccharides show the greatest chemical and structural variety. The nucleic acids are constructed from a handful of nucleotide bases so that the polymeric structure obtained is invariably linear. The number of amino acid building blocks used to construct the proteins is approximately twenty but, again, the proteins are always linear polymers. On the other hand, polysaccharides display a wide chemical and structural variability that is not found among the polypeptides and polynucleotides mainly due to the multiple hydroxyl functionality of the five- and six-carbon sugars. The replacement of one or more of such sugar hydroxyl functionalities by amine, ester, carboxylate, phosphate or sulfonate groups, leads to the frequent occurrence of tree-like branching and to the huge number of possible polymeric conformations of different solution behavior. For these reasons carbohydrate analysis involves, after isolation and purification, many steps, i.e. the determination of individual monosaccharides, of anomeric linkages, of branching and sequence, of anomeric configuration and, finally, of the chain conformation [19].

2.1. Isolation, purification and separation by ultra-filtration or by solvent precipitation

In the last two decades the filter (ultra-)fractionation technology has highly improved the methodologies for isolation and purification of polydisperse biopolymers, consisting of macromolecules, like algal polysaccharides, that present very often a large number of size fractions, going from oligomers of few sugar residues up to several million dalton of molecular weight. The tangential-flow filtration (known also as crossflow filtration) is one of the most useful tools for biopolymer separation and purification, both from seawater and culture medium. The principal advantage of this technique is that the residue which can obstruct the filter is substantially washed away during the filtration process by a tangential flow along the surface of the membrane. Depending on the biopolymer to be retained, membrane cut-off ranges from 1 kD to 1000 kD are used.

For large volumes, as seawater samples, a polysulfone multi-fiber system (hollow fiber tangential flow columns) is useful technique for simultaneous dia-filtration and concentration of samples given the large surface area available (on m^2 scale).

The addition of a non-solvent (or a bad-solvent for polysaccharides) to a given sample containing dissolved polysaccharides (algal cultures or seawater samples) allows the separation of the carbohydrate fraction by precipitation. This is a very common method for many advantages. It is non-destructive, inexpensive and relatively fast allowing also a fractionation in terms of polysaccharide molecular weight. Cold ethanol, isopranol or acetone are often used and added to the cell-free supernatant of cultures or filtered samples in an appropriate volume to volume ratio (about 4:1). The precipitate is usually re-dissolved in pure water and the solution dialysed exhaustively against EDTA (0.01-0.1 M) and Milli-Q water [6,7,20,21] The precipitation/re-dissolution treatment is commonly performed three-four times depending on the purity to be achieved.

2.2. Gas-chromatography of alditol acetates of neutral monosaccharides

Although the excretion of photosynthetic compounds is recognised as the major source of carbohydrates in seawater [10], so far there are not many papers reporting on the molecular composition of carbohydrates in the exudates from diatom cultures [see for examples 6-7,10,22-28]. The molecular-level characterization of dissolved polysaccharides may provide basic information on the origin, the bioreactivity and the fate of these biopolymers. For example, after monosaccharide analysis Hama and Yanagi reported that the turnover rate of dissolved glucose was the highest among dissolved neutral aldoses, while turnover rates of galactose, mannose, xylose, rhamnose and fucose were similar to each other and markedly lower than glucose [29]. This was a significant finding suggesting that the degree of degradability of autotrophic DOM depends mainly on the relative percentage of glucose with respect to other monosaccharides.

Gas-chromatographic (GC) methodologies for neutral monosaccharide analysis used to characterize the primary structure of marine polysaccharides were reported in several papers [5-7,20,30-39]. In general, the methodology is based on the acid hydrolysis of the polysaccharides and the suitable derivatisation of the saccharidic matter in order to obtain volatile compounds [40]. Thus, hydroxyl groups are subjected to chemical modifications obtaining silylated, acetylated, trifluoroacetylated, methylated or ethylated derivatives. Neutral and amino sugars are commonly analyzed after exhaustive hydrolysis with trifluoroacetic acid and the Neeser acetylation method [41]. The Neeser method provides a simple, rapid, and sensitive analytical method, which has been successfully used on glycoproteins and on plant and microbial cell-wall polysaccharide fractions in order to ensure complete release of amino sugars from glycoproteins, together with minimum losses of neutral sugars with an improved derivatization procedure by treatment with $CH_3ONH_2{\cdot}HCl$ in pyridine. In addition the occurrence of uronic acids requires a preliminar reduction process with carbodiimide and $NaBH_4$ [42]. The N-acetylated form of amino groups, that often occurs in marine polysaccharides, are removed by hydrolysis.

A very popular method of neutral carbohydrate analysis is the alditol acetate method originally described by Blakeney et al.[43], based on the four-step reaction described as:

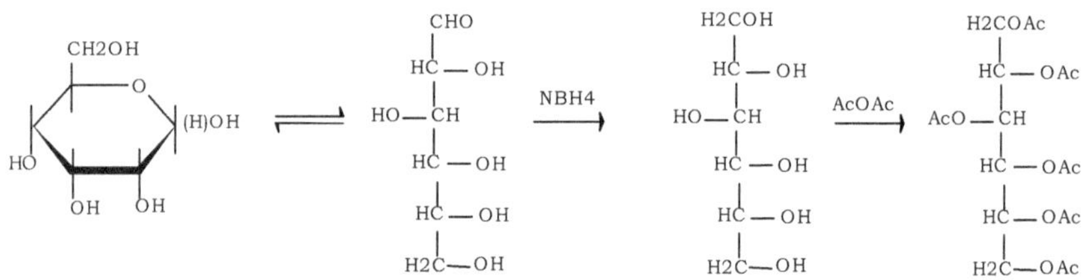

Scheme 1. Scheme 1

The complete separation of the mixture of neutral and amino sugars is usually obtained on polar capillary column (fused-silica coated with methyl silicone fluid).

As an example, the composition of extracellular polysaccharides produced by marine diatom *Chaetoceros decipiens* at the later exponential growth phase is presented in Figure 2. The exopolysaccharide fraction was isolated and purified by precipitation from bulk solution with isopropanol. The hydrolysis and the gas-chromatographic analysis yielded a suite of six neutral monosaccharides: glucose (glc), galactose (gal), mannose (man), xylose (xyl), rhamnose (rha) and fucose (fuc), present in different amount.

The molar ratio of monosaccharide presented in Table 1 shows rhamnose and fucose as the major components followed by galactose residue. The comparison with the composition of the exopolysaccharides obtained by Myklestad from *Chaetoceros decipiens* culture [5] shows a good agreement even at different growth conditions.

Exopolysaccharide molar composition	Rha	Fuc	Xyl	Man	Gal	Glc
Chaetoceros decipiens	7	7	1	4	5	1
Chaetoceros decipiens [5]	7	7	0.5	1	3	0.5

Table 1. Relative molar composition of exopolysaccharides from *Chaetoceros decipiens* cultures.

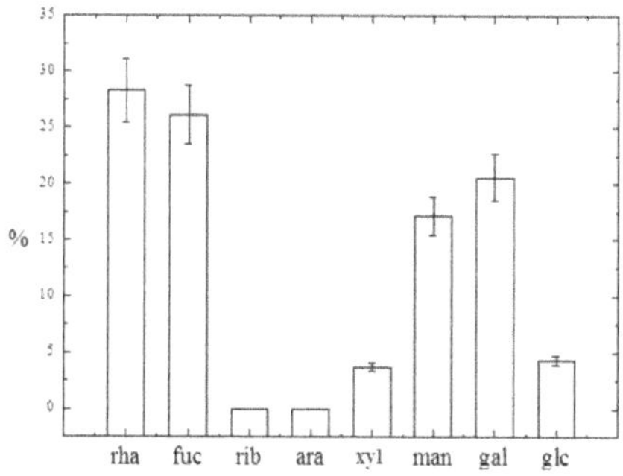

Figure 2. Monosaccharide pattern of the exopolysaccharides from cultured diatom *Chaetoceros decipiens* [44].

2.3. Anionic chromatography for charged and neutral monosaccharides

Carbohydrates in seawater include neutral sugars, aminosugars and acidic sugars, mainly uronic acids, phosphorylated and sulphated sugars [45-47]. By using high performance anion exchange chromatography (HPAEC) coupled with pulsed amperometric detector (PAD) [38] the simultaneous analysis of mixture of such substituted carbohydrates is possible after acid hydrolysis and neutralization. Chemical derivatization is not required so that time consuming gas-chromatographic methods for uronic residue analysis are avoided.

Monosaccharides are completely separated by an isocratic elution [48-50] or in addition with a gradient course of two mobile eluent phases [51]. Seawater samples and culture media require a desalting step preceding the acid hydrolysis and the use of membrane dialysis (of about 1kDa) instead of resins is strongly recommended [51,52] avoiding the losses in carbohydrate yield.

2.4. Methylation analysis and mass spectroscopy

The most widely used method for determining anomeric linkage structure of a polysaccharidic chain is the methylation analysis. The polysaccharide is partially methylated, then hydrolyzed and the resulted partially methylated monosaccharides are acetylated. These methylated alditol acetate sugars allow to establish which carbons are involved in the anomeric linkage. The advent of combined GC and mass spectrometry allows the identification of monosaccharides and provides linkage information on complex polysaccharides. The methylation reaction is commonly performed using Harris' method [53].

Unpublished data (P. Sist) on axenic culture of *Chaetoceros decipiens* are presented in Table 2. The result of the methylation analysis of the exopolysaccharide allowed to identify the linkages among monosaccharides along the polymeric chain. The results showed that fucose and rhamnose were present mainly as terminal residues (t-Rha and t-Fuc) but a lower percentage of rhamnose (8.3%) was linked in the chain backbone (2-Rha) and 5.4% of fucose

Methylation derivative	Linkage	%
2,3,4-Me_3 Rha	t-Rha	21.4
2,3,4-Me_3 Fuc	t-Fuc	10.7
2,3,4,6-Me_4 Man	t-Man*p*	3.2
3,4-Me_2 Rha	2-Rha	8.3
2,3,4,6-Me_4 Glc	t-Glc*p*	4.0
2,3,4,6-Me_4 Gal	t-Gal*p*	3.2
3,4-Me_2 Fuc	2-Fuc	2.7
3,4,6-Me_3 Man	2-Man*p*	2.3
2,3,4-Me_3 Man	6-Man*p*	3.3
3,5-Me_2 Xyl	2-Xyl*f*	1.0
Fuc	2,3,4-Fuc	5.4
3,4,6-Me_3 Gal	2-Gal*p*	13.4
2,3,6- Me_3 Gal	4-Galp	10.5
2,3,6-Me_3 Glc	4-Glc*p*	2.4
3-Me Xyl	2,4-Xyl*p*	1.8
3,6-Me_2 Man	2,4-Man*p*	1.9
4,6-Me_2 Man	2,3-Man*p*	1.0

Table 2. Methylation derivatives of monosaccharidic units of exopolysaccharide from *Chaetoceros decipiens* culture.

represented branched residues (2,3,4-Fuc). Galactose residues which were linked at carbon 2 and 4 (23.9% of 2-Galp and 4-Galp) resided predominantly in the backbone, while mannose was both a branched residue (2.9% of 2,4-Man and 2,3-Man) and mono-substituted in a linear chain (5.6% of 2-Man and 6-Man).

The relatively high percentage of galactose which could be present as α or as β anomeric configuration in the chain backbone suggested a possible extended and rigid chain conformation of the *D. decipiens* polysaccharide as also found for model polysaccharides in solution [54,55].

2.5. Gel permeation chromatography

The heterogeneity and the molecular weight polydispersity of polysaccharide fractions can be analyzed by gel permeation chromatography (GPC) [6,37,56]. This analytical technique is also called size-exclusion chromatography because the fractionation occurs according to the molecular size, V (not molecular weight!), of the polymer which easily or not permeates pores of a suitable dimension of the gel matrix packed in the column which has no specific, or weak, interactions with eluted polymers.

Unpublished data (R. Urbani) of exopolysaccharides obtained from mucilage marine samples are presented in Figure 3. The purified polysaccharides were dissolved in 0.05M NaCl and solutions were filtered on 0.45 μm filters and injected in a conditioned GPC column (1.5 m length, 310 mL internal volume), packed with Sepharose CL6B (10^4-10^6 Da). The chromatographic apparatus was pre-equilibrated using 0.05M NaCl at 25°C at a flow rate of 10 mL·h^{-1} and the column was calibrated with standard dextran solutions ($M_w = 2\cdot10^4$ Da and $M_w = 5\cdot10^5$ Da). A differential refractive index instrument was used as detector.

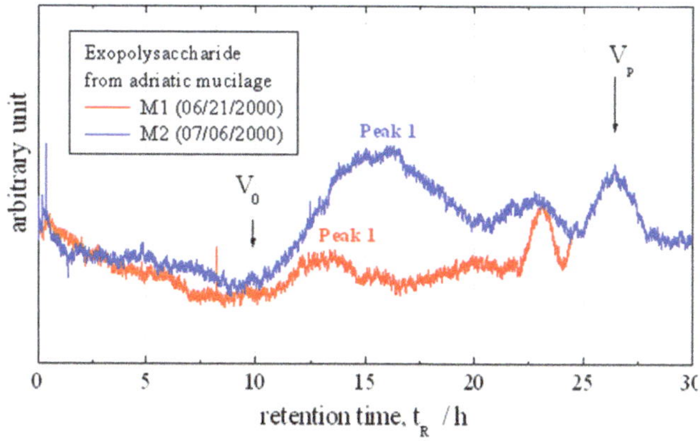

Figure 3. GPC of purified exopolysaccharides from Adriatic mucilage.

The small difference in the retention time (volumes) between peak 1 of M1 and M2 samples (t_R about 15 hours) is related to different stage (or age) of the mucilage aggregates with different degraded state [20]. The other peaks (t_R about 23 hours) very close to the (total) permeation volume, Vp, correspond to the low molecular weight fractions, very likely derived from hydrolytic activities of the bacterial pool on the native aggregates. High molecular weight fractions (peak 1 of M1 and M2) were collected, purified and derivatized for gas-chromatographic analysis. Their monosaccharidic patterns shown in Table 3 are quite similar suggesting the same origin of the high molecular weight polysaccharides even for aggregates of different degraded state [20].

Sample	Fraction	Rha	Fuc	Rib	Ara	Xyl	Man	Gal	Glc
M1	Peak 1	11.8	20.1	--	1.9	8.3	12.2	31.1	14.6
M2	Peak 1	9.3	15.8	--	2.7	13.7	13.3	29.7	15.5

Table 3. Relative monosaccharide composition (%) of unfractionated exopolysaccharide and corresponding fraction at high molecular weight.

A rapid and powerful non-preparative technique for the characterization of molecular dimensions of marine dissolved organic matter and exopolymeric material from diatoms cultures is the high performance size-exclusion chromatography (HPSEC) [57]. By using a hydrogel column system (cut-off 50.000-1.000.000 Da) HPSEC experiments were performed on purified exopolysaccharides from *Chaetoceros decipiens* cultures [44] and results are shown in Figure 4.

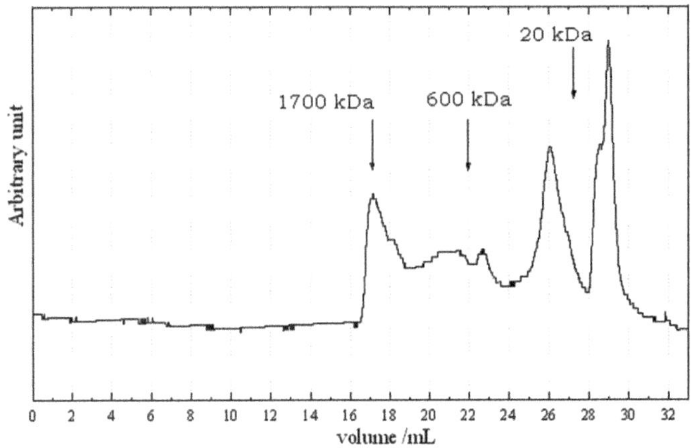

Figure 4. HPSEC chromatogram of purified exopolysaccharide from cultured *Chaetoceros decipiens*.

Two distinct peaks were observed: at lower (V=16-20 mL) and at higher retention volume (V=25-30 mL) which corresponded to about 1700 kDa and 20 kDa molecular weight substances, respectively. The low molecular size peaks were near the permeation volume (V_P) very likely due to oligosaccharides with an average number of monosaccharidic residues of about 110 units. The authors (R. Urbani, P. Sist) hypothesized that *Chaetoceros decipiens* produces a high molecular weight exopolysaccharide and some oligomers at low molecular weight probably derived from the release of storage carbohydrates by cultured diatom in given culture condition [44].

3. Solution and aggregation properties

In the macromolecular science it is well documented that going from flexible polysaccharides toward more rigid and extended chains, as well as with the presence of an ordered sequence in the primary structure, there is a general propensity of polymer chains to form chain associations/aggregations or multiple helical structures. Even if that sequence specificity were present, the absence of pronounced amphiphilic character in the sugar building blocks is not favorable for the formation of some kind of globular folding that occurs in many proteins and some nucleic acids [58]. In other words such globular structures are unknown in polysaccharides. Nonetheless, ordered conformations were proposed to represent the structure of many polysaccharides, both in solid state and in solution. In some cases the functionality is most closely associated with the occurrence of a randomly coiling polymeric character, that is, a propensity for the chain to move continuously through a vast range of nearly equally energetic conformations.

The presence of charged groups on the polysaccharidic chain confers peculiar properties to the macromolecules favoring, for example, the solubility or the association/dissociation processes in solution. On the polymer side the presence of the charged groups influences strongly all the conformational properties by enhancing the chain dimensions and increasing the hydrodynamic volume. Knowledge of polysaccharidic chain structures, from single chain up to the three-dimensional molecular shape and chain association, is essential to understand their capability to form supra-molecular structures including physical gels [54].

3.1. Viscometry and light scattering characterization

Physico-chemical techniques are widely used in order to estimate the biopolymer features such as chain stiffness and chain dimensions which are strictly related to the polymer propensity to give aggregation and highly structured systems like gels. The Smidsrød-Haug parameter B [59] measured by using capillary viscosity techniques as a function of ionic strength is related to chain stiffness: the more flexible the chain, the higher the response of the intrinsic viscosity on the ionic strength variation and the higher the B value.

Intrinsic viscosities, [η], of exopolysaccharides from Adriatic mucilage (M1 and M2) were measured in Cannon-Ubbelohde suspended-level capillary viscometers at different ionic strength and obtained by linear regression of reduced specific viscosities, η_{sp}/C (dL·g^{-1}), as a function of polymer concentrations. By plotting the intrinsic viscosity (the polymer hydrodynamic volume) as a function of the inverse of the root of ionic strength (Figure 5), both exopolysaccharides show a linear reduction of [η] at increasing of ionic strength due to screening effect of the salt on polymer charges. From the Smidsrød-Haug equations [59]:

$$\left[\eta\right] = A + S'\underline{k}^{-1} \approx A + S\left(\sqrt{I}\right)^{-1}$$

the stiffness parameter B is obtained where $[\eta]_{0.1}$ is the intrinsic viscosity at 0.1 M ionic strength:

$$B = \frac{S}{\left(\left[\eta\right]_{0.1}\right)^{1.3}}$$

It is interesting to compare the B stiffness parameter of exopolysaccharides with those reported in the literature for other polysaccharides of similar molecular weight. This is shown in Table 4 where dextran sulphate and polyphosphate represent rather flexible polymers with a high B value and, on the other hand, DNA and xanthan represent typical stiff polymers exhibiting a low B parameter. The M1/M2 exopolysaccharides exhibited a value similar to that of alginate considered as a semiflexible polymer.

Static and dynamic laser light scattering (SLLS and DLLS, respectively) are techniques widely used for polymer characterization, measuring average chain properties and thermodynamic quantities in solution. These properties are related to the propensity of the polymer system to give elongated and stiff chain: the radius of gyration (R_G), the weight-average molecular weight (M_w), the second virial coefficient (A_2) and the hydrodynamic diameter (d_h).

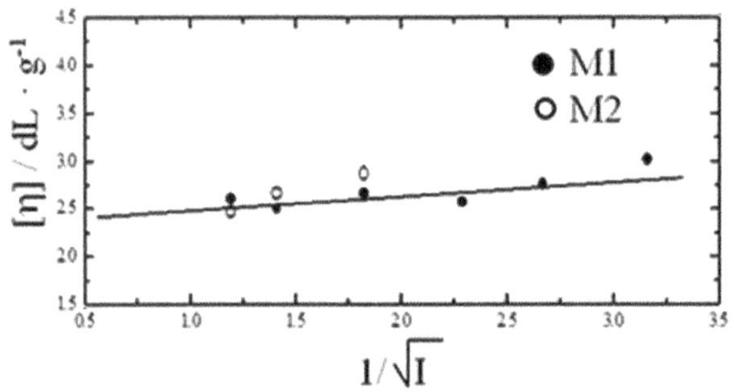

Figure 5. Plot of intrinsic viscosity *vs.* inverse of the square root of ionic strength for .M1 and M2 polysaccharides.

Polymer	B
Polyphosphate	0.44
Dextran sulphate	0.23
Hyaluronic acid	0.065
Alginate (mannuronic rich)	0.040
M1, M2	0.036
Alginate (guluronic rich)	0.031
DNA	0.0055
Xanthan	0.0053

Table 4. Stiffness parameters (B) of different model polymers.

Measurement of the scattering intensity at many angles allows the evaluation of the radius of gyration R_G, while by measuring the scattering intensity for many samples of various concentrations, the coefficient A_2 is obtained. Simultaneous linear least squares fits to both the angular and concentration dependence of scattering intensities are employed for the properties determination.

In Figures 6a and 6b the two Zimm plots of the M2 exopolysaccharide in 0.3M and 0.7M NaCl solutions, respectively, are presented. The results show a constancy of R_G (and M_W) at different salt concentrations, while A_2 becomes more negative for higher salt concentration as a consequence of the screening effect on polymer charges which may lead to an extensive degree of aggregation. In general, with respect to the thermodynamic stability of the polymer solution, negative A_2 values are a clear indication of the tendency of polysaccharide solution to undergo a phase separation to form a amorphous carbohydrate solid phase or a gel-like structure.

With comparison to model polysaccharides of different stiffness having the same M_w (Table 5), M2 polysaccharide possess higher R_G value than the flexible and coiled pullulan but also than the semi-rigid chain of the wellan and alginate biopolymers.

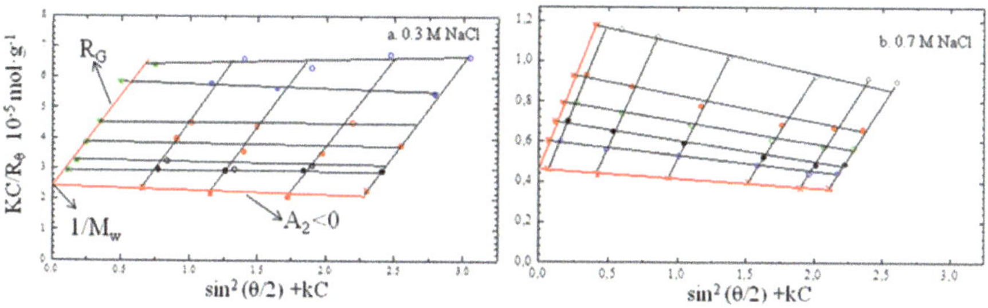

Figure 6. Zimm plot of M2 at two ionic strength (a) 0.3M and (b) 0.7M NaCl.

Sample	R_G /nm
Pullulan	24
Wellan	68
Alginate	73
M2	155[a]

[a] in 0.3M NaCl

Table 5. Radius of gyration of purified polysaccharides M1 and M2 compared to model polysaccharides with M_w=220,000 dalton in 0.1M NaCl.

Finally, taking into account all these results one may say that the M2 polysaccharide shows a marked polyelectrolytic behavior and an intrinsic chain stiffness, favoring at higher salt concentrations chain-to-chain association and/or gel formation. Thus, coupled to other favorable environmental conditions, the salinity gradient of halocline in the seawater column could play an important role in the first stage of aggregation of dissolved biopolymers.

Dynamic laser light scattering (DLLS), also called photon correlation spectroscopy, is a technique for the determination of hydrodynamic diameter of macromolecules and particulate matter in solution. In DLLS the fluctuation of the scattered light due to the Brownian motion of the molecules is detected by a photon counting detector. This technique is used nowadays to measure particle sizes at the nanometer scale and to follow the kinetics of particle formation from dissolved EPS material. As reported by Verdugo [60,61] marine aggregates may self-assemble from free DOM biopolymers and this process was easily followed by DLLS.

Following the same procedure the authors (R. Urbani, P. Sist) measured hydrodynamic diameters using Adriatic seawater samples collected during a three-years monitoring activity (1999-2001). Polymer assembly was monitored for 8 days by analyzing the scattering fluctuations detected at a 45° scattering angle. The autocorrelation function of the scattering intensity fluctuations was averaged over a 1-min sampling time and the particle size distribution calculated by the CONTIN method. Figure 7 shows an example of time dependence of hydrodynamic diameters in three seawater samples taken at different depths.

Although the aggregation kinetics was the same (asymptotic shaped curve in Figure 7) the highest value (1 μm) of hydrodynamic diameter was significantly lower than values reported by Verdugo group (several micrometers) for oceanic samples [60,61] and cultured diatoms, as well [62].

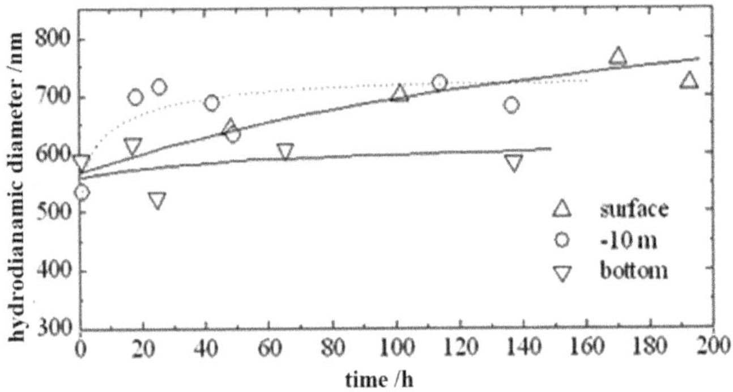

Figure 7. Hydrodynamic diameter in dependence on time for Adriatic seawater samples.

4. The supramolecular organization of polysaccharide fibrils: from cell release to giant gel networks

Diatoms excrete large quantities of extracellular polymeric substances, both as a function of their motility system and as a response to environmental conditions [63,64]. The natural occurrence of EPS is closely linked to diatom biomass, a pattern consistent over both macro (km) and micro (μm) scales. We will focus on diatom exopolysaccharides produced in cell culture and from large aggregates they form in marine environment [3]. Those polysaccharides stand out by their supramolecular organization and gelling capacity due to hydrogen bonding and electrostatic interaction resulting in formation of reversible physical gel. Unlike chemical gels that are formed by chemical reaction using a crosslinking agent, the characteristics of polysaccharide electrolytes is to form gels by physical bonds through intermolecular forces among polymer chains [65,66].

Intermolecular interactions are the basis of life, and an extremely important part of biological research, so an enormous range of techniques have been applied to their study. Among the new methodologies (experimental and theoretical) developed and applied to polysaccharide conformation and dynamics, solution properties, chain aggregation and gelation, the results obtained using atomic force microscopy (AFM) have been pointed out among those giving the most striking results [67]. AFM connects the nanometer and micrometer length scales utilizing a sharp probe tip that senses interatomic forces acting between the surface of a sample and the atoms at the apex of the tip. The physical basis behind AFM and its ability to „feel" the surface, make AFM a versatile tool in biophysics allowing high resolution imaging, nanomechanical characterization and measurements of inter and intramolecular forces in living and non-living structures [68]. Thanks to the simple principle on which it is based, the AFM is a surprisingly small and compact instrument. Its use includes electronic control unit, computer and usually two monitors for simultaneous checking of image and imaging parameters. The probe which scans the sample surface consists of a cantilever and the tip located at the free end of a cantilever. The deflection of the cantilever is measured by an optical detection system. Registered values of cantilever deflection are electronically converted into pseudo 3D image of a sample. AFM is a non-destructive method which gives real 3D images of the sample with a vertical resolution of 0.1 nm and lateral resolution of 1 nm. Measured forces range from 10^{-6} N to 10^{-11} N.

Here we will cover the recent achievements using AFM as the principal method in revealing the supramolecular organization of diatom exopolysaccharide fibrils beyond the chemical composition.

Polysaccharide samples for AFM imaging [69-72] and polysaccharide gels [70,73-77] are usually spread on freshly cleaved mica surface. The imaging of hydrated samples is preferably conducted in air to inhibit the unfavourable motion of polysaccharides in liquid medium. Such AFM studies have been validated against data obtained directly under buffers, transmission electron microscopy (TEM) studies and cryo-AFM. Balnois and Wilkinson [78] showed that when AFM is operated under ambient conditions, the thin water layer both sorbed to the biopolymers and present on the mica surface maintains molecular structure during AFM imaging.

Protocols for marine sample AFM imaging have been developed only recently: single diatom cells and released polymers [79]; isolated polysaccharides from diatom cultures [3,79]; marine gel polymers and networks [80]. The samples were prepared using the drop deposition method (5-10 μL aliquots) and mica as a substrate. Mica sheets were placed in enclosed Petri dishes for approximately 30–45 min to allow biopolymers and cells to settle and attach to the surface. Samples were then rinsed in ultrapure water (three times for 30 s) and placed in enclosed Petri dish to evaporate the excess of water on the mica. The rinsing step was necessary to remove the salt crystals that would hamper imaging. AFM imaging was performed in air at room temperature and 50–60% relative humidity. In contact mode the force was kept at the lowest possible value in order to minimize the forces of interaction between the tip and the surface. For imaging in tapping mode the ratio of the set point amplitude was maintained to the free amplitude (A/A_0) at 0.9 (light tapping).

4.1. Exopolysaccharide production at a single cell level

The ubiquitous marine diatom *Cylindrotheca closterium*, isolated from the northern Adriatic Sea, was used in AFM studies of exopolysaccharide production at a single cell level.

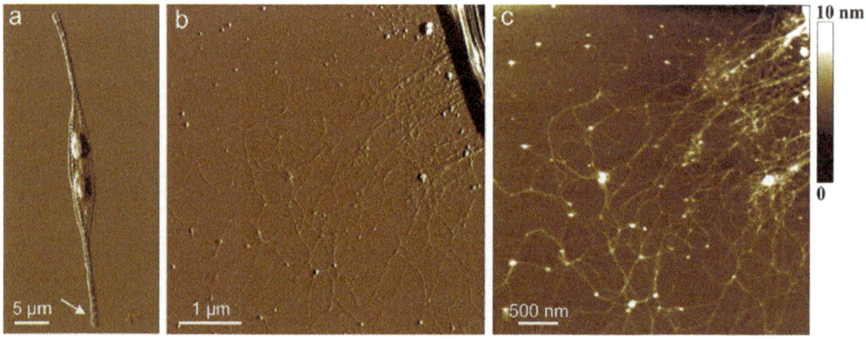

Figure 8. (**a**) AFM image of the whole *C. closterium* cell presented as deflection data. Arrow indicates the position of the polymer excretion site; (**b**) The released polymers still attached to the apex of the cell rostrum, deflection data, scan size 5 µm × 5 µm; (**c**) Released polymers presented as height data, scan size 4 µm × 4 µm and vertical scale shown as the color bar [3].

Figure 8a revealed the general features of a live *C. closterium* cell with the two chloroplasts and its drawn-out flexible rostra. Arrow indicates the position of polymer release shown in Figures 8b and 8c. The bundles of polymer fibrils extended up to10 µm from cell surface. Their heights are 5–7 nm at the position close to the site of excretion. At a distance of 1 µm the dense network is observed with fibril heights of 2–3 nm. At even larger distances the network is less dense with the fibril heights in the range of 0.4 to 1.2 nm. The lower value of fibril height corresponds to the single monomolecular polysaccharide chains [71]. At this

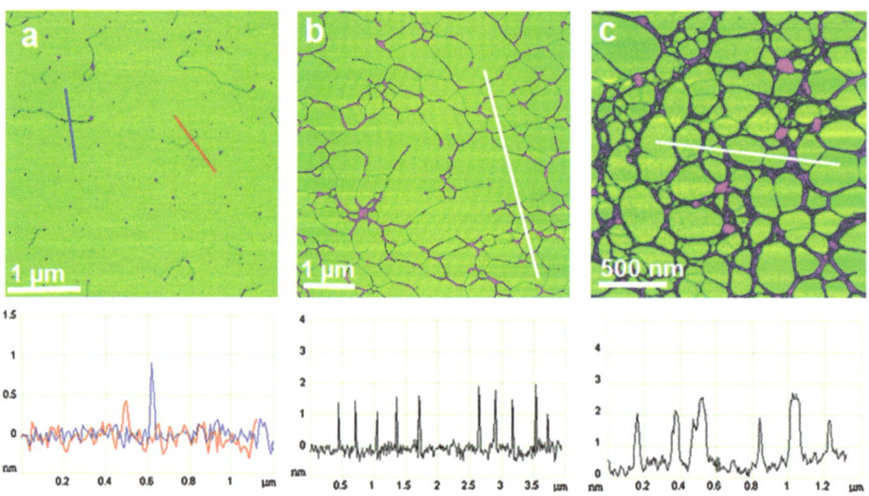

Figure 9. AFM images (tapping mode) of polysaccharides isolated from the *C. closterium* culture medium and dissolved in ultrapure water: (a) single fibrils (concentration 5 mg/L) vertical scale 2.5 nm; (b,c) fibril networks (concentration 10 mg/L). Vertical scales: 5 nm (b) and 10 nm (c) [79].

larger distance, the network appeared with incorporated spherical nanoparticles–globules. The globules are found to interconnect two or more fibrils. The globules may represent positively charged proteins whose function before the release is efficient intracellular packing of negatively charged polysaccharide fibrils, in line with molecular crowding in living cells [81].

4.2. Extracted and purified polysaccharides of C. closterium

The polysaccharide fraction isolated from the axenic *C. closterium* culture medium [7] was used to test the capacity of photosynthetically produced polymers to self-assemble into gel phase.

The isolated polysaccharides for AFM imaging were prepared in two concentrations (5 and 10 mg/L). Single fibrils prevail in samples prepared with polysaccharide concentration of 5 mg/L. Dissolved polysaccharides resumed a flexible fibrillar structure (Figure 9a) with fibril heights of 0.4 and 0.9 nm. The value of 0.4 nm corresponds to polysaccharide single molecular chain. AFM imaging of samples prepared from the solution containing 10 mg/L revealed fibrillar networks varying in the degree of fibril associations (Figures 9b,c). The fibril heights in the networks span over the same range (0.9–2.6 nm) and the mode of fibril association follows the same pattern. However, the segments forming the network shown in Figure 9c are significantly wider (50 vs. 140 nm), suggesting side-by-side associations once the maximum height of individual fibrils is reached. If we take the value of 2.6 nm observed using AFM analysis as the maximum fibril height, then up to six monomolecular fibrils can constitute a single polysaccharide fibril [79]. The absence of globules in the AFM images of isolated polysaccharides also indicates that the globules, which appeared as a constitutive component in the EPS of single cells (Figures 9b,c), are not polysaccharides.

4.3. Evolution of marine gel polymer networks

The fact that the isolated polysaccharide fraction has the capacity to self-assemble into a gel network in pure water is an important finding with implications on the mechanism of the macroscopic gel phase formation in marine systems. The marine gel is characterized as a thermoreversible physical gel and the dominant mode of gelation as crosslinking of polysaccharide fibrils by hydrogen bonding which results in helical structures and their associations [80]. This mechanism contrasts a more generally established view [82,83] that marine gel phase formation proceeds via cross-linking of negatively charged biopolymers (namely polysaccharides) by Ca^{2+} ions. Only recently, Ding *et al.* [62] reported that diatom EPS can spontaneously self-assemble in calcium-free artificial sea water, forming microscopic gels of 3–4 μm. They pointed out an overlooked issue of crosslinkers other than calcium ions in the formation of marine polymer networks.

Figure 10 represents the evolution of polymer networks of the macroscopic gel phase in the northern Adriatic Sea [84,85]. Samples were prepared from the macroaggregates with different residence time in the water column, from early stage of gel phase formation to the

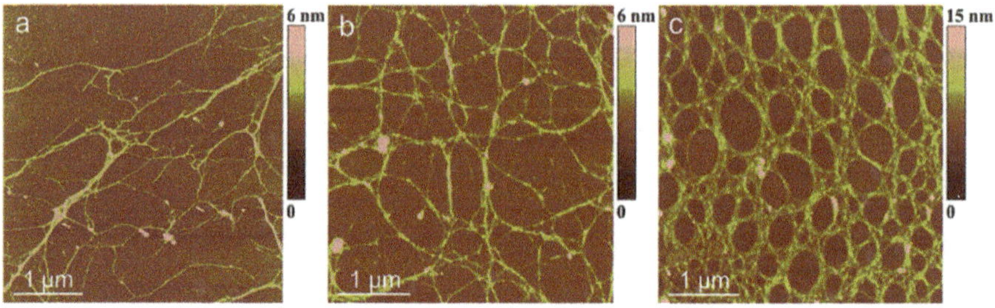

Figure 10. Evolution of polymer networks in the macroscopic gel phase from (**a**) to (**c**): early stage of gel phase formation to condensed gel network of older macroaggregate. AFM images are acquired in contact mode and presented as height data, scan size 4 μm×4 μm [3].

condensed (mature) gel network of an older macroaggregate. The long polymer strands with small patches of initial fibril associations (Figure 10a) coexisted with the continuous gel network shown in Figure 10b. With the prolonged residence time (one month) the more condensed network is formed as presented in Figure 10c. The analysis of fibril heights for early and mature gel state is given in Table 6. The fibril heights for the early stage of gel network correspond to the fibril heights produced by Adriatic *C. closterium*.

Polysaccharide fibrils	**Number of fibrils analyzed**	**Fibril height /nm**	
		Mean value	Range
Attached to the diatom cell	120	0.85 ± 0.32	0.4 - 1.8
Marine gel network: early stage	189	0.92 ± 0 .40	0.4 - 2.0
Marine gel network: mature gel	178	3.58 ± 0.76	1.6 - 5.0

Table 6. Comparison of polysaccharide fibril heights [3].

4.4. Exopolysaccharides interactions with nanoparticles

Diatom EPS production increases as a feedback response to the presence of NPs and may thus contribute to detoxification mechanisms [86,87]. Specifically, in the study of Ag NPs toxicity to the marine diatoms *C. closterium* and *C. fusiformis* [88] an increase of EPS production was documented and the incorporation of Ag NPs was clearly demonstrated. Although Ag NPs integrated in EPS–gel network are beneficial to the diatom cell (detoxification), their accumulation and persistence in microenvironments prolong their presence in the water column and make NPs available to higher organisms. Residing in a gel environment, particles are prevented from aggregation and export from the water column.

The diatom EPS component responsible for the NPs interaction was identified by bringing in direct contact polysaccharide fraction isolated from *C. closterium* EPS with Ag and SiO_2 NPs.

The polysaccharide fibril network was prepared by dissolving the polysaccharide fraction isolated from the *C. closterium* culture medium in ultrapure water at concentration of 20 μg/mL and stirred for 45 minutes before adding NPs, 10 μg/mL Ag-citrate coated NP, nominal size ~ 25 and SiO_2 NP (LUDOX® HS-40, Sigma), nominal size 15nm were used.

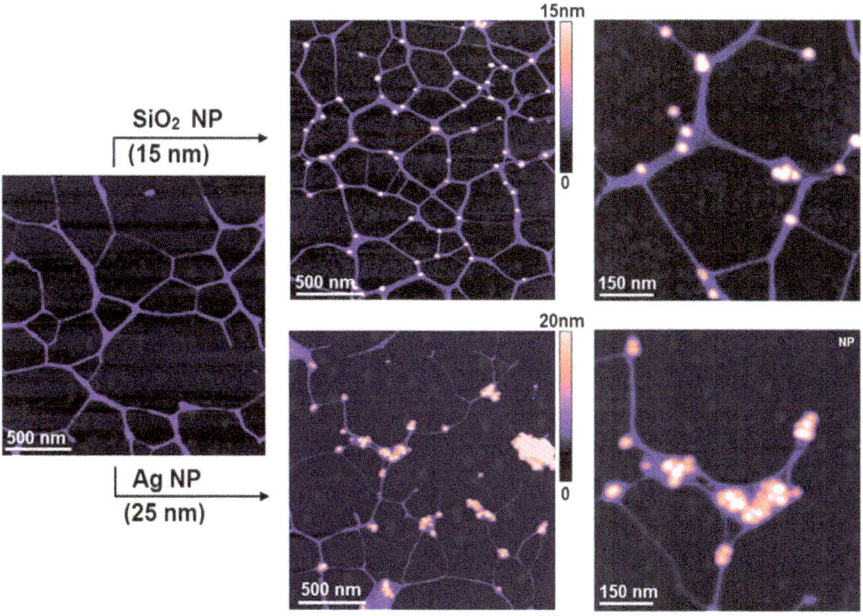

Figure 11. Interaction of polysaccharide network and nanoparticles visualized by AFM. The polysaccharide network was preformed in 20 mg/L of isolated polysaccharides in ultrapure water before addition of 10 mg/L NPs.

The Ag and SiO_2 NPs were detected exclusively on polysaccharide fibrils as single spherical particles or their agglomerates as shown by AFM images (Figure 11). The NPs did not induce cross-linking of fibrils nor change the fibril heights. Rather, particles are imbedded into the preexisting polysaccharide network. Besides the significance for the environment, such interaction of NPs and diatom polysaccharides can be applied for the design of composite materials, such as biocompatible gels with new properties [89]. Diatom extracellular polysaccharides could be also used as a capping agent giving rise to the stability of NPs in liquid environments over a broad range of ionic strength and pH.

4.5. Force spectroscopy as future prospective

In AFM force spectroscopy mode a single molecule or fiber is stretched between the AFM flexible cantilever tip and a flat substrate mounted on a highly accurate piezoelectric positioner. Polysaccharide molecule, protein or other biopolymer, is either adsorbed to the substrate or linked to it through the formation of covalent bonds. When the tip and substrate are brought together and then withdrawn, one or more molecules can attach to the tip by adsorption. As the distance between the tip and substrate increases, extension of the molecule generates a restoring force that causes the cantilever to bend. The deflection of the cantilever measures the force on the polymer with an accuracy of ~5 pN, while the piezoelectric positioner records the changes in the molecule's end-to-end length with an accuracy of 0.1nm.

AFM force spectroscopy is widely used method in polymer biophysics allowing measuring mechanical properties of single molecules, and with a possibility to directly quantify the forces involved in both intra- and inter-molecular polymer interactions [90-93]. It is also adopted in advancing diatom research into the nanotechnology era [94]. Most of the work done so far on measuring forces with AFM has distinguished non adhesive and adhesive EPS components and discovered the adhesive properties and designs that give explanation as to why diatoms have the great tendency to attach to surfaces. However, force spectroscopy has not yet been performed on single diatom polysaccharide fibrils or their networks. The data that follows are the results of exploratory experiments conducted on diatom polysaccharide molecules (single fibrils, Figure 12) and on marine gel polysaccharide network (Figure 13) in filtered seawater.

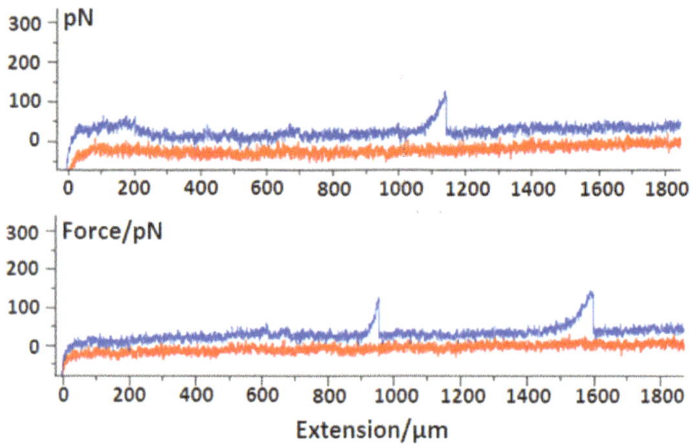

Figure 12. Force approach (in red) and extension (in blue) curves acquired for polysaccharide single fibrils in filtered seawater. Two typical curves are shown.

The extension curve with the two rupture events in Figure 12 could result from the two individual fibrils of different length (0.9 and 1.5 µm) simultaneously attached to the tip. The force spectrum signature for polysaccharide fibrils assembled in marine gel network is by far more complex (Figure 13) [95]. Assigning the underlying disentanglement events is in progress.

The force spectra can provide the critical piece of information that will allow us to characterize and quantify physical forces in polysaccharide network assemblies. Further developments will contribute to the new field of nanoecology and open the possibilities for rational design of polysaccharide gels with desired properties.

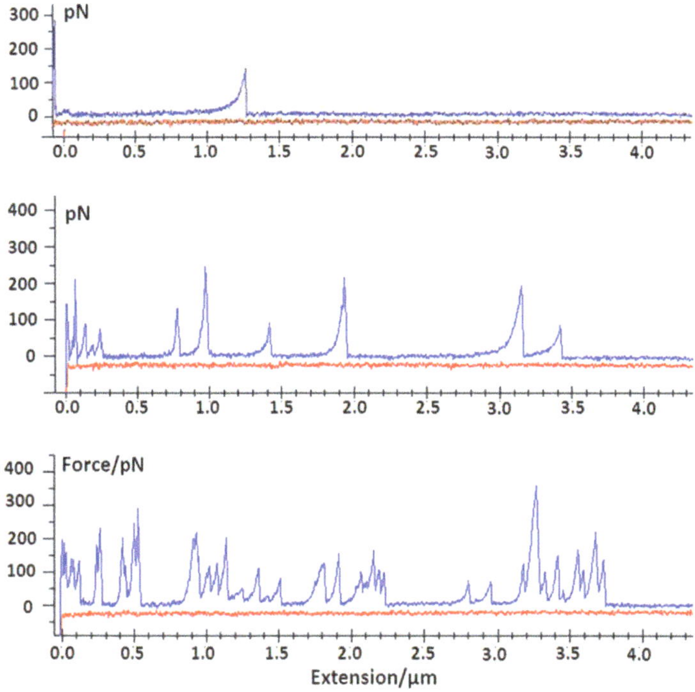

Figure 13. Force approach (in red) and extension (in blue) curves of polysaccharide fibrils assembled in marine gel network acquired in filtered seawater.

Author details

Ranieri Urbani and Paola Sist
Department of Life Sciences, University of Trieste, Trieste, Italy

Galja Pletikapić, Tea Mišić Radić, Vesna Svetličić * and Vera Žutić
2Division for Marine and Environmental Research, Ruđer Bošković Institute, Zagreb, Croatia

Acknowledgement

The work reported in the section "The supramolecular organization of polysaccharide fibrils" was performed within the scope of the project "Surface forces on atomic scale applied in marine science and nanotechnology", Croatian Ministry of Science and the project "Development of a nanotechnology-based targeted recognition system for bio-macromolecules", Croatian Science Foundation. AFM force spectroscopy of isolated polysaccharides was performed during G.P. stay in Jasna Brujić Group, Department of

* Corresponding Author

Physics, New York University. G.P. stay was financed through Croatian Science Foundation Doctoral Fellowship Award. Participation of Miklós Kellermayer was made possible through the networking of the COST Action "AFM for Nanomedicine and Life Science".

5. References

[1] David L Kirchman (2008) Microbial Ecology of the Oceans. New York: John Wiley & Sons. 2nd edition. 512 p.

[2] Zambianchi E, Calvitti C, Cecamore P, D'Amico F, Ferulano E, Lanciano P (1992) The mucilage phenomenon in the Northern Adriatic sea summer 1989: a study carried out with remote sensing techniques. Sci. Total Environ. (Suppl.) 581–598.

[3] Svetličić V, Žutić V, Mišić Radić T, Pletikapić G, Hozić Zimmerman A, Urbani R (2011) Polymer Networks Produced by Marine Diatoms in the Northern Adriatic Sea. Mar. Drugs 9: 666-679.

[4] Myklestad SM, Haug A (1972) Production of carbohydrate by the marine diatom *Chaetoceros affinis*: I. Effect of the conc. of nutrients in the culture medium. J. Exp. Mar. Biol. Ecol. 9: 125-136.

[5] Myklestad SM (1995) Release of extracellular products by phytoplankton with special emphasis on polysaccharides. Sci. Tot. Environ. 165: 155-164.

[6] Magaletti E, Urbani R, Sist P, Ferrari CR, Cicero AM (2004) Abundance and Chemical Characterization of Extracellular Carbohydrates Produced by the Marine Diatom *Cylindrotheca fusiformis*. Eur. J. Phycol. 39: 133-142.

[7] Urbani R, Magaletti E, Sist P, Cicero AM (2005) Extracellular Carbohydrates Released by the Marine Diatoms *Cylindrotheca closterium, Thalassiosira pseudonana* and *Skeletonema costatum* : effect of P-depletion and growth status. Sci. Tot. Environ. 353: 300-306.

[8] Santschi PH, Balnois E, Wilkinson KJ, Zhang J, Buffle J (1998) Fibrillar polysaccharides in marine macromolecular organic matter as imaged by atomic force microscopy and transmission electron microscopy. Limnol. Oceanogr. 43: 896–908.

[9] Biersmith A, Benner R (1998) Carbohydrates in phytoplankton and freshly produced dissolved organic matter. Mar. Chem. 63: 131-144.

[10] Aluwihare L, Repeta DJ (1999) A comparison of the chemical characteristics of oceanic DOM and extracellular DOM produced by marine algae. Mar. Ecol. Prog. Ser. 186: 105-117.

[11] Myklestad S (1974) Production of carbohydrates by marine planktonic diatoms I: Comparison of nine different species in culture. J. Exp. Mar. Biol. Ecol. 15: 261–74.

[12] Myklestad S (1977) Production of carbohydrates by marine planktonic diatoms II: Influence of the N/P ratio in the growth medium on the assimilation ratio, growth rate, and production of cellular and extracellular carbohydrates by *Chaetoceros affinis* var *willei* (Gran) Hustedt and *Skeletonema costatum* (Grev) Cleve. J. Exp. Mar. Biol. Ecol. 29: 161–79.

[13] Chen W, Wangersky PJ (1996) Rates of microbial degradation of dissolved organic carbon from phytoplankton cultures. J Plankton Res. 18: 1521–33.

[14] Alldredge AL (1999) The potential role of particulate diatom exudates in forming nuisance mucilaginous scums. Ann. Ist. Super Sanita` 35: 397– 400.

[15] Alcoverro T, Conte E, Mazzella L (2000) Production of mucilage by the Adriatic epipelic diatom Cylindrotheca closterium (Bacillariophyceae) under nutrient limitation. J. Phycol. 36: 1087– 1095.

[16] Staats N, Stal LJ, Mur LR (2000) Exopolysaccharide production by the epipelic diatom Cylindrotheca closterium: effects of nutrient conditions. J. Exp. Mar. Biol. Ecol. 249: 13–27.

[17] Granum E, Kirkvold S, Myklestad SM (2002) Cellular and extracellular production of carbohydrates and aminoacids by the marine diatom *Skeletonema costatum*: diel variations and effects of N depletion. Mar. Ecol. Prog. Ser. 242: 83– 94.

[18] Guerrini F, Cangini M, Boni L, Trost P, Pistocchi R (2000) Metabolic responses of the diatom *Achnanthes brevipes* (Bacillariophyceae) to nutrient limitation. J. Phycol. 36: 882–890.

[19] Gerald O Aspinall (1982) The Polysaccharides. New York: Academic Press. 248 p.

[20] Giani M, Berto D, Zangrando V, Castelli S, Sist P, Urbani R (2005) Chemical Characterisation of Different Typologies of Mucilaginous Aggregates in the Northern Adriatic Sea. Sci. Tot. Environ. 353: 232-246.

[21] Urbani R, Brant DA (1992) Kelco microbial polysaccharides S-130 (Welan) and S-657 display similar dilute aqueous solution behaviour. Carbohydr. Polym. 11: 169–191.

[22] Khodse VB, Fernandes L, Gopalkrishna VV, Bhosle NB, Fernandes V, Matondkar SGP, Bhushan R (2007) Distribution and seasonal variation of concentrations of particulate carbohydrates and uronic acids in the northern Indian Ocean. Mar. Chem. 103: 327–346.

[23] Yang GP, Zhang YP, Lu XL, Ding HB (2010) Distributions and seasonal variations of dissolved carbohydrates in the Jiaozhou Bay, China. Estuar. Coastal Shelf Sci. 88: 12-20.

[24] Khodse VB, Fernandes L, Bhosle NB, Sardessai S (2008) Carbohydrates, uronic acids and alkali extractable carbohydrates in contrasting marine and estuarine sediments: Distribution, size fractionation and partial chemical characterization. Org. Geochem. 39: 265-283.

[25] Lai J, Yu Z, Song X, Cao X, Han X (2011) Responses of the growth and biochemical composition of *Prorocentrum donghaiense* to different nitrogen and phosphorus concentrations. J. Exp. Mar. Biol. Ecol. 405: 6-17.

[26] Meon B, Kirchman DL (2001) Dynamics and molecular composition of dissolved organic material during experimental phytoplankton blooms. Mar. Chem. 75: 185-199.

[27] Hama T, Yanagi K, Hama Y (2004) Decrease in molecular weight of photosynthetic products of marine phytoplankton during early diagenesis. Limnol. Oceanogr. 49: 471-481.

[28] McCarthy M, Hedges J, Benner R (1996) Major biochemical composition of dissolved high molecular weight organic matter in seawater. Mar. Chem. 55: 281-297.

[29] Hama T, Yanagi K (2001) Production and neutral aldose composition of dissolved carbohydrates excreted by natural marine phytoplankton populations. Limnol. Oceanogr. 46: 1945 – 1955.

[30] Allan GG, Lewin J, Johnson PG (1972) Marine polymers. IV Diatoms polysaccharides. Bot. Mar. 15: 102-108.

[31] Aluwihare LI, Repeta DJ, Chen RF (1997) A major biopolymeric component to dissolved organic carbon in surface seawater. Nature 387: 166-169.

[32] Skoog A, Benner R (1997) Aldose in various size fractions of marine organic matter: Implications for carbon cycling. Limn. Oceanogr. 42: 1803-1813.

[33] Biersmith A, Benner R (1998) Carbohydrates in phytoplankton and freshly produced dissolved organic matter. Mar. Chem. 63: 131-144.

[34] Benner R, Opsahl S (2001) Molecular indicators of the sources and transformation of dissolved organic matter in the Mississippi river plume. Org. Geochem. 32: 597-611.

[35] Cotrim da Cunha L, Serve L, Blazi JL (2002) Neutral sugar as biomarkers in the particulate organic matter of a French Mediterranean river. Organic Geochem. 33: 953-964.

[36] Repeta D, Quan TM, Aluwihare L, Accardi AM (2002) Chemical characterization of high molecular weight dissolved organic matter in fresh and marine waters. Geochim. Cosmochim. Acta 66: 955-962.

[37] Sartoni G, Urbani R, Sist P, Nuccio C, Berto D, Giani M (2008) Benthic mucilaginous aggregates in the Mediterranean Sea: Origin, chemical composition and polysaccharides characterization. Mar. Chem. 111: 184-198.

[38] Kaiser K, Benner R (2009) Biochemical composition and size distribution of organic matter at the Pacific and Atlantic time-series stations. Mar. Chem. 113: 63-77.

[39] Giani M, Sist P, Berto D, Serrazanetti G, Ventrella V, Urbani R (2012) The organic matrix of pelagic mucilaginous aggregates in the Tyrrhenian Sea (Mediterranean Sea). Mar. Chem. 132-133: 83-94.

[40] Christopfer J Biermann, Gary D McGinnis (1988) Analysis of Carbohydrates by GLC and MS. Boca Raton: CRC Press. 304 p.

[41] Neeser JR (1985) G.l.c. of O-methyloxime and alditol acetate derivatives of neutral sugars, hexosamines, and sialic acids: "one-pot" quantitative determination of the carbohydrate constituents of glycoproteins and a study of the selectivity of alkaline borohydride reductions. Carbohydr. Res.138: 189-198.

[42] Taylor RL, Conrad HE (1972) Stoichiometric depolymerization of polyuronides and glycosaminoglycuronans to monosaccharides following reduction of their carbodiimide-activated carboxyl group. Biochemistry 11: 1383-1388.

[43] Blakeney AB, Harris PJ, Henry RJ, Stone BA (1983) A simple and rapid preparation of alditol acetates for monosaccharide analysis. Carbohydr. Res. 113: 291-299.

[44] Sist P (2005) Ciclo biogeochimico della sostanza organica in ambiente marino costiero: caratterizzazione e ruolo dei biopolimeri marini. PhD Thesis, University of Trieste, Italy.

[45] Mopper K (1977) Sugars and uronic acids in sediment and waters from the Black Sea and North Sea with emphasis on analytical techniques. Mar. Chem. 5: 585-603.

[46] Kaiser K, Benner R (2000) Determination of amino sugars in environmental samples with high salt content by high-performance anion-exchange chromatography and pulsed amperometric detection. Anal. Chem. 72: 2566–2572.
[47] Leppard GG (1995) The characterization of algal and microbial mucilage's and their aggregates in aquatic ecosystems. Sci. Tot. Environ. 165: 103–131.
[48] Metaxatos A, Panagiotopoulos C, Ignatiades L (2003) Monosaccharide and aminoacid composition of mucilage material produced from a mixture of four phytoplanktonic taxa, J. Exp. Mar. Biol. Ecol. 294: 203– 217.
[49] Goldberg SJ, Carlson CA, Hansell DA, Nelson NB, Siegel DA (2009) Temporal dynamics of dissolved combined neutral sugars and the quality of dissolved organic matter in the northwestern Sargasso Sea. Deep-Sea Res. 56: 672–685.
[50] Goldberg SJ, Carlson CA, Brock B, Nelson NB, Siegel DA (2010) Meridional Variability in Dissolved Organic Matter Stocks and Diagenetic State Within the Euphotic and Mesopelagic Zone of the North Atlantic Subtropical Gyre. Mar. Chem. 119: 9-21.
[51] Engel A, Händel N (2011) A novel protocol for determining the concentration and composition of sugars in particulate and in high molecular weight dissolved organic matter (HMW-DOM) in seawater. Mar. Chem. 127: 180-191.
[52] Giroldo D, Ortolano PIC, Vieira AAH (2007) Bacteria–algae association in batch cultures of phytoplankton from a tropical reservoir: the significance of algal carbohydrates. Freshwater Biol. 52: 1281–1289.
[53] Harris PJ, Henry RJ, Blakeney AB, Stone BA (1984) An improved procedure for the methylation analysis of oligosaccharides and polysaccharides. Carbohydr. Res. 127: 59-73.
[54] Urbani R, Cesàro A (2001) Chain Conformation of Polysaccharides in Different Solvents. In: Wypych G, editor. Handbook of Solvents. Toronto: Chem Tech Publishing. pp. 706-735.
[55] Ruggiero JR, Urbani R, Cesàro A (1995) Conformational Features of Galacturonans: II. Configurational Statistics of Pectic Polymers. Int. J. Biol. Macromol. 17: 213-218.
[56] Mandal SK, Singh RP, Patel V (2011) Isolation and Characterization of Exopolysaccharide Secreted by a Toxic Dinoflagellate, *Amphidinium carterae* Hulburt 1957 and Its Probable Role in Harmful Algal Blooms (HABs). Microb. Ecol. 62: 518–527.
[57] Minor EC, Simjouw JP, Boon JJ, Kerkhoff AE, Van der Horst J (2002) Estuarine/marine UDOM as characterized by size-exclusion chromatography and organic mass spectrometry, Mar. Chem. 78: 75–102.
[58] Gerald O Aspinall (1982) The Polysaccharides. New York: Academic Press. 248 p.
[59] Smidsrød O, Haug A (1971) Estimation of the relative stiffness of the molecular chain in polyelectrolytes from measurements of viscosity of different ionic strengths. Biopolymers 1: 1213-1227.
[60] Chin WC, Orellana MV, Verdugo P (1998) Spontaneous assembly of marine dissolved organic matter into polymer gels. Nature 391: 568–72.

[61] Verdugo P, Santschi PH (2010) Polymer dynamics of DOC networks and gel formation in seawater. Depp-Sea Res. II 57: 1486-1493.
[62] Ding YX, Hung CC, Santschi PH, Verdugo P, Chin WC (2009) Spontaneous Assembly of Exopolymers from Phytoplankton. Terr. Atmos. Ocean. Sci. 20: 741-747.
[63] Underwood GJC, Paterson DM (2003) The importance of extracellular carbohydrate production by marine epipelic diatoms. Adv. Bot. Res. 40: 184-240.
[64] Underwood GJC, Boulcott M, Raines CA, Waldron K (2004) Environmental effects on exopolymer production by marine benthic diatoms: dynamics, changes in composition, and pathways of production. J. Phycol. 40: 293-304.
[65] Rinaudo M (2005) Advances in characterization of polysaccharides in aqueous solution and gel state. In: Dumitriu S, editor. Polysaccharides-Structural Diversity and Functional Versatility. New York: Marcel Dekker. pp. 237–252.
[66] Jacob N Israelachvili (2010) Intermolecular and surface forces. London: Academic Press.3rd edition. 700 p.
[67] Cesàro A, Bellich B, Borgogna M (2012) Biophysical functionality in polysaccharides: from Lego-blocks to nano-particles. Eur. Biophys. J. 41(4): 379-95.
[68] Peter Eaton, Paul West (2010) Atomic Force Microscopy. New York: Oxford University Press. 248 p.
[69] Kirby AR, Gunning AP, Morris VJ (1996) Imaging polysaccharides by atomic force microscopy. Biopolymers 38: 355–366.
[70] Victor J Morris, Andrew R Kirby, A Patrick Gunning (1999) Atomic Force Microscopy for Biologists. London: Imperial College Press. 332 p.
[71] Abu-Lail NI, Camesano TA (2003) Polysaccharide properties probed with atomic force microscopy. J. Microsc. 212: 217–238.
[72] Abu-Lail N.I, Camesano T.A (2004) Atomic force microscope and single-molecule force microscopy studies of biopolymers. In: Schwarz J, Contescu C, Putyera K, editors. Dekker Encyclopedia of Nanoscience and Nanotechnology. London: Taylor & Francis. pp. 119–131.
[73] McIntire TM, Brant DA (1997) Imaging of individual biopolymers and supramolecular assemblies using noncontact atomic force microscopy. Biopolymers 42: 133–146.
[74] Ikeda S, Morris VJ, Nishinari K (2001) Microstructure of aggregated and nonaggregated kappa-carrageenan helices visualized by atomic force microscopy. Biomacromolecules 2: 1331–1337.
[75] Keresztes Z, Rigó T, Telegdi J, Kálmán E (2001) Investigation of biopolymer networks by means of AFM. Appl. Phys. A. 72: 113–116.
[76] Funami T, Hiroe M, Noda S, Asai I, Ikeda S, Nishimari K (2007) Influence of molecular structure imaged with atomic force microscopy on the rheological behavior of carrageenan aqueous systems in the presence or absence of cations. Food Hydrocolloid. 21: 617–629.
[77] Noda S, Funami T, Nakauma M, Asai I, Takahashi R, Al-Assaf, S, Ikeda S (2008) Molecular structures of gellan gum imaged with atomic force microscopy in relation to

the rheological behavior in aqueous systems. 1. Gellan gum with various acyl contents in the presence and absence of potassium. Food Hydrocolloid. 22: 1148–1159.

[78] Balnois E, Wilkinson KJ (2002) Sample preparation techniques for the observation of environmental biopolymers by atomic force microscopy. Colloid. Surf. A 207: 229–242.

[79] Pletikapić G, Mišić Radić T, Hozić Zimmermann A, Svetličić V, Pfannkuchen M, Marić D, Godrijan J, Žutić V (2011) Extracellular polymer release AFM imaging of extracellular polymer release by marine diatom *Cylindrotheca closterium* (Ehrenberg) Reiman& J.C. Lewin. J. Mol. Recognit. 24: 436–445.

[80] Mišić Radić T, Svetličić V, Žutić V, Boulgaropoulos B (2011) Seawater at the nanoscale: Marine gel imaged by atomic force microscopy. J. Mol. Recognit. 24: 397-405.

[81] Falciatore A, Ribera d'Alcala M, Croot P, Bowler C (2000) Perception of environmental signals by a marine diatom. Science 288: 363–2365.

[82] Verdugo P, Alldredge AL, Azam F, Kirchmand DL, Passow U, Santschi PH (2004) The oceanic gel phase: a bridge in the DOM–POM continuum. Mar. Chem. 92: 67– 85.

[83] Verdugo P, Orellana MV, Chin WC, Petersen TW, van den Eng G, Benner R, Hedges J (2008) Marine biopolymer self-assembly: implications for carbon cycling in the ocean. Faraday Discuss. 139: 393–398.

[84] Stachowitsch M, Fanuko N, Richter M (1990) Mucous aggregates in the Adriatic Sea: an overview of stages and occurrence. Mar. Ecol. 11: 327–350.

[85] Precali R, Giani M, Marini M, Grilli F, Ferrari CR, Pećar O, Paschini E (2005) Mucilaginous aggregates in the northern Adriatic in the period 1999–2002: typology and distribution. Sci Total Environ. 353: 10–23.

[86] Navarro E, Baun A, Behra R, Hartmann NB, Filser J, Miao AJ, Quigg A, Santschi P, Sigg L (2008) Environmental behavior and ecotoxicity of engineered nanoparticles to algae, plants, and fungi. Ecotoxicology 17: 372-386.

[87] Miao AJ, Schwehr KA, Xu C, Zhang SJ, Luo ZP, Quigg A, Santschi P (2009) The algal toxicity of silver engineered nanoparticles and detoxification by exopolymeric substances. Environ. Pollut. 157: 3034–3041.

[88] Pletikapić G, Vinković Vrček I, Žutić V, Svetličić V (2012) Atomic force microscopy characterization of silver nanoparticles interactions with marine diatom cells and extracellular polymeric substance. J. Mol. Recognit. 25: 309-317.

[89] Boissière M, Tourrette A, Devoisselle JM, Di Renzo F, Quignard F (2006) Pillaring effects in macroporous carrageenan-silica composite microspheres. J Colloid Interface Sci. 294: 109-16.

[90] Bustamante C, Macosko JC, Wuite GJ (2000) Grabbing the cat by the tail: manipulating molecules one by one. Nat. Rev. Mol. Cell Biol. 1: 130-6.

[91] Bhanu P. Jena, J. K. Heinrich Hörber (2006) Force Microscopy: Applications in Biology and Medicine. New Jersey: John Wiley & Sons. 300 p.

[92] Yves Dufrene (2011) Life at the Nanoscale: Atomic Force Microscopy of Live Cells. Singapore: Pan Stanford Publishing. 460 p.

[93] Alexander Noy (2007) Handbook of Molecular Force Spectroscopy. Livermore: Springer. 300 p.
[94] Higgins M.J, Wetherbee R (2011) The Role of Atomic Force Microscopy in Advancing Diatom Research into the Nanotechnology Era. In: Dufrene Y, editor. Life at the Nanoscale: Atomic Force Microscopy of Live Cells. Singapore: Pan Stanford Publishing. pp. 405-420.
[95] Pletikapić G, Mišić Radić T, Berquand A, Murvai U, Kellermayer MSZ, Vinković Vrček I, Žutić V, Svetličić V (2012) Marine diatom cells and extracellular polymers: nanostructure, nanomechanics and interactions with nanoparticles. International Conference: From Solid State to BioPhysics VI: From Physics to Life Sciences, 9th-16th June 2012, Dubrovnik, Croatia.

Chapter 13

Exopolysaccharides of the Biofilm Matrix: A Complex Biophysical World

Pierre Lembre, Cécile Lorentz and Patrick Di Martino

Additional information is available at the end of the chapter

1. Introduction

Microbial biofilm development is observed on virtually all submerged surfaces in natural and industrial environments. Biofilms are also observed at interfaces as pellicles, or in the bulk of aquatic environments as flocs or granules [1, 2]. A biofilm is a complex structure made of aggregates of microbial cells within a matrix of extracellular polymeric substances (EPS) (Figure 1). The matrix structure constitutes the elastic part of the biofilm. Interstitial voids and channels separating the microcolonies contain a liquid phase, mainly constituted by water. This liquid phase is the viscous part of the biofilm. The EPS matrix provides the biofilm with mechanical stability through these viscoelastic properties [3].

All major classes of macromolecule, i.e., polysaccharides, proteins, nucleic acids, peptidoglycan, and lipids can be present in a biofilm. Although extracellular polysaccharides are considered as the major structural components of the biofilm matrix, extracellular DNA plays an important role in the establishment of biofilm structure [4]. Moreover, nucleases can be regulators of biofilm formation [5]. To get a better understanding of the role of extracellular polysaccharides in the biofilm architecture and mechanical properties, it is necessary to take a look at the properties of a limited number of components, which can be isolated. Most microbial exopolysaccharides are highly soluble in water or dilute salt solutions, and capsule-forming polysaccharides are attached to the cells surface through covalent bonds to other surface polymers. Many of the extracellular polysaccharides produced in biofilms are insoluble and not easily separated from the cells, complicating the precise determination of their chemical structures and physical properties. Jahn et al. extracted a mixture of polymers from *Pseudomonas putida* biofilm material and found it to be very heterogeneous [6]. Most bacterial exopolysaccharides can exist either in ordered or disordered forms. Elevated temperatures and extremely low ionic concentrations favour the disordered forms. Polysaccharide molecules can interact with themselves or with heterologous ions and molecules to yield gels, often with multivalent cations playing a significant role in the process. Polysaccharides also interact with proteins molecules both as solutes and when attached to the surface of the microbial cells. The polysaccharide - protein interactions in the matrix induce both structural and functional properties. Indeed, some of these proteins are enzymes constituting an external digestion system [7].

Biofilms in differing environments can be exposed to a very wide range of hydrodynamic conditions, which greatly affect the matrix and the biofilm structure [8]. The shear rate determines the rate of erosion of cells and regions of the matrix from the biofilm. Polysaccharides of the matrix exhibit flow and elastic recovery; because of the flexibility of the matrix its shape can change in response to an applied force. The shear stress to which a biofilm is exposed also affects the physical morphology and dynamic behaviour. Biofilms grown under higher shear are more strongly adhered and have a stronger EPS matrix than those grown under lower shear [9]. Biofilm density can be influenced by the fluid shear during growth [10]. *Pseudomonas* biofilms grown under laminar flow generally consist of hemispherical mound-shaped microcolonies, which form an isotropic pattern on the surface [9]. The biofilm microcolonies grown in turbulent flow are elongated in the downstream direction to form filamentous streamers. The streamers are attached to the glass substratum by an upstream head while the downstream tails are free to oscillate in the flow. Thus, hydrodynamics conditions influence both the structure and the material properties of biofilms [9]. This may be related to the physical arrangement of individual polymer strands in the biofilm EPS matrix [11]. The constitution of the biofilm matrix of *S. enteritidis* varies with pressure forces applied to the biofilm. Indeed in the absence of pressure, the sugars in the biofilm matrix are mainly composed of glucose and very little fucose. However in the presence of power flow, the share of fucose in the biofilm matrix is increased from 11% to about 30% [12].

In this chapter, after the presentation of exopolysaccharides extraction and purification from the biofilm matrix, the structural and physical properties of bacterial alginates, cellulose and other exopoysaccharides related to biofilm formation are discussed. An illustration of the complexity of the biofilm matrix architecture and the role of exopolysaccharides in the properties of the matrix is given through biofilms formation at the surface of nanofiltration membranes used for drinking water production.

2. Exopolysaccharides extraction and purification from the biofilm matrix

This section focuses on specific extraction methods targeting exopolysaccharides. General extraction methods for exopolysaccharides are first presented, followed by a presentation of the corresponding exopolysaccharides properties and carbohydrate contents.

2.1. Methods for exopolysaccharides extraction

Exopolysaccharides constitute the main EPS in many biofilms. They form the backbone of a network where other EPS components can be included. The stability of the biofilm matrix is dominated by entanglement of EPS and weak physicochemical interactions between molecules. These interactions correspond to various binding forces such as electrostatic attractive forces, repulsive forces (preventing collapsing), hydrogen bonds, van der Waals interactions and ionic attractive forces [13].

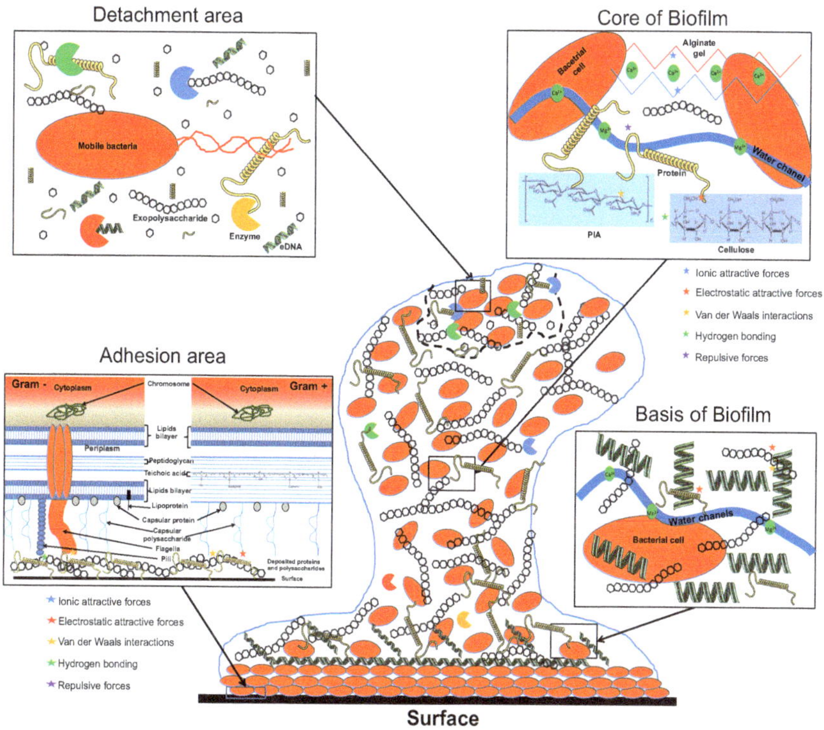

Figure 1. Schematic representation of a mature biofilm. In the centre, overall diagram of the structure of a biofilm to an interface solid / liquid: bacteria are attached to the solid surface and included in a self-induced polymer matrix. In the area of contact between bacteria and surface, the microbial cells can interact with the surface via several protein and polysaccharide appendages (pili, flagella, LPS, capsular polysaccharides) depending on the type of bacteria. On the basis of the biofilm, bacterial cells are embedded in a matrix containing high eDNA concentrations, in addition to proteins and polysaccharides. The eDNA plays a major role in early biofilm formation. In the core of the biofilm, channels of water carrying ions and nutrients cross the biofilm matrix containing high concentrations of exopolymeric substances. All these exocellular compounds form a protective gel around the microorganisms. In the biofilm detachment area, microbial enzymes destroy the exopolymeric matrix and release the cells that regain mobility, to be able to colonize new surfaces.

The exopolysaccharides recovery from the biofilm matrix in order to get a better understanding of their nature, requires to break down the interactions between EPS and selectivity separate them from other EPS and from matrix cells without cell lysis. The evaluation of cell lysis can be performed by measuring activity of the intracellular marker enzyme glucose-6-phosphate dehydrogenase (G6PDH; EC 1.1.1.49). Thus, substantial cell lysis occurring during the EPS extraction is commonly observed [14]. Regarding extraction methods, publications dealing with selective extraction of exopolysaccharides are missing, as already reviewed by Denkaus *et al.* [15]. Indeed, procedures described in the literature mainly deals with EPS extraction.

Physical and/or chemical methods are used to extract EPS from biofilms. Some EPS are tightly associated to the biofilm structure, sometimes through covalent bounds to the cells surface and are not directly extracted. Others free EPS are directly released. The easily released EPS can be separated using physical methods such as high-speed centrifugation and ultrasonication. Indeed, centrifugation is often used to separate soluble EPS from bacterial cells from pure cultures. Firmly cells-associated EPS require chemical methods of extractions. EPS cross-linked by divalent cations can be released from the biofilm matrix by complexing agents such as ethylenediamine tetraacetic acid (EDTA), by cation-exchange resins such as Dowex or by a formaldehyde treatment with or without sodium hydroxide [14, 16].

Various methods used to extract EPS can be applied to the extraction of exopolysaccharides as illustrated on Figure 2.

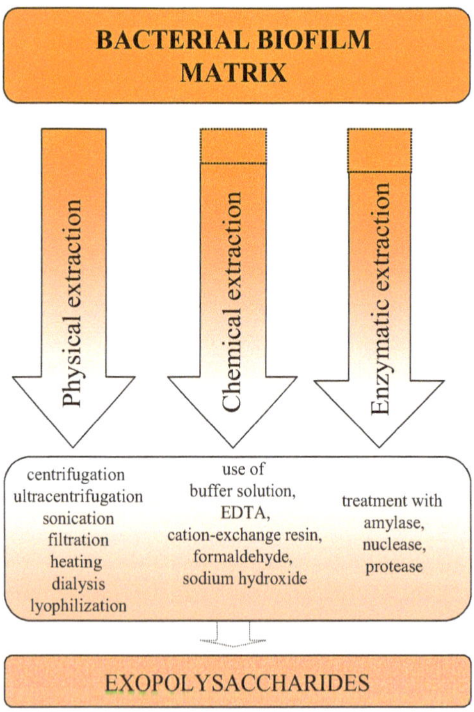

Figure 2. Pathways of exopolysaccharides extraction methods from biofilms

EPS extraction can be done from pure cultures of from complex microbial communities. For example, EPS material can be removed from *Pseudomonas aeruginosa* bacteria by centrifugation at 40 000g for 2 hours at 10°C. Purification of alginate is mostly obtained by precipitation in the presence of organic solvents from culture supernatants and treatment with enzymes such as nucleases and proteases to remove contaminating nucleic acids and proteins [14]. EPS from activated sludge samples can also be extracted by a centrifugation protocol. Then, residual bacteria can be removed from the supernatant by filtration on cellulose acetate membranes (0.2 µm). For removal of low molecular weight material, the supernatants can be dialyzed against deionised water. Depending on the studies, dialysis tubings can have various molecular weight cut-off. Then, dialysate is concentrated by lyophilisation. Other processes of exopolysaccharides extraction from various biofilm species are presented in Table 1.

Biofilm source/ support	Sample preparation	Method of determination	Reference
Pseudomonas fluorescens / polymethyl methacrylate plates	▪ extraction by EDTA 1.5%, 5°C, 3 h ▪ dialyse against deionized water (14 kDa) ▪ precipitation of proteins by pH adjsutement	phenol-sulfuric acid method	Oliveira *et al.*, [17]
Enterobacter cloacae /zinc selenide crystal	▪ extraction by EDTA or NaOH	phenol-sulfuric acid method	Boualam *et al.*, [18]
Leuconostoc mesenteroides strains /stainless steel	▪ scraping and washing biofilm material ▪ extraction by deionized water ▪ centrifugation at 10 000 rpm, 30 min ▪ hydrolysis (endodextranase) ▪ methylation	GC-MS	Leathers and Bischoff [19] Leathers and Cote [20]
Thermus aquaticus YT- /cellophane membrane	▪ extraction by NaCl 0.9% ▪ centrifugation, 8 000g, 15 min ▪ precipitation with alcohol 95 %, 4°C, one night ▪ centrifugation, 2500g, 0°C, 15 min ▪ dissolution in double deionized water and lyophilization	size-exclusion chromatography, GC-MS, HPAEC-PAD*, MS/MS, NMR	Lin *et al.*, [21]
aerobic activated sludge from a wastewater treatment	Comparison of five extraction processes : ▪ extraction by - EDTA 2 %, 4 C, 3 h - cation exchange resin, 4°C, 1 h - formaldehyde, 4°C,1 h - formaldehyde plus NaOH 1 N, 4 C, 3 h - formaldehyde plus ultrasonication (60 W), 2.5 min ▪ high-speed centrifugation 20 000g, 20 min, ▪ filtration (0.2 µm) ▪ dialysis (3,5 kDa), ▪ lyophilization, -50 C, 48 h	anthrone method	Liu and Fang, [16]

*HPAEC-PAD: High Performance Anion Exchange Chromatography with Pulsed Amperometric Detection

Table 1. Extraction condition and determination of exopolysaccharides from various biofilm species

The content of the EPS extracts is done by chemical analyses. The exopolysaccharide content of EPS can be determined by the phenol-sulphuric acid method described by Dubois *et al.* [22] or by using the anthrone method according to Dreywood [23], with glucose as the standard. The protein content of EPS can be determined by the Bradford method [24] or by using the bicinchoninic acid reagent [20] with bovine serum albumin as the standard.

As mentioned by several authors, yields of EPS extracted from biofilms depend on the extraction method used. Pan *et al.* [26] reported that chemical methods significantly increased the extraction yield from natural biofilm compared to physical methods. Nevertheless, the use of chemical methods can modify the composition of the EPS extracted. Indeed, added chemical extractants such as EDTA or formaldehyde can be present in the EPS extracts as contaminants and then can modify the EPS quantification efficiency. Moreover, chemicals can induce the release of intracellular components from treated cells and contaminate extracellular substances by intracellular material. These contaminants may be eliminated.

2.2. Exopolysaccharides of the biofilm matrix

The ability to synthesize exopolysaccharides is widespread among microorganisms, and microbial exopolysaccharides play important roles in biofilm formation, pathogen persistence, and have several applications in the food and medical industries. Exopolysaccharides are considered to be important components of the biofilms matrix [27]. However, some studies suggest that exopolysaccharides may not always be essential for biofilm formation [28]. Most of the matrix exopolysaccharides are very long with a molecular weight of 500-2000 kDa. They can be homo-polymers such as cellulose, curdlan or dextran, or hetero-polymers like alginate, emulsan, gellan or xanthan. Exopolysaccharide chains can be linear or branched. They are generally constituted by monosaccharides and some non-carbohydrate substituents such as acetate, pyruvate, succinate, and phosphate [29]. Various examples of exopolysaccharides encountered in bacterial biofilm are presented in Table 2.

2.3. Carbohydrate content of exopolysaccharides

Composition as well as conformation of sugar monomers may modify the properties of the exopolysaccharides and thus of the biofilm matrix. Mono-carbohydrate constituted exopolysaccharides are often D-glucose, D-galactose, D-mannose, L-fucose, L-rhamnose, L-arabinose, N-acetyl- D-glucose amine and N-acetyl-D-galactose amine as well as the uronic acids D-glucuronic acid, D-galacturonic acid, D-manuronic acid and L-guluronic acid. Other sugar monomers less frequently occurring are D-ribose, D-xylose, 3-keto-deoxy-D-mannooctulosonic acid and several hexoseamineuronic acids [29]. Some examples of carbohydrate content in biofilm are presented in Table 3.

In conclusion of this section, it is clear that the extraction of exopolysaccharides from biofilms usually require a multi-method protocol. Furthermore, there is no standard extraction procedure established, making difficult the meaning, comparison and interpretation of published results. However, recent studies tend to evaluate whether molecular diversity of EPS are potential markers for biofilm macro-scale characteristics [40].

Exopolysaccharides	Biofilm species	Reference
alginate	*Azotobacter vinelandii* *Pseudomonas aeruginosa* *Pseudomonas fluorescens*	Gorin and Spencer [30] Sabra *et al.*, [31] Donati and Paoletti [32]
curdlan 1,3-β-curdlan-type curdlan-type	*Agrobacterium sp.* ATCC 31749 *Cellulomonas flavigena* *Cellulomonas sp.*	Ruffing and Chen [33] Kenyon *et al.*, [34] Young and Reguera [35]
xanthan	*Xanthomonas citri* subsp. citr	Guo *et al.*, [36]
1,6-α-glucan 1,3-α- glucan	*Streptococcus mutans*	Aires *et al.*, [37]
glucan alternan dextran	*Leuconostoc mesenteroides* strain NRRL B-1355	Cote and Leathers [38]
pyruvated galactan	*Methylbacterium* sp. (isolated from a Finnish paper machine)	Verhoef *et al.*, [39]

Table 2. Examples of exopolysaccharides of bacterial biofilms

Biofilm source/ original source	Carbohydrate nature (concentration)	Method of carbohydrate characterization	Reference
European intertidal mudflat (Marennes-Oléron Bay, France)	galacturonic acid (20%) mannose (19,5%) glucose (19%) arabinose (15%) xylose (8%) galactose (7%)	GC-MS	Pierre *et al.*, [41]
Pseudomonas fluorescence Biovar II	mannose (14%) glucose (<5%) arabinose (28%) xylose (<5%) galactose (45%) fucose (6%) rhamnose (<5%) ribose (<5%)	GC-MS	Hung *et al.*, [42]
Pseudomonas aeruginosa PAO1, PDO300, *alg*D, PA14	mannuronic acid (0-100) % glucose 0-56 % rhamnose (0- 20.7 %) galactose (0-12.4 %) mannose (0-13.9 %) xylose (0-9.7 %) ketodeoxyoctulosonate (0-9.1%) *N*-acetyl galactosamine (0-1.9 %) *N*-acetyl fucosamine (0-7.5 %) *N*-acetyl glucosamine 0-3.8 % *N*-acetyl quinovosamine 0- 18.1% unknown amino sugar 0-5 %	GC-MS	Wozniak *et al.*, [28]
Membrane bioreactor/fouling	uronic acid* glucose* mannose* fructose* xylose* ribose* arabinose* *N*-acetylglucosamine* glucuronic acid* galacturonic acid* maltose* saccharose*	HPLC-SM	Al-Halbouni *et al.*, [43]

(*) not indicated

Table 3. Carbohydrate content of various biofilms

3. Structure and function of different polysaccharides from the biofilm matrix

The most famous exopolysaccharides present inside biofilms are alginate, cellulose and poly-N-acetyl glucosamine. This section focuses on their structures and their function inside biofilms.

3.1. Bacterial alginates

Alginate, a polysaccharide which occurs in brown algae and in different bacteria like *Azotobacter vinelandii* [30] and *P. aeruginosa* [44] has been extensively studied. Alginate is an exopolysaccharide with a relatively high molecular mass (10^4-10^6 g/ml). It consists of the uronic acid residues β-D-mannuronate (M) and its C-5 epimer, α-L-guluronate (G) [45] (Figure 3).

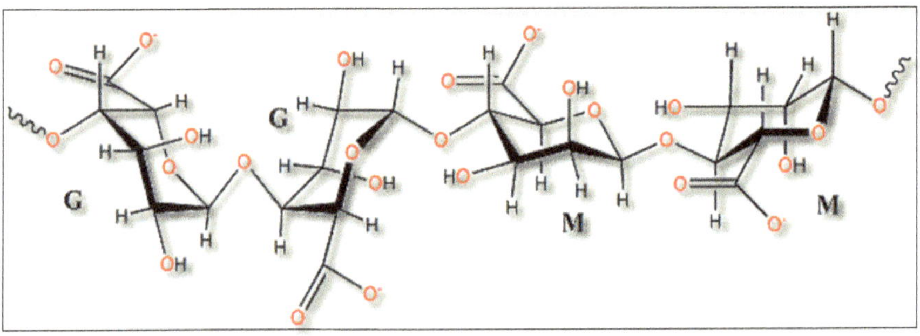

Figure 3. Structure of alginate

Generally, the monomers form a block copolymer with homopolymeric regions of poly-β-D-mannuronate (M-blocks) and poly-α-L-guluronate (G-blocks) as well as heteropolymeric regions (MG-blocks). The absence of G-blocks differentiates alginates produced by *P. aeruginosa* from alginates expressed by algae or by *A. vinelandii* [46]. The functional properties of the alginates strongly correlate with the composition (M/G ratio) and with the uronic acid sequence.

There are 24 genes located on the bacterial chromosome, involved in the production and secretion of alginate in *P. aeruginosa* [45]. Eight genes are implicated in the exportation of alginic acids (Figure 4), and twelve in the biosynthesis of the polysaccharide [47]. The four remaining genes are involved in the regulation of the synthesis.

Alginates can form a gel in the presence of chelating divalent cations. This structure formed is called a Grant "egg-box" [47]. The alginate gel is formed by ionic bonds between the G-rich blocks and divalent cations. The mechanical properties of alginate gels can vary depending on the amounts of guluronic acid present in the polymer. Moreover, alginate gels can be formed in vitro in the presence of proteins such as gelatin [48].

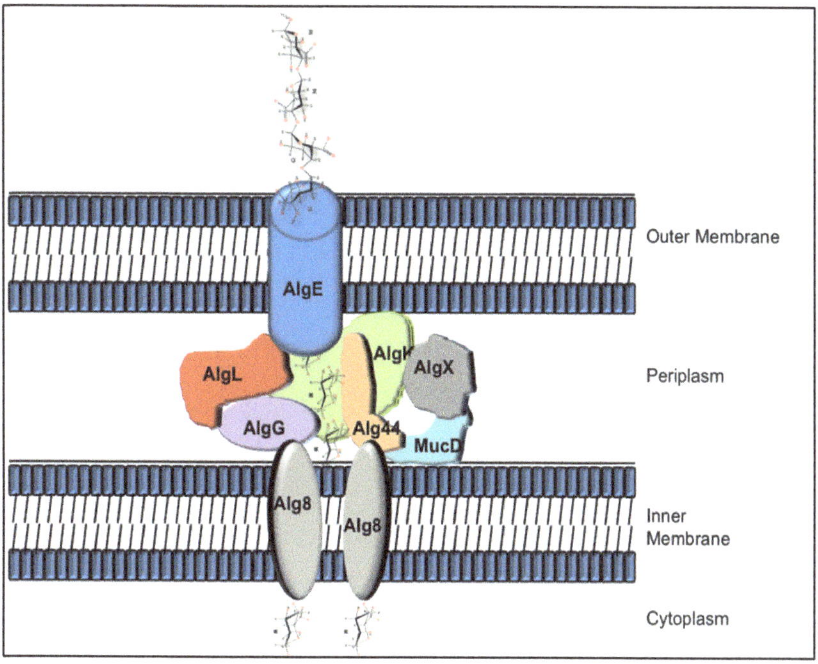

Figure 4. Biosynthesis of bacterial alginate

3.2. Cellulose

Cellulose is the most abundant sugar polymer found on the surface of the planet. It is found throughout the living world: in plants, animals, fungi and in bacteria such as *Salmonella, E. coli, Acetobacter, Agrobacterium* and *Rhizobium* [49].

Salmonella and *E. coli* produce cellulose as a crucial component of the extracellular matrix [50]. Cellulose consists of a β-1-4 linked linear glucose (Figure 5). The formation of cellulose fibers is provided by hydrogen bonds between the chains of glucose. These formed sheets are very stable and their number varies depending on the nature of the environment.

Cellulose has a crystalline structure. Each crystal of cellulose contains numerous glycan chains in parallel orientation. The reducing ends are at one terminus while the non-reducing ends are at the opposite terminus. The structure is not uniform and amorphous regions cohabit with highly crystalline regions.

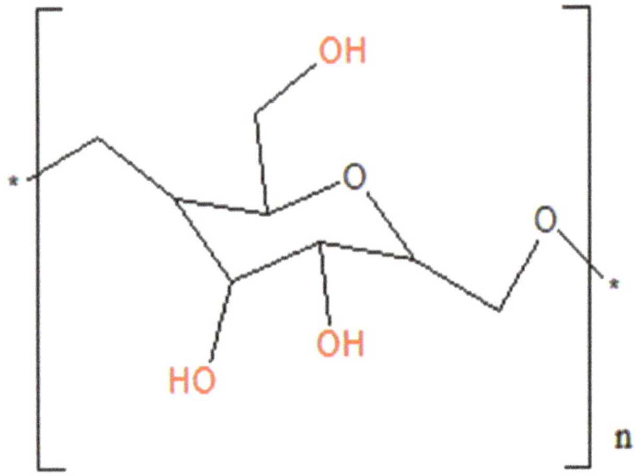

Figure 5. Structure of cellulose

Genes involved in the production of cellulose in *E. coli* and *S. typhimurium* are called *bcs* for bacterial cellulose synthesis (Figure 6). The four *bcs* genes called *bcsA, bcsB, bcsZ* and *bcsC* are organized as an operon. The *bcs* operon is partially regulated by AgfD, a thin aggregative fimbriae which increases the production of cellulose and curli [49].

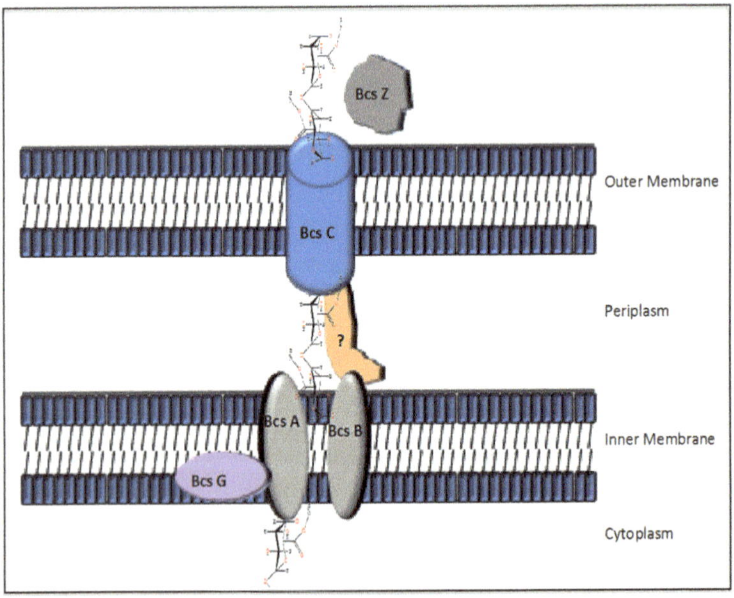

Figure 6. Cellulose biosynthesis

Cellulose can form a gel at adequate temperatures. Cellulose solutions are liquid at room temperature. Gels can form in a cellulose solution at either high temperature (above 50 °C) or low temperature (less than 10 °C). After gelification, cellulose solutions remain more or less stable in the gel state at room temperature [51, 52]. The gel structure of cellulose may explain the mechanical properties of biofilms formed by bacterial species producing this polymer.

3.3. Poly-N-acetyl glucosamine

The polysaccharide intercellular adhesin (PIA) or the related poly-N-acetyl glucosamine (PNAG) polymer is required for bacterial adherence and biofilm formation of some bacterial species. This polysaccharide family was first described in *Saphylococcus* species [53], and further in *E. coli* [54]. PNAG is a positively charged linear homoglycan composed of β-1,6-N-acetylglucosamine residues with approximately 20% deacetylated residues [55] (figure 7).

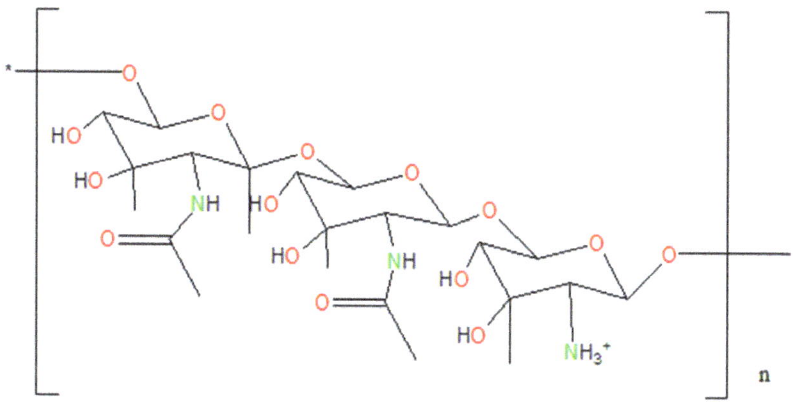

Figure 7. Structure of PNAG

The genes involved in the biosynthesis of PIA are named *ica* for intercellular adhesion. This operon is composed of a regulation gene *icaR* and four biosynthetic genes: *icaADBC* [56, 57] (Figure 8).

PNAG forms a protective matrix around bacterial cells that is also involved in cell-to-cell interactions [53, 54]. PNAG can also interact with eDNA, reinforcing the biofilm matrix structure [58].

3.4. Other polysaccharides involved in biofilm formation

Individual strains or one strain put in different environmental conditions, are able to produce several different extracellular polysaccharides. In mucoid strains of *P. aeruginosa* isolated from patients with cystic fibrosis, mucoidy is due to the overproduction of alginate that is the major constituent of the biofilm matrix. Nevertheless, biofilms formed by non mucoid strains do not contain alginate. A glucose-rich polymer named PEL (pellicle), and a mannose-rich polymer named PSL (polysaccharide synthesis locus), makes significant contribution to *P. aeruginosa* biofilm structure [59]. The *pel* genes are present in several strains but some commonly used reference strains, such as PAK and PAO1 do not express strongly these genes under common laboratory conditions. The *psl* genes are only present in some strains, but not in the well known PA14 laboratory strain. In some instance, PEL, and PSL can be present together in the biofilm matrix of *P. aeruginosa*.

Other polymers are present in the matrix of the biofilm of *S. epidermidis* for example: teichoic acid [60]. There are two types of teichoic acid in *S. epidermidis*: teichoic acid associated with the bacterial membrane (CW TA) and extracellular teichoic acid (EC TA). The EC TA is responsible for the increased viscosity of the colony. The EC TA is a (1-3)-linked poly(glycerol phosphate), substituted at the 2-position with α-glucose, α-N-acetylglucose, D-alanine and α-glucose-6-alanine (Figure 9).

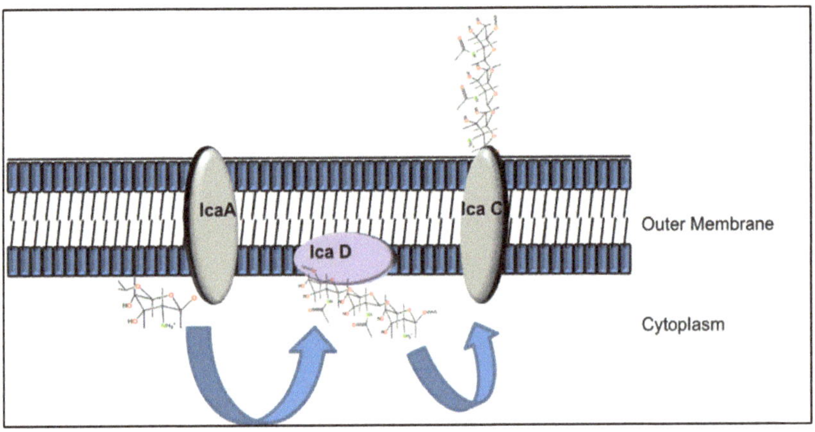

Figure 8. PNAG biosynthesis

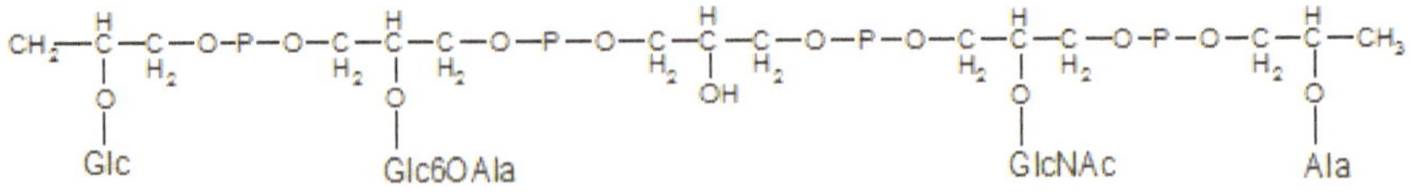

Figure 9. Structure of teichoic acid

In *E. coli*, colanic acid, a sugar polymer composed of galactose, fructose and glucose, is regularly found in the biofilm matrix [61] (Figure 10).

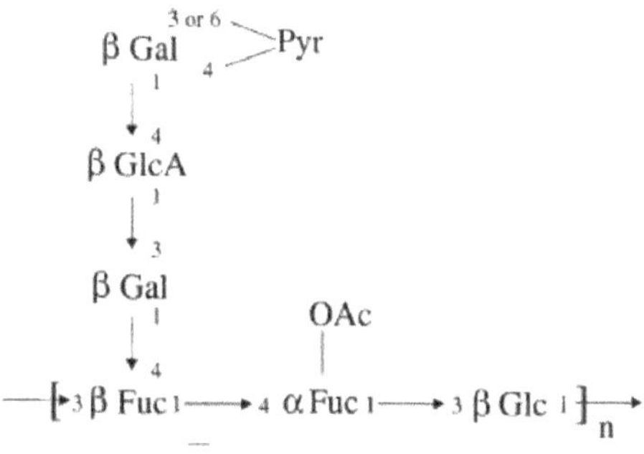

Figure 10. Structure of colanic acid

It must be remembered that although different strains can apparently synthesize the same EPS, there can be differences in physical properties especially with respect to viscosity and gel formation. Several biofilm studies have used colanic acid-producing *E. coli* [61]. Prigent-Combaret, C. et al. [62], yet this polymer can vary greatly in mass and viscosity, as can bacterial alginates.

4. An example of complex biofilm: biofilm formation at the surface of nanofiltration membranes used for drinking water production.

We and others have previously studied very complex biofilms formed on nanofiltration (NF) membranes during surface water filtration in drinking water production processes [63, 64]. After several years of filtration, the foulant consists in a brown viscous layer covering the entire surface of the membrane [65] (Figure 11).

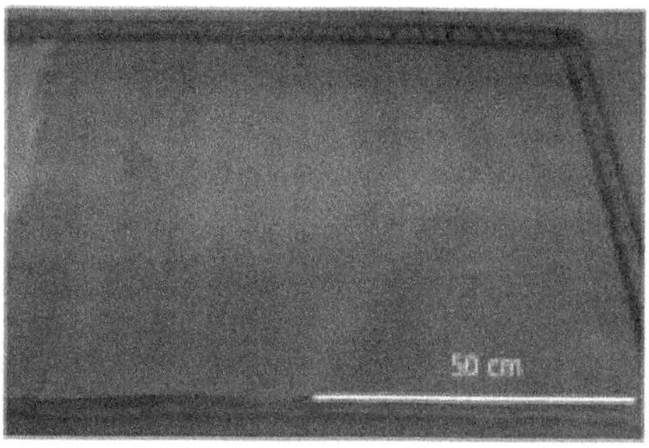

Figure 11. Visual examination of a fouled NF membrane

Dry weight of the foulant is about 2 g/m^2. The NF biofilms harbours mainly exopolysaccharides and proteins, as shown by characteristic ATR-FTIR signals near 1650 cm^{-1} (amide I), 1550 cm^{-1} (amide II), 1450 cm^{-1} (due in part to C-H deformation), 1400 cm^{-1} (due in part to symetric stretch for the carboxylate ion), 1250 cm^{-1} (P=O and C-O-C stretching and/or amide III), and in a broad complex region from 1250 to 900 cm^{-1} (due in part to C-O-

C, C-O, ring-stretching vibrations of polysaccharides and the P=O stretch of phosphodiesters) (Figure 12).

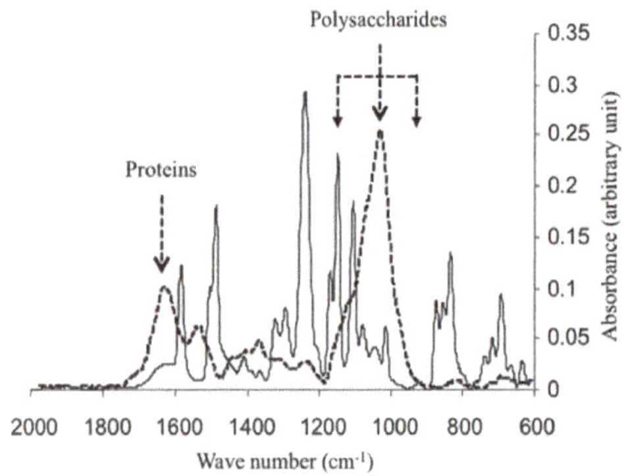

Figure 12. ATR-FTIR spectra of a virgin membrane (plain line) and of a fouled membrane (dotted line)

Fluorescence microscopy observations after nucleic acid staining with DAPI and polysaccharides staining with lectins labelled with fluorescein isothiocyanate or tetramethylrhodamine isothiocyanate indicate a high spatial heterogeneity inside the foulant matter with a mean thickness of 32.5 ± 17.7 μm [66] (Figure 13). Examples of lectins that can be used for such polysaccharides staining experiments are peanut agglutinin (PNA) targeting β-gal(1->3)galNAc residues, wheat germ agglutinin (WGA) targeting $(glcNAc)_2$ and NeuNAc residues, *Bandeiraea simplicifolia* (BS-1) agglutinin targeting α-gal and α - galNAc residues and Concanavalin A (ConA) targeting α-man and α-glc residues.

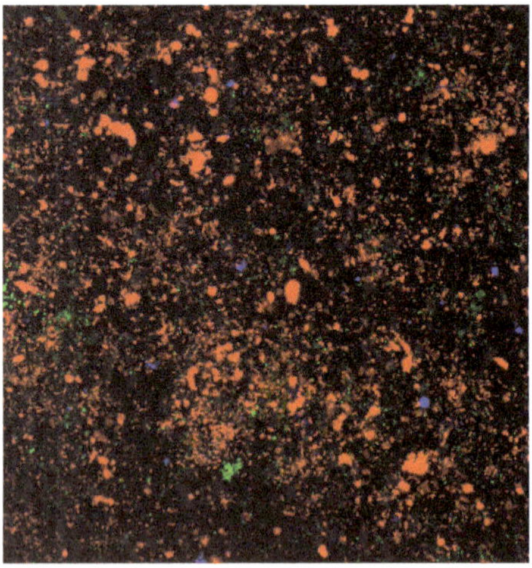

Figure 13. CLSM visualization of the heterogeneity of a NF biofilm after staining with DAPI, TRITC and TITC-labelled lectins. Magnification x630

The microbial cells, mainly composed of bacteria, are localized in the superficial layer of the fouling material and are organized as microcolonies interspersed at the membrane surface. Some algae are also present, as shown by autofluorescence properties. The presence of a dense and wide polysaccharide matrix harbouring few microbial cells at the NF membrane surface has been associated with differences in the efficiency of cleaning procedures against different foulants categories [65, 67]. Polysaccharide residues are found in areas where microcolonies are present and in areas devoid of microbial cells. This polysaccharide organization has been previously observed with environmental biofilms grown in vitro with river water as the sole source of carbon and nutrients [68]. High staining with PNA and BS-1, respectively reveals high occurrence of galactosides residues in the polysaccharide components of the foulants. The BS-1 lectin staining pattern indicates a high degree of spatial organisation with the observation of long and entangled fibers. WGA staining shows short fibers and cloud stained areas. PNA and ConA lectin staining are more interspersed. The polysaccharide composition of the fouling layer changes quantitatively and qualitatively during spring and summer [64]. Lectin staining increases from March to September for all the lectins used. Staining with BS-1 increases constantly in March, June and September. A high increase of binding with PNA, and ConA is observed between March and June, but the binding of these two lectins does not change between June and September. Staining with the WGA is weak in March and June and is higher in September. The lectin-binding changes with time may be linked to an increase of the biomass attached at the membrane surface and to changes among the populations of attached cells. Nutrients, oxygen level and the concentration of metals can influence the exopolymer abundance of environmental model biofilms grown in vitro with river water as the sole source of carbon and nutrients [69]. The modification of these parameters leads to a shift in the glycoconjugate makeup of the biofilms.

Biofilms may be considered to be highly porous polymer gels [70] and diffusion studies demonstrate gel-like characteristics [71]. Previous work has suggested that laboratory-grown and some natural biofilms are viscoelastic in nature [3, 8, 72]. During rotation analysis, a rheofluidification behaviour is observed for NF biofilms [66]. Different mechanisms can explain shear thinning of a biofilm. Break down of links between polymers in the biofilm matrix or deflocculation of particles corresponding to an irreversible modification of the biofilm structure can occur. Such irreversible modifications are unlikely in the experimental conditions published because of the reversibility of viscosity changes with shear rate [66]. Shear thinning of NF biofilms may be related to the polymeric composition of the biofilm matrix. With shear acceleration, polymers may follow the direction of the flow leading to viscosity decrease. This has been previously observed with purified polysaccharides like cellulose [73]. Moreover, bending of biofilm structures in the shear direction during the application of shear stress has been mentioned to explain the viscoelastic response of a mixed culture biofilm [72]. NF biofilms have been submitted to oscillation analysis with a cone-plate rheometer [66]. In such experiments, a sinusoidal oscillation of defined maximum strain and oscillatory frequency is applied to a sample and the storage (G') and lost (G'') modulus are measured. The storage modulus characterizes the ability of the material to store energy, whereas the loss modulus characterizes energy dissipation in the material under dynamic excitations. If the material is perfectly elastic then the resultant stress wave is exactly in phase with the strain wave. By contrast, when the rate of change of the sinusoidal oscillation is a maximum and the strain is zero, for purely viscous systems, the resultant stress wave will be exactly 90° out-of phase with the imposed deformation. For NF biofilms, during oscillation analysis, values of storage modulus (G') stay higher than values of loss modulus (G'') over the entire range of frequencies covered, indicating that the NF biofilm behave like a highly elastic physical gel [74]. Polysaccharides alone, like alginates, are known to realize a sol-gel transition under adequate physicochemical conditions [75]. The physicochemical microenvironment inside NF biofilms may be permissive to exopolysaccharides sol-gel transition. The gel state is resistant enough and presents a micro porosity favourable for resistance to flow forces, microcolonies development and cell nutrition inside the biofilm structure. This model of sol-gel transition of polysaccharides inside biofilms is consistent with rheological properties previously demonstrated for other biofilms: *Streptococcus mutans* biofilms have elasticity and viscous behaviour analogous to NF biofilms for a range of frequencies between 0.1 to 20 Hz [76]. The rheofluidification behaviour and gel-type rheological properties shared by different type of biofilms and purified polysaccharides suggest that the critical components of the biofilm matrix determining the biofilm texture are polysaccharides.

The time-dependent strain response observed in the creep curves clearly indicated that NF biofilms exhibited viscoelastic behaviour. Viscoelasticity is thought to be a general mechanical property of biofilms. A very wide range of elasticity and viscosity values has been previously observed for a wide sample of biofilms formed artificially in laboratory experiments or coming from natural aquatic environments [4, 72, 76]. Thus, it wasn't surprising to observe that the rheological properties of NF biofilms are different from the ones of natural biofilms from different aquatic environments like Nymph Creek (Yellowstone National Park) and Chico Hot springs (Montana) algal biofilms [4]. These differences in viscosity and elasticity between biofilms can be related to different exopolysaccharide contents and to different shearing strains. Bacterial and algal alginates are known to have different monomeric composition leading to a stronger binding of cations for bacteria, a property involved in the formation of a stable gel in the presence of ambient Ca^{2+} cations [77].

The specificity of NF biofilms may be the necessity to resist shear forces applied to the membrane during the filtration process. In the Méry-sur-Oise plant, NF membranes are operated at feed pressure of approximately 10 bars [78]. The high membrane feed pressures may influence the rheological properties of NF biofilms by increasing cohesive forces in the biofilm bulk, increasing forces, which keep the exopolymers to the membrane surface, and thus strengthening the mechanical stability of the biofilm. This may explain at least in part the NF biofilms resistance to industrial cleaning [65].

Shaw et al. have previously shown that the elastic relaxation time varied much less between biofilms of different origins. λ was estimated to be the time required for viscous creep length to equal elastic deformation length (so that memory of initial conditions is lost), i.e., $\lambda \approx \eta / G$.

The elastic relaxation time of about 30 minutes lies within the range previously determined for various biofilms [4]. The universality of the viscoelastic transition of biofilms has been suspected to have critical survival impact [4]. The ability of biofilms to deform in response to mechanical stress may be a conserved strategy of defence to enable persistence on surfaces in different flow conditions.

5. Conclusion

Extracellular polysaccharides are considered as the major structural components of the biofilm matrix. A large variety of polysaccharides required for bacterial adherence and biofilm formation have been described. Polysaccharide molecules can interact with themselves or with ions and proteins. These interactions result from electrostatic attractive forces, repulsive forces, hydrogen bonds, van der Waals interactions and ionic attractive forces. All these forces influence the structure and the stability of the biofilm matrix and the way EPS and polysaccharides can be extracted from the biofilm bulk. A universal protocol for extracellular polysaccharide extraction from the biofilm matrix does not exist. Each study may adapt usual extraction procedures to biofilm specificities and to the nature of the polysaccharide studied. The viscoelasticity nature of biofilms is universal but biofilms in differing environments exposed to different hydrodynamic conditions will encounter changes in the structure, composition and then physical properties of their matrix. Biofilm science is highly exciting since it is a mixture of biology, biophysic, chemistry and much more.

Author details

Pierre Lembre, Cécile Lorentz and Patrick Di Martino*
Groupe Biofilm, Laboratoire ERRMECe-EA1391, Institut des Matériaux-FD4122,
Institut Universitaire de Technologie, Université de Cergy-Pontoise, Cergy-Pontoise, France

6. References

[1] Marti S, Nait Chabane Y, Alexandre S, Coquet L, Vila J, Jouenne T, Dé E (2011) Growth of *Acinetobacter baumannii* in pellicle enhanced the expression of potential virulence factors. PLoS One 6(10), e26030.
[2] Wan J, Bessière Y, Spérandio M (2009) Alternating anoxic feast/aerobic famine condition for improving granular sludge formation in sequencing batch airlift reactor at reduced aeration rate. Water Res. 43(20), 5097-108.
[3] Shaw T, Winston M, Rupp C J, Klapper I & Stoodley P (2004) Commonality of elastic relaxion times in biofilms. Phys rev lett 93: 1-4.
[4] Whitchurch C B, Tolker-Nielsen T, Ragas P C, Mattick J S (2002) Extracellular DNA required for bacterial biofilm formation. Science 295, 1487.

* Corresponding Author

[5] Kiedrowski M R, Kavanaugh J S, Malone C L, Mootz J M, Voyich J M, Smeltzer M S, Bayles K W, Horswill A R (2011) Nuclease modulates biofilm formation in community-associated methicillin-resistant *Staphylococcus aureus*. PLoS One. 6(11), e26714.
[6] Jahn A, Griebe T, Nielson P H (2000) Composition of *Pseudomonas putida* biofilms: accumulation of protein in the biofilm matrix. Biofouling 14, 49–57.
[7] Wicker-Böckelmann U, Wingender J, Winkler U K (1987) Alginate lyase releases cell-bound lipase from mucoid strains of *Pseudomonas aeruginosa*. Zentralbl Bakteriol Mikrobiol Hyg A. 266(3-4), 379-89.
[8] Stoodley P, Dodds I, Boyle J D, Lappin-Scott H M (1999) Influence of hydrodynamics and nutrients on biofilm structure. J. Appl. Microbiol. 85, S19–S28.
[9] Stoodley P, Cargo R, Rupp C J, Wilson S, Klapper L (2002) Biofilm material properties as related to shear-induced deformation and detachment phenomena. J Industrial Microbiol. Biotechnol. 29, 361-367.
[10] Vieira M J, Melo J F, Pinjeiro M M (1993) Biofilm formation: hydrodynamic effects on internal diffusion and structure. Biofouling 7, 67-80.
[11] Flemming H C, Wingender J, Mayer C, Koerstgens V, Borchard W (2000) Cohesiveness in biofilm matrix polymers. In: Allison D., Gilbert P., Lappin-Scott H.M., Wilson M. (Eds), Community structure and cooperation in biofilms. SGM Symposium Series 59. Cambridge Univ. Press, Cambridge, 87-105.
[12] Mangalappalli-Illathu A K, Lawrence J R, Swerhone G D W, Korber D R (2007) Architectural adaptation and protein expression patterns of *Salmonella enterica* serovar enteritidis biofilms under laminar flow conditions. International journal of food Microbiology, 123: 109-120.
[13] Flemming H C and Wingender J (2010) The biofilm matrix. Nature Reviews Microbiology 8(9): 623-633.
[14] Wingender J, Strathmann M, Rode A, Leis A and Flemming H C (2001) Isolation and biochemical characterization of extracellular polymeric substances from *Pseudomonas aeruginosa* Methods in Enzymology 336:302-14.
[15] Denkhaus E, Meisen S, Telgheder U and Wingender J (2007).Chemical and physical methods for characterisation of biofilms. Microchimica Acta. 158: 1-27.
[16] Liu H, Fang H H P (2002) Extraction of extracellular polymeric substances (EPS) of sludges. J. Biotechnol. 95, 249-256.
[17] Oliveira R, Marques F and Azeredo J (1999) Purification of polysaccharides from a biofilm matrix by selective precipitation of proteins. Biotechnology Techniques. 13: 391-393.
[18] Boualam M, Quiles F, Mathieu L, Block J C (2002) Monitoring the effect of organic matter on biofilm growth in low nutritive waters by ATR-FT-IR spectroscopy. Biofouling 18(1): 73-81.
[19] Leathers T D, Bischoff K M (2011) Biofilm formation by strains of *Leuconostoc citreum* and L. mesenteroides. Biotechnology letters 33(3): 517-523.
[20] Leathers T D and Cote G L (2008) Biofilm formation by exopolysaccharide mutants of *Leuconostoc mesenteroides* strain NRRL B-1355. Applied Microbiology and Biotechnology 78(6): 1025-1031.

[21] Lin M H, Yang Y L, Chen Y P, Hua K F, Lu C P, Sheu F, Lin G H, Tsay S S, Liang S M and S H Wu (2011) A novel exopolysaccharide from the biofilm of *Thermus aquaticus* YT-1 induces the Immune response through toll-like receptor 2 Journal of Biological Chemistry J Biol Chem. 286(20): 17736-45.
[22] Dubois M, Gilles K A, Hamilton J K, Rebers PA, Smith F (1956) Colorimetric method for determination of sugars and related substances. Analytical Chemistry 38: 350–356.
[23] Dreywood R (1946) Qualitative test for carbohydrate material. Ind. Eng. Che. (Anal. ed.), 18, 8, 499.
[24] Bradford M M (1976) A rapid and sensitive method for the quantitation of microgram quantities of protein utilizing the principle of protein-dye binding. Anal. Biochem. 72, 248 - 256.
[25] Smith P K, Krohn R I, Hermanson G T, Mallia A K, Gartner F H, Provenzano M D, Fujimoto, E K, Goeke N M, Olson B J and Klenk D C (1985) Measurement of protein using bicinchoninic acid. Anal. Biochem. 150 (1): 76–85.
[26] Pan X, Liu J, Zhang D, Chen X, Li L, Song W and Yang J (2010) A comparison of five extraction methods for extracellular polymeric substances (EPS) from biofilm by using three-dimensional excitation-emission matrix (3DEEM) fluorescence spectroscopy. Water SA 36 111-116.
[27] Sutherland I W (2001) Biofilm exopolysaccharides: a strong and sticky framework. Microbiology 147(1): 3-9.
[28] Wozniak D J, Wyckoff T J O, Starkey M, Keyser R, Azadi P, O'Toole G A and Parsek M R (2003) Alginate is not a significant component of the extracellular polysaccharide matrix of PA14 and PAO1 *Pseudomonas aeruginosa* biofilms Proceedings of the National Academy of Sciences 100 (13): 7907-7912.
[29] Andersson S (2009) Characterization of bacterial biofilms for. wastewater treatment. Royal Institute of Technology. Thesis of Stockholm University, School of Biotechnology.
[30] Gorin J P A, Spencer T J F (1966) Exocellular alginic acid from *Azotobacter vinelandii*. Canadian journal of chemistry, 44: 993-998.
[31] Sabra W, Zeng A P, Deckwer W D (2001) Bacterial alginate: physiology, product quality and process aspects, Applied Microbiology and Biotechnology, 56: 315-325.
[32] Donati I, and Paoletti S (2009) Material Properties of Alginates. Alginates: Biology and Applications, Microbiology Monographs 13, 1-54, Springer.
[33] Ruffing, A M and Chen R R (2012) Transcriptome profiling of a curdlan-producing *Agrobacterium* reveals conserved regulatory mechanisms of exopolysaccharide biosynthesis. Microbial cell factories 11.
[34] Kenyon W J, Esch S W.and.Buller C S (2005) The curdlan-type exopolysaccharide produced by *Cellulomonas flavigena* KU forms part of an extracellular glycocalyx involved in cellulose degradation. International Journal of General and Molecular Microbiology 87(2): 143-148.
[35] Young, J. M., S. B. Leschine and G. Reguera. (2012). Reversible control of biofilm formation by *Cellulomonas* spp. in response to nitrogen availability. Environmental microbiology 14(3).

[36] Guo Y P, Sagaram U S, Kim J S and Wang N (2010) Requirement of the galU Gene for Polysaccharide Production by and Pathogenicity and Growth In Planta of *Xanthomonas citri* subsp *citri*. Applied and Environmental Microbiology 76(7): 2234-2242.

[37] Aires C P, Tenuta L M, Carbonero E R, Sassaki G L, Iacomini M and Cury J A (2011) Structural characterization of exopolysaccharides from biofilm of a cariogenic *streptococci*. Carbohydrate Polymers 84(4): 1215-1220.

[38] Cote G L and Leathers T D (2009) Insoluble glucans from planktonic and biofilm cultures of mutants of *Leuconostoc mesenteroides* NRRL B-1355. Applied Microbiology and Biotechnology 82(1): 149-154.

[39] Verhoef R, Waard P D, Schols H A, Siika-aho M and Voragen A G J (2003) *Methylobacterium* sp. isolated from a Finnish paper machine produces highly pyruvated galactan exopolysaccharide. Carbohydrate Research 338(18): 1851-1859.

[40] Ras M, Lefebvre D, Derlon N, Paul E, Girbal-Neuhauser E (2011) Extracellular Polymeric Substances diversity of biofilms grown under contrasted environmental conditions. Water Res. 45 (4), 1529-38.

[41] Pierre G, Graber M, Rafiliposon BA, Dupuy C, Orvain F, De Crignis M, Maugard T (2012) Biochemical composition and changes of extracellular polysaccharides (ECPS) produced during microphytobenthic biofilm development (Marennes-Oléron, France). Microb Ecol. 63(1): 157-69.

[42] Hung C C, Santschi P H, Gillow J B (2005) Isolation and characterization of extracellular polysaccharides produced by *Pseudomonas fluorescens* Biovar II, Carbohydrate Polymers, Volume 61(2): 141-147.

[43] Al-Halbouni D, Dott W and Hollender J (2009) Occurrence and composition of extracellular lipids and polysaccharides in a full-scale membrane bioreactor. Water Research 43(1): 97-106.

[44] Linker A, Jones R S (1964) A Polysaccharide resembling Alginic Acid from a *Pseudomonas* Micro-organism, Nature, 204: 187-188.

[45] Gacesa P (1998) Bacterial alginate biosynthesis-recent progress and future prospects, Microbiology, 144: 1133-1143.

[46] Remminghorst U, Rehm B H A (2006) Bacterial alginates: from biosynthesis to applications, Biotechnology letters, 28:1701-1712.

[47] Stokke B T, Draget K I, Smidsrod O, Yuguchi Y, Urakawa H, Kajiwara K (2000) Small-Angle X-ray scattering and rheological characterization of alginate gels. Macromolecules, 33: 1853-1863.

[48] Doumèche B, Picard J, Larreta-Garde V (2007) Enzyme-catalyzed phase transition of alginate gels and gelatin-alginate interpenetrated networks. Biomacromolecules, 8: 3613-3618.

[49] Zogaj X, Nimtz M, Rhode M, Bokranz W, Romling U (2001) The multicellular morphotypes of *Salmonella typhimurium* and *Escherichia coli* produce cellulose as the second component of the extracellular matrix. Molecular Microbiology, 39: 1452-1463.

[50] Solano C, Garcia B, Valle J, Berasain C, Ghigo J M, Gamazo C, Lasa I (2002) Genetic analysis of *Salmonella enteritidis* biofilm formation : critical role of cellulose, Molecular Microbiology, 43 : 793-808.

[51] Weng L, Zhang L, Ruan D, Shi L, Xu J (2004) Thermal gelation of cellulose in a NaOH/thiourea aqueous solution, Langmuir, 20: 2086-2093.
[52] Cai J, Zhang L (2006) Unique gelation behavior of cellulose in NaOH/Urea aqueous solution, Biomacromolecules, 7: 183-189.
[53] Mack D, Fischer W, Krokotsch A, Leopold K, Hartmann R, Egge H, Laufs R (1996) The intercellular adhesion involved in biofilm accumulation of *Staphylococcus epidermidis* is a linear β-1,6-linked glucosaminoglycan: purification and structural analysis. Journal of Bacteriology, 178: 175-183.
[54] Cerca N, Maira-Litran T, Jefferson K K, Grout M, Goldmann D A, Pier G B (2007) Protection against *Escherichia coli* infection by antibody to the *Staphylococcus aureus* poly-N-acetylglucosamine surface polysaccharide. PNAS, 104:7528-7533.
[55] Gotz F (2002) *Staphylococcus* and biofilm, Molecular Microbiology, 43: 1367-1378.
[56] Heilmann C, Gerke C, Perdreau-Remington F, Gotz F (1996) Characterization of Tn*917* insertion mutants of *Staphylococcus epidermidis* affected in biofilm formation. Infection and immunology, 64: 277-282.
[57] Gerke C, Kraft A, Sussmuth R, Schweitzer O, Götz F (1998) Characterization of the N-acetyglucosaminyl-transferase activity involved in the biosynthesis of the *Staphylococcus epidermidis* polysaccharide intercellular adhesion. The journal of biological chemistry, 273: 18586-18593.
[58] Izano E A, Amarante M A, Kher W B, Kaplan J B (2008) differential roles of poly-N-acetylglucosamine surface polysaccharide and extracellular DNA in *Staphylococcus aureus* and *Staphylococcus epidermidis* biofilms. Applied and environmental microbiology, 74: 470-476.
[59] Ryder C, Byrd M, Woznik D J (2007) Role of polysaccharides in *Pseudomonas aeruginosa* biofilm development. Current opinion in Microbiology, 10: 644-648.
[60] Sadovskaya I, Vinogradov E, Li J, Jabbouri S (2004) Structural elucidation of the extracellular and cell-wall teichoic acids of *Staphylococcus epidermidis* RP62A, a reference biofilm-positive strain. Carbohydrate research, 339: 1467-1473.
[61] Danese P N, Pratt L A, Kolter R (2000) Exopolysaccharide production is required for development of *Escherichia coli* K-12 biofilm architecture. Journal of bacteriology, 182: 3593-3596.
[62] Prigent-Combaret C, Prensier G, Le Thi T T, Vidal O, Lejeune P, Dorel C (2000) Developmental pathway for biofilm formation in curli-producing *Escherichia coli* strains: role of flagella, curli and colanic acid. Environ. Microbiol. 2, 450–464.
[63] Liikanen R, Yli-Kuivila J & Laukkanen R (2002) Efficiency of various chemical cleanings for nanofiltration membrane fouled by conventionally-treated surface water. J Membr Sci, 195: 265-276.
[64] Doumèche B, Galas L, Vaudry H, Di Martino P (2007) Membrane Foulants Characterization in a Drinking Water Production Unit. Food and Bioproducts Processing, 85, 42-48.
[65] Di Martino P, Doumeche B, Galas L Vaudry H, Heim V, Habarou H (2007). Assessing chemical cleaning of nanofiltration membranes in a drinking water production plant: a

combination of chemical composition analysis and fluorescence microscopy. Water Science & Technology. 55, 219–225.
[66] Houari A, Picard J, Habarou H, Galas L, Vaudry H, Heim V, Di Martino P (2008) Rheology of biofilms formed at the surface of NF membranes in a drinking water production unit. Biofouling. 24(4): 235-240.
[67] Houari A, Seyer D, Couquard F, Kecili K, Démocrate C, Heim V, Di Martino P (2010) Characterization of biofouling and cleaning efficiency of nanofiltration membranes. Biofouling. 26 (1): 15 - 21.
[68] Neu T R, Swerhone G D W, Lawrence J R, (2001), Assessment of lectin-binding analysis for in situ detection of glycoconjugates in biofilm systems. Microbiology, 147, 299-313.
[69] Lawrence J R, Chenier M R, Roy R, Beaumier D, Fortin N, Swerhone G D W, Neu T R, Greer C W (2004) Microscale and molecular assessment of impacts of nickel, nutrients, and oxygen level on structure and function of river biofilm communities. Appl. Environ. Microbiol., 70, 4326-4339.
[70] Christensen B E, Characklis W W (1990) Physical and chemical properties of biofilm. In: Characklis W.G., Marshall, K.C., editors. Biofilms. New York: Wiley, 93-130.
[71] de Beer D, and Stoodley P (1995) Relation between the structure of an aerobic biofilm and transport phenomena. Water Sci Technol 32(8): 11-18.
[72] Towler B W, Rupp C J, Cunningham Al B & Stoodley P (2003) Viscoelastic properties of a mixed culture biofilm from rheometer creep analysis. Biofouling 19: 279-285.
[73] Cancela M A, Alvarez E, and Maceiras R (2005) Effects of temperature and concentration on carboxymethylcellulose with sucrose rheology. J Food Engineer 71, 419-424.
[74] Kavanagh G N, Ross-Murphy S B (1998) Rheologicals characterisation of polymer gels. Prog Polymer Sci. 23: 533-562.
[75] Draget KI, Steinsvå K, Onsøyen E, Smidsrød O (1998) Na- and K-alginate; effect on Ca^{2+}-gelation *Carbohyd Polymers* 35: 1-6.
[76] Vinogradov A M, Winston M, Rupp C J & Stoodley P (2004) Rheology of biofilms formed from dental plaque pathogen *Streptococcus mutans*. Biofilms 1: 49-56.
[77] Schürks N, Wingender J, Flemming H C & Mayer C (2002) Monomer composition and sequence of alginates from *Pseudomonas aeruginosa*. Int J Biol Macromol 30: 105-111.
[78] Cyna B, Chagneau G, Bablon G and Tanghe N (2002) Two years of nanofiltration at the Méry-sur-Oise plant, France. Desalination 147: 69-75.

Permissions

The contributors of this book come from diverse backgrounds, making this book a truly international effort. This book will bring forth new frontiers with its revolutionizing research information and detailed analysis of the nascent developments around the world.

We would like to thank Professor Desiree Nedra Karunaratne, for lending her expertise to make the book truly unique. She has played a crucial role in the development of this book. Without her invaluable contribution this book wouldn't have been possible. She has made vital efforts to compile up to date information on the varied aspects of this subject to make this book a valuable addition to the collection of many professionals and students.

This book was conceptualized with the vision of imparting up-to-date information and advanced data in this field. To ensure the same, a matchless editorial board was set up. Every individual on the board went through rigorous rounds of assessment to prove their worth. After which they invested a large part of their time researching and compiling the most relevant data for our readers. Conferences and sessions were held from time to time between the editorial board and the contributing authors to present the data in the most comprehensible form. The editorial team has worked tirelessly to provide valuable and valid information to help people across the globe.

Every chapter published in this book has been scrutinized by our experts. Their significance has been extensively debated. The topics covered herein carry significant findings which will fuel the growth of the discipline. They may even be implemented as practical applications or may be referred to as a beginning point for another development. Chapters in this book were first published by InTech; hereby published with permission under the Creative Commons Attribution License or equivalent.

The editorial board has been involved in producing this book since its inception. They have spent rigorous hours researching and exploring the diverse topics which have resulted in the successful publishing of this book. They have passed on their knowledge of decades through this book. To expedite this challenging task, the publisher supported the team at every step. A small team of assistant editors was also appointed to further simplify the editing procedure and attain best results for the readers.

Our editorial team has been hand-picked from every corner of the world. Their multi-ethnicity adds dynamic inputs to the discussions which result in innovative outcomes. These outcomes are then further discussed with the researchers and contributors who give their valuable feedback and opinion regarding the same. The feedback is then collaborated with the researches and they are edited in a comprehensive manner to aid the understanding of the subject.

Apart from the editorial board, the designing team has also invested a significant amount of their time in understanding the subject and creating the most relevant covers. They scrutinized every image to scout for the most suitable representation of the subject and create an appropriate cover for the book.

The publishing team has been involved in this book since its early stages. They were actively engaged in every process, be it collecting the data, connecting with the contributors or procuring relevant information. The team has been an ardent support to the editorial, designing and production team. Their endless efforts to recruit the best for this project, has resulted in the accomplishment of this book. They are a veteran in the field of academics and their pool of knowledge is as vast as their experience in printing. Their expertise and guidance has proved useful at every step. Their uncompromising quality standards have made this book an exceptional effort. Their encouragement from time to time has been an inspiration for everyone.

The publisher and the editorial board hope that this book will prove to be a valuable piece of knowledge for researchers, students, practitioners and scholars across the globe.

List of Contributors

Stefan Kwiatkowski
Alltech Inc. Nicholasville KY, USA

Stefan Edgar Kwiatkowski
University of Kentucky, USA

Susana P. Miranda Castro and Eva G. Lizárraga Paulín
Laboratorio de Biotecnología, Facultad de Estudios Superiores Cuautitlan, UNAM, Cuautitlan Izcalli, Estado de México, México

Shauna L. Reckseidler-Zenteno
Athabasca University, Athabasca, Alberta, Canada

Rosa Eugenia Reyes, Alejandra Aquino Andrade, Rafael Coria Jiménez, Carolina Romo González and Maribel Ortiz Herrera
Departamento de Medicina Experimental, Instituto, Nacional de Pediatría, México D.F., México

Rosa Eugenia Reyes
Departamento de Sistemas Biológicos, División CBS, Universidad Autónoma Metropolitana Unidad, Xochimilco, Coyoacán, México D.F., México

Tanya V. Ivashina
Skryabin Institute of Biochemistry and Physiology of Microorganisms, Russian Academy of Sciences, Pushchino, Moscow Region, Russia

Vladimir N. Ksenzenko
Institute of Protein Research, Russian Academy of Sciences, Pushchino, Moscow Region, Russia

Natalya Nikolaevna Trofimova, Elena Nikolaevna Medvedeva, Nadezhda Viktorovna Ivanova, Yuriy Alekseevich Malkov and Vasiliy Anatolievich Babkin
Laboratory of Wood Chemistry, Federal Research Budget Institution, A.E. Favorsky Institute of
Chemistry, Siberian Branch, Russian Academy of Sciences, Irkutsk, Russia

Desiree Nedra Karunaratne, R.G.U. Jayalal and Veranja Karunaratne
University of Peradeniya, Peradeniya, Sri Lanka

Mona A. Esawy, Eman F. Ahmed and Wafaa A. Helmy
Department of Chemistry of Natural and Microbial Products, National Research Centre, Dokki, Giza, Egypt

Nahla M. Mansour
Gut Microbiology and Immunology Lab, Central of Excellence for Advanced Sciences (CEAS), National Research Centre, Dokki, Giza, Egypt

Waled M. El-Senousy
Water Pollution Research Department, National Research Centre, Dokki, Giza, Egypt

Mounir M. El-Safty
Central Laboratory for Evaluation of Veterinary Biologics (CLVB), Abbassia, Cairo, Egypt

Richard A. Cunha and Eduardo F. Franca
Institute of Chemistry, Federal University of Uberlândia, Uberlândia, MG, Brazil

Thereza A. Soares, Victor H. Rusu, Frederico J.S. Pontes and Roberto D. Lins
Department of Fundamental Chemistry, Federal University of Pernambuco, Recife, PE, Brazil

V. Poinsot, M.A. Carpéné and F. Couderc
Laboratoire IMRCP, UMR 5623, Université Paul Sabatier, Toulouse, France

Aleksandr N. Zimnitskii
«NanoDerm-profi» Ltd., Moscow, Russia

Ranieri Urbani and Paola Sist
Department of Life Sciences, University of Trieste, Trieste, Italy

Galja Pletikapić, Tea Mišić Radić, Vesna Svetličić and Vera Žutić
2Division for Marine and Environmental Research, Ruđer Bošković Institute, Zagreb, Croatia

Pierre Lembre, Cécile Lorentz and Patrick Di Martino
Groupe Biofilm, Laboratoire ERRMECe-EA1391, Institut des Matériaux-FD4122, Institut Universitaire de Technologie, Université de Cergy-Pontoise, Cergy-Pontoise, France